Fish Ecology in Arctic North America

The Alaska Chapter, American Fisheries Society
thanks the following sponsors of the symposium
Fish Ecology in Arctic North America

Alaska Cooperative Fish and Wildlife Research Unit
Arctic Institute of North America, U.S. Corporation
Inuvialuit Fisheries Joint Management Committee (Canada)
LGL Alaska Research Associates, Inc.
North Slope Borough, Alaska
School of Fisheries and Ocean Sciences, University of Alaska

The Chapter also thanks the following organizations
for support of this symposium publication

BP Exploration (Alaska) Inc.
Canada Department of Fisheries and Oceans
U.S. Fish and Wildlife Service
U.S. Minerals Management Service

Fish Ecology in Arctic North America

EDITED BY

James B. Reynolds
Alaska Cooperative Fish and Wildlife Research Unit
Biological Resources Division
U.S. Geological Survey
University of Alaska Fairbanks

American Fisheries Society Symposium 19

Proceedings of the *Fish Ecology in Arctic North America* Symposium

Held at Fairbanks, Alaska, USA
19–21 May 1992

American Fisheries Society
Bethesda, Maryland
1997

The American Fisheries Society Symposium series is a registered serial. Suggested citation formats follow.

Entire book

Reynolds, J., editor. 1997. Fish ecology in Arctic North America. American Fisheries Society Symposium 19, Bethesda, Maryland.

Article within the book

Colonell, J. M., and B. J. Gallaway. 1997. Wind-driven transport and dispersion of age-0 Arctic ciscoes along the Alaska Beaufort coast. Pages 90–103 *in* J. Reynolds, editor. Fish ecology in Arctic North America. American Fisheries Society Symposium 19, Bethesda, Maryland.

Library of Congress Catalog Card Number: 97-70116

ISBN 0-913235-96-2

ISSN 0892-2284

Printed in the United States of America on acid-free paper

American Fisheries Society
5410 Grosvenor Lane, Suite 110
Bethesda, Maryland 20814-2199, USA

Dedication

To Rebecca (Becky) Everett who lost a one-year battle with cancer on 26 November 1993. Becky's expertise was in the field of fishery genetics. She was a fishery biologist with the U.S. Fish and Wildlife Service in Anchorage, Alaska, where she directed the Regional Office program on char and salmon genetics. Becky was a dedicated and meticulously honest researcher with a passionate desire to preserve and protect our fish and wildlife resources. This symposium proceedings honors her dedication to providing the very best data possible.

Contents

Preface and Acknowledgements

The legacy of arctic fisheries science, especially in Alaska, is found mostly in the gray literature—agency and consultant reports dealing with management issues such as fish harvest and environmental impact assessment. Much of this documentation has been driven by the development of mineral, oil, and gas resources in northern Alaska and Canada. Although this information, mostly generated since the 1970s, has been useful for management and potentially valuable to science, much of it has not been subject to peer review and faces the uncertain future of permanent loss in the untraceable literature (Norton 1989), despite enormous costs of acquisition and analysis.

Two approaches have been used to reverse the trend toward gray literature and establish a new professional legacy. One method has been the integration of multiple data sources into a unified, user-friendly database (e.g., Slaybaugh et al. 1989). The other approach has been more direct: a conscious effort to publish findings in peer-reviewed outlets such as reviews of past, primarily unpublished, works (e.g., Armstrong 1986); collections of theme-related papers (e.g., Norton 1989); and symposia designed to encourage managers, as well as scientists, to take their work a step beyond the gray literature (e.g., this volume).

The professional environment for publishing by arctic fisheries scientists and managers, especially in government agencies and consulting firms, has moved from either a passive ("on-your-own-time") or obstructive ("not-even-on-your-own-time") one to a more positive, supportive environment created by supervisors and administrators. Leaders of agencies and organizations are demonstrating, through action, their recognition of the benefits of publication: the stronger legal standing of peer-reviewed literature; higher credibility for their personnel and organization; and better accountability of progress, ultimately to the stockholder or taxpayer. For example, the Alaska Department of Fish and Game recently established its own journal, the *Alaska Fisheries Research Bulletin*, accompanied by a policy of peer review and high editorial standards. Arctic scientists with Alaskan-oriented findings now have an alternative to national or international journals.

This symposium, *Fish Ecology in Arctic North America*, was conceived as a program objective by the Alaska Chapter of the American Fisheries Society (AFS) at its 1990 annual meeting in Homer, Alaska. A symposium steering committee—R. Bailey, W. Barber, C. George, R. Meyer, J. Reynolds, L. Thorsteinson, A. Wertheimer (chair), and W. Wilson, all of Alaska, and J. Reist of Canada—agreed to broadly define "arctic" to include subarctic regions; to focus on arctic North America because of the expected "draw" to a meeting sponsored by a state-based fisheries organization; and to include all arctic North American fishes and invertebrates, and their habitats, as potential topics for the symposium. Thus, the purpose of the symposium, as given by the committee, was to "summarize current knowledge of the biology and ecology of freshwater, anadromous, and marine fishes in arctic Alaska and Canada." Although this purpose represented a formidable goal, committee members assumed it was approachable if contributors would include summaries of previous work reported in the gray literature, as well as the peer-reviewed literature, in the interpretation of their own works. This approach was seen as another means of escape from dependence on the gray literature. Contributors would acknowledge the traceable gray literature to a substantial degree, incorporating its value into their papers for exposure to peer review. The call for papers from the committee included an informal effort to encourage the participation of managers and others not usually involved in scientific proceedings.

The symposium was held 19–21 May 1992, on the Fairbanks campus of the University of Alaska. Sixty papers (50 oral, 10 poster) were presented to the 112 registrants. As expected, most attendees were from Alaska and Canada. Fittingly, a blizzard blanketed Fairbanks with snow during the symposium, leaving no doubt among the participants that they were at an arctic conference!

Out of the symposium arose the 34 peer-reviewed papers included in this publication. I extend sincere thanks to those who gave their time and talent as peer reviewers. In England: H. Parker; in Norway, H. Hop; in Sweden, J. Hammar; and in Canada, T. Beacham, D. Bodaly, W. Bond, K. Chang-Kue, E. Crossman, W. Gazey, J. Green, W. Griffiths, E. Gryselman, M. Healey, J. Johnson, L. Johnson, A. Kristofferson, P. McCart, K. Mills, D. Morris, J. Nelson, D. Noakes, A. Peden, J. Reist, R. Tallman, D. Taylor, R. Waples, and H. Welch.

In the United States: J. Adams, M. Alexandersdottir, K. Alt, W. Barber, R. Barry, W. Bechtol, R. Behnke, D. Bernard, J. Bickham, A. Bingham, J. Bryan, C. Burger, J. Burr, M. Castellini, R. Clark, J. Cole, J. Colonell, K. Coyle, D. Daum, A. DeCicco, R. Dillinger, W. Donaldson, R. Fechhelm, B. Gallaway, J. George, A. Gharrett, J. Gordon, N. Gotell, L. Haldorson, C. Hemming, N. Hughes, S. Jewett, W. Johnson, C. Kaya, S. Kelsch, T. Kline, J. LaPerriere, and B. Lubinski.

Also in the United States: R. Marshall, O. Mathisen, D. McBride, M. McDonald, R. Meyer, S. Meyer, A. Moles, L. Moulton, S. Nichols, B. Norcross, D. Norton, B. Osborne, A. Paul, M. Philo, T. Quinn, J. Raymond, T. Royer, J. Seeb, T. Shirley, R. Smith, W. Smoker, S. Sonnichsen, A. Springer, L. Thorsteinson, K. Troyer, A. Tyler, T. Underwood, D. Vincent-Lang, G. Wedemeyer, T. Weingartner, A. Wertheimer, R. Wilmot, W. Wilson, K. Winemiller, D. Wiswar, and W. Wooster.

In the AFS editorial office, Heather Boyd provided considerable editorial help and Janet Harry handled production of the manuscripts into typeset pages.

In Alaska, Betsy Sturm assisted with production of figures and the cover design.

For several reasons, mostly to do with being a volunteer editor paid to do other things full time, I was responsible for this symposium being published at least two years behind schedule. I offer heartfelt thanks to all those who were so patient and supportive during this painfully long process: the steering committee, especially its chair Alex Wertheimer; the Alaska Chapter leadership, particularly Kate Wedemeyer and Dana Schmidt; the AFS editorial office, most particularly Beth Staehle; and those most affected by the delay, the authors. Thank you, one and all.

Was the goal of the symposium—"to summarize current knowledge of . . . fishes in arctic Alaska and Canada"—achieved? No, strictly speaking. Most of the papers in this volume are reports of current research, not exhaustive reviews. Nevertheless, this volume contains valuable information for arctic fisheries scientists and managers, a point well summarized by Lionel Johnson in his symposium review (Johnson 1997, this volume). The "arctic fish" symposium will serve as a key fisheries reference for many years to come.

JAMES B. REYNOLDS
Editor

References

Armstrong, R. H. 1986. A review of Arctic grayling studies in Alaska, 1952–1982. Biological Papers of the University of Alaska 23:3–17.

Johnson, L. 1997. Living with uncertainty. Pages 340–345 *in* J. Reynolds, editor. Fish ecology in Arctic North America. American Fisheries Society Symposium 19, Bethesda, Maryland.

Norton, D. W., editor. 1989. Research advances on anadromous fish in arctic Alaska and Canada. Biological Papers of the University of Alaska 24.

Slaybaugh, D. K., B. J. Gallaway, and J. S. Baker. 1989. The databank for arctic anadromous fish: description and review. Biological Papers of the University of Alaska 24:4–26.

Symbols and Abbreviations

The following symbols and abbreviations, as well as others approved for the Systeme International d'Unites (SI), may appear in this book without definition.

A	ampere
AC	alternating current
°C	degrees Celsius
cal	calorie
cm	centimeter
Co.	Company
Corp.	Corporation
cov	covariance
DC	District of Columbia
D	dextro (as a prefix)
d	day
df	degrees of freedom
dL	deciliter
E	east
E	expected value
e	base of natural logarithm (2.71828 . . .)
e.g.	(exempli gratia) for example
eq	equivalent
et al.	(et alii) and others
etc.	et cetera
F	filial generation; Farad
°F	degrees Fahrenheit
ft	foot (30.5 cm)
ft^3/s	cubic feet per second (0.0283 m^3/s)
g	gram
G	giga (10^9, as a prefix)
gal	gallon (3.79 L)
h	hour
ha	hectare (2.47 acres)
in	inch (2.54 cm)
Inc.	Incorporated
i.e.	(id est) that is
J	joule
k	kilo (10^3, as a prefix)
kg	kilogram
km	kilometer
L	levo (as a prefix)
L	liter (0.264 gal, 1.06 qt)
lb	pound (0.454 kg, 454 g)
log	logarithm
Ltd.	Limited
M	mega (10^6, as a prefix) molar (as a suffix or by itself)
m	meter (as a suffix or by itself) milli (10^{-3}, as a prefix)
mi	mile (1.61 km)
min	minute
mol	mole
N	normal; north; newton
N	sample size
NS	not significant
n	ploidy; nanno (10^{-9}, as a prefix)
o	ortho (as a chemical prefix)
oz	ounce (28.4 g)
P	probability
p	para (as a chemical prefix)
p	pico (10^{-12}, as a prefix)
pH	negative log of hydrogen ion activity
ppm	parts per million
qt	quart (0.946 L)
R	multiple correlation or regression coefficient
r	simple correlation or regression coefficient
rad	radian
S	siemens; south
SD	standard deviation
SE	standard error
s	second
UK	United Kingdom
U.S.	United States (adjective)
USA	United States of America (noun)
V	volt
V,	Var variance (population)
var	variance (sample)
W	watt; west
yd	yard (0.914 m, 91.4 cm)
α	probability of type I error
β	probability of type II error
Ω	ohm
μ	micro (10^{-6}, as a prefix)
′	minute (angular)
″	second (angular)
°	degree (temperature or angular)
%	per cent (per hundred)
‰	per mille (per thousand)

American Fisheries Society Symposium 19:1–3, 1997

Circumpolar Perspective on the Importance of Fish Resources within the Arctic Ecosystem and Environmental Changes That Could Affect These Resources

CHRISTOPHER J. HERLUGSON
BP Exploration (Alaska) Inc.
Post Office Box 196612, Anchorage, Alaska 99519, USA

Ladies and Gentlemen, I'm here this morning to talk about the importance of your research. I'd like to thank the Alaska Chapter of the American Fisheries Society for the opportunity to address this symposium. I believe it is important to promote scientific research in forums such as this so as to bring together those working in the field and to summarize the current knowledge on the biology and ecology of arctic fish as we move forward into the next century.

The next century holds many new challenges for research and resource management in the Arctic. We are faced with a world in which interest groups are arguing about whose reports are printed on recycled paper, who is more in favor of animal rights, and even who has filed the most lawsuits against federal or state agencies. And the "environment" for research in the Arctic is going to be no less hostile in the coming decades. You might think that you are working in a remote environment, but the world is getting smaller. These factors make research on Arctic wildlife increasingly difficult—as if research in the Arctic weren't already difficult.

More and more, we are also faced with political challenges. The Arctic is a very "polarized" place. We in Alaska are blessed with extremes of climate—although some of you may not consider snow in the middle of May a blessing. We are also blessed with extremes of personalities with respect to development and conservation: Some people view the Arctic as a vast, underdeveloped stockpile of resources, while others see a vanishing wilderness, desperately in need of protection from all human activity. We know that the Arctic is an incredible laboratory where one can observe and thereby learn about the biological, chemical, and physical components and cycles of the ecosystem.

But if you look into the future, you see applied research, because the arctic ecosystem is faced with a myriad of environmental issues. Most of these issues will propel your research into an arena that is more public than you've ever experienced. For example, one issue that is particularly important in polar regions is the greenhouse effect; many researchers believe the poles will be more severely affected by global warming than will the lower latitudes. Increases in greenhouse gas emissions could result in climate change and temperature fluctuations. They could alter water circulation and breakup and freeze-up patterns. This issue could potentially change the very fabric of the arctic ecosystem. Depletion of the ozone layer is another major environmental issue, with continuing speculation on the associated risks of skin cancer for northern residents or even the possibility of genetic mutations in other species. Arctic haze is another concern challenging the ecosystem. Changes in the atmosphere caused by industrial developments far beyond the boundaries of the Arctic have brought outsiders into the local environmental issues. Acid rain and heavy-metal deposition from outside the Arctic also have the potential to influence arctic systems and lifestyles.

All these buzzwords—global warming, climate change, acid rain, arctic haze—are making arctic biology more international and making applied research just as important as basic research. For example, the Arctic Environmental Protection Strategy, signed in 1991, represents the first time eight nations with territory above the Arctic Circle have joined in cooperative scientific activity to study acidification, chlorinated organics, oil, radioactivity, heavy metals, and noise. And this type of agreement is more than just political posturing. It shows the importance research will have in international resource management decisions.

Another example is provided by the former Soviet Union. This region is beginning to deal with its changing role and influence in global politics and is coming to grips with widespread environmental problems. Such problems include the dumping of nuclear waste into the Barents Sea, pollution from badly managed oil and gas operations, and scars from mining. The impacts of political changes in that country—the one with the most land above the Arctic Circle—cannot be ignored by the scientific community. We are faced with a situation where different nations have different priorities and different needs. One government may place high value on a particular resource, while another does not. We are dealing with international resources and international ecosystems.

The Arctic, with its changing physical characteristics, is the best example of the universality of the natural world, the concept of transboundary pollution—not just because the problems may have either international beginnings or international ramifications, but because the biological resources themselves are international: Arctic cisco *Coregonus autumnalis*, polar bear *Ursus maritimus*, bowhead whale *Balaena mysticetus*, beluga (white whale) *Delphinapterus leucas*, Arctic char *Salvelinus alpinus*, and Arctic cod *Boreogadus saida*. We cannot protect one part of the system because it has economic value or intrinsic value without protecting the parts of the ecosystem that sustain it. Research on key species of this ecosystem can help settle issues of policy and conservation. It can help manage resource development, including commercial and subsistence fishing, oil and gas exploration, mining, communications, even tourism. Within the Arctic, resource development will continue or even increase. The Arctic has a wealth of natural resources, and people will not just walk away from that wealth. These studies presented here today and others planned for the future are not just academic exercises; they are providing us with analytical tools to answer real questions.

The key to keeping up with environmental issues will be applied research. Basic research is important, but applied research will provide the answers to the environmental questions of the future. The problem is how to respond to those questions.

Biology is obviously one focus. Many of the environmental issues that are gaining the attention of the public require basic biological knowledge. One of the overriding concerns with issues such as global warming, acid rain, or even pollution in the oceans is that while impacts seem to be accelerating, our knowledge to deal with these impacts is inadequate.

The protection of biological diversity will be a key point in the future. To many people, biological diversity is the key to the health of this planet. The distribution and abundance of species may be viewed as a vital component of any study. An international treaty on protecting biological diversity may emerge from the Earth Summit in Brazil.

Your research will have to answer some new, probing questions. Is each ecosystem protected? How do you protect an ecosystem? How many protected areas do we need? Which species are important and why? How do we deal with the natural evolution of ecosystems that may in fact decrease diversity?

Fish are viewed as indicator species—keys to understanding environmental trends—forming a connection between man and his environment. Research can provide data for understanding and anticipating population booms and crashes, data necessary for assigning value to various habitats and species, and data for assessing the biological importance of change. Variations in numbers, diversity, changes in abundance—these are issues of more than academic interest. They are vital to our continued ability to harvest biological resources, to drill for petroleum, and to mine for minerals.

People who live in the Arctic are looking to science more and more for information on how to manage their resources—to ensure that the productivity is there for future generations and to ensure access to traditional subsistence resources. They are also looking to move into the twentieth century with modern conveniences that we often take for granted in Fairbanks and Anchorage but that have become available to many arctic residents only recently. Commercial fishers need to know the cycles and fluctuations of fish populations to control their activities and to assure sustainable development. And they are asking tough questions: Can we manipulate a biological system to enhance it in some way to meet specific management objectives? Can we manage development to minimize change and reduce effect? This research is relatively new and thus perhaps different from the kind of research that you may have done in the past.

The environmental impacts of development are being evaluated to see if there have been biological benefits, not just biological impacts. The north slope is a prime example. Research has shown that gravel mine sites, because of their depth, are now used as overwintering habitat for anadromous fish. Research is being conducted in the Beaufort Sea to evaluate the migration patterns of Arctic ciscoes across international boundaries and the effects of temperature fluctuations on growth potential and body condition. Cooperative monitoring programs are being conducted between industry and agencies and between neighboring countries. In the future, arctic biological research cannot afford to be isolated.

In the coming decades, researchers will need to understand the political as well as the biological importance of their work in responding to environmental issues. Whether we like it or not, biology has entered the world of politics—entering Congress and state legislatures, where people get up to speak and say very little, nobody listens, and then everyone disagrees. One problem with Congress is that law is the number one profession for today's legislators. These are people who make decisions based on exact answers. And science is being called on to produce those facts so that members of Congress can make scientifically sound decisions. We must all take responsibility for what that legislation entails. It must be based on sound knowledge and understanding of the real environmental issues.

We expect reauthorization of the Marine Mammal Protection Act (16 U.S.C. §§ 1361 to 1407 [1976 & Supp. V 1981]), the Endangered Species Act (16 U.S.C. §§ 1531 to 1543 [1976 & Supp. V 1981]), and the Clean Water Act (33 U.S.C. §§ 1251 to 1376 [Supp. V 1981]). All of these measures could significantly influence the type of information politicians will ask for, such as data to establish cause-and-effect relationships. But these data are useless unless we have people who can understand and interpret it. The results of your research should educate our children, our lawmakers, and ourselves about the interrelationships between humans and the natural world. And the public needs educating. I read that on a recent science exam a student wrote "H_2O is hot water and CO_2 is cold water." That statement is almost as worrisome as former President Reagan's statement that "trees cause pollution." These are the renewed challenges for biological research in the Arctic.

But these challenges also offer countless new opportunities for applied research: "resource management," "sustainable development," "biological diversity." These are not just biological or political buzzwords. These are not just the emotional campaigns of extreme environmentalists. These are the questions of your future. Your research must be viewed in light of these environmental issues, in the context of increasing internationalism, less interest in pure academic disciplines, and an increasing role of politics. We may not be able to answer these questions completely or in ways to which we have been accustomed. We will never be able to answer the long-term, burning questions—that is the nature of science. There are always more questions. But we must be able to address the broader environmental issues.

Some people might look at your agenda for the next few days and see a collection of boring and irrelevant academic papers presented by academics, for academics. It is your responsibility to wake those people up—show them that ichthyologists are worth listening to. Give them information they need to look at real problems. Show them how the work you do will help address and solve the problems of our coexistence with nature.

American Fisheries Society Symposium 19:4–12, 1997

The Canadian Perspective on Issues in Arctic Fisheries Management and Research

JAMES D. REIST
Department of Fisheries and Oceans
501 University Crescent, Winnipeg, Manitoba, Canada R3T 2N6

Abstract.—The nature of fisheries management and research in the Canadian Arctic is undergoing significant evolution. This is driven by diverse outside factors such as increasing human population, empowerment of aboriginal groups through settlement of land claims, renewable and nonrenewable resource development, global concerns with environmental issues and conservation, and reform in the roles and responsibilities of government. These changes present significant challenges for management and research pertaining to arctic fish. These challenges are further complicated by the practical problems faced by arctic resource managers and researchers—poor or nonexistent scientific databases, too many management units spread over a large geographical area, limited (and decreasing) program resourcing, and logistical problems and high costs associated with arctic work. Clearly, fisheries management and research in the Canadian Arctic must change to keep pace with these issues. Management structures have evolved to the concept of comanagement organizations, which have been formalized in many cases as legislated comanagement boards. Research structures have lagged behind somewhat but are also evolving. The concept of cooperative research projects that are designed to address localized, site-specific fisheries management issues but also are components of longer-term, more global research programs is advanced to meet this need. In this way, the shorter-term needs of resource users and managers can be met while greater fundamental understanding of biological processes can also be achieved. Together, these activities will allow us to meet the goals of conservation and sustainable development for arctic fisheries.

Recently, the arctic regions of the globe have experienced increased development of both renewable and nonrenewable resources. Coupled with this has been increased public interest and awareness of environmental issues and conservation, especially in arctic regions. These factors will continue to influence the management of arctic renewable resources, and this will significantly influence the direction of research upon such resources. I will provide an overview from the Canadian perspective of the changes anticipated in arctic fisheries management and research in the near future.

I have organized the presentation into four sections. First, I will give some background on the Canadian Arctic, the responsibility of the Department of Fisheries and Oceans (DFO) with respect to it, and some general thoughts on the needs of fisheries management and research. Second, I will address a few of the major issues surrounding resource management in the region. Third, I will return to a discussion of DFO's role and responsibility and how these are presently evolving and may evolve over the next decade. Finally, I will propose a general framework for fisheries management and research designed to meet future challenges.

The Canadian Arctic and the Department of Fisheries and Oceans

Although numerous definitions of the Arctic are possible, in Canada this area is typically viewed as that region north of 60°N latitude—the provincial-territory northern boundary. From a biological standpoint, however, the region encompassed by this definition includes a substantive portion of the subarctic biome consisting of boreal forest. The Arctic, as defined herein, consists of all lands north of 60°N latitude and west of Hudson Bay, the Arctic Islands, Hudson and James bays and contained islands, as well as the marine waters east and west to the international boundaries with Greenland and Alaska, respectively. Areas of northern Quebec and Labrador and the adjacent marine areas are also included in this definition of the Arctic.

Administratively, the responsibility for the management of fisheries resources and habitat within this area rests primarily with the Central and Arctic Region of the Department of Fisheries and Oceans, Canada. The administrative division is somewhat different from the area defined above as the Arctic. In particular, the Central and Arctic Region is responsible for management within the North Slope (Beaufort Sea drainages) of the Yukon Territory, all of the Northwest Territories, and adjacent marine waters including Hudson Bay. Excluded areas are the remainder of the Yukon Territory (Pacific Region responsibility) and arctic areas of Quebec and Labrador (responsibility of Quebec and Newfoundland Regions respectively).

This article limits discussion to that part of the Arctic managed by the Central and Arctic Region. This area of responsibility contains about 38% (3.8 million square kilometers) of the land area of Canada and about 20% of its freshwaters. The arctic coastline is 173,000 km long (71% of Canadian total) and the arctic continental shelf is 1.1 million square kilometers. Marine areas constitute about 67% of Canadian coastal waters (3.2 million square kilometers). The land masses in this area cover about 23 degrees of latitude (60°N to northern Ellesmere Island at 83°N) and 81 degrees of longitude (60°W at Ellesmere to 141°W at the international boundary with Alaska). The Canadian Arctic consists of two major drainage basins, the Arctic basin and Hudson Bay basin, within which are one of the world's largest rivers, the MacKenzie River; two of the world's great lakes, Great Slave Lake and Great Bear Lake; and a major inland sea, Hudson and James bay.

This area is characterized by extremes of remoteness, cold temperatures, perennial ice presence in many locations, and lack of daylight for much of the year. Extremes in interannual variation in climate, ice-out, freeze-up, and shifts in polar pack ice also occur. The wide latitudinal range results in extremes of seasonality, with relatively long, summer growing seasons and typical light–dark regimes in the south grading to short, intense growing seasons and 24-h daylight in the North. All these factors affect biological production, the nature and degree of anthropogenic impacts on the ecosystems, and thus the capacity to compensate for impacts.

The land is sparsely populated by about 50,000 people comprised of about 16,000 Dene and Metis, 18,000 Inuit, and 16,000 people of other ethnic origins. In general, Inuit occupy coastal and more northerly regions whereas Dene and Metis are found in interior and southerly regions. Native populations can be widely dispersed, at least seasonally, on the land, but the recent trend has been for centralization in communities. People of other ethnic origins live primarily in the larger communities.

In the Canadian Arctic there are 224 marine fish species (27% of 834 Canadian total) and 42 freshwater fish species (23% of 181). Sixteen of the freshwater species belong to one family (Salmonidae) and about half exhibit some degree of anadromy. Despite these numbers, diversity is generally low, especially in the northern locations. In addition, invertebrate species (e.g., shrimp and scallops), marine plants, birds, and mammals (whales, walrus, and seals) occur in marine waters throughout the area.

Exploitation of fishery resources has been a dominant feature of human activity especially among Native populations both throughout prehistory and presently. Fishing serves important economic, cultural, and social functions in the lives of Inuit, Dene, and Metis. Three types of fish exploitation are distinguished: (1) Domestic fishing is conducted by Natives or by persons of mixed ancestry (at least one-fourth Native) to obtain food. The aboriginal domestic fishery is unregulated, and unlicensed access is viewed as an inherent right (Yaremchuk and Wong 1989). Non-Natives wishing to fish domestically require an appropriate license. Per capita consumption of fish is estimated at 85 kg/year. The geographic location, cultural tradition, and traditions and habits of individuals determine which species are targeted. (2) Commercial fishing—defined as all fishing for sale or barter—must be licensed and is regulated by species, mesh size, gear, closures, and quotas. Several significant commercial fisheries have been operating for many years in the Canadian Arctic, including Great Slave Lake for lake whitefish *Coregonus clupeaformis*, lake trout *Salvelinus namaycush*, and inconnu *Stenodus leucichthys*; Cambridge Bay for Arctic char *Salvelinus alpinus*; numerous inland lake fisheries primarily for lake whitefish; and scattered and periodic coastal fisheries primarily for Arctic char and coregonids. Coregonids, especially broad whitefish *C. nasus* in the lower Mackenzie River, have been the subject of several attempts at commercial exploitation, but these efforts have proved unsustainable primarily as a result of high transportation costs to the south and stiff price competition from fisheries for the closely related lake whitefish. Efforts aimed at reestablishing viable commercial fisheries (e.g., lower Mackenzie River coregonids) or at developing new fisheries (e.g., nearshore exploitation of Greenland halibut *Reinhardtius hippoglossoides* in the eastern Arctic) are continuing. In theory, domestic and commercial fishing are separate and subject to different regulation and management practices. However, throughout much of the area aboriginal domestic and commercial fishing are intertwined, often being conducted by the same people at the same time and place. The destination of the fish determines the fishery to which it should be apportioned. This is particularly true for small commercial fisheries but is less relevant for the larger commercial fisheries. (3) Sportfishing is an activity regulated by license, gear (angling), size, quotas, and closures. Areas that are close to population centers or are road accessible experience substantive pressure, which is often undefined and uncontrolled with respect to the number of anglers. In contrast, remote sites have a well-defined and for the most part well-controlled sport fishery. Targeted species include salmonids—especially lake trout, Arctic char, and Arctic grayling *Thymallus arcticus*—as well as walleye *Stizostedion vitreum* and northern pike *Esox lucius*.

Currently, DFO manages the fishery resources within this area from a hierarchy of priorities. The overriding priority is conservation of the stock. Subject to this, the second priority is the right of indigenous peoples to harvest for subsistence purposes. The third priority, pursued only after the previous two have been met, is to use the resources to enhance economic and social well-being, usually through the development of commercial or sport fisheries.

Recent court challenges (Sparrow Case; Usher 1991) have altered the interpretation of the constitution of Canada. Aboriginals are now given the first rights to the resources, thus ensuring the possibility for such peoples to maintain distinctive Native lifestyles and culture. In view of this, the general principle followed by DFO for management of arctic fisheries is to achieve an equitable distribution of benefits among all Canadians while respecting the rights and obligations resulting from aboriginal land claims, prior use of resources, and needs of those living near the resource.

Ideally, effective scientifically based fisheries management is predicated upon sound knowledge of the biology of the exploited species. Knowledge of life history, response to various levels of exploitation, effects (both direct and indirect) of other anthropogenic influences, and levels of reproduction, growth, recruitment, mortality, abundance, and harvest must be accrued for each biological population that is exploited. Fisheries research accumulates this knowledge base and fisheries management applies it. From such knowledge, and with the underlying principle of sustainability (that is, conservation) guiding the process, an idealized model for the management of a particular fishery is developed and applied. For commercial fisheries, quotas are set that account for food and recreational fisheries; harvest is monitored over time, and effects of the harvest on the populations are noted. The limit to the combined harvest must be below that which the population can sustain indefinitely. Thus, a feedback loop is necessary to adjust the harvest depending upon the direction and degree of effect.

Of course, day-to-day fisheries management also does or should encompass many other aspects. These include (1) allocation among types of fisheries or among various groups using the resource, (2) habitat management, (3) coordination and development of new fisheries, (4) regulation and compliance, (5) assurance of fish quality from capture to consumption, (6) development of human resources related to fisheries, and (7) education and communication regarding conservation and fisheries management practices.

Issues Affecting Arctic Fisheries Management and Research

Two sets of issues will continue to affect fisheries management and research in the Canadian Arctic for the foreseeable future: the practical issues of fisheries management and the public issues that impinge on renewable resources in the North. Practical issues include items such as (1) poor or nonexistent scientific data bases, (2) too many units (stocks) requiring management distributed over too large an area handled by too few fisheries managers, (3) the sporadic nature of many fisheries, and (4) the development and application of practical but comprehensive management and research models. With respect to public issues, four major groups are pertinent: (1) aboriginal land claims, (2) economic development, (3) environmental concerns, and (4) government deregulation and restructuring. As all of these issues evolve, profound changes will occur in the management of the resources as well as in the research underpinning such management.

Practical Issues

Perhaps the biggest single practical problem is the dearth of scientific data and information bases upon which fisheries management can be founded. Annually, the fisheries managers from the Central and Arctic Region are called upon to set commercial quotas for approximately 300 stocks of anadromous fish (particularly Arctic char and coregonids), many stocks of freshwater fish (particularly lake whitefish and walleye), and several marine stocks (Greenland halibut). Marginally adequate databases and biological understanding exist for only a handful of these stocks. Thus, the typical management conundrum exists of providing necessary advice without an adequate information base. The solutions are to use existing knowledge despite its lack of completeness, to set quotas by rules of thumb (e.g., exploitation rate of 10% of the fishable stock), to extrapolate from more to less well known situations (transfer of knowledge), or to monitor the fisheries where possible and control harvests as necessary. In specific situations of either a completely unknown stock or especially important stocks, monitoring of both the fishery and biological parameters of the stock occurs where resources permit.

For most fish populations in arctic regions, both the experiential history and actual biological database are much more limited than in similar southern situations. For example, the history of management for the largest commercial Arctic char fishery in the Northwest Territories (Cambridge Bay) is a little more than 20 years long. This compares poorly with similar experience for more southerly fisheries (e.g., on the order of 50 to more than 100 years for coastal, anadromous, and Great Lakes fisheries). In some selected situations, the biological understanding is reasonably good, but in others we know virtually nothing. For example, in many situations the delineation of taxa at the species level either is not adequately known or has not penetrated to the level of the management model. The determination and delimitation of the basic units of biological management (biological stocks), long considered fundamental to proper sustainable fisheries management, is in most cases absent or at best rudimentary. Studies of taxonomic variability at levels (e.g., life history types) between the stock and species and the implications of this to management are virtually nonexistent, despite the general knowledge of a large diversity at this level. The biological structure (e.g., age, size, etc.) of unexploited populations, the dynamics of this structure over time as a result of natural variability in the biotic and abiotic environments, and the effects of exploitation on this structure are virtually unknown for any arctic fish species.

The stability and resilience of arctic fish populations to perturbations, whether simple, additively cumulative, or synergistic, are completely unknown. It is also not known whether arctic fish populations are fundamentally or qualitatively (as opposed to quantitatively) different from southern fish populations. Thus, the effectiveness of transfer of information and experience from southern situations to the Arctic is unknown.

Clearly one of the most basic and pressing practical issues affecting arctic fisheries is to redress this information gap, which can be achieved only by a directed, longer-term effort in which management and research needs and objectives are identified clearly. Specific needs are (1) development of comprehensive workable models for fisheries management from which clear research needs are identifiable, (2) development of associated research plans, (3) prioritization (and proper sequencing) of the information needs (components) of such plans, and (4) implementation of research effort. The commitment to this process must include the recognition that the accumulation of such information is, of necessity, a long-term approach. Short-term benefits can definitely result from this process, but a long-term commitment is needed to produce the greatest benefit.

The second significant practical problem is one of scale—that is, too many stocks, too large an area, and too few arctic fisheries researchers and managers. This is coupled with an anticipated increase in both the number and scale of the fisheries, brought about by factors such as development of northern fisheries in response to population needs as well as by failures in more southerly fisheries. The acquisition of adequate knowledge and its application to achieve sustainable fisheries for more than 300 arctic stocks is a Herculean task. This can never be achieved for every stock without a major outlay of resources (which is unlikely to occur in the current fiscal climate). Thus, both fisheries research and management must have as an inherent component some assessment of the transferability of knowledge from well-studied to less well-studied or even to unknown situations. A peculiarity, and perhaps an attribute that contributes significantly to the sustainability of several arctic fisheries, is the sporadic nature of the fisheries themselves. In part, this may be driven by lifestyles, climatic vagaries, market forces, or other factors. Regardless of the cause, from the perspective of fisheries management and assuming the periodicity is sufficiently long, the effect is generally beneficial in that pressure on the population is relaxed and recovery presumably is effected to some degree. However, from the perspective of fisheries research, inconsistent exploitation results in complicating factors for the interpretation of effects.

The development and application of comprehensive yet practical and workable fisheries management and research models for arctic fish populations is a significant need. A clearly defined management model is necessary that outlines the mechanism(s) by which particular types of fisheries are to be managed. In addition to the biology of the exploited resource, three factors complicate the development of a management model: (1) the type(s) of fishery executed on a particular stock, (2) the number of units (stocks, species, etc.) upon which the fishery operates, and (3) the spatial and temporal extent of the fishery (i.e., multiple sites or times at which the fishery occurs). Thus, more than one general management model may be necessary. From the general model, specific management models applicable to the particular fisheries can be developed and implemented. Managers of arctic fisheries in the Central and Arctic Region have made good progress in the development of fisheries management plans for specific fisheries. From these, the elements of a more general management model can be developed. From the general management model and the detailed specific management plans, both general and specific research needs can be clearly defined and programs to meet the needs can be implemented.

Public Issues

Perhaps the largest single public issue affecting the Arctic is the settlement of comprehensive aboriginal land claims. From the mid-1970s to the mid-1990s, issues regarding Native land claims, right to self-determination, and self-government have become the focus of both aboriginal and nonaboriginal portions of the Canadian public.

In Canada, the proprietary right of Natives to an interest in the land they inhabit has been implicitly recognized since 1763. Formalization of this occurred to some extent with the treaty process that proceeded into the twentieth century. However, throughout the Arctic, vast areas and the Native groups therein were not covered by any treaty. These factors, as well as a supportive stance on the part of the Canadian public, have resulted in the government entering into a series of negotiations with Native groups to settle Comprehensive Land Claims. Several land claims have been settled and fully implemented, including those with the Inuit of northern Quebec and the Inuvialuit of the western coastal arctic. Other claims have been settled and are in the initial stages of implementation—Gwich'in of the inland lower Mackenzie Valley region and Inuit of the remainder of the Northwest Territories (Nunavut). Most remaining claims in the arctic area are in various stages of negotiation.

Although details differ, all claim settlements for the Arctic have common trends that include the creation of private land ownership by the Native groups, exclusive rights to harvest in certain areas, preferential rights to harvest in other areas, equal representation on decision-making groups concerning resources, as well as financial compensation packages for giving up rights to other lands. All claim settlements establish wildlife management boards consisting of equal numbers of Native and government appointees. In general, such boards are charged with the responsibility of establishing total harvest levels within the limits of conservation, allocation of resources, and establishment of the subsistence levels that must be met. Coupled with all of this is the development of a corporate structure to administer the land claim settlement.

The second major public issue affecting the Arctic is economic development. This occurs at both the level of the Native groups using monies from the land claim process for development and of other groups, including governments, who wish to develop northern resources. The types of development include renewable resource development, such as commercial fisheries; tourism; and nonrenewable resource development, such as mines, hydrocarbon extraction, and hydroelectric development on rivers draining to Hudson and James bays. Such activity also includes all related infrastructure development, which in many cases is substantive.

The third major public issue, which is intimately intertwined with economic development, is the environment. Environmental issues fall into two areas: mitigation of negative effects from future activities, and cleanup and mitigation of the current legacy. Recent initiatives by the eight arctic countries have seen the development of a formal agreement on the arctic environment—the Arctic Environmental Protection Strategy. Initially, four major components of this were (1) Arctic Monitoring and Assessment Program (AMAP), (2) Conservation of Arctic Flora and Fauna (CAFF), (3) Protection of the Arctic Marine Environment (PAME), and (4) Emergency Prevention, Preparedness, and Response. Recently, a fifth component, Sustainable Development, has been added.

Although AMAP will address specific contaminants issues, other more general issues are also of concern. These include widespread effects such as long-range transport of atmospheric pollutants, contaminants issues associated with development as well as cleanup of previous problems (e.g., military base decommissioning and cleanup), climate change, and ozone depletion. Local-scale effects, such as mercury contamination resulting from water impoundment, depletion of animal populations, and alteration of habitats are also of concern.

Conservation is a major issue of local and global concern. Initiatives under the CAFF and PAME components of the Arctic Environmental Protection Strategy will address this problem specifically for the Arctic. There is more to the concept of conservation than simply preserving the last few organisms of a particular taxon. The habitat, ecosystem structure and function, genetic potential of the population, as well as an ecologically viable minimum population must also be preserved. A significant problem herein is the understanding necessary to link biological conservation limits with the environment in which the organisms exist. Although new, the Sustainable Development initiative will connect resource development, conservation, and protection issues. To address these strategies, substantive research at levels beyond the immediate population or taxon of concern must be designed and implemented. That is, arctic fisheries management and research cannot only focus on the exploited population but must also address wider issues of ecosystem structure, function, and preservation.

Finally, the fourth major public issue is the current trend in government for substantive reform of many aspects of its role and delegation of authority. This is manifested in response to the public desire for greater accountability and participation in decision-making processes. Related to this issue are the economic and fiscal management reforms being undertaken by all levels of government. These issues directly affect the fiscal resources available to organizations, such as DFO, that are charged with the responsibility for managing and researching arctic fish populations. Such issues also substantively alter how management and research activities will be conducted in the future. For example, in combination with the Native land claim settlements, fiscal reform means that fishery managers and researchers will have to develop mechanisms that are accountable and allow for direct participation in the process by the affected groups (or their representatives). Generally, such reform also means that managers and researchers will be asked to do more with less—thus, increased efficiency will be required.

Intimately related to and in part driven by all of the above is the more general issue of the increasing human population in the Arctic. In Canada, through immigration from the south and through very high birth rates among Native groups, the population in northern areas is burgeoning. In and of itself, such population growth creates increased demand on fisheries as well as acting as a positive feedback mechanism on the aforementioned major issues. This, in turn, further increases the demands for effective fisheries management and research.

The Future Challenge

Arctic fisheries management and research face substantive challenge in the future. Increased development and environmental concern will increase the demand for all aspects of fisheries management and research. This increased demand will be accommodated by fewer managers and researchers working from limited databases. The research and management activities must take place in open forums with direct participation by those affected by management decisions. Strategies to develop and implement arctic fisheries management and research programs must take these factors into account yet also accommodate the realities of data collection, population modeling, understanding cause–effect relationships, and developing predictive power for both natural and anthropogenic effects. Furthermore, such strategies must also accommodate short-term immediate needs and situations as well as develop long-term understanding of the forces shaping the evolution and continued existence of the resources and their ecosystems.

The Changing Role and Responsibility of the Department of Fisheries and Oceans

Comanagement

In the future, the role and responsibility of DFO will entail greater responsiveness and applicability of the products of both management and research.

The continued refinement of the process of comanagement involving government and the resource users will occur primarily within the framework of land claim settlements. Substantive general input from all stakeholders must be considered by the comanagement boards, which, as established under land claim settlements, have real decision-making ability backed by legislated responsibility. They must also have the commitment of all of the involved groups to the process. They must be open and be independently funded in order to function properly. Such boards will establish total harvest levels subject to conservation issues, allocate subsistence quotas, and control fishing on certain lands. Obviously, fishery comanagement boards will also have to interact efficiently with DFO or other agencies addressing environmental and development issues, as well as with adjacent boards addressing transboundary stocks.

Thus, the management role of the government must change from that of controlling to one of advising, licensing, monitoring, facilitating, and implementing. Fishery managers must become true partners in the overall comanagement process.

Cooperative Research

Where does this leave research? Individuals and groups responsible for research on arctic fish, their ecology, and the habitat face perhaps the greatest challenge. Currently, in Canada, research tends to be holistic in its approach, attempting to address widespread questions in a comprehensive manner usually from a long-term perspective. Often such research, although intrinsically interesting in its own right, may not be particularly germane to the immediate problems facing the fisheries, the resource users, or the managers. Furthermore, such research is often conducted from the perspective of the interest and background of the individual scientists, and benefits or application may take years to occur. The reality of the changing management role for arctic fisheries, increasing trends toward regional autonomy, greater power of self-determination on the part of aboriginal groups, as well as the continuation of fiscal restraint policies will greatly affect how and what research is conducted over the next decade.

Researchers and research managers must enter into their own cooperative research process that includes both the resource users and the resource managers. Such a process would allow these groups to achieve a mutually agreed-upon research plan tailored to the specific needs of the fisheries as well as meeting the longer-term goal of improved management of the fisheries. This process would also provide a mechanism whereby the substantive traditional knowledge of the resource user could be incorporated into the research program.

This may sound like the advocacy of a policy of strict, short-term, applied research only. This is not so, because if carried to an extreme such a policy would ultimately operate to the detriment of the resource in that lack of theoretical and widespread understanding will result in narrowly focused and perhaps inappropriate management. Rather, what is being advocated is a mechanism of component or compartmentalized research in which the results and understanding gained serve both to address the shorter-term needs of client groups such as resource users and managers, as well as to contribute to longer-term research plans and programs that will achieve greater fundamental understanding of processes, cause–effect relationships, and mechanisms by which conservation and sustainable development can be achieved. Researchers must also be in the forefront of the development and application of new technology in order to increase the efficiency of research and management activities. Thus, we must be proactive in conducting three types of research: (1) the development of appropriate methodology, (2) the site-specific planning and assessments necessary as the basis for management decisions, and (3) the longer-term biological research necessary to achieve better understanding of and predictability for the systems with which we work. Researchers must also become better communicators of the relevance of our research and do so in clear terms understandable to the lay public, the resource managers, and the comanagement boards.

Future Management and Research Approaches

Fisheries Management

In recognizing the changing role for arctic fisheries management over the past several years, the staff of the DFO Central and Arctic Region have developed and are presently implementing a process for fisheries management planning. This effort, outlined below, is extracted from de March (1991). The steps of the fisheries management process are (1) identifying the need for a particular management plan, (2) establishing priorities, (3) preparing a status report for the fishery stock, (4) establishing a safe harvest level, deciding whether need and information base permit preparation of a formal management plan, (5) developing such a plan, (6) approval and publication of the plan, (7) implementation of the plan, and (8) review and revision on an ongoing basis as more information and experience is gathered. The core of this exercise is the management plan itself, which consists of several component plans not all of which are necessarily present: fishing plan (e.g., safe harvest level, allocation), compliance and harvest statistics plan, inspection plan (as necessary for local or domestic fisheries; a mandated requirement for commercial export fisheries), research plan to define research needs and priorities, monitoring plan to provide feedback on stock status, habitat plan to identify and mitigate existing or potential threats, economic development plan, human resources development plan, and communication and education plan.

Several steps are necessary for applying this process to a particular fishery. The managed entity may consist of one or more biological stocks, occupy a greater or lesser geographical area during the life cycle, and may also cross jurisdictional boundaries. Thus, a reasonable level of biological understanding is a prerequisite to applying the aforementioned process. If this understanding is not available, appropriate research is necessary. For example, the biological units being managed and their spatial–temporal distributions must be defined, which requires increased levels of communication and cooperation between the various stakeholders. By the year 2000 these probably will include resource harvesters, other members of the public, formal comanagement boards, various federal government departments, and the organizational units within these departments (e.g., fisheries managers, economists, researchers).

Fundamental to the entire process is concurrence achieved between all the relevant parties, both with respect to the goals of the entire management process and with the various processes by which such goals are to be achieved. Given the different levels of knowledge and different constituencies of interested parties for particular resources, both concurrence and goals can only be established on a case-by-case basis and probably only after considerable consultation.

Fisheries Research

Fisheries research in all its forms is either explicit or implicit in the aforementioned management approach and it is intimately interrelated to proper fisheries management. Perhaps because of the inherently complex nature of research, the multiplicity of definitions and connotations possible for research, or the focus of researchers more on the problems rather than on the need for change in the process, similar processes and plans for research to meet future and changing needs are not as well formulated as those outlined above for fisheries management. The following is an initial and general attempt to redress this absence and to develop the idealized attributes of a research plan.

Short-term research focused on the information needs of a particular fishery as well as long-term strategic research of a more general nature are needed. The former meets the immediate needs of the resource comanagers. The latter develops knowledge, methodology, theory, and the principles of transferability on a wider basis, as well as addressing questions of a more global perspective.

Ideally, research should be applicable at various levels. For example, research results should be applicable to a specific stock management problem, a more general management problem, and development of new theoretical import. A case in point is that of data collected for genetic stock discrimination and identification. Proper structuring of sampling and analytical programs can result in the same data being useful to answer the following types of questions: (1) the immediate delineation of the unit for management and for organizing research results—that is, practical criteria for basic stock discrimination; (2) the degree of mixing of migratory stocks (i.e., identification of the contributions of unit stocks to mixed groups); (3) the delineation and identification of life history types (e.g., migratory and nonmigratory) that may complicate management scenarios; (4) the delineation and identification of taxa at the species level and whether hybridization or introgression is occurring; (5) understanding patterns of relationship and evolutionary history of the taxa at various levels; (6) global uniqueness of taxa at various taxonomic levels, thus implying the various levels of conservation that may be required; (7) development and application of new techniques to address problems; and (8) general understanding of biotic and abiotic factors that promote or maintain structuring of fish into biological units at various levels. All interested partners (fishery comanagers, research managers, and researchers) must recognize the validity of such a multilevel approach and promote and fund it where possible. Researchers themselves must structure their programs to achieve multiple end-uses for their data. In this way, the efficiency is increased for the high expenditures necessary for short-term, site-specific and long-term, strategic research in the Arctic.

Research must be more product-oriented in the sense of meeting short-term goals, be as timely as possible to reduce the lag time between understanding and implementation, and be proactive in its approach rather than reactive. That is, research should anticipate problems and address them before they become of substantive management significance. To achieve this, there must be strong functional and cooperative connections between research units, management units, the comanagers, and the resource users. Also, clear lines of communication and a high degree and frequency of information exchange are required.

An additional aspect of idealized arctic fisheries research is the involvement of aboriginal peoples as both proponents of research and as researchers in their own right, for example, in the development and implementation of monitoring programs. Research on arctic fisheries can benefit from the traditional ecological knowledge available from aboriginal peoples. A major focus of effort should be the development of mechanisms by which the results of western-based science and traditional knowledge can be meshed to increase understanding. Traditional knowledge often can provide the background knowledge and the starting point for fisheries research, thus focusing the scientific problem more clearly and increasing the efficiency of the research.

Because of government restructuring and the general decline of research funding, other mechanisms must also be found to increase efficiency of use of the resources devoted to research. Such mechanisms could include multiple end-users for specimens collected (e.g., biological data for fisheries management, tissues for stock identification, tissues for contaminants analyses, etc.). This would result in enhanced fiscal efficiency and more coherent understanding of the biological problems. With respect to increased efficiency of conducting research, all parties involved in managing, impacting, exploiting, or conducting research on the Arctic and its biota must realize that they have an obligation to participate in and communicate to the public the results of their research. For example, research should not be conducted simply to address local environmental impacts from industrial development but rather should be part of a more general research plan designed to increase both local biological knowledge as well as to contribute to general biological understanding and theory concerning arctic fisheries resources. Implicit in this is the need to make all research results readily available to the public, both as interpreted reports and as raw data deposited in appropriate archives.

With these considerations in mind, research regarding arctic fishery resources and their environments should primarily be grounded in a thorough understanding of those resources and the forces that impinge upon them. Thus, rather than being narrowly focused on particular issues such as fisheries management or a particular environmental impact, a significant component of research also must address the problem of more general understanding. With such a basis, the development of better cause–effect understanding can result and be applicable to issues of fisheries exploitation, local environmental impacts, global environmental modification, and perhaps most importantly to the additive and synergistic effects of these issues upon particular resources. In this way, the ideal goals of conservation and sustainability may be achievable for arctic fishery resources.

Conclusion

Throughout the preparation of this presentation, I kept thinking about the ancient Chinese curse, that is, may you live in interesting times. This is a curse because interesting times are usually times of great change. Globally, regionally, and certainly within the Canadian Arctic, we are in a period of substantive change. Such change and the transition time between stable structures often lead to conflict. However, times of change present both challenges and opportunities. Our challenge is to meet the change appropriately; at the same time, we have the opportunity to create cooperatively a better situation than that of the past. Hopefully, in the year 2000 or beyond, it will be said that we rose to the challenge and met it.

Acknowledgments

I thank B. Ayles, R. Clarke, and J. Mathias for critical review of earlier versions of this article as well as for access to unpublished documents outlining change in the Central and Arctic Region's fisheries management approach. I have also drawn upon ideas of G. Yaremchuk (Department of Fisheries and Oceans, Ottawa), as communicated directly to me and as embodied in his unpublished manuscript, The Ice Goes Out: The Conservation and Management of the Fish and Marine Mammal Resources of the Northwest Territories and North Slope of the Yukon Territory, A Discussion Paper.

References

de March, L. 1991. A process for developing fishery management plans in the Central and Arctic region. Canadian Manuscript Report of Fisheries and Aquatic Sciences 2094.

Usher, P. J. 1991. Some implications of the *Sparrow* judgement for resource conservation and management. Alternatives 18(2):20–21.

Yaremchuk, G. C. B., and B. Wong. 1989. Issues in the management of native domestic fishing in the Northwest Territories. Canadian Manuscript Report of Fisheries and Aquatic Sciences 2010.

American Fisheries Society Symposium 19:13–39, 1997

A Review of Fish Ecology in Arctic North America

G. Power
University of Waterloo, Department of Biology
Waterloo, Ontario, Canada N2L 3G1

Abstract.—Aquatic environments in the Arctic offer some protection against the extremes of the climate. The fish fauna is low in diversity because arctic ecosystems are still in the early stages of establishment, and environmental conditions are unsuitable for many species. Temperatures are low, with a narrow range, especially in marine habitats. The extreme seasonal variation in light affects primary production and energy supply. There is a marked pulse in freshwater discharge, sediment, and nutrient export from rivers, but there are long periods of stability in all aquatic environments. Ice is a significant component of the environment, particularly in marine, shallow lake, and riverine waters. The arctic environment still retains the remains of a colder past in the form of permafrost, which greatly influences freshwater habitats. Arctic fishes have many morphological, behavioral, and physiological adaptations for survival. Morphological adaptations can involve sensory organs, coloration, and metamerism. Species must tolerate long periods of low light and temperature, and their seasonal patterns of activity, growth, and reproduction must be matched to the seasons. Anadromy is common among salmonids and advantageous for energy accumulation. Migrations and movements must be precise and timed so that fishes arrive where they can best exploit the productive potential of their environment and avoid its worst hazards. Physiological adaptations, such as increases in metabolic rate, may not occur in arctic fishes but species are adapted to low temperatures. Many species have evolved antifreeze proteins; others live in undercooled states in deeper marine waters. These mechanisms may have evolved separately in antarctic and arctic regions, with a possible south-to-north transfer in one or more families. The arctic fish resource has been used by people for subsistence since their arrival in arctic regions. Harvests vary greatly between groups but today may average 60 kg per capita per year in arctic North America. Harvests are substantial compared with sport and commercial catches, and maintaining them is a priority. Slow growth, longevity, and iteroparity of arctic fish stocks may be adaptive but lead to low sustainable yields in fisheries. The fragmenting of freshwater and anadromous species into small, reproductively isolated stocks has implications for survival, stock genetics, and conservation. In North America most recent information on arctic fishes has been obtained as a necessary adjunct to development. Little attention has been paid to marine species of no economic or cultural importance. Many questions remain to be answered about the ability of arctic fishes to tolerate exploitation and industrial intrusions into their habitat. Threats to their environment originating in the south and global change pose other problems that could affect their longer-term survival. Much work remains to be done to fully understand and appreciate the ecology of arctic fishes.

According to Cohen (1970), there are at least 19,700 species of extant fish-like vertebrates; these are grouped into Agnatha (50), Chondrichthyes (515–555), and Osteichthyes (19,135–20,980). Nelson (1984) estimated there were 21,723 species but thought the real number might be as high as 28,000 if all species were described. Of these, rather few are found in the Arctic. The number depends on the definition adopted for the arctic region and whether totals are inclusive or exclusive of species with centers of distribution farther south. Dunbar (1968) commented that fishes were poorly represented and that there were probably fewer than 50 arctic marine species. Recent work has not changed that figure greatly.

Hunter et al. (1984) provide maps showing the sites of capture for 64 marine species from Canadian arctic waters that are in the collections of the National Museums of Canada and the former Fisheries and Oceans, Canada, Arctic Biological Station. They also map sampling sites of 21 anadromous species. Some of the marine species are recorded only at the western or eastern edge of the Canadian arctic sector and are more properly Pacific or Atlantic species. Forty-two of the species are widely distributed in Canadian arctic marine waters.

McAllister et al. (1987) list 135 fish species taken in Canadian arctic marine waters in a useful publication that provides English, French, and Inuktitut vernacular names. The list includes two freshwater, plus anadromous, Atlantic, and Pacific species. For freshwater species, it is difficult to define the area which is properly arctic. McPhail and Lindsey (1970) estimated that there were only about 60 species of freshwater and anadromous species in Alaska and northwestern Canada. Lindsey and McPhail (1986) include 69 species in their discussion of the same area; only 7 of these occur in the

Pacific drainage of Alaska, 3 are introduced species, and 1 is marine. If species occurring only in Great Slave Lake and its tributaries (which may be more properly considered subarctic) are removed, the list is reduced to 51 species. In the arctic archipelago and eastern Arctic the number of freshwater and anadromous species may be quite small. There are only 8 species from the archipelago and 18 species from the Ungava Bay drainage, including the fourhorn sculpin *Myoxocephalus quadricornis*, which is of marine origin (Crossman and McAllister 1986). The Hudson Bay drainage, on the other hand, has quite a rich fish fauna with 101 species, but much of the drainage cannot be considered arctic. The exact number of fish species living in arctic North America is not important—they account for less than 1% of known species. The Arctic Ocean and Hudson Bay represent about 4.3% of the area of all marine waters and support about 0.8% of marine fish species. Arctic freshwaters support a similar proportion of the known species of freshwater and anadromous fishes.

Why do arctic waters support so few species? The obvious answer that they are too cold may be too facile. Except in a few circumstances, aquatic habitats are protected from the lowest arctic temperatures, the minima being 0°C in freshwaters and −1.9°C in marine waters. It is true that low temperatures may slow physiological reactions and limit biological production, but this presupposes no adaptation to life at low temperatures. Dunbar (1968), in his critical analysis of ecological development in polar regions, suggested three factors that must be considered in addition to temperature. These are (1) the large seasonal oscillations in light and nutrient supply, (2) the generally low productivity, and (3) the immaturity of arctic ecosystems. Immaturity encompasses factors such as the pattern of deglaciation, access from glacial refugia, stage of succession, and selection of genotypes, all of which have temporal components. These considerations apply as much to fish as to the rest of the arctic biota.

This review addresses these questions and others that can be asked about fish in the North American Arctic. A description of the habitats used by fish provides a background for the discussion. Morphological and physiological adaptations that have enabled fish to colonize arctic waters are considered. The direct and indirect use of the fish resource raises questions about the size of harvestable surpluses, management options, resource allocation, and the sustainability of fisheries. Finally, other effects of human activity are discussed. Development projects such as hydroelectric power plants, mining, and oil extraction are modifying arctic environments, increasing the amount of direct pollution, and causing other more irreversible changes. Atmospherically transported pollutants and climate change probably will have other long-term effects. Appreciation of the ecology of arctic fish requires a better understanding of these topics.

The Environment of Arctic Fishes

The general features of the arctic environment are well known; they result from the large seasonal variation in the amount and quality of solar radiation reaching the ground. This is caused by the annual progression of the Earth's orbit around the sun and its axis of rotation, which controls temperature and the energy available to primary producers. Water offers considerable protection from temperature extremes but less shielding from the effects of seasonal light cycles. The three major aquatic environments—marine, lacustrine, and fluvial—present a different set of problems to their biota because of their geography, their relative sizes, and the properties of their water. These descriptions provide an environmental setting for the discussions that follow. Hammar (1989 and the references therein) gives a broad review of arctic freshwaters and Herman (1989) does so for arctic oceanography.

Marine

Biologists view arctic marine waters as stable and predictable because of the area of the Arctic Ocean (14×10^6 km^2), its large volume, its density structure, and the ice cover, which reduces mixing. Temperatures are consistently low. In general, seawater freezes at between −1.7 and −1.9°C, depending on the salinity. Arctic seawater is usually somewhat warmer, between −1.0 and +0.5°C depending on the depth and the influence of water exchanges with the Atlantic and Pacific oceans. Only in estuaries, coastal fjords, bays, and lagoons do temperatures rise above this, and here, in localized areas, they may reach 10°C (Fechhelm et al. 1992). Surface salinities of the Arctic Ocean are somewhat lower than those of warmer seas for two reasons: the Arctic basin receives a considerable amount of freshwater run-off, particularly from the Eurasian continent, and evaporation is low. Except in estuary plumes, salinities range between 28 and 35‰, slightly higher at depths. During freezing of tidal pools, Davenport (1992) found salinities up to

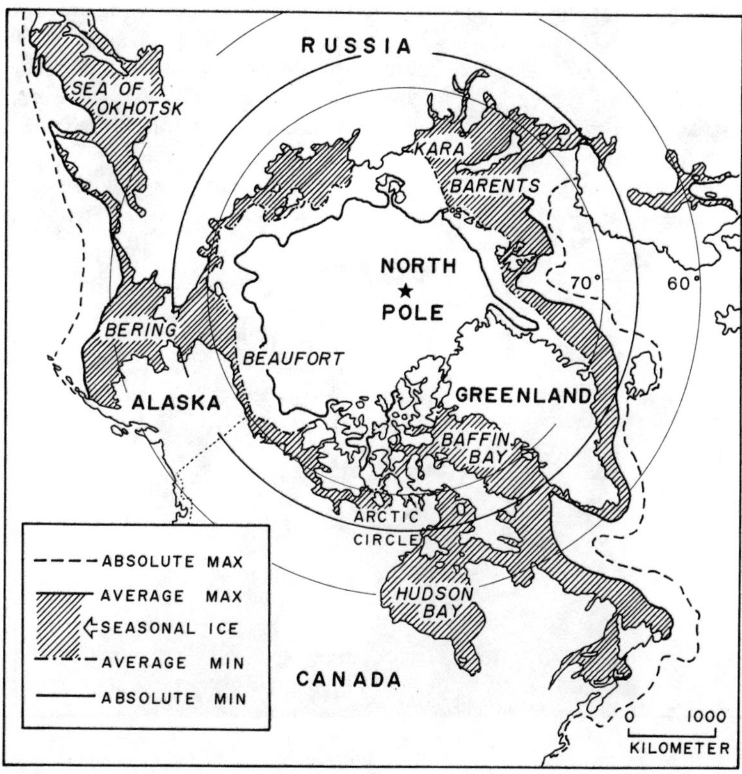

FIGURE 1.—The average and extreme seasonal limits (max = maximum, min = minimum) of arctic sea ice of more than one-eighth coverage (simplified after Barry 1989).

40‰; H. E. Welch (Freshwater Institute, personal communication) measured 70‰ in small ponds.

Sea ice is an important part of the environment. It consists of a mixture of new ice less than 1 year old and ice up to 4 years old; mixed with the latter is some glacial ice (Barry 1989). It is in constant motion, which leads to the formation of open leads, ridges, and fractures. Although new ice can grow to about 2 m in thickness during winter, floes that do not melt and have rafted upon other floes produce a rough texture with ice keels extending as deep as 30 m. The seasonal extent of the ice and the general circulation are shown in Figures 1 and 2. At the shore, the ice freezes to the shoreline as landfast ice, which extends seaward as a shelf for a variable distance. There is a zone of shear between the offshore pack ice and the landfast ice, and in some places zones of open water called polynyas are maintained. The open leads and polynyas are essential to the survival of arctic marine mammals and birds and contribute significantly to Arctic Ocean productivity (Stirling 1980; Bradstreet and Cross 1982). Tides cause the ice to rise and fall twice daily, and the landfast and nearshore ice must flex or move with this motion. A consequence of all this activity is considerable scraping, gouging, and rearranging of the shore and the shallow coastal seabed. This results in an impoverished nearshore benthos. The ice creates a new environment, generally termed the cryopelagic zone. On the undersurface of the ice, a cryopelagic community is formed of ice algae, zooplankton, and fish. This community has attracted a lot of attention in recent years because it may account for a significant fraction of the energy fixed in arctic seas (Nemoto and Harrison 1981; Carey and Montagna 1982; Grainger et al. 1985). These organisms are especially well adapted to life at low temperature, and the algae fix carbon at very low light intensities. They appear to benefit from nutrients concentrated in the bottom layer of the ice during seawater freezing. For a good general account of the marine environment, consult Percy et al. (1985) or Craig (1989a). More specialized treatments can be found in Dunbar (1977), Barnes et al. (1984), Herman (1989), and Smith (1990).

In North America the generally accepted limits of

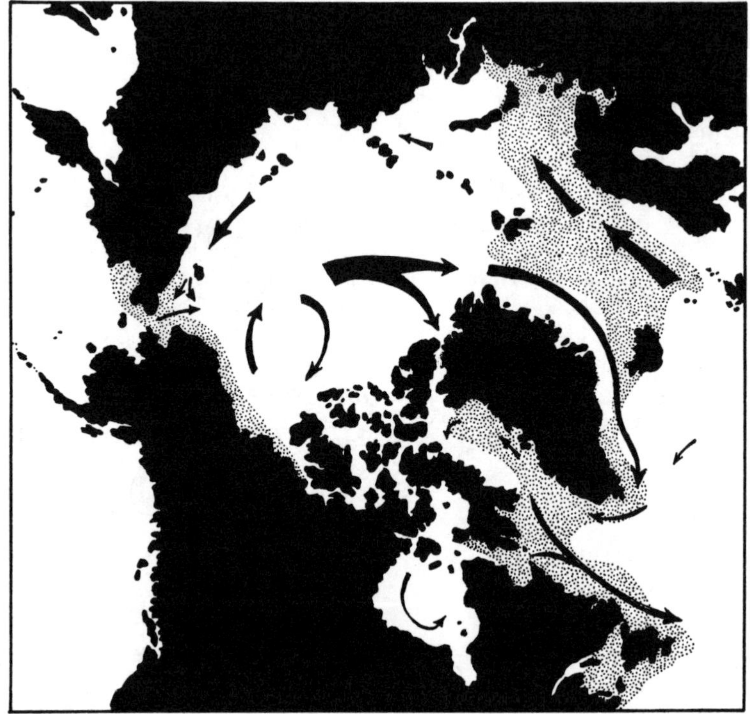

FIGURE 2.—The extent of the arctic (white area in the center) and subarctic (stippled areas) marine zones. Arrows indicate the general pattern of surface circulation (based on Dunbar 1968).

the marine arctic and subarctic are those proposed by Dunbar (1968), as shown in Figure 2. Many of the areas of interest for fisheries discussed in this review are in the subarctic zone. These include the continental shelf of the Beaufort Sea, Ungava Bay, part of Hudson Strait, and the southeast coastal waters of Baffin Island. Fish living in arctic marine waters have to adapt to low temperature, in most places below the usual freezing point of teleost blood; seasonally constrained low productivity; perpetual or long periods of darkness; and an ice-affected shoreline.

Lacustrine

When Rawson (1954) wrote his account of limnological investigations in arctic and subarctic North America, not much work had been done. His account identified several opportunities, including clearing up the problems of lake classification and understanding arctic lake production dynamics. Progress has been made, but the same questions, and others, remain to be answered. The thermal classification of polar lakes proposed by Forel included those whose surface temperatures did not rise above 4°C (Rawson 1954). The choice of 4°C coincides with the temperature at which freshwater is at its lowest density, which is important in terms of freezing because cooler, lighter water stratifies on the surface where ice forms on still, clear, fall nights with a high radiant heat loss. The high specific heat of water and its latent heat of fusion result in a considerable exchange of energy when water changes state. More recent thermal classifications (Reid 1961) are based on Hutchinson's terminology (Hutchinson 1957) and include amictic lakes for those with permanent ice covers and cold monomictic for those with waters that never exceed 4°C and that experience an annual period of mixing during the ice-free season. Alternative classifications consider productivity, which is biologically more meaningful.

Oligotrophic lakes fix less than 25 g $C \cdot m^{-2} \cdot year^{-1}$ in the open water season, eutrophic lakes fix greater than 75 g $C \cdot m^{-2} \cdot year^{-1}$, while mesotrophic lakes fall in between. Most, but not all, arctic lakes would fall into the first category. Other classifications deal with the origins of lake basins, and in the Arctic these usually have been formed by glacial action, erosion of valleys, glacial debris dams, and ice dams. Poor drainage in permafrost, melt kettles, slumping, and melting of permafrost are common origins of smaller shallow lakes in deposits removed from bedrock (Livingstone 1966). No classification satisfactorily identifies arctic lakes because area, volume, climate, and exposure all determine actual conditions. Arctic lakes are cold, isothermal in the ice-free season, oligotrophic, and experience a long period of ice cover each year. For any locality the duration of ice cover varies with the size of the lake, exposure to wind, and water exchange rate.

Ice is probably the most characteristic feature of arctic lakes and it affects several processes. Its presence does not imply complete stability, although it does limit gas exchange with the air, which, in shallow lakes, can lead to reduced oxygen levels and winterkill of fish. Brewer (1958) studied a small lake near Point Barrow, Alaska, that froze when the water was isothermal and almost 0°C. Under an ice cover, the water temperature rose quickly to about 2°C because of radiant heating from above (except close to the ice). When a snow cover was established, the water cooled again, only to slowly warm as heat diffused from the bottom. It cooled toward spring as this heat dissipated. Char Lake, Cornwallis Island, fits the cold monomictic classification and, as the focus for the Canadian International Biological Programme effort, is one of the most intensively studied arctic lakes. It cools to 0.2°C or less before freezing if it is windy, to 0.5°C under calm conditions (Schindler et al. 1974a). There is inverse stratification over winter and a suggestion of slight cooling. Concentrations of most substances in the water increase during winter as a result of freezing-out. Nutrients are low and precipitation adds little phosphorus or nitrogen. Welch and Kalff (1974) and Rigler et al. (1974) provide an introduction to primary production and zooplankton in Char Lake. MacCallum and Regier (1984) report on the only fish species, Arctic char *Salvelinus alpinus*.

Perhaps the most complete studies of arctic lakes were carried out in the Saqvaqjuac lakes on the northwest shore of Hudson Bay. In a series of articles, Welch (1985), Welch and Bergmann (1985a, 1985b), Bergmann and Welch (1985), Welch et al. (1987), and Jorgenson et al. (1992) describe the physical and chemical processes in a group of natural and fertilized lakes of different sizes formed in glacial depressions in Precambrian granitic gneiss bedrock. Annual heat budgets, patterns of light attenuation through various forms of ice and snow, and a general model for winter circulation are given. Midwinter water movement is slow, about 10 m/d, radially shoreward at the surface, down the slope of the basin to the deepest depression and presumably upwelling from there to complete the circulation. The forces involved are stored heat in the bottom, which warms the adjacent water and increases its density, and freeze-out of solutes from the ice, which increases water density as it flows shoreward. Heat gained from the bottom is balanced by heat lost through the ice. An interesting result of the work at Saqvaqjuac lakes was the delayed response time of the invertebrate community to increased primary production after lake fertilization. *Gammarus* biomass increased quickly, Trichoptera larvae more slowly. Neither group had reached equilibrium 3 years after the experiments began. It was suggested that the long life cycles of arctic species account for the responses observed and that macroinvertebrate biomass would eventually decrease in response to fish predation (Jorgenson et al. 1992).

A characteristic of arctic lakes is that a considerable proportion of the annual heat budget is used to melt the ice (Livingstone 1966; Schindler et al. 1974a; Welch et al. 1987). Food chains are also remarkably simple, as is evident in Char Lake. Arctic lakes, especially small lakes with little water exchange, are sensitive to eutrophication by sewage, which increases production and makes them extremely vulnerable to oxygen depletion in winter (Schindler et al. 1974b). Arctic lakes are relatively benign habitats for fish, providing they are deep enough (>3–4 m) and not polluted. Fish living in them must adapt to low temperature, low productivity, and a long period of ice cover and darkness.

Meromictic lakes represent special habitats found in previously glaciated coastal regions. They result from sea level changes and isostatic rebound of the land. These lakes are chemically stratified with deep saline bottom water and they support a relict marine fauna that has diverged from the present fauna. An unusual lake of this type on Little Cornwallis Island is being used for tailings deposit from a heavy metal mine. It has anoxic hypersaline bottom water, which may have been concentrated by freezing-out from marine permafrost. Floating on this is a layer of anoxic photosynthetic sulphur bacteria. The lake is heliothermic and the deeper waters are above 5°C. The fourhorn sculpin in the lake have atrophied head spines (Ouellet and Pagé 1988), which is not unusual in landlocked populations of this species. Such populations were considered vulnerable and genetically distinct enough to warrant special protection (Houston 1990).

Fluvial

Lotic habitats may be the most difficult of all places in the Arctic for fish (Power et al. 1993). Small streams freeze solid in winter and can serve only as temporary habitats for fish that move into them and as sources of drift. Spring-fed streams and rivers, especially those with large lakes upstream, may maintain flows all winter or be reduced to a series of isolated pools in the deepest parts of the channel. In North America the largest mainland rivers have their sources south of the Arctic and flow throughout the year, with lowest flows occurring in winter. They provide vital habitat for spawning and overwintering of anadromous fish.

The North American Arctic is a polar desert with total precipitation being less than 400 mm/year, except in parts of eastern Baffin Island and northern Labrador. Snow cover lasts for more than 6 months. Except at high latitudes and in the mountains, most precipitation falls as rain in the summer. River discharge is greatly controlled by temperature and rainfall events. In the high arctic region, glacier and snowmelt is the source of most stream water. Farther south, surface runoff from rain is important, the response time depending on the state of the ground. In spring, when the surface is frozen, runoff is rapid; later in the year more of the rain can infiltrate the thawed active layer above the permafrost (Prowse and Ommanney 1990).

Permafrost is continuous throughout the Arctic and can reach 1,000 m in thickness and penetrate under the continental shelf. To the south it thins and becomes discontinuous; pockets of frozen ground depend on insulation and exposure. Water absorbs a considerable amount of solar radiation, and its thermal properties cause a considerable amount of heat to be stored during summer, some of which is transferred to the ground through the beds of lakes and streams, thus often maintaining patches and corridors of unfrozen ground, called taliks, beneath water bodies. Taliks connect rivers to groundwater flows, allowing recharge and discharge in springs, which assume particular importance in the Arctic as sites for fish spawning and overwintering. A more complete description of this subject can be found in van Everdingen (1987) and of the effects climate change may have on this regime in Ripley (1987).

About half of the runoff from Canada flows north, transporting a considerable amount of heat and nutrients to the arctic seas. Large northern rivers support a fauna that contains many elements from farther south (Lindsey and McPhail 1986). On the north slope of Alaska, Craig (1989a) distinguished between mountain streams whose hydrograph is controlled by melting snow and ice; slightly warmer, lower velocity, coastal plain streams draining the tundra-covered slopes of the northwest; smaller tundra streams; and small streams. The fish fauna is most developed in the coastal plain streams, with fewer species in tundra streams. In the small streams, there are few fish and no anadromous stocks. On the other side of the continent, estuaries and lower reaches of small streams entering Hudson Bay are colonized each summer by euryhaline threespine sticklebacks *Gasterosteus aculeatus* and fourhorn sculpin.

Several factors make arctic rivers difficult environments for fish. The annual hydrographs include a period each year when discharge is reduced to almost zero (Figure 3). The hydrographs include periods of very high discharge during which up to 75% of the annual runoff can occur in less than 1 month. These discharges fall outside the range of flows that provide good fish habitat. In addition, stream volume tends to be very unstable and dependent on weather. Overwintering habitat is critical, and Craig (1989a) makes this point very strongly in his discussion of critical habitat. In order to survive, fish must use specific overwintering areas that are limited in capacity. These locations are occupied for many months and all life history stages may be confined to them.

Ice is important in arctic rivers and its role has been reviewed by Prowse and Gridley (1993). When first forming along the banks, it provides cover for fish, but as it expands and thickens, it encroaches on living space and impedes movement (Figure 4). Even small increases in ice thickness can have a pronounced effect (Craig 1989a). Anchor ice, which forms on the bed of streams in shallow riffle areas, can cause considerable changes in flow, scouring, backing up of water, and diversions. At certain times this may form and release daily. Frazil ice, which forms as small crystals in turbulent water, may irritate fish, and where it accumulates downstream it can form curtain dams. Ice dams can lead to surface icings and downstream dewatering. Where surface ice-cover is complete and flows are reduced, oxygen depletion can occur. This has been observed in several arctic rivers, including the Sagavanirktok and Colville in Alaska, and in tributaries of the Yukon in the Northwest Territories (Schreier et al. 1980; Schmidt et al. 1989).

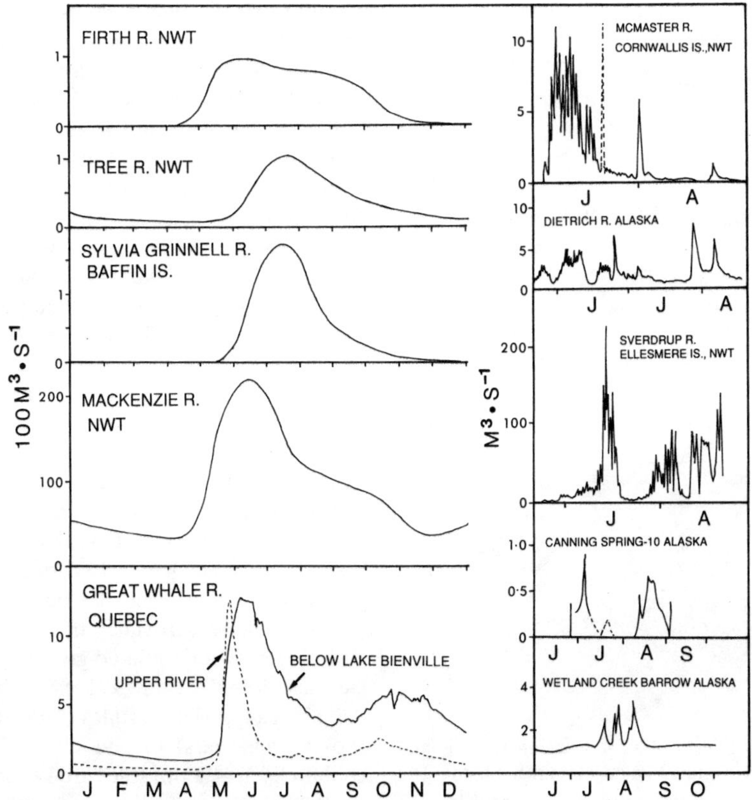

FIGURE 3.—Hydrographs of a variety of arctic streams and rivers. The Firth (69°N, 140°W), Tree (68°N, 112°W), and Sylvia Grinnell (64°N, 69°W) rivers have annual discharges close to 1×10^6 dam^3 (cubic decameter). Winter discharge in the Tree is maintained by many lakes. The Mackenzie River at the point of gauging (67.5°N, 134°W) has an annual discharge of about 275×10^6 dam^3. Size and southern headwaters maintain winter flows. The Great Whale River (54.5°N, 73°W) is farther south and its hydrographs show the influence of the large Lake Bienville. The McMaster, Dietrich, and Sverdrup rivers show marked responses to snow and ice melt and summer precipitation, as do the small Alaskan streams. Canning Spring-10 maintains a fairly even base-flow typical of springs (based on Environment Canada 1990; Prowse and Ommanney 1990; Sparks 1991.) Abbreviations are R. = River, NWT = Northwest Territories, and IS. = Island.

The breakup of ice on streams and rivers can cause severe disruptions to the flow. Ice can move and form dams during freeze-up, during periods of thaw in winter, and at breakup. Dams cause rapid changes in discharge, both during their formation and upon failure. They are studied mainly because of the damage they can cause by flooding, bridge collapse, and damage to other river works (Beltaos et al. 1990). Dams pose a hazard to fish and other stream organisms. Mainland arctic rivers are prone to this problem because of the ice thickness and the tendency to break up from south to north.

In the process of finding suitable overwintering sites, strong selective pressures must have been (and are) exerted on fluvial arctic fish stocks. These fish must be able to tolerate long periods in a dark, restricted space at near 0°C and must live on their energy reserves because there is little or no opportunity to feed and respond appropriately to sudden changes in their environment. For these reasons, the genetic integrity of the stocks is vital and, if the habitat becomes unsuitable, the small isolated units of the population can be easily lost.

Survival in an Arctic Environment

A distinction must be made between adaptation and acclimation. Adaptation is a consequence of natural selection, and over time it increases the fitness of the genome of a species for the environment in which it lives. Acclimation involves adjustments in an individual's life support processes that

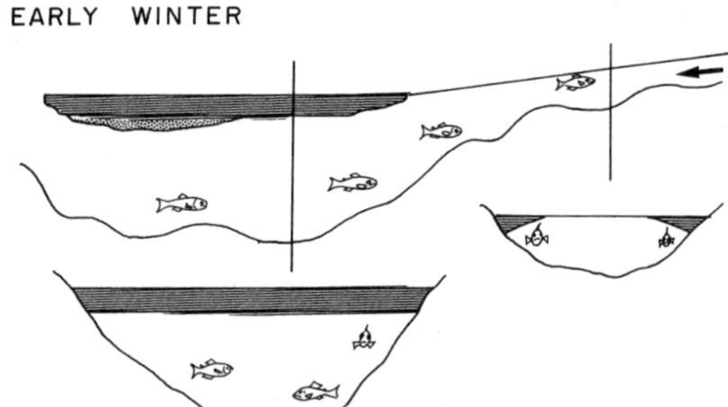

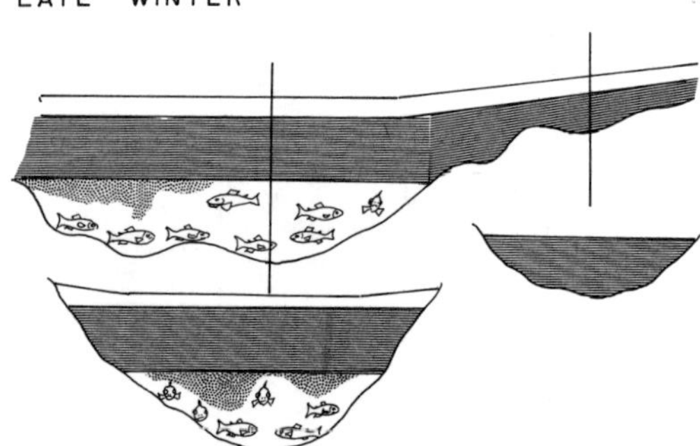

FIGURE 4.—The growth of ice in a schematic riffle and pool of an arctic stream and its effect on fish habitat. The clear top layer is snow, the horizontally shaded layer is firm ice, and the stippled lower layer is slush or frazil ice.

enable it to cope with seasonal (temporary) conditions. Both adaptation and acclimation can involve changes in morphology, behavior, and physiology. Adaptation is a genotypic response, acclimation a phenotypic response. Alternatively, the two can be thought of as ultimate and proximate levels of response, according to Dunbar (1968, 1970). Adaptation and acclimation are important in understanding the abundance and survival of fish in the Arctic.

Morphological Responses

Form and function are inextricably connected, to the extent that authors trying to describe the particular mix of morphological characters that might have survival value in arctic environments usually refer to features of the environment for clues. The large seasonal amplitude of illumination, the seasonal pulse in productivity, ice as a substrate, low temperatures, high water viscosities, stability of the Arctic Ocean and large lacustrine environments, and variability of smaller freshwater environments might all have provided selection pressures leading to adaptational morphological responses. Proximate morphological responses might also be superimposed. Slow growth and longevity or low developmental temperatures can modify the anatomy, but such changes result from acclimation and are modified if conditions change. For morphological features it may be difficult to separate adaptation from acclimation, except using controlled experiments. The age of the arctic ecosystems implies that there may have been very little time for evolution of elaborate specializations with respect to the environment, unless species were preadapted to move into the area from elsewhere. The mesopelagic and deep oceanic ecosystems are the most likely sources of marine colonizers, but fish also could have moved into the region from cold temperate refugia, as did anadromous and freshwater species.

Terrestrial ecologists considered the question of adaptation to arctic environments much sooner than did fish biologists. Two rules that grace the pages of many ecology textbooks are (1) Bergmann's Rule, which states that for warm-blooded endotherms, within related groups, the largest species are found in the coldest habitats; and (2) Allen's Rule, which states that the extremities (noses, tails, ears) of endotherms are smaller in cooler climates. Both rules are generalizations, and many exceptions can be found, especially to Bergmann's Rule. Attempts to apply these rules to ectotherms have not been very productive, although Dunbar (1968) has commented on the large size of arctic plankton and related it, among other things, to better flotation in more viscous water and long life cycles resulting from low energy. He called it the Bergmann Effect. The only comparable rule applying to fish is the so-called Jordan's Rule, which can be stated as follows: fish in polar and cold temperate waters tend to have more vertebrae and other meristic parts than do related species from warmer waters. Lindsey (1975) extended the discussion of Jordan's Rule when he examined the relationship between vertebral number and maximum body length in fish. Over a large number of families, including fish of diverse shapes and swimming methods, he found a significant correlation between the number of vertebrae and maximum length. He termed the phenomenon *pleomerism* and noted that latitudinal gradients in vertebral number were often the additive effect of pleomerism and latitudinal gradients in maximum length. He pointed out that rules in ecology were statements of trends rather than laws and that exceptions were to be expected. He offered no explanation for pleomerism but made several suggestions. One was that intervertebral joints in larger fish tend to be less flexible, therefore more joints are required to maintain mobility. It could be argued that in colder waters, fish need greater muscle mass to overcome increased water viscosity. Doubling the length cubes the muscle mass, but unless adjustments are made, the area of articulation between vertebrae is only doubled. To strengthen vertebrae against compression stresses, they would have to increase in diameter and the consequence of this would be decreased flexibility. More evidence should be collected because an alternative suggestion is that lower temperatures prolong the stage of development when the myomeres are established, and this increases vertebral number (Ali and Lindsey 1974).

The polar night and life at low light intensities under an ice cover for much of the year may be responsible for the observation that, when pairs of species, one arctic and one boreal, are compared, the arctic species generally has the larger eyes. McAllister (1977) listed five such pairs but also pointed out that fish living at greater depths also live at low light intensities, so that in making comparisons, the depths normally occupied should be taken into account. He provided two examples of pairs in which the boreal species has larger eyes but lives at greater depths. It is assumed that large eyes are more sensitive. Other sensory organs may assume increased importance under conditions in which vision becomes unreliable. Tactile, chemosensory, pressure-sensory, and electrosensory organs are all used by fish under appropriate conditions. McAllister (1977) pointed out that the nasal flaps on the anterior nostrils of Greenland cod *Gadus ogac* are twice as large as those on Atlantic cod *G. morhua* and associated this with greater olfactory sensitivity. He also suggested that the tactile sense of Greenland cod might be better because the mental barbel was relatively long and that other marine species benefit from elongated pelvic and pectoral fin rays, which assume a tactile function. Again, the assumption is that larger size confers greater sensitivity, but this needs to be confirmed. Perhaps better evidence of adaptive morphological features that would increase sensitivity is the structure of the lateral-line sensory canals. Andriashev (1970) pointed out that in cryopelagic fishes of the Arctic and Antarctic the usual system of canals is absent, and neuromast cells are exposed on the surface. McAllister (1977) listed the genera *Lota*, *Eleginus*, *Boreogadus*, *Arctogadus*, and the small-eyed boreal *Microgadus* as having free neuromasts on the head. He also pointed out that the fourhorn sculpin has its preopercular neuromasts exposed on filaments, as in deep sea fishes. There is obviously need for more research on the sensory adaptations of arctic fishes.

Color was another item discussed by McAllister (1977). Arctic fish species, with few exceptions, are dull in coloration, especially in comparison with tropical species. The only species showing bright colors belong to the genus *Salvelinus*, some of which are brilliantly colored orange-red during spawning. A few marine species have adopted contrasting white and dark markings, which would tend to be visible at low light intensities. Perhaps with fewer species there is less need for species and sexual recognition displays. Dull coloration has cryptic value and is likely to be selected for the protection it affords against predation.

Behavioral Responses

Behavior is most likely to be thought of as a proximate response to conspecifics, food availability, other biological factors, or immediate environmental conditions. The repertoire of behavior displayed by any species is, however, adaptive and likely to be that which increases fitness in the species. An example is the annual cycle of movements from feeding to spawning to rearing to overwintering areas carried out by many arctic fish species. The importance of the overwintering migration was recognized early by Russian workers (Nikolsky 1963) and has since been widely acknowledged on this continent (Armstrong 1986; Winters et al. 1988; Craig 1989a; Schmidt et al. 1989; West et al. 1992). Because of the nature of the arctic habitat, there is a strong selection for being in certain places at certain times to take advantage of temporary opportunities to feed or to avoid mortalities because of the risks of freezing, dewatering, or being washed away. Larkin (1956) made the point that most temperate freshwater species are generalists in their diets and take advantage of any food source they can consume. Dunbar (1968) has pointed out that a characteristic of arctic ecosystems is that species fluctuate in abundance, and predatory species must be flexible in their requirements if they are to persist. Thus, there exists the dual requirement for precision in timing and navigation where movements are concerned and for flexibility in energy intake and sources.

Movements and migrations have been investigated in detail in arctic and subarctic anadromous salmonids. Fortunately, several recent books and symposia have focused on the topic or included contributions (Balon 1980; Johnson and Burns

1984; McCleave et al. 1984; McKeown 1984; Kawanabe et al. 1989). The American Fisheries Society symposium on anadromous and catadromous fishes (Dadswell et al. 1987) included articles on marine migrations of Arctic char and Atlantic salmon *Salmo salar*, as well as a discussion on the prevalence of diadromous behavior worldwide. The latter topic was expanded into a book (McDowall 1988). Anadromous behavior is clearly more prevalent in the North, and coastal and marginal arctic seas contain a high proportion of anadromous species belonging to the families Petromyzontidae, Acipenseridae, Salmonidae, Osmeridae, Cottidae, and Gasterosteidae. In arctic seas Salmonidae is the dominant family, with 11 anadromous species reaching at least 71°N. These species are particularly important because their annual migrations bring them into estuaries and rivers where they are concentrated and available to predation by marine mammals and to capture by coastal human populations. The question can be asked, Why is the salmonid family so successful in the North? It is considered a primitive teleost family, as are the smelts (Osmeridae). The lampreys (Petromyzontidae) and the sturgeons (Acipenseridae) are even more ancient groups of fish. In fact, as McDowall (1988) points out, diadromous behavior is concentrated among primitive groups and families of fishes, and because these tend to be most varied and abundant in the cool temperate northern hemisphere, that alone explains the high proportion of anadromous species in arctic waters. Facts about the prevalence of anadromy are beyond refute. How, when, where, and why this particular life history pattern arose are subjects of debate (McDowall 1988). Many theories have been proposed with different authors displaying a preference for either marine or freshwater origin of the groups concerned and a distant past or recent origin of the habit. Each assembles an array of evidence and argument to support the proposed thesis; none is able to offer an irrefutable hypothesis.

Of more immediate interest for this review are events that occurred during the last 200,000 years in Alaska and the last 12,000–15,000 years in the rest of the North American Arctic; at the beginning of this period it can be safely assumed that the Salmonidae (and other fish families) had habits similar to those at present. The question of interest then becomes why anadromous groups have been so successful establishing themselves in arctic waters. The simple and probably correct answer is that their migratory habits and their ability to tolerate saline conditions enabled them to move into habitats as soon as they became suitable and accessible during glacial retreat. A more penetrating analysis of the question might ask whether maintaining an anadromous habit confers any special advantages. It might allow them to grow and benefit from the pulse of secondary and tertiary production as it is distributed down current from its source in estuaries and coastal upwellings (Skreslet 1985). This pattern of movement is demonstrated by outmigrants from the Mackenzie River, the largest North American source of freshwater entering the Arctic basin. Craig (1989a) estimated the gain in food available to anadromous species in Beaufort Sea coastal waters. He concluded that coastal waters contained almost 10 times the quantity of food available in freshwater streams when the areas of the different habitats were factored into the calculation. Only spring-fed streams match coastal waters in productivity, but they provide only a fraction of the habitat.

The distribution of coastal biomass is patchy; thus, fish must locate areas of sufficient biomass concentration to use as an energy source. This is precisely the advantage of an anadromous life cycle, and these pelagic species range widely in the shallow, warmer, productive coastal waters. Along the coast of the Beaufort Sea, marine species such as Arctic cod *Boreogadus saida* and fourhorn sculpin share this band of coastal water with anadromous salmonids (Craig 1984). In reviewing the life history patterns of the coregonids, Reist and Bond (1988) observed that most fish move along coastal corridors within the 5-m depth contour. This corridor may be closed when onshore winds push the more-saline, cold, offshore surface water to the coast or when causeway construction blocks the coastal corridor (Fechhelm et al. 1989; Griffiths et al. 1992).

Lower salinities and higher temperatures of coastal, estuary, and fjord surface waters appear to suit many anadromous species. The estuarine and coastal fish communities of James and Hudson bays include several anadromous species of coregonids and chars. In Richmond Gulf these species tend to be concentrated in the warmer, less saline, surface waters of the eastern basin, which receives most of the freshwater input (Dutil and Power 1980; Boivin and Power 1985). In James and Hudson bays the situation is less clear, but all species are associated with estuaries and estuarine plumes at least during winter and early spring (Morin et al. 1982; Morin and Dodson 1986; Kemp et al. 1989).

Characteristics of arctic species that must have had considerable survival value are flexibility in life history and in migratory behavior. Many species include strict lacustrine or riverine resident stocks.

There are freshwater stocks that migrate into streams in summer for feeding, as do many lake trout *Salvelinus namaycush* in the Arctic, returning to lakes to overwinter. Mostly streams are selected for spawning and initial rearing habitat, but lakes may also serve in these capacities. Arctic grayling *Thymallus arcticus* spawn in inlet or outlet streams of lakes in which they live; they may live, spawn, and overwinter in large rivers; they use a variety of other freshwater streams as migration routes, for feeding, for reproduction, and for overwintering (Armstrong 1986). Coregonids display multiple life history patterns including lacustrine, riverine, migrations between these, movements between freshwater estuaries, and fully anadromous (Reist and Bond 1988).

The genus *Salvelinus* is best known for its flexible life history strategies. The brook trout *S. fontinalis* in Hudson Bay, Ungava Bay, and Northern Labrador occurs in a diversity of forms, from those living in small streams and maturing at 10–15 cm in length to large anadromous stocks (Power 1980). Arctic char occurs as a variety of stocks inhabiting small streams, rivers, lakes, and as anadromous fish (Johnson 1980; Johnson and Burns 1984). Distributions of the three *Salvelinus* species overlap, and they often share similar habitats and life histories. Barriers to reproduction between species are not complete. Splake, the hybrid of brook trout and lake trout, was bred to replace lake trout decimated by predation by sea lampreys *Petromyzon marinus* in the Great Lakes (see Berst et al. 1980 for a review). Fertile natural hybrids of Arctic char and brook trout and of lake trout and Arctic char have recently been discovered in Labrador (Hammar 1989; Hammar et al. 1991). The possibility of hybridization added to the natural variability of the char species; their survival in and dispersal from several glacial refugia has led to some difficult taxonomy. Discussions about the Arctic char complex (Nyman 1989) probably will continue as new methods are applied to solving taxonomic puzzles. Atlantic salmon in Ungava Bay exhibit a wide range of life history patterns, although in more temperate regions the life cycle is shifted toward anadromy (Robitaille et al. 1986). Thorpe (1987) balanced the reproductive advantages of spawning in freshwater against the physiological costs and risks of anadromy. He concluded that maturation and physiological preparation for seawater were mutually incompatible. In consequence, rapid growth, causing early maturity, should inhibit smolting and result in an entirely freshwater life cycle. Both sockeye salmon *Oncorhynchus nerka* and Atlantic salmon responded in this direction after severe reduction in numbers caused by exploitation. Whether the prevalence of alternate life history strategies displayed by arctic salmonids is the consequence of periodic catastrophic reductions in numbers is not certain, but it is a possible explanation.

The described examples indicate that successful arctic salmonids are varied and flexible in their life cycles. Such adaptability coupled with longevity and iteroparity, which allows maturation in years when energy reserves are adequate, is part of the suite of characteristics that ensures long-term survival. Such species can tolerate limited food supplies, blockages to migration routes, winterkills that may annihilate segments of their populations, and other extremes that would eliminate more specialized species. Behavior accounts for some of their tenacity.

Physiological Considerations

An article by Scholander et al. (1953) excited fish ecologists by suggesting that the metabolism of high-latitude species was somehow adjusted to compensate for the low temperature. Arctic species were reported to have elevated metabolic rates compared with those of temperate or tropical species extrapolated to the same temperature. They functioned normally at temperatures that induced torpor or death in other species and were said to exhibit cold adaptation or temperature compensation. The idea was appealing, it seemed to fit the neo-Darwinian logic, and it led to many studies that collected evidence either in support of or critical of the concept. The debate was about issues such as methodology, the conditions under which measurements were made, the advisability of extrapolation beyond the data, the validity of making measurements outside normal temperatures, the wisdom of interspecies comparisons, and even the question of what was actually being measured. This work was largely field based and was subject to many of the problems inherent in field measurements. Set up of controls and replication were difficult; ages, health, and condition of experimental animals were unknown. During the same period, laboratory-based studies of fish metabolism (actually, oxygen consumption) were being done, and these established the protocol and rigor required for this type of measurement (Job 1955; Fry 1957, 1971; Beamish 1964). This work led to a reexamination of the question of cold adaptation. Holeton (1974), working at Resolute Bay, Northwest Territories, measured the oxygen consumption of 11 marine species, representing 4 families. He used a flow-through respirometer (except for some smaller specimens), made repeat measurements for several days, and selected the lowest oxygen consumption rate measured to avoid errors caused by handling-stress or meal digestion. Because his fish could move, he measured routine metabolism. Even so, except for Arctic cod, which remained quite active in the respiration chamber, rates were well below levels considered cold adapted. He showed that arctic fish required about 48 h before their oxygen consumption rates stabilized, and he suggested that errors of methodology had led to earlier erroneous conclusions about cold adaptation. Clarke (1980) supported Holeton's interpretation and extended it to polar marine invertebrates. He agreed that it was illogical to argue that increased basal metabolism was adaptive in animals living in an environment where the productivity was low or that it would be coupled with increased ability to be active. His analysis pointed in the opposite direction. Reduced basal metabolism and scope for activity was usual, coupled with deferred maturity, iteroparity, large eggs, reduced reproductive costs, and increased longevity. He viewed this as *K*-strategy (as in MacArthur and Wilson [1967]) suited to an environment where abiotic factors were stable and predictable and where the seasonal availability of food varied greatly.

The results of Wohlschlag's work on antarctic nototheniids are usually cited in defense of cold adaptation, but a study by Forster et al. (1987) using more recent methodology has provided routine estimates lower than Wohlschlag's basal rate. Although the *Pagothenia borchgrevinki* used displayed too much spontaneous activity to estimate a basal rate, the rate was probably elevated. Active oxygen consumption measurements were more consistent, and aerobic scope was similar to that of other teleosts. Sustained swimming speeds of small fish (~20 g) were 2.5 lengths/s, of larger fish (~100 g) 1.8 lengths/s. These authors felt that comparisons between fish from different latitudes would be more meaningful if maximum O_2 consumption rates and swimming speeds were compared. However, interpretation of data from fish of different shapes and habits still might be complicated. Montgomery and Macdonald (1984) studied escape response and eye movement, both assumed to be selected for maximum speed, in one cryopelagic and two benthic antarctic species. Maximum swimming speed showed only slight temperature compensation, whereas eye movements (saccade velocities) were more fully compensated.

In arctic waters, the basal metabolic rate of the Greenland cod was measured by Mikhail and Welch (1989), and their results did not suggest any cold adaptation. Ursin (1984) approached the question from a different perspective by using growth parameters instead of oxygen consumption to estimate metabolic rates. He makes the point that the Q_{10} estimated by Scholander et al. (1953) over the range from polar to tropical (0–30°C) was 1.6. Using data on maximum length and sea surface temperatures in a growth equation and three groups of fish (clupeids, large demersal, and small demersal), he arrives at the same average Q_{10} value. Small coral reef fishes did not fit the trend, having low metabolic rates perhaps associated with their habits. Large tropical fishes exhibit routine metabolic rates at ambient temperatures, which are about three times those of cold temperate species.

One way to avoid handling-stress associated with measurements on live fish is to use excised tissue. The contraction time of white myotomal muscle varies with temperature and fish size and can be used to predict maximum swimming speed. McVean and Montgomery (1987) compared a temperate species, *Girella tricuspidata*, with an antarctic species, *P. borchgrevinki*, of similar form. For fish of similar size, muscle of the antarctic species had a shorter contraction time, especially at low temperature. Despite this clear temperature compensation, at their respective ambient temperatures muscle contraction in the temperate species was fastest.

Hochachka (1988) offers an explanation for the confusion and differences in some of the results by considering metabolic adjustments to temperature at the cellular level. To remain viable, cells must balance the passive diffusion of ions through transmembrane channels with active transport against concentration gradients. One reason some cells are cold-sensitive is that temperature affects the physical diffusion process, $Q_{10} = 1.2 - 1.4$, differently from the biological transport, $Q_{10} > 2$.

Antarctic fish maintain channel to pump flux ratio by increasing the pump density, which occurs at the metabolic cost of approximately doubling their basal metabolism. Deep-sea fishes living at stable low temperatures have adopted the alternate strategy of downward adjustment in functional channel densities. The cost of this is drastically reduced metabolic rate, reduced scope for activity, and impaired osmoregulatory capacity. If Hochachka is correct, then arctic fish probably did not arise from deep-sea fish. Because they do not appear to exhibit high basal metabolic rates like the antarctic Nototheniidae, they must have solved the problem of maintaining activity and osmoregulation independently.

TABLE 1.—Arctic fish species in which antifreeze glycopeptides (AFGP) and peptides (AFP) have been reported (sources: Davies et al. 1988; Ewart and Fletcher 1990; Cheng and DeVries 1991; Davenport 1992).

Family	Species	Common name	Antifreeze	Type
Clupeidae	*Alosa pseudoharengus*	Alewife	AFP	II
	Clupea pallasi	Pacific herring	AFP	II
	C. harengus harengus	Atlantic herring	AFP	II
Osmeridae	*Osmerus mordax*	Rainbow smelt	AFP	II
Gadidae	*Boreogadus saida*	Arctic cod	AFGP	
	Eleginus gracilis	Saffron cod	AFGP	
	Gadus morhua	Atlantic cod	AFGP	
	G. ogac	Greenland cod	AFGP	
	Microgadus tomcod	Atlantic tomcod	AFGP	
Zoarcidae	*Lycodes polaris*	Canadian eelpout	AFP	III
Cottidae	*Hemitripterus americanus*	Sea raven	AFP	II
	Myoxocephalus aenaeus	Grubby	AFP	I
	M. verrucosus	Warty sculpin	AFP	I
	M. scorpius	Shorthorn sculpin	AFP	I
Pleuronectidae	*Pleuronectes ferrugineus*	Yellowtail flounder	AFP	I
	P. americanus	Winter flounder	AFP	I
	P. quadrituberculatus	Alaska plaice	AFP	I

Teleost fish in arctic seas live continuously or seasonally at temperatures that are below the freezing point of their body fluids. They can do this by avoiding contact with ice and existing in a supercooled (undercooled) state or by relying on antifreeze proteins to depress their freezing point below that of seawater. Scholander and his colleagues were the first to recognize supercooling in arctic fish (Scholander et al. 1957). Experimenting with the Arctic cod and other species, they showed that some species could survive, apparently indefinitely, in an aquarium held at −1.5°C provided no ice was present. If ice, or a partly frozen fish, was introduced, the fish froze instantly and died. The freezing point of the blood of these fish was −0.9 to −1.0°C, which is slightly below that of most teleosts. In the absence of ice nuclei, these fish can exist in a supercooled state. Fish that are dependent on this method of avoiding freezing must live continuously in deeper waters, or move there in winter, to avoid risk of contact with ice; this conclusion does not seem to fit the cryopelagic habits of Arctic cod.

The presence of antifreeze solutes in the serum, along with slightly higher concentrations of sodium chloride and other body fluid constituents, is the most common way of avoiding freezing (Table 1). These lower the freezing point to about −2.2°C, which is below that of seawater, −1.9°C. The antifreeze solutes are relatively large molecules and have been identified as glycopeptides (DeVries 1971; DeVries et al. 1971) and peptides (Duman and DeVries 1976). Several reviews on the topic have appeared (DeVries 1982, 1983; Feeney et al. 1986; Davies et al. 1988; Cheng and DeVries 1991; Davenport 1992). The structure and biochemistry of the molecules have been described (DeVries 1984, 1986; Davies and Hew 1990) and the site of synthesis, seasonal cycle, and evolution discussed (Scott et al. 1986; Davies et al. 1988).

Antifreeze proteins are surprisingly ubiquitous. They promote undercooling in polar fish species and they have been shown to modify the growth of extracellular ice crystals in a variety of freezing-tolerant organisms, including plants (Griffith et al. 1992), intertidal invertebrates (Davenport 1992), and insects (Duman et al. 1991). The aquaculture industry, which is interested in protecting salmonid fish, caged at the surface, from the lethal effects of low winter temperatures along the North American East Coast, has provided considerable stimulus for work on fish antifreeze proteins. Attempts are underway to transfer the genes coding for antifreeze production into Atlantic salmon, but a freeze-resistant strain is not yet commercially available. When antifreeze protein from winter flounder *Pleuronectes americanus* is injected into rainbow trout *Oncorhynchus mykiss*, it confers freezing resistance to this species, showing that the action of the proteins is not species-specific (Fletcher et al. 1986).

The following summary is based largely on the reviews by Cheng and DeVries (1991) and Davenport (1992). About half of the observed depression of the freezing point of body fluids of fish in polar seas is caused by an increase in dissolved sodium chloride and other small molecular weight solutes. The remainder of the depression is caused by antifreeze proteins, which are believed to function by adsorbing to ice crystals—the molecular structure of antifreeze proteins allows them to bond to the ice crystal lattice. This alters the pattern of water molecule attachment to the crystal and inhibits its growth until the temperature is reduced to −2.2°C. The crystals then grow rapidly, assuming long spicule forms resembling glass wool. Freezing and death ensue at a fraction of a degree below this. To demonstrate the presence of antifreeze proteins in fish blood, the freezing and melting points of the plasma are determined. If antifreeze is present, the plasma freezing-point is about 1°C below its melting point, a condition termed thermal hysteresis. This phenomenon is attributable to macromolecular components in the plasma. When these are removed by dialysis, the plasma freezes and melts at the same temperature. The freezing point then depends on the concentration of solutes remaining. The antifreeze proteins act noncolligatively, meaning their action is not proportional to the number of molecules and, therefore, they lower the freezing point without changing osmotic balance. Concentrations are often about 30–40 mg/mL, above which level they have little further effect on the freezing point.

Protection against freezing involves more than just preventing the blood from freezing. The gut contents, excretory fluids, ocular fluids, endolymph, neural tissue, and cell contents are all at risk, and various explanations have been proposed about how they are protected. To maintain osmotic balance, marine fish imbibe seawater, absorb sodium chloride and water from the gut, and secrete excess sodium chloride via the gills. The fluid in the gut becomes increasingly dilute as salts are removed and a gradient is produced for water uptake. Antarctic nototheniids secrete antifreeze proteins in the bile so that the gut contents maintain a freezing point below that of the environment (O'Grady et al. 1983). These glycopeptides are not digested, so they represent a metabolic cost to the fish (Davenport 1992). The seasonal cycle of antifreeze concentration increasing in winter and decreasing in summer would help reduce these costs. Similar protection is not afforded urine, as no antifreeze has been found in this fluid. Cheng and DeVries (1991) suggest that because it is important for polar fish to conserve their antifreeze, rapid loss in the urine would be energetically costly. The only explanation offered as to why urine does not freeze is that the urethra contains mucus and the sphincter is tight enough to prevent ice nucleation. Ocular fluids may contain small amounts of antifreeze; none have been detected in endolymph. These fluids, and perhaps the brain and nervous tissues, are thought to be protected because they are internal and surrounded by tissues that are cryoprotected and act as barriers to ice propagation. The skin over the cornea is fortified with antifreeze glycopeptides and can prevent nucleation by ice crystals (Turner et al. 1985). A similar explanation was invoked by Fletcher et al. (1988) to account for the greater freezing-resistance of Arctic char compared with brook trout and Atlantic salmon. They thought that the epidermis was acting as a barrier to ice propagation.

Several authors have remarked on the general similarity of glycopeptide and peptide antifreezes found in arctic and antarctic fish and regarded this as an example of convergent evolution. Scott et al. (1986) interpreted the data differently. They used the four main identified types (Table 1) to reconstruct the relationships and origins of the fish possessing them. The main teleost groups had arisen by the Eocene (40 million years ago), when the climate was considerably warmer than at present. The antarctic cooling (30 million years ago) first exposed teleosts to temperatures below their freezing point. The Nototheniidae developed an antifreeze that is essentially the same across three families. The Zoarcidae include both arctic and antarctic representatives, and evidence suggests that this family evolved in the south, the area of highest endemism, and spread north in a transhemisphere dispersal. The Cottidae possess two types of antifreeze, and these are believed to have originated independently during the arctic cooling (2.5 million years ago). The similarity between the antifreeze peptide of the shorthorn sculpin *Myoxocephalus scorpius* and that of the distantly related winter flounder is taken as an example of convergence. The Gadidae require special consideration because the glycopeptide they possess is almost identical to that of the Nototheniidae. Scott et al. (1986) think it improbable that such a complex molecule could have evolved twice, and therefore, postulate a relationship between these groups that dates back at least 30 million years. If this is true, the Gadidae, which are generally thought of as a northern hemisphere group, must have dispersed north after their antifreeze system was established. Perhaps its presence gave them a competitive edge with the onset of arctic cooling (2.5 million years ago), an advantage they have clearly maintained. Davies et al. (1988) agree with this interpretation, believing that the four basic types of antifreeze protein are sufficiently distinct that they must have arisen independently and that their origin probably coincided with the recent antarctic and arctic glaciations. Using antifreeze-protein gene detection techniques, Davis et al. (1988) have examined the distribution of gene signals within fish families. They concluded that the suborder Blennioidei should probably be separated from the suborder Zoarcoidei, as in Nelson's (1984) classification, because of the absence of a DNA sequence that hybridizes to the type III antifreeze gene sequence. They relate ability to produce higher concentrations of antifreeze to gene amplification, in which the functional units are organized in direct tandem repeats. They postulate that the climate during glaciation provided the selective pressure for gene amplification in species living in freezing, ice-laden seas. The trigger that initiates seasonal production of antifreezes is the autumn reduction in daylength. Once production is started, the concentration in circulation is temperature dependent. Synthesis is most rapid at low temperatures, clearance at high temperatures.

In addition to strictly arctic fish, cold temperate fish that inhabit or move into icy waters to spawn may exhibit protective adaptations against freezing. Supercooling has been reported in the cunner *Tautogolabrus adspersus*, a north temperate species that overwinters in a dormant state in rock crevices in inshore waters of Newfoundland and dies if touched by ice. Mortalities of this species in winter coincide with turbulent conditions (Green 1974). An antifreeze protein found in the skin of this species (Valerio et al. 1990) shows a seasonal cycle of activity. This must afford some winter protection against freezing because the skin is the first tissue likely to come into contact with ice. The capelin *Mallotus villosus* has a circumpolar distribution and occurs in the Bering Sea, Hudson Bay, and Hudson Strait. It lays its eggs in early spring in the breaker zone, even on shore, and at low tide its eggs are often exposed and subjected to freezing air temperatures. Davenport (1992) studied capelin development and found that the eggs could survive up to 6 h at temperatures down to −5.2°C. Some eggs survived longer, and others tolerated even lower temperatures for short periods. The mechanism depends on an intact chorion and a large perivitelline space. On exposure to freezing temperatures, the water film around the egg starts to freeze, thus raising the osmolarity of the unfrozen water, which results in water being drawn out of the large perivitelline space, and the chorion shrivels. The embryo is unaffected inside the vitelline membrane. The increased salinity of the residual perivitelline fluid prevents freezing and produces a barrier to ice nucleation in the embryo. Upon a rise in temperature, the elastic chorion becomes spherical and the perivitelline fluid is replaced. The Atlantic tomcod *Microgadus tomcod* spawns in winter, the peak spawning time for the Sainte-Anne River stock in Quebec being mid-January (Fortin et al. 1990). On route to spawning sites in the river, the fish migrate through curtains of slush and frazil ice that accumulate in the estuary. The grubby *Myoxocephalus aenaeus* and the winter flounder also spawn in winter in shallow inshore waters where they may come into contact with ice. Reisman et al. (1987) suggested that the presence of antifreeze proteins in these species during winter has obvious survival value.

Fisheries

Any discussion of arctic fisheries must begin by clearly distinguishing between subsistence, domestic, sport- or recreational, and commercial fishing. Subsistence fishing has been carried out for as long as people have inhabited the Arctic. The other fisheries are recent developments. Through the eyes of indigenous residents, the subsistence fisheries are the most important because they help to feed people. Subsistence fisheries are part of the culture and have a significance above and beyond the food they provide. Masty (1991) makes the point strongly: Many Cree names relate to the quality and type of fish that can be caught, reliable fishing spots enable nomadic families to travel light, and fish have saved many Cree families from starvation. He cited many uses for fish, including fish fat used as ointment for skin problems and fat from boiled lake trout stomachs as a substitute for milk of a mother who died in childbirth. Inuit and other Indian groups have similar cultural ties to fish. Subsistence fisheries are generally acknowledged as the most important use of fish and take precedence over other uses where conflicts arise (Clarke et al. 1989).

Sportfishing may be practiced by wage-earning residents purely for entertainment and it may provide a food supplement. When tourists fish, it provides revenue and the type of employment locals appreciate. Commercial fishing is a source of jobs and income. The catch may or may not be processed, and usually it is sold outside the region. The importance of these fisheries is that they provide opportunities for seasonal, sometimes year-round, work in a region where there are few or no alternatives outside government-supported employment.

Berkes (1990) drew together much of the gray literature on subsistence fisheries in Canada and compiled a list of 96 studies representing 93 communities that satisfied his criteria with respect to data and methodology. Many of these were aboriginal communities in subarctic or arctic Canada. The range of estimates of fish harvest was substantial, 6–600 kg per capita per year. There were no obvious differences with latitude. Almost half of the harvest values were between 43 and 91 kg, including 22 of 35 arctic locations. His data were approximately lognormal, with a mean of 60 kg, which provided about 42 kg per capita per year of edible fish. When there were estimates from a community over several years, they varied with fish availability and employment opportunities. Seasonal jobs often coincided with the fishing season. Berkes felt that the subsistence harvest represented at least 10% of the reported Canadian inland commercial catch, and this was probably a gross underestimate. Crawford (1989), in reviewing the state of arctic fisheries, quoted estimates of the subsistence catch at 40% of the total catch in the Northwest Territories. Corkum and McCart (1981) estimated it at 80% of the catch in the Mackenzie River delta. Crawford (1989) emphasized the incomplete and often unreliable statistics on fish harvest for personal consumption. It is difficult to determine whether subsistence fisheries in Canada are increasing in importance or are in decline. The Inuit, Indian, and Metis populations are growing with a doubling time of between 20 and 25 years, but so is the wage economy. Dogs have been replaced by snowmobiles and outboard engines have replaced paddles. Food from the south is shipped into northern communities by air on a regular basis. Subsistence fishing is being replaced to some extent by sportfishing for food on weekends and holidays. In a study of an Ojibwa community in northern Ontario, Hopper and Power (1991) noted that older males did most of the net fishing. In the Inuit community of Kangiqsualujjuac, northern Quebec, most of the harvest of Arctic char was taken by a few experienced hunters (Boivin et al. 1989). In both communities, the catch was distributed throughout the community so that the elderly, incapacitated, and other community members who were unable to fish benefited from the catch.

In Alaska, more effort has been put into studies of subsistence fishing, partly because of conflicts between domestic and commercial use of salmon species (Wheeler 1987; Andrews 1988; Walker et al. 1989). Inland Indian communities are at the end of the line of harvesters, so downstream catches have to be regulated to allow sufficient escapement for their fisheries to be maintained. With respect to the arctic coast, Craig (1989b) has reviewed the available information. The fisheries are based on about 19 species of anadromous, marine, and freshwater fishes, the most important being pink salmon *Oncorhynchus gorbuscha*, Arctic char, Arctic cisco *Coregonus autumnalis*, least cisco *C. sardinella*, and Arctic grayling. Most species are taken in gill nets in the vicinity of villages during the open-water season. Although fishing is culturally less respected than hunting, it nonetheless contributes a substantial amount of food. In 1985 the fish harvest for six North Slope villages was estimated at 21 kg per capita. This estimate is on the low end of estimates for the Canadian North, but as Craig points out, the data upon which it was based were not all recent and may have been incomplete.

In an attempt to estimate potential fish yields from lakes, Schlesinger and Regier (1982) investigated the importance of the morphoedaphic index (M.E.I.) and mean annual temperature (climate). On a global scale, climate had the greatest effect. They produced a map showing theoretical upper limits to fish yield. Arctic lakes fell into the range of 2 to less than 1 kg/ha. In subarctic lakes the limit rose to 5 kg/ha. If subsistence fisheries are based entirely on lacustrine stocks and Berkes's (1990) estimate of 60 kg/year harvest is accepted, it implies each person in the arctic harvests the yield from 50–75 ha of lakc. A consequence is that the fuel required to catch fish will increase northward. Hopper and Power (1991) commented on the area required and the need for cheap fuel to maintain fish harvests. The hunter support program initiated in northern Quebec after the James Bay and Northern Quebec Agreement provides a subsidy to full-time hunters and fishers to help maintain their activity.

Fortunately, the dependence on anadromous species in most coastal communities means that the fish can move and they may be caught locally during migration. The fish are then harvested intensively during a short season, and means of preservation become important. Boivin et al. (1989) noted that the lengths, weights, and ages of Arctic char increased with distance of the fishing sites from an Inuit community. Near sites experienced the greatest fishing pressure. It certainly cannot be assumed that subsistence fisheries will not affect the resource.

Commercial fisheries in the Arctic are relatively small and local in extent. They are based on easily accessible, high-value species that are worth transporting south. These are mainly anadromous whitefish, salmon, Arctic char from coastal waters, or lake trout and whitefish from inland waters. Demand for quality fish has increased as stocks farther south have been damaged, and this has certainly helped to promote and maintain commercial harvests in the North. The Canadian government embarked on an intensive promotional campaign in the 1960s to establish a high-price gourmet market for Arctic char; other species did not require this.

There are several problems with arctic commercial fisheries. Local markets are limited, although this is changing as the wage economy grows. To cover the costs of handling and shipping, wages to fishers are kept low. In many cases the costs of gear and fuel exceed revenue and subsidies are required, which often have come from the government in the form of loans. Management of fisheries is sometimes complex, especially when there is a mixture of unregulated subsistence and commercial fishing. On the one hand, there are gear and season regulations; on the other, no regulations apply. In a transitional economy, a need for cash for a specific purchase is often the stimulus to fish. Once the need has been satisfied, interest declines. Other problems relate to the sparsity of the resource, and its slow growth and rate of replacement. This is particularly true of lake trout, some whitefish stocks, and Arctic char, which may be 7–12 years old before they are large enough to harvest. These problems are more fully discussed by Crawford (1989) and in reports prepared for the Canadian Arctic Fisheries Scientific Advisory Committee (Clarke et al. 1989; Cosens et al. 1990).

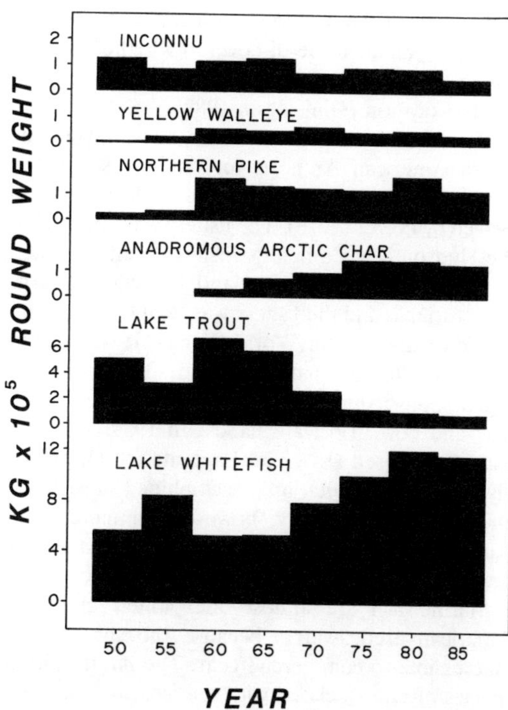

FIGURE 5.—Five-year mean commercial catches of the principal species taken in the Northwest Territories between 1948 and 1987 (data from Yaremchuck et al. 1989). Round weight is in 100,000 kgs. Fish species are inconnu *Stenodus leucichthys*, yellow walleye *Stizostedion vitreum*, northern pike *Esox lucius*, Arctic char *Salvelinus alpinus*, lake trout *S. namaycush*, and lake whitefish *Coregonus clupeaformis*.

Catches in the Canadian Northwest Territories of the six principal species between 1948 and 1987 are given in Figure 5. The figure shows 5-year mean harvests, so as to remove the rather large annual variations in landings, which possibly reflect influences other than stock size on the catches. After an initial buildup, there has been a marked decline in lake trout harvests, which have fallen to less than one-fifth of the best 5-year period. Catches of lake whitefish *Coregonus clupeaformis* have increased, and the growth of the Arctic char fishery—which, while low in weight, is high in value—has helped stabilize commercial fishing revenues. The Arctic char fishery has the highest profile and much effort has gone into maintaining it (Dempson et al. 1986; Kristofferson et al. 1986; Gillis 1988; Crawford 1989). After a number of collapses of ill-planned, badly regulated Arctic char fisheries, operations are now carefully monitored. Quotas are set, and reduced if there is any indication of overfishing. New fisheries are evaluated and the effects of a limited fishery on stock parameters monitored. Only when it is determined that the stock can be exploited profitably is commercial fishing allowed. Otherwise, the stock is reserved for subsistence use.

Commercial exploitation of marine fish or anadromous species in the sea is limited to the continental edge of the Bering Sea and to Hudson Strait and northern Labrador. The target species, such as Pacific herring *Clupea pallasi* or Atlantic cod, have their centers of distribution to the south. The eastern Arctic shows most promise, but so far the greatest potential seems to be in the harvest of the northern shrimp *Pandalus borealis* or the striped pink shrimp (Aesop shrimp) *P. montagui*. The Iceland scallop *Chlamys islandica* has also shown some harvest potential (Crawford 1989). Cosens et al. (1990) noted Ungava Bay and Cumberland Sound as possible sites for scallop fisheries, but scallop growth rate was slow in Cumberland Sound. Killiniq, at the northern tip of Labrador, was the site of an experimental fishery for Atlantic cod and Greenland halibut *Reinhardtius hippoglossoides* in the mid-1980s; this followed several earlier attempts to promote a commercial fishery. The catches were low and unpredictable, and weather and ice conditions hampered the operation and damaged the gear.

Returns from the fishery were insufficient to cover costs. It remains to be discovered whether there is any real possibility of a commercial operation based on this location (Gillis et al. 1987).

The longest running commercial fishery in the North American Arctic is for Atlantic salmon in Ungava Bay. The history of this fishery was reviewed by Power (1976). The fishery was initiated by the Hudson Bay Company. Fish were either salted or frozen and sent to England for sale. Returns were variable and the fishery was temporarily abandoned at the start of World War II. It was reactivated in 1961 and has continued since then, first under sponsorship of the Inuit Cooperative at Kujjuac and now in private hands. In 1989, 1990 the fishery collapsed (Roy and Laplante 1991), probably from overexploitation in a combined subsistence and commercial fishery; however, an increase in arctic water outflow past the northern part of Labrador may have closed the migration window.

Marine fish are almost unexploited in North American arctic waters because they are mostly inaccessible to commercial gear. The Arctic cod, a species often associated with sea ice, may be quite abundant (Frost and Lowry 1984; Percy et al. 1985; Welch et al. 1992; Hop et al. 1997, this volume). Acoustic techniques are now being tested as a means of estimating biomass of fish species under the ice (Crawford 1988) and there has been some success recording Arctic cod. It is unlikely that the Canadian government would allow a fishery for Arctic cod because of the key role of this fish in arctic marine food webs upon which the Inuit subsistence economy depends.

Sportfishing is an important recreational activity in North America. In Canada alone, direct spending by anglers in 1985 was estimated at US$2.5 billion and their total economic impact was estimated at $4.4 billion (Pearse 1988). In the United States in 1989, marine recreational anglers spent an estimated $7.2 billion (O'Bannon 1990). Participation rate in recreational fishing is likely to be higher than average for northern residents. Visitors are attracted by the chance to catch trophy fish and experience the North. Angling provides jobs to guides, outfitters, and fishing-lodge operators and their staff. It generates business in the air charter, travel, accommodation, clothing, fishing equipment, craft, and souvenir parts of the economy. Some money stays in the North, while much is transferred south for materials and supplies.

In Alaska, pink salmon, Dolly Varden *Salvelinus malma*, Arctic char, and Arctic grayling dominate the angling catch. In the interior, Arctic grayling is the preferred species (Armstrong 1986). In the Northwest Territories, Arctic char, lake trout, northern pike *Esox lucius*, and walleye *Stizostedion vitreum* are popular in the western mainland, Arctic char in the archipelago (Crawford 1989). In northern Quebec, Atlantic salmon and Arctic char are mainly sought; 28 fishing lodges were open in that area in 1974 (Power and Le Jeune 1976). In the Northwest Territories, 41 lodges were operating in 1982. Growth in sportfishing in the territories can be seen in the numbers of licenses sold. These tripled between 1968–1969 and 1985, to just over 16,000. The number of visitor licenses was rather stable at about 3,500. The growth in the resident sector was largely the result of growth of communities like Yellowknife, Hay River, and Fort Smith (Crawford 1989). It should be realized that many southerners are attracted to northern communities by the recreational opportunities they offer, and quality angling is perhaps the most important. For this reason alone, special attention needs to be paid to maintaining a quality resource near communities. A conflict arises between subsistence fisheries and recreational use of fish near communities, especially if there is a large indigenous population. Land claim settlements and priority use of certain species as was agreed to in the James Bay and Northern Quebec Agreement (Anonymous 1976), complicate the management issues. As Pearse (1988) has emphasized, governments have been slow to recognize the importance of sportfishing, but it is an important consideration in the development of the North.

Issues in Science, Conservation, and Development

Like all scientific literature, that on arctic fish ecology has proliferated. The geographical boundaries of the Arctic are not always explicitly defined, and this applies to titles, abstracts, and even contents of articles on marine and freshwater fishes. An added complication is that most recent research has been funded as a consequence of development and the results may be confidential or published without peer review in the gray literature (Collette 1990).

There must be several thousand English-language titles dealing with North American arctic fishes. There have been several attempts to make the information more available by producing reviews (Magnin 1977; Corkum and McCart 1981; Barnes et al. 1984; Norton 1989), symposia (Balon 1980; Johnson and Burns 1984; Kawanabe et al. 1989), popular accounts (Percy et al. 1985), and bibliographies (Pfeifer 1977; Scott 1979; Armstrong et al. 1986; McAllister and Steigerwald 1986; Winters et al. 1988; Marshall et al. 1990; Heuring et al. 1991). Unfortunately not all these documents are readily accessible. The importance of fish as food, for recreation, and as monitors of the state of the environment has ensured that at least some species have been studied in detail. Others have been virtually ignored because they are small, apparently rare, or inaccessible. These species need some attention, and a strong case has been made for a comprehensive volume on arctic fishes (McAllister 1990).

Conservation of arctic fish resources is an issue that needs to be addressed before it is too late (Hammar 1989). We have assumed that the remoteness of the region is sufficient protection from the effects of humans, but this is no longer acceptable. Freshwater and anadromous species are at most risk because, for part of their lives, they occupy habitats that are used by humans. More than one-third of the world's freshwater fish fauna is considered threatened today (Nyman 1991). Arctic species are not likely to be spared extinction; fluvial species are especially vulnerable because of the limited space for spawning and overwintering. A difficulty with many arctic fish is defining the unit that should be preserved. Species are ill-defined, many exist in several forms with varied life histories, growth, longevity, and reproductive patterns. Reist (1988) discussed this as it applies to anadromous arctic species and raised the question with respect to marine and lacustrine species, about which much less stock-specific information exists. He also emphasized the complexity of conserving stocks that are spatially and temporally separated and between which the rates of genetic exchange are unknown. Managers and conservationists need to understand the complexity of population structuring to deal with fish resources. Nyman (1991) has addressed the question of size of unit that must be preserved to ensure long-term survival. His focus was conservation of biodiversity in fish resources as identified in recent mission statements of organizations such as the United Nations Environment Program, the International Union for Conservation of Nature, and the World Wildlife Fund. The subject has propagated a whole alphabet soup of acronyms: MVP (minimum viable population), MDA (minimum dynamic area), SLOSS (single large or several small reserves), and PVA (population viability analysis). A recent development has been PVA risk analysis, which tries to incorporate chance random events into conservation strategy. This could be very important in the Arctic, where much of a species reserve reproductive potential might be needed to compensate for random catastrophic losses caused by abiotic events.

This raises the question of sustainability of arctic fisheries. Ursin (1984) observed that marine fisheries in the Arctic were based on demersal species, tropical on pelagic species, and temperate on both. This was because cycling of primary production in the tropics was so fast that none of it fell to the bottom. The reverse was true in the Arctic, at least in coastal waters; but would it be true under the polar pack?

Subsistence fisheries are certain to continue in the foreseeable future. The debatable question is to what extent should recreational and commercial fisheries be encouraged in the arctic? Unfortunately, many fishery positions are based on these activities and it could be argued that it is in the profession's interests to promote them. Conservation, defined as "wise use of," does not conflict with these uses of the resource. What should our position be if technology develops to catch Arctic cod under the arctic sea ice? We know that this fish plays a key role in arctic food chains (Frost and Lowry 1984; Norton and Weller 1984; Finley et al. 1990; Welch et al. 1992; Hop et al. 1997). What happens if Greenland cod becomes of interest for conversion into fish meal? Rather little is known of its role in the eastern Arctic (Mikhail and Welch 1989). Should we wait until more is known, or should we allow a fishery to proceed and monitor it? Hammar (1989) pointed out the disastrous consequences of overfishing capelin and other small fishes in the Barents Sea on predators dependent on them, such as seabirds and seals. What is the potential for aquaculture in the Arctic? The Inuit and Cree have expressed considerable interest in aquaculture in northern Quebec. Arctic char, with its ability to grow rapidly in cool water, has potential in the aquaculture industry. This species has been the subject of a recent technology transfer mission to Europe and an economic analysis of interaction between farmed and wild produce (Smith 1989). It is hard to believe it can be farmed competitively in the Arctic, because of the costs of fish food and exporting the produce, but it might be better to subsidize the industry than to pay welfare. It is doubtful whether attempts to transfer antifreeze genes to salmonid species will change the situation, because, except in indoor closed-culture systems, ice will create too many hazards. The possibility of increasing natural production of Arctic char has been suggested in Quebec (Power and Barton 1987), and some projects have already been carried out (Dumas 1990). There must be many possibilities across the North American Arctic to increase stocks of anadromous Arctic char and possibly other species by removing barriers to migration and making overwintering lakes accessible.

Several topics relating the growth of human populations in the North to requirements for water supplies, sewage, and garbage disposal have been the subject of reviews (Healey and Wallace 1987). Increasing numbers of inhabitants will put added stress on fish stocks. This expansion is caused by the natural growth of indigenous populations, by industrial development, and by increased government presence. Thus far, industrial development has only breached the Arctic in a few locations and is associated with oil, gas, and mineral extraction. This development brings its own risks to the environment. To what extent this activity is compatible with the idea of sustainable development and with the Canadian government policy of no net loss of habitat for fish remains to be seen. In the Arctic, there was a tendency to compensate for the high costs of doing business by relaxing environmental standards. This has now ended and may slow the pace of development, especially in Canada where land settlement agreements have and will give the native population more voice in decision making. Hydroelectric development is still largely a subarctic problem because of the energy losses in transmission. In Quebec it is moving into fringe arctic regions, with the proposed Great Whale River scheme. This is meeting heavy opposition, and is temporarily on hold. If it proceeds, it will have effects on anadromous and fluvial fish stocks in the area. Much more insidious than any of the hydroelectric developments are the proposed major water diversion schemes (Bocking 1987). These envisage turning south the headwaters of many major rivers to feed the industrial heartland and the southwestern United States. The impact of such schemes on arctic fish resources is hard to imagine. They will be complex and devastating to many stocks.

Related problems that are appearing, and that may threaten arctic fish stocks, are the long-range transport of atmospheric pollutants and climate change. The former problem is of concern to northern communities (which consume a lot of "country food" from higher trophic levels) because of possible human health effects (Wagemann 1989; Anonymous 1992a). Long-range transport of atmospheric pollutants contributes to increased mercury, heavy metals, and other contaminants in northern fish stocks (Brouard et al. 1990; Muir et al. 1990; Oehme 1991; Anonymous 1992b). These contaminants may spoil fish taste (Lockhart et al. 1987), but effects on the viability of fish stocks have not been demonstrated. There is evidence that more rigorous emissions controls since the 1970s have led to a reduction of lead, cadmium, and zinc deposited in the Greenland snows (Boutron et al. 1991). Climate change is also a product of global activities, and while there may not yet be irrefutable proof that the climate is warming, the evidence is mounting. In a response to problems likely to be created by global change, the Royal Society of Canada (Mungall and McLaren 1991) released a book that was written by 33 authors, all acknowledged experts in their fields. It reports some alarming observations, several of which are described in the following paragraph so as to underscore the uncertainty concerning the future of arctic fish resources.

The average expansion and melt back of the sea ice over the Arctic involves an area equal to 70% of the area of Canada. In the east, heavy ice-years occur roughly every 10–14 years. The position of the ice margin each season means the difference between near-starvation and survival for Inuit dependent on marine resources. It also influences navigation, nutrient supply, and fisheries. Since 1976 the average ice thickness near the pole has fallen 40%. Runoff controls salinity in the Arctic basin; increased runoff leads to more ice. In dealing with an issue as complex as global change, there are feedback loops that produce unexpected results; warmer temperatures lead to increased glacier melting and more sea ice! In freshwater, the effects might be more beneficial. We can expect increased fire frequency and evaporation, increased dissolved solids and nutrients, and higher productivity. There also can be problems caused by permafrost melting, erosion, and changed drainage patterns. There are, and there will be, many unresolved questions about arctic fish ecology—enough to keep fishery biologists busy for a long time.

Postscript

Writing a review on arctic fishes is a challenge because of the large amount of literature (primary, secondary, and gray), much of which cannot be incorporated. Each of the referees made useful suggestions that improved the original manuscript, but I believe each would have written a review with a very different emphasis. The following list is based on questions and issues they raised that could not be adequately integrated into the review but that should stimulate further debate on arctic fish ecology.

1. More emphasis should have been placed on the long-term prospects for arctic ecosystems and the fish resources they support. Pollution from north-flowing rivers, arctic marine dumping, and atmospheric transport is probably more significant than most of us realize! Arctic ecosystems are threatened.
2. An opportunity was missed to use information from simple arctic ecosystems to explore and test general ecological concepts. One advantage of most arctic ecosystems that should be exploited is that they are still essentially undisturbed, which means they are ideal reference points for comparison with more complex systems elsewhere.
3. Why have marine mammals succeeded in penetrating the Arctic Ocean when large fishes have generally failed?
4. Discussion of the physiological and ecological processes leading to longevity in arctic fishes might contribute to an understanding of aging and theories of senility.
5. A more detailed discussion of population dynamics of arctic fish—particularly how stable population structures are maintained in the face of a variable, abiotically dominated environment and how arctic stocks respond to exploitation—might help us understand the dynamics of other exploited populations. Data on some exploited Arctic char stocks suggest there are few changes in the parameters normally measured (age, growth, mean weight) to indicate the level of exploitation.

I hope these points help to stimulate interest in arctic fish ecology. One referee thought a book or books could be written on the topic, another thought arctic fishes had been reviewed too often. It is obvious there are many opinions to be discussed and many questions still to be answered.

Acknowledgments

I thank Jim Reist and Jim Johnson, Department of Fisheries and Oceans, Freshwater Research Institute, Winnipeg, for their help in finding and assembling a considerable part of the literature used to write this review. I also thank Buster Welch, Lionel Johnson, and Johan Hammar for the effort they put into reviewing my original manuscript and making many helpful suggestions.

References

Ali, M. Y., and C. C. Lindsey. 1974. Heritable and temperature-induced meristic variation in the medaka, *Oryzias latipes*. Canadian Journal of Zoology 52:959–976.

Andrews, E. F. 1988. The harvest of fish and wildlife for subsistence by residents of Minto, Alaska. Alaska Department of Fish and Game, Division of Subsistence, Technical Paper 137, Juneau.

Andriashev, A. P. 1970. Cryopelagic fishes of the Arctic and Antarctic and their significance in polar ecosystems. Pages 297–304 *in* M. W. Holdgate, editor. Antarctic ecology, volume 1. Academic Press, London.

Anonymous. 1976. The James Bay and Northern Québec Agreement. Editeur officiel du Québec, Montreal.

Anonymous. 1992a. Contaminants in the marine environment of Nunavik. Collection Nordicana 56, Centre d'études nordiques, Université Laval, Québec.

Anonymous. 1992b. Proceedings of the eighteenth annual aquatic toxicity workshop. Canadian Technical Report of Fisheries and Aquatic Sciences 1863.

Armstrong, R. H. 1986. A review of arctic grayling studies in Alaska, 1952–1982. Biological Papers of the University of Alaska 23:3–17.

Armstrong, R. H., H. Hop, and J. H. Triplehorn. 1986. Indexed bibliography of the holarctic genus *Thymallus* (grayling) to 1985. Biological Papers of the University of Alaska 23:18–110.

Balon, E. K., editor. 1980. Charrs, salmonid fishes of the genus *Salvelinus*. Dr. W. Junk, The Hague, Netherlands.

Barnes, P. W., D. M. Schell, and E. Reimnitz, editors. 1984. The Alaskan Beaufort Sea, ecosystems and environments. Academic Press, Orlando, Florida.

Barry, R. G. 1989. The present climate of the Arctic Ocean and possible past and future states. Pages 1–46 *in* Y. Herman, editor. The Arctic Seas, climatology, oceanography, geology, and biology. Van Nostrand Reinhold, New York.

Beamish, F. W. H. 1964. Respiration of fishes with special emphasis on standard oxygen consumption. II. Influence of weight and temperature on respiration of several species. Canadian Journal of Zoology 42:176–188.

Beltaos, S., R. Gerard, S. Petryk, and T. D. Prowse. 1990. Working group on river ice jams. Field studies and research needs. National Hydrology Research Institute Science Report 2. Environment Canada, Saskatoon, Saskatchewan.

Bergmann, M. A., and H. E. Welch. 1985. Spring meltwater mixing in small arctic lakes. Canadian Journal of Fisheries and Aquatic Sciences 42:1789–1798.

Berkes, F. 1990. Native subsistence fisheries: a synthesis of harvest studies in Canada. Arctic 43:35–42.

Berst, A. H., P. E. Ihssen, G. R. Spangler, G. B. Ayles, and G. W. Martin. 1980. The splake, a hybrid charr *Salvelinus namaycush* × *S. fontinalis*. Pages 841–887 *in* Balon (1980).

Bocking, R. C. 1987. Canadian water: a commodity for export. Pages 105–135 *in* Healey and Wallace (1987).

Boivin, T., and G. Power. 1985. An assessment of the fisheries resources of Richmond Gulf, P. Q. Makivik Corporation, Renewable Resource Development, Lachine, Québec.

Boivin, T. G., G. Power, and D. R. Barton. 1989. Biological and social aspects of an Inuit winter fishery for Arctic char (*Salvelinus alpinus*). Pages 653–672 *in* H. Kawanabe, F. Yamazaki, and D. L. G. Noakes, editors. Biology of charrs and Masu salmon. Physiology and Ecology Japan Special Volume 1, Kyoto.

Boutron, C. F., U. Görlach, J.-P. Candelone, M. A. Bolshov, and R. J. Delmas. 1991. Decrease in anthropogenic lead, cadmium and zinc in Greenland snows since the late 1960s. Nature 353:153–156.

Bradstreet, M. S. W., and W. E. Cross. 1982. Trophic relationships at high arctic ice edges. Arctic 35:1–12.

Brewer, M. C. 1958. The thermal regime of an arctic lake. Transactions of the American Geophysical Union 39:278–284.

Brouard, D., C. Demers, R. Lalumière, R. Schetagne, and R. Verdon. 1990. Evolution of mercury levels in fish of the La Grande hydroelectric complex Québec (1978–1989). Hydro-Québec, Centre de Documentation, Montreal, Québec.

Carey, A. G., Jr., and P. A. Montagna. 1982. Arctic sea ice faunal assemblage: first approach to description and source of the underice meiofauna. Marine Ecology Progress Series 8:1–8.

Cheng, C. C., and A. L. DeVries. 1991. The role of antifreeze glycopeptides and peptides in the freezing avoidance of cold-water fish. Pages 1–14 *in* G. di Prisco, editor. Life under extreme conditions: biochemical adaptations. Springer-Verlag, Berlin.

Clarke, A. 1980. A reappraisal of the concept of metabolic cold adaptation in polar marine invertebrates. Biological Journal of the Linnean Society 14:77–92.

Clarke, R. McV., and five coauthors. 1989. Report of the Arctic Fisheries Scientific Advisory Committee for 1986/87 and 1987/88. Canadian Manuscript Report of Fisheries and Aquatic Sciences 2015.

Cohen, D. M. 1970. How many recent fishes are there? Proceedings of the California Academy of Sciences 38:341–346.

Collette, B. B. 1990. Problems with gray literature in fishery science. Pages 27–31 *in* J. Hunter, editor. Writing for fishery journals. American Fisheries Society, Bethesda, Maryland.

Corkum, L. D., and P. J. McCart. 1981. A review of the fisheries of the Mackenzie Delta and nearshore Beaufort Sea. Canadian Manuscript Report of Fisheries and Aquatic Sciences 1613.

Cosens, S. E., J. F. Craig, and T. A. Shortt. 1990. Report of the Arctic Fisheries Scientific Advisory Committee for 1988/89. Canadian Manuscript Report of Fisheries and Aquatic Sciences 2063.

Craig, P. C. 1984. Fish use of coastal waters of the Alaskan Beaufort Sea: a review. Transactions of the American Fisheries Society 113:265–282.

Craig, P. C. 1989a. An introduction to anadromous fishes in the Alaskan Arctic. Biological Papers of the University of Alaska 24:27–54.

Craig, P. C. 1989b. Subsistence fisheries at coastal villages in the Alaskan Arctic, 1970–1986. Biological Papers of the University of Alaska 24:131–152.

Crawford, R. E. 1988. Fisheries acoustics in the high arctic: under-ice applications. Pages 95–98 *in* Hydroacoustics workshop proceedings. Canadian Technical Report of Fisheries and Aquatic Sciences 1641.

Crawford, R. 1989. Exploitation of Arctic fishes. Canadian Manuscript Report of Fisheries and Aquatic Sciences 2002.

Crossman, E. J., and D. E. McAllister. 1986. Zoogeography of freshwater fishes of the Hudson Bay drainage, Ungava Bay and the Arctic archipelago. Pages 53–104 *in* C. H. Hocutt and E. O. Wiley, editors. The zoogeography of North American freshwater fishes. Wiley, New York.

Dadswell, M. J., and five coeditors. 1987. Common strategies of anadromous and catadromous fishes. American Fisheries Society Symposium 1, Bethesda, Maryland.

Davenport, J. 1992. Animal life at low temperature. Chapman and Hall, London.

Davies, P. L., and C. L. Hew. 1990. Biochemistry of fish antifreeze proteins. FASEB (Federation of American Societies for Experimental Biology) Journal 4:2460–2468.

Davies, P. L., C. L. Hew, and G. L. Fletcher. 1988. Fish antifreeze proteins: physiology and evolutionary biology. Canadian Journal of Zoology 66:2611–2617.

Dempson, J. B., L. J. LeDrew, and G. Furey. 1986. Summary of catch statistics by sub-area and assessment unit for the northern Labrador Arctic char fishery in 1985. Report to Canadian Department of Fisheries and Oceans by Canadian Atlantic Fisheries Scientific Advisory Committee (Research document 86/26), Ottawa.

DeVries, A. L. 1971. Glycoproteins as biological antifreeze agents in Antarctic fishes. Science 172:1152–1155.

DeVries, A. L. 1982. Biological antifreeze agents in coldwater fishes. Comparative Biochemistry and Physiology A 73:627–640.

DeVries, A. L. 1983. Antifreeze peptides and glycopeptides in cold-water fishes. Annual Review of Physiology 45:245–260.

DeVries, A. L. 1984. Role of glycopeptides and peptides in inhibition of crystallization of water in polar fishes. Philosophical Transactions of the Royal Society of London B Biological Sciences 304:575–588.

DeVries, A. L. 1986. Antifreeze glycopeptides and peptides: interactions with ice and water. Methods in Enzymology 127:293–303.

DeVries, A. L., J. Vandenheede, and R. E. Feeney. 1971. Primary structure of freezing point depressing proteins. Journal of Biological Chemistry 246:305–309.

Duman, J. G., and A. L. DeVries. 1976. Isolation, characterization and physical properties of protein antifreezes from the winter flounder, *Pseudopleuronectes americanus*. Comparative Biochemistry and Physiology B 53:375–380.

Duman, J. G., L. Xu, L. G. Neven, D. Tursman, and D. W. Wu. 1991. Hemolymph proteins involved in insect subzero-temperature tolerance. Ice nucleators and antifreeze proteins. Pages 94–127 *in* R. E. Lee, Jr. and D. L. Denlinger, editors. Insects at low temperature. Chapman and Hall, New York and London.

Dumas, R. 1990. Arctic charr stream enhancement in

Nunavik: summary of activities in 1989. Makivik Corporation, Renewable Resource Development, Lachine, Québec.

Dunbar, M. J. 1968. Ecological development in polar regions. Prentice-Hall, Englewood Cliffs, New Jersey.

Dunbar, M. J. 1970. Ecosystem adaptation in marine polar environments. Pages 105–111 *in* M. W. Holdgate, editor. Antarctic ecology, volume 1. Academic Press, London.

Dunbar, M. J., editor. 1977. Polar oceans—proceedings of the polar ocean conference. McGill University, Montreal, and Arctic Institute of North America, Calgary, Alberta.

Dutil, J.-D., and G. Power. 1980. Coastal populations of brook trout, *Salvelinus fontinalis*, in Lac Guillaume-Delisle, (Richmond Gulf) Québec. Canadian Journal of Zoology 58:1828–1835.

Environment Canada. 1990. Surface water data Yukon and Northwest Territories 1989. Inland Waters Directorate, Water Resources Branch, Water Survey of Canada, Ottawa.

Ewart, K. V., and G. L. Fletcher. 1990. Isolation and characterization of antifreeze proteins from smelt (*Osmerus mordax*) and Atlantic herring (*Clupea harengus harengus*). Canadian Journal of Zoology 68: 1652–1658.

Fechhelm, R. G., J. S. Baker, W. B. Griffiths, and D. R. Schmidt. 1989. Localized movement patterns of least cisco (*Coregonus sardinella*) and Arctic cisco (*C. autumnalis*) in the vicinity of a solid-fill causeway. Biological Papers of the University of Alaska 24:75–106.

Fechhelm, R. G., R. E. Dillinger, Jr., B. J. Gallaway, and W. B. Griffiths. 1992. Modeling of in situ temperature and growth relationships for yearling broad whitefish in Prudhoe Bay, Alaska. Transactions of the American Fisheries Society 121:1–12.

Feeney, R. E., T. S. Burcham, and Y. Yeh. 1986. Antifreeze glycoproteins from polar fish blood. Annual Review of Biophysics and Biophysical Chemistry 15: 59–78.

Finley, K. J., M. S. W. Bradstreet, and G. W. Miller. 1990. Summer feeding ecology of harp seals (*Phoca groenlandica*) in relation to Arctic cod (*Boreogadus saida*) in the Canadian high arctic. Polar Biology 10:609–618.

Fletcher, G. L., M. H. Kao, and R. M. Fourney. 1986. Antifreeze peptides confer freezing resistance to fish. Canadian Journal of Zoology 64:1897–1901.

Fletcher, G. L., M. H. Kao, and J. B. Dempson. 1988. Lethal freezing temperatures of Arctic char and other salmonids in the presence of ice. Aquaculture 71:369–378.

Forster, M. E., C. E. Franklin, H. H. Taylor, and W. Davidson. 1987. The aerobic scope of an antarctic fish, *Pagothenia borchgrevinki* and its significance for metabolic cold adaptation. Polar Biology 8:155–159.

Fortin, R., M. Léveillé, P. Laramée, and Y. Mailhot. 1990. Reproduction and year-class strength of the Atlantic tomcod (*Microgadus tomcod*) in the Sainte-Anne River, at La Pérade, Quebec. Canadian Journal of Zoology 68:1350–1359.

Frost, K. J., and L. F. Lowry. 1984. Trophic relationships of vertebrate consumers in the Alaskan Beaufort Sea. Pages 381–401 *in* P. W. Barnes, D. M. Schell, and E. Reimnitz, editors. The Alaskan Beaufort Sea. Ecosystems and environments. Academic Press, Orlando, Florida.

Fry, F. E. J. 1957. The aquatic respiration of fish. Pages 1–63 *in* M. E. Brown, editor. The physiology of fishes, volume I. Academic Press, New York.

Fry, F. E. J. 1971. The effect of environmental factors on the physiology of fish. Pages 1–98 *in* W. S. Hoar and D. J. Randall, editors. Fish physiology, volume VI. Academic Press, New York.

Gillis, D. J. 1988. Arctic char (*Salvelinus alpinus*) fisheries management: an evaluation of current methods and of the information pertinent to the northern Québec resource. Ministère du Loisir, de la Chasse et de la Pêche, Direction régionale du Nouveau-Québec, Service de l'aménagement et de l'exploitation de la faune, Technical Report, Québec.

Gillis, D. J., M. R. Allard, and F. Axelsen. 1987. Killiniq fisheries project. Phase III. Makivik Corporation Research Department, Lachine, Québec.

Grainger, E. H., A. A. Mohammed, and J. E. Lovrity. 1985. The sea ice fauna of Frobisher Bay, arctic Canada. Arctic 38:23–30.

Green, J. M. 1974. A localized mass winter kill of cunners in Newfoundland. Canadian Field-Naturalist 88:96–97.

Griffith, M., P. Ala, D. S. C. Yang, W.-C. Hon, and B. A. Moffatt. 1992. Antifreeze protein produced endogenously in winter rye leaves. Plant Physiology 100: 593–596.

Griffiths, W. B., B. J. Gallaway, W. J. Gazey, and R. E. Dillinger. 1992. Growth and condition of Arctic cisco and broad whitefish as indicators of causeway-induced effects in the Prudhoe Bay region, Alaska. Transactions of the American Fisheries Society 121: 557–577.

Hammar, J. 1989. Freshwater ecosystems of polar regions: vulnerable resources. Ambio 18(1):6–22.

Hammar, J., J. B. Dempson, and E. Verspoor. 1991. Natural hybridization between Arctic char (*Salvelinus alpinus*) and brook trout (*S. fontinalis*): evidence from northern Labrador. Canadian Journal of Fisheries and Aquatic Sciences 48:1437–1445.

Healey, M. C., and R. R. Wallace, editors. 1987. Canadian aquatic resources. Canadian Bulletin of Fisheries and Aquatic Sciences 215.

Herman, Y., editor. 1989. The Arctic seas, climatology, oceanography, geology and biology. Van Nostrand Reinhold, New York.

Heuring, L. G., J. A. Babaluk, and K. E. Marshall. 1991. A bibliography of the Arctic charr, *Salvelinus alpinus* (L.) complex: 1985–1990. Canadian Technical Report of Fisheries and Aquatic Sciences 1775.

Hochachka, P. W. 1988. Channels and pumps—determinants of metabolic cold adaptation strategies. Comparative Biochemistry and Physiology B 90:515–519.

Holeton, G. F. 1974. Metabolic and cold adaptation of polar fish: fact or artefact? Physiological Zoology 47:137–152.

Hop, H., H. E. Welch, and R. E. Crawford. 1997. Population structure and feeding ecology of Arctic cod schools in the Canadian high Arctic. Pages 68–80 *in* J. Reynolds, editor. Fish ecology in Arctic North America. American Fisheries Society Symposium 19, Bethesda, Maryland.

Hopper, M., and G. Power. 1991. The fisheries of an Ojibwa community in northern Ontario. Arctic 44: 267–274.

Houston, J. 1990. Status of the fourhorn sculpin, *Myoxocephalus quadricornis*, in Canada. Canadian Field-Naturalist 104:7–13.

Hunter, J. G., S. T. Leach, D. E. McAllister, and M. B. Steigerwald. 1984. A distributional atlas of records of the marine fishes of Arctic Canada in the National Museums of Canada and Arctic Biological Station. Syllogeus 52.

Hutchinson, G. E. 1957. A treatise on limnology, volume 1. Geography, physics, and chemistry. Wiley, New York.

Job, S. V. 1955. The oxygen consumption of *Salvelinus fontinalis*. University of Toronto Studies, Biology Series 61:1–39.

Johnson, L. 1980. The arctic charr, *Salvelinus alpinus*. Pages 15–98 *in* Balon (1980).

Johnson, L., and B. Burns, editors. 1984. Biology of the Arctic charr: proceedings of the international symposium on Arctic charr. University of Manitoba Press, Winnipeg.

Jorgenson, J. K., H. E. Welch, and M. F. Curtis. 1992. Response of amphipoda and trichoptera to lake fertilization in the Canadian Arctic. Canadian Journal of Fisheries and Aquatic Sciences 49:2354–2362.

Kawanabe, H., F. Yamazaki, and D. L. G. Noakes, editors. 1989. Biology of charrs and Masu salmon. Proceedings of the international symposium on charrs and Masu salmon. Physiology and Ecology Japan Special Volume 1, Kyoto.

Kemp, A., L. Bernatchez, and J. J. Dodson. 1989. A revision of coregonine fish distribution and abundance in eastern James–Hudson Bay. Environmental Biology of Fishes 26:247–255.

Kristofferson, A. H., D. K. McGowan, and W. J. Ward. 1986. Fish weirs for the commercial harvest of searun Arctic charr in the Northwest Territories. Canadian Industry Report of Fisheries and Aquatic Sciences 174.

Larkin, P. A. 1956. Interspecific competition and population control in freshwater fish. Journal of the Fisheries Research Board of Canada 13:327–343.

Lindsey, C. C. 1975. Pleomerism, the widespread tendency among related fish species for vertebral number to be correlated with maximum body length. Journal of the Fisheries Research Board of Canada 32: 2453–2469.

Lindsey, C. C., and J. D. McPhail. 1986. Zoogeography of fishes of the Yukon and Mackenzie basins. Pages 639–674 *in* C. H. Hocutt and E. O. Wiley, editors. The zoogeography of North American freshwater fishes. Wiley, New York.

Livingstone, D. A. 1966. Alaska, Yukon, Northwest Territories and Greenland. Pages 559–574 *in* D. G. Frey, editor. Limnology in North America. University of Wisconsin Press, Madison.

Lockhart, W. L., D. A. Metner, D. A. J. Murray, and D. C. G. Muir. 1987. Hydrocarbons and complaints about fish quality in the Mackenzie River, Northwest Territories, Canada. Water Pollution Research Journal of Canada 22:616–628.

MacArthur, R. H., and E. O. Wilson. 1967. The theory of island biogeography. Princeton University Press, Princeton, New Jersey.

MacCallum, W. R., and H. A. Regier. 1984. The biology and bioenergetics of Arctic charr in Char Lake, N.W.T., Canada. Pages 329–340 *in* L. Johnson and B. L. Burns, editors. Biology of the Arctic charr. Proceedings of the international symposium on Arctic charr. University of Manitoba Press, Winnipeg.

Magnin, E. 1977. Ecologie des eaux douces du territoire de la Baie James. Société d'énergie de la Baie James, Montreal, Québec.

Marshall, K. E., M. Layton, and C. Stobbe. 1990. A bibliography of the lake trout *Salvelinus namaycush* (Walbaum), 1984 through 1990. Canadian Technical Report of Fisheries and Aquatic Sciences 1749.

Masty, D. 1991. L'utilisation des ressources fauniques (poisson). *in* Forum Québécois pour l'examen public du complexe Grande-Baleine. 2ième Colloque University Laval, St. Foy, Québec.

McAllister, D. E. 1977. Ecology of the marine fishes of Arctic Canada. Pages 49–65 *in* Section II marine ecology. Circumpolar conference on northern ecology September 1975. National Research Council of Canada, Ottawa.

McAllister, D. E. 1990. Diversity and biology of invertebtrates and fish. Pages 120–126 *in* C. R. Harington, editor. Canada's missing dimension—science and history in the Canadian Arctic islands, volume II. Canadian Museum of Nature, Ottawa.

McAllister, D. E., V. Legendre, and J. G. Hunter. 1987. List of Inuktitut (Eskimo), French, English and scientific names of marine fishes of Arctic Canada. Canadian Manuscript Report of Fisheries and Aquatic Sciences 1932.

McAllister, D. E., and M. B. Steigerwald. 1986. Bibliography of the marine fishes of Arctic Canada, 1771–1985. Canadian Manuscript Report of Fisheries and Aquatic Sciences 1909.

McCleave, J. D., G. P. Arnold, J. J. Dodson, and W. H. Neill, editors. 1984. Mechanisms of migration in fishes. Plenum, New York.

McDowall, R. M. 1988. Diadromy in fishes: migrations between freshwater and marine environments. Croom Helm, London.

McKeown, B. A. 1984. Fish migration. Croom Helm, London.

McPhail, J. D., and C. C. Lindsey. 1970. Freshwater fishes of northwestern Canada and Alaska. Bulletin of the Fisheries Research Board of Canada 173.

McVean, A. R., and J. C. Montgomery. 1987. Temperature compensation in myotomal muscle: Antarctic versus temperate fish. Environmental Biology of Fishes 19:27–33.

Mikhail, M. Y., and H. E. Welch. 1989. Biology of

Greenland cod, *Gadus ogac*, at Saqvaqjuac, northwest coast of Hudson Bay. Environmental Biology of Fishes 26:49–62.

Montgomery, J. C., and J. A. Macdonald. 1984. Performance of motor systems in Antarctic fishes. Journal of Comparative Physiology A 154:241–248.

Morin, R., and J. J. Dodson. 1986. The ecology of fishes in James Bay, Hudson Bay and Hudson Strait. Pages 293–325 *in* I. P. Martini, editor. Canadian inland seas, Elsevier oceanography series 44. Elsevier, New York.

Morin, R., J. J. Dodson, and G. Power. 1982. Life-history of anadromus cisco (*Coregonus artedii*), lake whitefish (*C. clupeaformis*), and round whitefish (*Prosopium cylindraceum*) populations of eastern James–Hudson Bay. Canadian Journal of Fisheries and Aquatic Sciences 39:958–967.

Muir, D. C. G., C. A. Ford, N. P. Grift, D. A. Metner, and W. L. Lockhart. 1990. Geographic variation of chlorinated hydrocarbons in burbot (*Lota lota*) from remote lakes and rivers in Canada. Archives of Environmental Contamination and Toxicology 19:530–542.

Mungall, C., and D. J. McLaren, editors. 1991. Planet under stress—the challenge of global change. The Royal Society of Canada, Oxford University Press, Don Mills, Ontario.

Nelson, J. S. 1984. Fishes of the world. Wiley, New York.

Nemoto, T., and G. Harrison. 1981. High latitude ecosystems. Pages 95–126 *in* A. R. Longhurst. Analysis of marine ecosystems. Academic Press, London.

Nikolsky, G. V. 1963. The ecology of fishes. Academic Press, New York.

Norton, D. W., editor. 1989. Research advances on anadromous fish in arctic Alaska and Canada. Nine papers contributing to an ecological synthesis. Biological Papers of the University of Alaska 24.

Norton, D., and G. Weller. 1984. The Beaufort Sea: background, history, and perspective. Pages 3–19 *in* P. W. Barnes, D. M. Schell, and E. Reimnitz, editors. The Alaskan Beaufort Sea—ecosystems and environments. Academic Press, Orlando, Florida.

Nyman, L. 1989. Why is there a "charr problem"? Pages 25–32 *in* H. Kawanabe, F. Yamazaki, and D. L. G. Noakes, editors. Biology of charrs and Masu salmon. Physiology and Ecology Japan, Special Volume 1, Kyoto.

Nyman, L. 1991. Conservation of freshwater fish: protection of biodiversity and genetic variability in aquatic ecosystems. Fisheries Development Series 56. World Wide Fund for Nature, Solna, Sweden.

O'Bannon, B. K., editor. 1990. Fisheries of the United States, 1989. Current Fishery Statistics 8900, U.S. Government Printing Office, Washington, DC.

Oehme, M. 1991. Further evidence of long-range air transport of polychlorinated aromates and pesticides: North America and Eurasia to the Arctic. Ambio 20:293–297.

O'Grady, S. M., J. C. Ellory, and A. L. DeVries. 1983. The role of low molecular weight antifreeze glycopeptides in the bile and intestinal fluid of Antarctic fishes. Journal of Experimental Biology 104:149–162.

Ouellet, M., and P. Pagé. 1988. Canada's most fascinating lake. Geos 17:1–7.

Pearse, P. H. 1988. Rising to the challenge. Report prepared for Canadian Wildlife Federation, Ottawa.

Percy, R., B. Smiley, and T. Mullen. 1985. Fishes, invertebrates and marine plants. The Beaufort Sea and the search for oil. Beaufort Sea Project, Department of Fisheries and Oceans, Sidney, British Columbia.

Pfeifer, W. E. 1977. Bibliography of the fishes of the Beaufort Sea. Biological Papers of the University of Alaska 17.

Power, G. 1976. History of the Hudson's Bay company salmon fisheries in the Ungava Bay region. Polar Record 18:151–161.

Power, G. 1980. The brook charr, *Salvelinus fontinalis*. Pages 141–203 *in* Balon (1980).

Power, G., and D. R. Barton. 1987. Some effects of physiographic and biotic factors on the distribution of anadromous Arctic char (*Salvelinus alpinus*) in Ungava Bay, Canada. Arctic 40:198–203.

Power, G., R. Cunjak, J. Flannagan, and C. Katopodis. 1993. Biological effects of river ice. Pages 97–119 *in* T. D. Prowse and N. C. Gridley, editors. Environmental aspects of river ice. National Hydrology Research Institute Science Report 5. Environment Canada, Saskatoon, Saskatchewan.

Power, G., and R. Le Jeune. 1976. Le potentiel de pêche du Nouveau-Québec. Cahiers de Géographie de Québec 20:409–428.

Prowse, T. D., and N. C. Gridley, editors. 1993. Environmental aspects of river ice. National Hydrology Research Institute Science Report 5. Environment Canada, Saskatoon, Saskatchewan.

Prowse, T. D., and C. S. L. Ommanney, editors. 1990. Northern hydrology: Canadian perspectives. National Hydrology Research Institute Science Report 1. Environment Canada, Saskatoon, Saskatchewan.

Rawson, D. S. 1954. Limnology in the North American Arctic and subarctic. Arctic 7:206–212.

Reid, G. K. 1961. Ecology of inland waters and estuaries. Van Nostrand Reinhold, New York.

Reisman, H. M., G. L. Fletcher, M. H. Kao, and M. A. Shears. 1987. Antifreeze proteins in the grubby sculpin, *Myoxocephalus aenaeus*, and the tomcod, *Microgadus tomcod*: comparison of seasonal cycles. Environmental Biology of Fishes 18:295–301.

Reist, J. D. 1988. Stock genetics in arctic anadromous fish: an organizational basis for biological research. Pages 17–24 *in* R. M. Meyer and T. M. Johnson, editors. Fisheries oceanography, a comprehensive formulation of technical objectives for offshore application in the Arctic. U.S. Department of the Interior, Minerals Management Service, Anchorage, Alaska.

Reist, J. D., and W. A. Bond. 1988. Life history characteristics of migratory coregonids of the lower Mackenzie River, Northwest Territories, Canada. Finnish Fisheries Research 9:133–144.

Rigler, F. H., M. E. MacCallum, and J. C. Roff. 1974. Production of zooplankton in Char Lake. Journal of the Fisheries Research Board of Canada 31:637–646.

Ripley, E. A. 1987. Climatic change and the hydrological regime. Pages 137–178 *in* Healey and Wallace (1987).

Robitaille, J. A., Y. Côté, G. Shooner, and G. Hayeur. 1986. Growth and maturation patterns of Atlantic salmon (*Salmo salar*) in the Koksoak River, Ungava, Quebec. Canadian Special Publication of Fisheries and Aquatic Sciences 89:62–69.

Roy, L., and M. Laplante. 1991. Bilan de la situation du Saumon atlantique (*Salmo salar*) au Nouveau-Québec en 1990. Ministère du Loisir, de la Chasse et de la Pêche du Québec. Direction régionale du Nouveau-Québec. Service de l'aménagement et de l'exploitation de la faune, Québec.

Schindler, D. W., and five coauthors. 1974b. Eutrophication in the high arctic—Meretta Lake, Cornwallis Island (75°N lat.). Journal of the Fisheries Research Board of Canada 31:647–662.

Schindler, D. W., H. E. Welch, J. Kalff, G. J. Brunskill, and N. Kritsch. 1974a. Physical and chemical limnology of Char Lake, Cornwallis Island (75°N lat.). Journal of the Fisheries Research Board of Canada 31:585–607.

Schlesinger, D. A., and H. A. Regier. 1982. Climatic and morphoedaphic indices of fish yields from natural lakes. Transactions of the American Fisheries Society 111:141–150.

Schmidt, D. R., W. B. Griffiths, and L. R. Martin. 1989. Overwintering biology of anadromous fish in the Sagavanirktok River delta, Alaska. Biological Papers of the University of Alaska 24:55–74.

Scholander, P. F., W. Flagg, V. Walters, and L. Irving. 1953. Climatic adaptation in arctic and tropical poikilotherms. Physiological Zoology 26:67–92.

Scholander, P. F., L. Van Dam, J. W. Kanwisher, H. J. Hammel, and M. S. Gordon. 1957. Supercooling and osmoregulation in Arctic fish. Journal of Cellular and Comparative Physiology 49:5–24.

Schreier, H., W. Erlebach, and L. Albright. 1980. Variations in water quality during winter in two Yukon rivers with emphasis on dissolved oxygen concentration. Water Research 14:1345–1351.

Scott, K. M. 1979. Arctic stream processes—an annotated bibliography. U.S. Geological Survey, Geological Survey Water-Supply Paper 2065. U.S. Government Printing Office, Washington, DC.

Scott, G. K., G. L. Fletcher, and P. L. Davies. 1986. Fish antifreeze proteins: recent gene evolution. Canadian Journal of Fisheries and Aquatic Sciences 43:1028–1034.

Skreslet, S. 1985. Freshwater outflow in relation to space and time dimensions of complex ecological interactions in coastal waters. Pages 3–12 *in* S. Skreslet, editor. The role of freshwater outflow in coastal marine ecosystems. NATO ASI (Advanced Science Institutes) Series Series G Ecological Sciences 7.

Smith, R. 1989. Market interaction of Canadian farmed and wild Arctic char. Economic and Commercial Analysis Report 22, Canada Department of Fisheries and Oceans, Ottawa.

Smith, W. O., Jr., editor. 1990. Polar oceanography. Part A, physical science. Part B, chemistry, biology and geology. Academic Press, San Diego, California.

Sparks, D. 1991. Regime fluvial et potentiel hydroelectrique de la Grande Riviere de la Baleine, la Petite Riviere de la Baleine et la riviere Nastapoka. *In* Forum Quebecois pour l'examen public du complexe Grande-Baleine. 2ieme Colloque University Laval, St. Foy, Quebec.

Stirling, I. 1980. The biological importance of polynyas in the Canadian Arctic. Arctic 33:303–315.

Thorpe, J. E. 1987. Smolting versus residency: developmental conflict in salmonids. Pages 244–252 *in* M. J. Dadswell and five coeditors. Common strategies of anadromous and catadromous fishes. American Fisheries Society Symposium 1, Bethesda, Maryland.

Turner, J. D., J. D. Schrag, and V. L. DeVries. 1985. Ocular freezing avoidance in antarctic fish. Journal of Experimental Biology 118:121–131.

Ursin, E. 1984. The tropical, the temperate and the arctic seas as media for fish production. Dana 3:43–60.

Valerio, P. F., M. H. Kao, and G. L. Fletcher. 1990. Thermal hysteresis activity in the skin of the cunner *Tautogolabrus adspersus*. Canadian Journal of Zoology 68:1065–1067.

van Everdingen, R. O. 1987. The importance of permafrost in the hydrological regime. Pages 243–276 *in* Healey and Wallace (1987).

Wagemann, R. 1989. Comparison of heavy metals in two groups of ringed seals (*Phoca hispida*) from the Canadian Arctic. Canadian Journal of Fisheries and Aquatic Sciences 46:1558–1563.

Walker, R. J., E. F. Andrews, D. B. Anderson, and N. Shishido. 1989. Subsistence harvest of Pacific salmon in the Yukon River drainage, Alaska, 1977–88. Alaska Department of Fish and Game, Division of Subsistence, Technical Paper 187, Juneau.

Welch, H. E. 1985. Introduction to limnological research at Saqvaqjuac, northern Hudson Bay. Canadian Journal of Fisheries and Aquatic Sciences 42:494–505.

Welch, H. E., and M. A. Bergmann. 1985a. Water circulation in small arctic lakes in winter. Canadian Journal of Fisheries and Aquatic Sciences 42:506–520.

Welch, H. E., and M. A. Bergmann. 1985b. Winter respiration of lakes at Saqvaqjuac, N.W.T. Canadian Journal of Fisheries and Aquatic Sciences 42:521–528.

Welch, H. E., and J. Kalff. 1974. Benthic photosynthesis and respiration in Char Lake. Journal of the Fisheries Research Board of Canada 31:609–620.

Welch, H. E., J. A. Legault, and M. A. Bergmann. 1987. Effects of snow and ice on the annual cycles of heat and light in Saqvaqjuac lakes. Canadian Journal of Fisheries and Aquatic Sciences 44:1451–1461.

Welch, H. E., and seven coauthors. 1992. Energy flow through the marine ecosystem of the Lancaster Sound region, Arctic Canada. Arctic 45:343–357.

West, R. L., M. W. Smith, W. E. Barber, J. B. Reynolds, and H. Hop. 1992. Autumn migration and overwintering of Arctic grayling in coastal streams of the Arctic National Wildlife Refuge, Alaska. Transactions of the American Fisheries Society 121:709–715.

Wheeler, P. 1987. Salmon fishing patterns along the middle Yukon River at Kaltag, Alaska. Alaska Department of Fish and Game, Division of Subsistence, Technical Paper 156, Juneau.

Winters, J. F., P. K. Weber, A. L. De Cicco, and N. Shishido. 1988. An annotated bibliography of selected references of fishes of the North Slope of Alaska, with emphasis on research conducted in National Petroleum Reserve—Alaska. Department of Wildlife Management, North Slope Borough, Barrow, Alaska.

Yaremchuck, G. C. B., and five coauthors. 1989. Commercial harvests of major fish species from the Northwest Territories, 1945–1987. Canadian Data Report of Fisheries and Aquatic Sciences 751.

American Fisheries Society Symposium 19:40–59, 1997

A Review of the Physical Oceanography of the Northeastern Chukchi Sea

THOMAS J. WEINGARTNER
*University of Alaska, Institute of Marine Science
Fairbanks, Alaska 99775, USA*

Abstract.—This article reviews the physical oceanography of the ice-free season (July to October) of the northeastern Chukchi Sea. The Chukchi ecosystem is unique among all Arctic Ocean shelf seas because its physical and biological characteristics are profoundly influenced by the northward transport of Pacific Ocean waters through Bering Strait. The salient circulation feature of the northeastern Chukchi Sea is the Alaska Coastal Current, which advects warm, dilute Alaska coastal water northward from Bering Strait into the Arctic Ocean through Barrow Canyon. The main core of this current coincides with a bottom temperature front observed in 1982, 1986, and 1990. Both the front and Alaska coastal water parallel the bathymetry and their intensities vary in proportion to the topographic relief. Benthic and fish community structures are substantially different on either side of this front. The hydrography and benthic biological data suggest that this bottom front is a quasi-permanent feature of the northeastern Chukchi Sea. However, neither Alaska coastal water nor this front were observed in 1981 when anomalously southward winds persisted over the Chukchi Sea from July through November. Such winds would have effected a westward displacement of Alaska coastal water from the northeastern Chukchi Sea and a reduction in the northward transport through Bering Strait. Although surface fronts are also observed, they are more variable and are not necessarily colocated with the bottom front. Their strength and position vary in response to fluctuations in ice-edge position and the winds. This review also discusses aspects of processes occurring at the ice edge and shelf break of the northern Chukchi Sea as well as the large interannual variations in ice cover, winds, and the transport through Bering Strait.

The Chukchi Sea lies at the northeastern end of the enormous Eurasian continental shelf system of the Arctic Ocean. These shelf seas—Barents, Kara, Laptev, East Siberian, and Chukchi—constitute 12% of the continental shelf area of the globe (2.6×10^7 km^2, Walsh 1988) and 22% of the total area of the Arctic Ocean (17×10^6 km^2, Carmack 1990). But the Chukchi Sea is unique among them in that waters of Pacific Ocean origin flow through it, influencing the regional circulation, sea ice conditions, and water-mass property and nutrient distributions. These waters are also an important source of plankton and carbon for both the Chukchi Sea and the Arctic Ocean (Walsh et al. 1989). Collectively, these features affect the distribution and abundance of the local marine biota within the Chukchi Sea and are critically important to populations of seasonally migrating marine organisms.

In this review of the salient physical oceanographic characteristics of the northeastern Chukchi Sea, the open-water season (mid-July through October) is emphasized because it is of primary concern to fisheries biologists, and relatively little data were collected during the ice-covered seasons. The northeastern portion of the basin is highlighted because this region has been studied to assess the potential effects of oil exploration. Moreover, because of limited access to the western Chukchi Sea, little information has been published on this region since the monograph by Coachman et al. (1975). Detailed information on the oceanography of Bering Strait and the southern Chukchi Sea are found in Coachman et al. (1975) and Walsh et al. (1989).

Bathymetry

The Chukchi Sea is bounded to the north by the shelf break (200-m isobath) of the Arctic Ocean and to the south by Bering Strait—a distance of more than 800 km (Figure 1). Its lateral extent varies from a minimum of about 85 km in Bering Strait to a maximum of about 900 km between Long Strait in the west and Point Barrow in the east.

The basin has a relatively flat bottom, with depths deepening gradually from 30 m in the east to 55 m in the west. However, subtle bathymetric features steer the currents, thus influencing sea ice and water property distributions: (1) Hope Sea Valley, a broad 55-m deep depression, extending northwest from the southern Chukchi Sea to the Herald Sea Valley; (2) Herald Shoal in the center of the basin; and (3) Hanna Shoal in the northeast. The latter two shoal regions have minimum depths of about 25 m.

Two shelf break bathymetric features facilitate

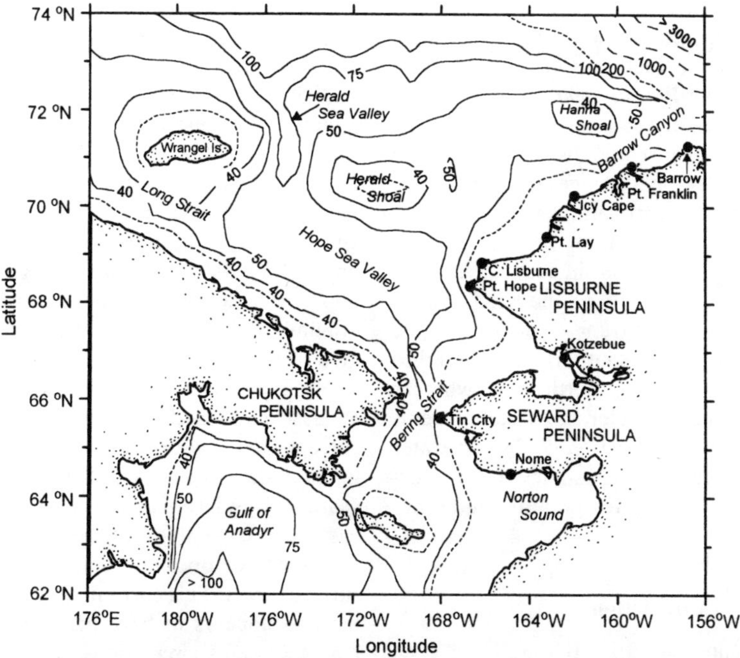

FIGURE 1.—Bathymetric map of northern Bering and Chukchi seas.

cross-shelf exchange of water by steering currents and channeling upwelled water onto the shelf: (1) Herald Sea Valley, east of Wrangel Island, is a relatively gentle, 50-km wide depression that terminates on the shelf at about 150-m depth, and (2) Barrow Canyon is a steep-sided, box canyon originating offshore of Point Franklin that cuts across the continental slope, terminating at depths greater than 3,000 m. In winter these canyons serve as likely conduits for the flow of dense water produced by brines distilled from growing sea ice. An additional feature of the shelf break's bathymetry is the westward divergence of isobaths, which results in a decrease in bottom slope from 10 m/km in the northeast to 1 m/km in the northwestern Chukchi Sea. This change in shelf break slope might have important implications with respect to the exchange of shelf waters with those of the Arctic basin.

Winds

Wind stress acting directly on the surface of the ocean or coupled to the ocean via a floating ice pack accelerates ocean currents. For shallow seas, such as the Chukchi, these wind fluctuations are especially important because they can induce rapid changes in the current structure over the whole water column and, consequently, redistribute water mass properties.

The winds over the Chukchi and northern Bering seas are strongly influenced by a polar high-pressure cell centered at about 79°N, 170°W (Pease 1987). On a mean annual basis this pressure system drives easterly winds over the northern Chukchi Sea and northeasterly winds over the southern Chukchi and northern Bering seas. North of Point Lay (~69.5°N), the monthly mean winds vary little throughout the year, and the variance about these monthly means is constant from February through July but increases through the fall. South of Point Lay, both the mean monthly winds and their variance vary seasonally. For example, mean July winds at Tin City (Figure 1) are from the south-southeast, and 55% of all recorded winds blow from southerly quadrants during this month. In February mean winds are from the north, and 78% of all recorded winds blow from the northerly quadrants (Brower et al. 1988). The temporal and spatial variability patterns are a consequence of low-pressure systems propagating eastward across the Bering Sea. In winter months, storms propagate mainly across the southern Bering Sea and into the Gulf of Alaska, whereas in summer and fall the relative frequency of storms propagating northeastward into the southern Chukchi Sea increases (Overland and Pease 1982; Brower et al. 1988; Aagaard et al. 1990). The storm track patterns and the complexity of the coastal topography result in relatively low correlations in wind stress fluctuations between coastal stations in the southern and the northern Chukchi Sea (Aagaard et al. 1990). Their correlations were computed using wind data sampled at 6-h intervals over record lengths of 12 or more months. Given the seasonal changes in storm tracks, it is possible that these correlations vary seasonally and that the spatial correlation of the winds is broader in winter months than in summer and fall months. Spatial heterogeneity of the wind field in autumn influences the current structure of the northeastern Chukchi Sea; southerly winds tend to accelerate coastal currents northward, whereas northerly winds tend to accelerate coastal currents southward. Differential advection of distinct water masses by these currents could establish strong horizontal gradients (fronts) in temperature and salinity.

Circulation

The salient circulation feature of the Chukchi Sea is the northward flow through Bering Strait, which has a mean annual transport of 0.82 Sverdrup (Coachman and Aagaard 1988; 1 Sverdrup = 10^6 m^3/s). This transport is driven by the 0.5-m drop in sea level between the Aleutian Basin of the Bering Sea and the Arctic Ocean (Overland and Roach 1987). Approximately 85% of this inflow originates in the deep Bering Sea and the Gulf of Anadyr (Kinder et al. 1986). The remainder is derived from the northeast shelf of the Bering Sea. Transport fluctuations, at time scales ranging from days to weeks, are of the same magnitude as the mean and are significantly correlated to the along-strait component of wind stress (Coachman and Aagaard 1988). Northward transport is minimum in winter when southward winds prevail and is maximum in summer when northward winds occur more frequently.

North of Bering Strait, this flow bifurcates offshore of the Lisburne Peninsula; one branch continues along the northeastern coast as the Alaska Coastal Current (ACC), and the remainder flows to the northwest. Both of these flows are sustained, in part, by the sea level difference between the Bering Sea and the Arctic Ocean. The northwestward current consists of a diffuse, slowly drifting transport through the Hope and Herald sea valleys and into the Arctic Ocean (Coachman et al. 1975). In contrast, the ACC is a swift (0.1 m/s) coastal current whose main axis is found within 20 to 50 km of the Alaskan coast and that flows into the Arctic Ocean through Barrow Canyon. There is also a southeastern-flowing current, the Siberian Coastal Current, which enters the Chukchi Sea through Long Strait and flows along the Chukotsk Peninsula before joining the Bering Strait inflow.

Relatively little is known about the circulation over the shelf break of the northern Chukchi Sea. However, hydrography (reviewed by Carmack 1990) and trajectories of buoys deployed in sea ice (Colony 1984) indicate that, on average, currents in the upper 50 m flow westward and constitute the southern limb of the wind-driven anticyclonic (clockwise) gyre of the Arctic Ocean (Coachman and Aagaard 1974). Beneath this surface layer and extending to at least 2,500-m depth, hydrography and moored current-meter observations collected along the continental slope of the Alaskan Beaufort Sea (Aagaard 1984; Aagaard et al. 1990) reveal a well-defined eastward current, the Beaufort Undercurrent, which flows opposite of the prevailing westward surface winds. This current has a width of about 70 km and appears to be a persistent feature of the Arctic Ocean's circulation. Mean speeds between 60- and 100-m depth are about 0.05 m/s, but fluctuations up to 1 m/s are common. Above 250-m depth, Chukchi source waters originate in the Bering Sea. Beneath 250-m depth, undercurrent waters originate in the Atlantic Ocean, entering the eastern Arctic Ocean through Fram Strait. Presumably, the undercurrent is much weaker west of Barrow Canyon because of the decrease in shelf break slope and because its volume increases after the ACC joins it at Barrow Canyon. Although the dynamics of this undercurrent awaits theoretical study, it must be related, in part, to these inflows. Insofar as the undercurrent flows in a direction opposite that of the wind, with shallow water to its right, it bears a strong similarity to the poleward-flowing undercurrents observed along the eastern boundaries of midlatitude oceans, and it is probable that these undercurrents share similar dynamics.

Water Masses

Based upon the mean circulation scenario, the various Chukchi Sea water masses and their origins can be better appreciated. The temperature–salinity characteristics of these waters are shown in Figure 2, which was constructed from data collected between Bering Strait and Barrow from 1981 to 1992. The water mass nomenclature of Coachman et al. (1975) and Coachman and Shigaev (1992) is adopted in this discussion.

The northward flow from Bering Strait consists of

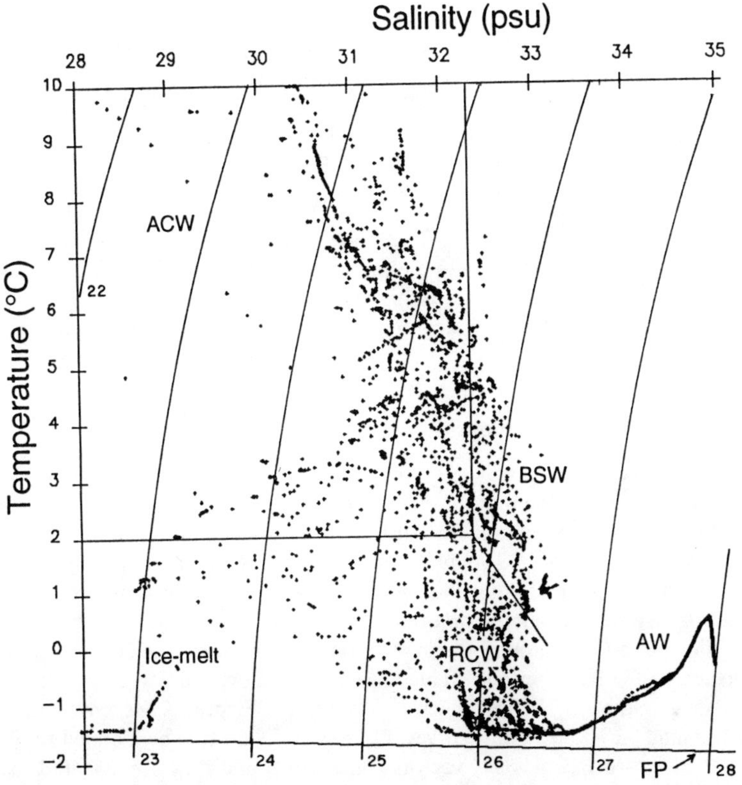

FIGURE 2.—Temperature–salinity diagram of major water masses of the Chukchi Sea. Acronyms are as follows: ACW is Alaska coastal water, BSW is Bering shelf water, RCW is resident Chukchi water, AW is Atlantic water, and FP indicates the freezing-point curve of seawater for a given salinity. Temperature is in degrees Celsius and salinity is in practical salinity units.

Alaska coastal water and Bering shelf water. A well-defined front, extending northward from Bering Strait to the Lisburne Peninsula, separates Alaska coastal water on the east side of the front from Bering shelf water (Coachman et al. 1975; Walsh et al. 1989). Water mass properties of Alaska coastal water vary broadly, with temperatures ranging from 2 to 13°C and with salinities less than 32.2 practical salinity units (psu). The low salinities are the result of the influx of freshwater from river discharge (primarily from the Yukon) along the Alaskan coast. Because Alaska coastal water is less dense than more-saline waters, it is confined to the surface layer and is warmed by solar radiation. Bering shelf water is colder (0–30°C) and more saline (32.5–33.0 psu) than Alaska coastal water. It consists of a blend of Bering Sea shelf water formed in winter and water flowing onto the shelf from the deep Bering Sea through the Gulf of Anadyr (Coachman and Shigaev 1992). There are also much higher concentrations of dissolved nutrients and chlorophyll in Bering shelf water than in Alaska coastal water (Walsh et al. 1989), and Bering shelf water is a major source of oceanic zooplankton for the Chukchi Sea (Springer et al. 1989).

Four additional water masses are observed on the northeastern Chukchi shelf. The first derives from melting sea ice and is characterized by low salinities and temperatures. The second, resident Chukchi water, is advected onto the shelf from the upper layers of the Arctic Ocean or is shelf water remnant from the previous winter. Its low temperatures (<1°C) and relatively high salinities (32–33 psu) reflect the effects of freezing and brine drainage from growing sea ice. The third water mass, Atlantic water (so named because of its origin), has very high salinities (~34.5 psu) and moderate temperatures (0.5°C) and is generally found at depths greater than 250 m along the continental slope of the Chukchi and Beaufort seas. Atlantic water is observed in

Barrow Canyon (Garrison and Paquette 1982), especially during upwelling events (Mountain et al. 1976; Aagaard and Roach 1990), but it is not clear how frequently these waters are advected into shallower depths. Bourke and Paquette (1976) reported sporadic cases of warm, saline water extending as far south as 70°N along the bottom of the northeastern Chukchi Sea. They hypothesized that Atlantic water upwelled through Barrow Canyon. The fourth water mass is indicated by the freezing-point curve in Figure 2. Winter waters on the Chukchi shelf are generally only a few tenths of a degree warmer than the freezing point.

Shelf Hydrography and Circulation

The observations reported by Johnson (1989) are a useful introduction to several of the more important oceanographic features of the region. His data were collected during a cruise from late August to early September 1986 that began near Point Barrow and concluded off Point Hope. During the first few days of the cruise, the winds (measured at Barrow) blew from the northeast and then reversed and blew from the southeast and southwest. The ice edge was located between 72.5 and 73°N during this survey.

Figure 3 shows the near-surface velocity vectors obtained throughout the cruise from a shipboard Acoustic Doppler Current Profiler. At the beginning of the survey, a vigorous southwestward flow (~0.75 m/s) was observed offshore of Barrow. (Simultaneous current-meter data from the same area showed that these currents reversed along with the wind and flowed toward the northeast for the duration of the survey.) The trajectory of the ACC is apparent in the vectors that indicate a swift northeastward and northward current to the west and northwest of the Lisburne Peninsula. The vectors veer eastward between 70.5 and 71.5°N and then northeastward approaching the coast. In the shallow (<25-m depth) embayment, to the north of the Lisburne Peninsula, the flow is weak and the vector pattern suggests a clockwise circulation cell. Coachman et al. (1975) inferred a similar circulation pattern here, and the observations are consistent with laboratory results showing the formation of eddies and stagnation points in the lee of bluff bodies protruding into a coastal flow (Boyer et al. 1987). Finally, the influence of Hanna Shoal is indicated by the meridional divergence of the flow at 72°N, 165°W and eastward flow to the north of this shoal region.

The hydrography from this cruise is presented in the form of maps showing contours of the surface and bottom isotherms and isohalines (Figure 4, A–D). The surface maps (Figure 4, A and B) indicate that Alaska coastal water extends over the whole region south of about 71°N. The tongue of cold, low-salinity, ice meltwater projecting southwest from Barrow is consistent with the observed surface flow at the beginning of the survey. These two water masses are separated by a surface front (delineated by the 3–6°C isotherms and the 31.5-psu isohaline) between 70.5 and 71°N that extends offshore to the west–southwest for about 200 km. The front weakens at the point where the 32-psu isohaline attains its easternmost position and then bends to the southwest.

Along the bottom (Figure 4, C and D), Alaska coastal water is observed inshore and to the south of the 32.5-psu and 3°C isopleths. A sharp thermal front (parallel to, but much stronger than, the surface thermal front) separates Alaska coastal water from the rest of the sampled domain, which consists of resident Chukchi water and mixtures of Alaska coastal water and resident Chukchi water. This bottom front parallels the 30- and 40-m isobaths and is strongest north of Icy Cape where these isobaths converge (Figure 1). Moreover, the velocity vectors (Figure 3) are roughly parallel to the isobaths (and the front) and they are oriented such that warmer and less-saline water is found to the right of the direction of flow. Both of these observations are consistent with theories pertaining to the dynamics of density-driven, coastal geostrophic flows over a sloping bottom (e.g., Csanady 1982). Geostrophic motion implies an equilibrium flow in which the pressure gradient is balanced by the Coriolis acceleration. For the case just discussed, the pressure gradient is perpendicular to the local orientation of the front and the isobaths and arises because of the density contrast between warm, dilute and cold, salty water.

By way of comparison, temperature and salinity at the surface (Figure 5, A and B) and bottom (Figure 5, C and D) maps are shown from data obtained from late August through mid-September 1990. Throughout August and most of September, winds at Barrow blew persistently and vigorously from the east, while winds at Kotzebue, although more variable, were primarily from the southwest and west. The ice edge in September 1990 was at about 74°N. Alaska coastal water was observed at the surface at all locations, and temperatures were higher and salinities lower relative to the 1986 survey. Although a strong front was observed running westward from the coast midway between Cape Lisburne and Point Lay, there is no indication of a

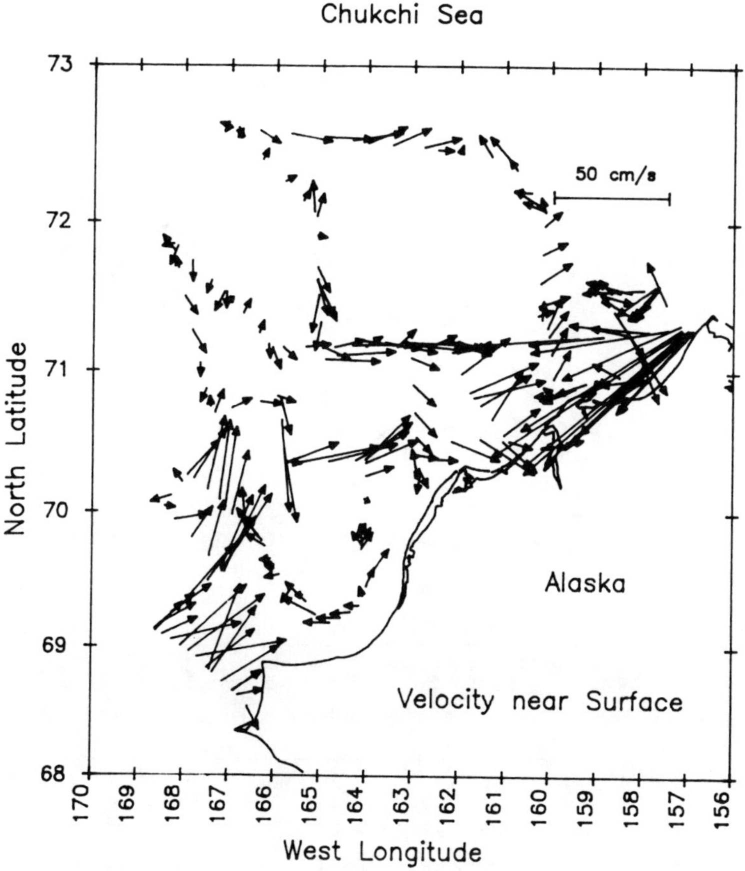

FIGURE 3.—Near-surface velocity vectors in August–September 1986 as measured with a vessel-mounted Acoustic Doppler Current Profiler (after Johnson 1989).

front intersecting the coast between Icy Cape and Point Franklin.

Alaska coastal water extends over a broad area of the bottom and is found everywhere inshore of the 32-psu isohaline (Figure 5, C and D). Alaska coastal water is also bounded by a weaker thermal front in a similar location and orientation as that observed in 1986. Resident Chukchi water lies north of this front, while west of Point Hope the bottom waters seaward of the front are characteristic of the colder, saltier Bering shelf water.

The relatively coarse station-spacing of the 1986 and 1990 surveys permit mapping large-scale hydrographic features, but they do not resolve adequately the strength of these fronts. However, Aagaard's (1988) cross-sectional profiles of temperature and salinity collected in late August 1982 from transects running northwestward from the coast off Point Lay (Figure 6A) and Point Franklin (Figure 6B) illustrate the intensity of these fronts. The Point Lay section consists wholly of Alaska coastal water and shows a well-mixed, 20-m-deep surface layer of uniform temperature (7°C) and salinity (~30 psu) extending over the entire transect. The surface layer is separated from the colder and saltier bottom waters by a 5-m thick thermocline across which the temperature decreases from 6 to 2°C. Approximately 40 km offshore, the thermocline intersects the bottom, forming a front along the 30-m isobath. Bottom temperatures decrease by 4°C over the 25-km width of the front. The Point Franklin transect is more complex. Alaska coastal water is confined to a band within 20 km of the coast and is separated from offshore waters by an intense surface-to-bottom-temperature front. Seaward of the front, waters in the uppermost 20 m are derived from ice melt, whereas those at greater

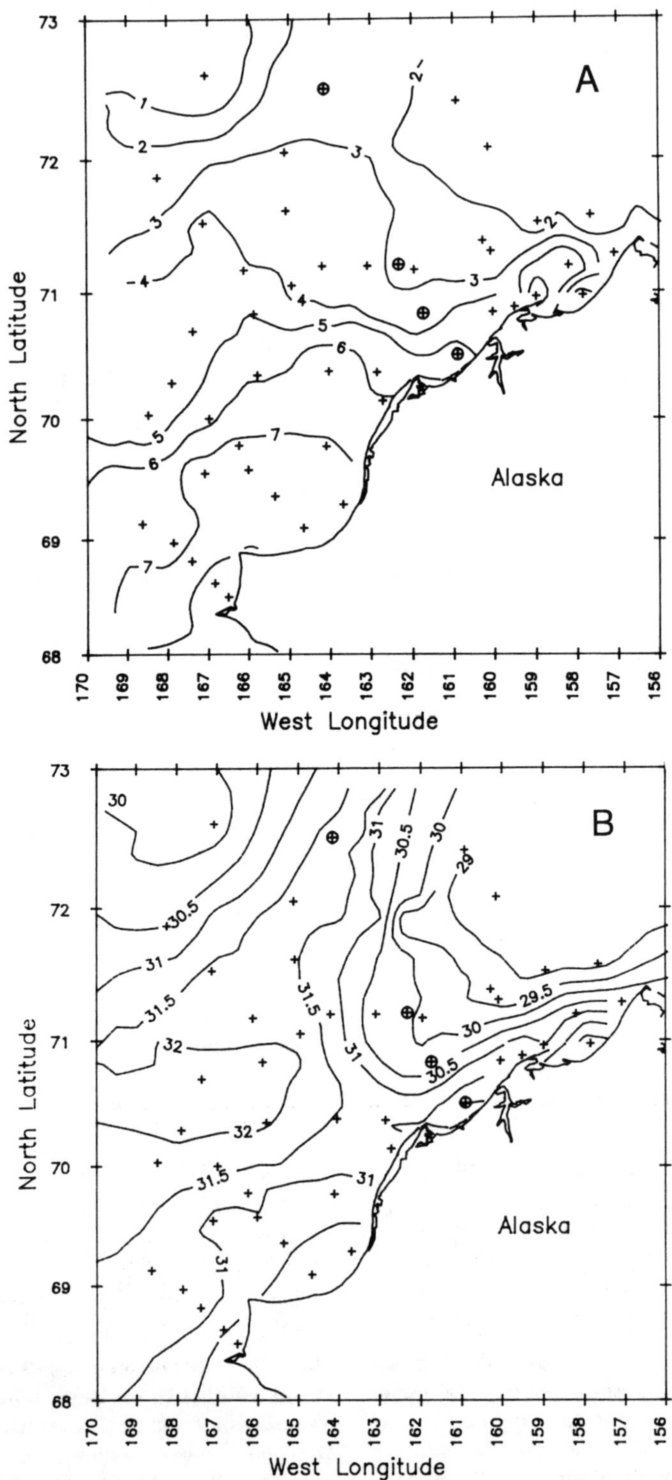

FIGURE 4.—Contour maps of (**A**) surface temperature, (**B**) surface salinity, (**C**) bottom temperature, and (**D**) bottom salinity in August–September 1986 (after Johnson 1989). Temperature is in degrees Celsius and salinity is in practical salinity units.

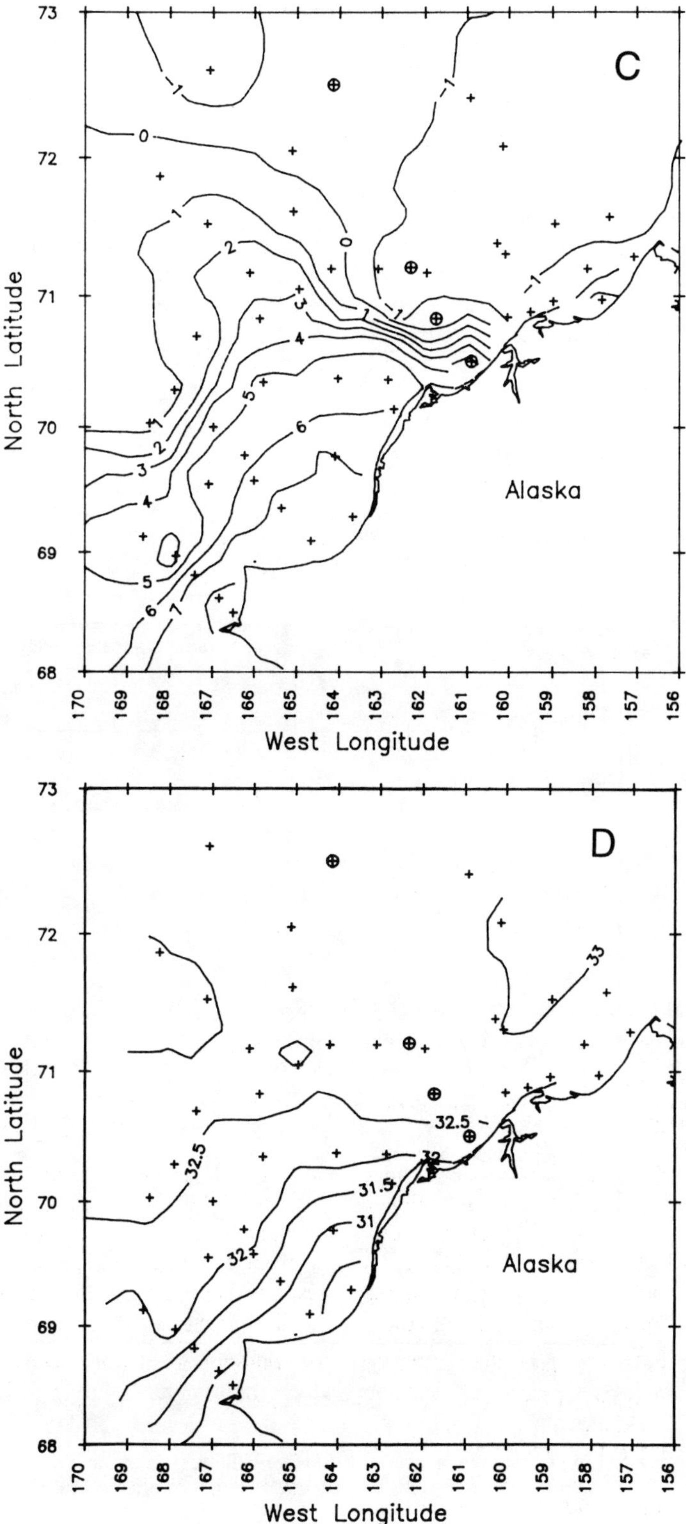

FIGURE 4.—Continued.

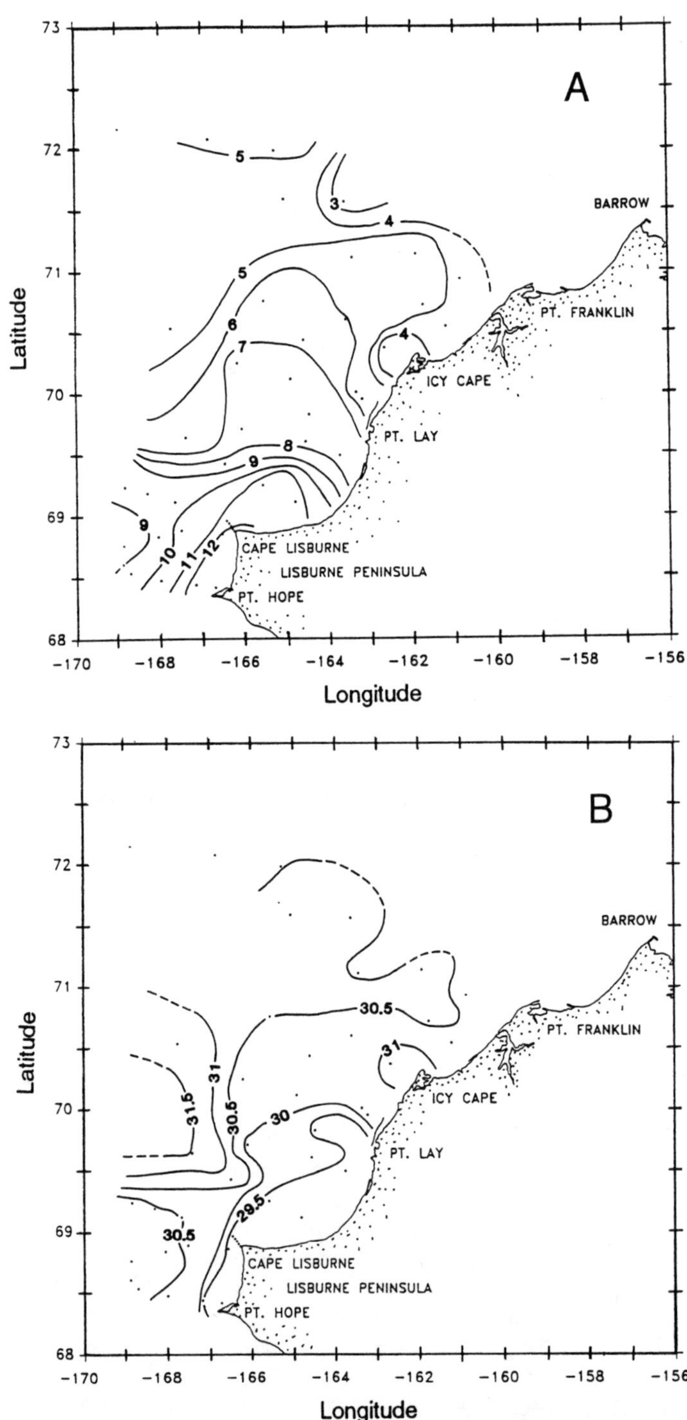

FIGURE 5.—Contour maps of (**A**) surface temperature, (**B**) surface salinity, (**C**) bottom temperature, and (**D**) bottom salinity in September 1990. Temperature is in degrees Celsius and salinity is in practical salinity units.

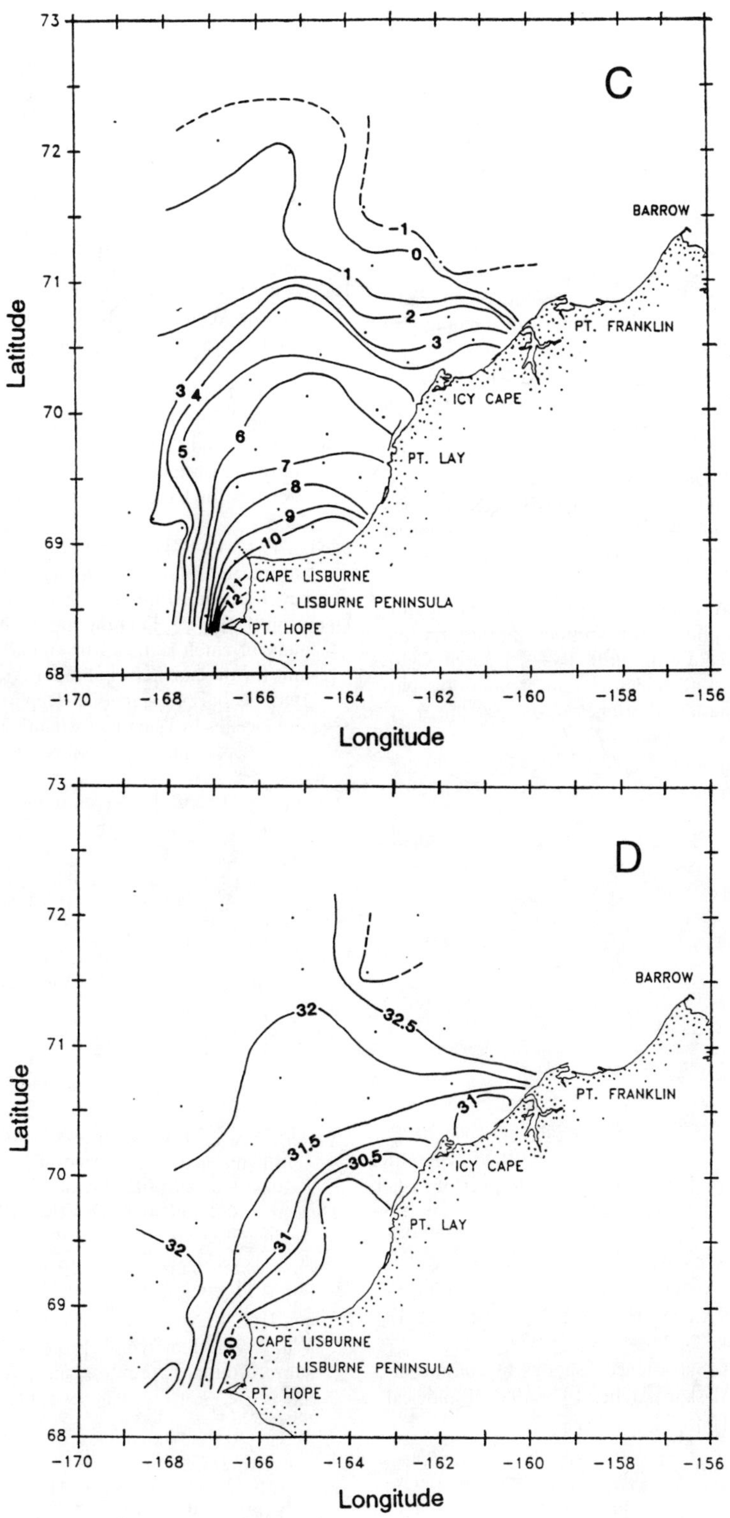

FIGURE 5.—Continued.

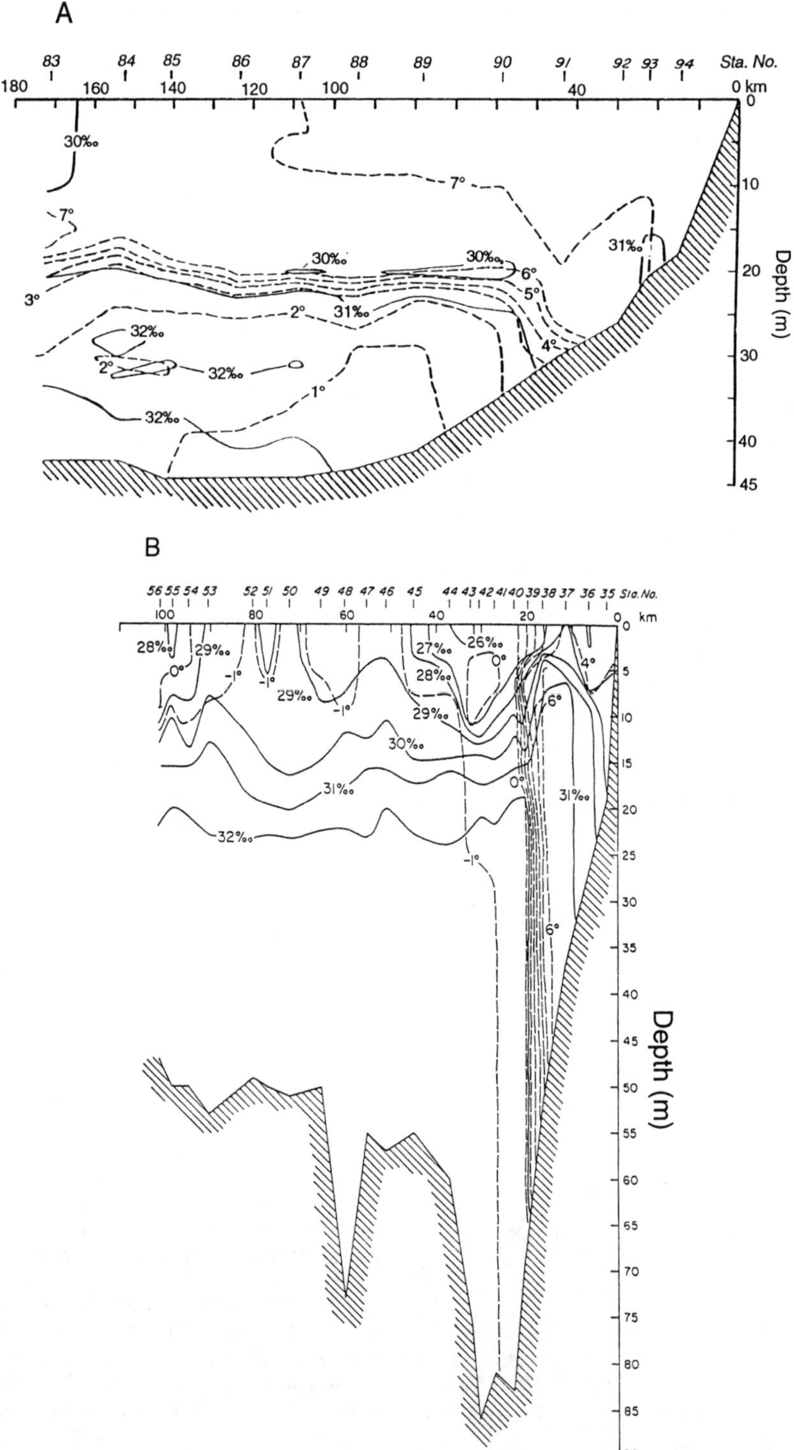

FIGURE 6.—Cross-sectional profiles of temperature (°C) and salinity (practical salinity units) from transects extending northwestward from (**A**) Point Lay and (**B**) Point Franklin in August–September 1982 (after Aagaard 1988). Station numbers are along the top of each graph. Temperature is in degrees Celsius and salinity is in practical salinity units.

depths are resident Chukchi water. Moreover, there is neither a well-mixed surface layer nor a thermocline. Rather, the surface layer is strongly stratified because of the salinity difference between the meltwater and resident Chukchi water. The most striking aspect of this transect, however, is the intensity of the thermal front across which the temperature changes by 6°C within 5 km.

Taken together, these results plus those shown by Fleming and Heggarty (1966) and Coachman et al. (1975) suggest that the bottom front between the 25- and 40-m isobaths is a quasi-permanent feature of the ice-free season of the northeastern Chukchi Sea. (However, Aagaard's [1988] data from September 1981 show no evidence of this front and reasons for its absence in that year are discussed under "Interannual Variability.") In examining the surface temperature distributions, it is evident that surface and bottom fronts are not necessarily colocated. In fact, the presence and the location of surface thermal fronts appear to be much more variable, and this variability is probably a consequence of the greater influence of wind forcing in the near-surface layer and the location of the ice edge. Paquette and Bourke (1981) also noted that the positions of the surface and bottom fronts associated with the marginal ice zone of the Chukchi Sea do not often coincide. One consequence of these results is that analysts cannot draw inferences on the absence or presence of bottom fronts based upon satellite thermal imagery.

Although the bottom front can be displaced in response to wind-induced current fluctuations (Johnson 1989), its mean position is most likely established where the bottom depth is equal to the mean depth of the mixed layer. In the Chukchi Sea, tidal velocities are small (~0.05 m/s) and the most important energy sources for mixing are cooling at the ocean surface and the winds. The mixed-layer depth represents an equilibrium depth over which mixing processes are balanced by stabilizing processes (atmospheric heating, freshwater influx from rivers, and melting sea ice). Hence, variability in these factors should also be reflected in the position of the bottom front.

Although the body of physical oceanographic data from the northeastern Chukchi Sea is small, further support for the contention that this bottom front is an annually recurring hydrographic feature is apparent in biological data. Feder et al. (1990) show large increases in abundance and biomass of benthic communities within the vicinity and north of the front. Grebmeier et al. (1988) show similar changes in benthic community structure crossing the front south of the Lisburne Peninsula. Moreover, large populations of walrus *Odobenus* and grey whales *Eschrichtius* are known to feed annually on the benthic communities within and north of the front between Icy Cape and Point Franklin (Moore and Clarke 1986; Phillips and Colgan 1987; Feder et al. 1990).

In summary, the hydrography shows that the flow of the ACC is augmented by horizontal pressure gradients established by the density contrasts between water masses on the shelf. The main axis of this current is parallel to a bottom front oriented along the bathymetry. Along the axis of the current, speeds vary in conjunction with the strength of the front and the magnitude of the bottom slope. The latter point is illustrated by the mean and maximum current speeds reported by Aagaard (1984) from yearlong current-meter records obtained offshore of Point Franklin (large bottom slope) and Cape Lisburne (smaller bottom slope). Mean and maximum speeds from the Point Franklin moorings were about 0.2 and 1.1 m/s, respectively. The corresponding values for the Cape Lisburne moorings were 0.05 and 0.62 m/s.

Johnson's (1989) and Aagaard's (1988) results show that superposed on the mean flow of the ACC are large temporal variations that occur on time scales of days to weeks and are correlated with the north–south wind component. (Northward winds accelerate the along-isobath flow component, i.e., in the same direction as the mean flow. Southward winds decelerate this flow component; sufficiently strong southward winds can lead to current reversals.) This current variability might play an important role in transporting shelf water into the interior of the western Arctic Ocean, given D'Asaro's (1988) hypothesis that, at the mouth of Barrow Canyon, these fluctuations generate the long-lived eddies observed throughout the interior of the western Arctic Ocean. Aagaard's (1988) results show that along-isobath current fluctuations at the Point Franklin moorings were also correlated with those at Cape Lisburne, implying that the velocity field of the northeastern Chukchi Sea is coherent over alongshore distances as large as 350 km. The current response to fluctuations in the wind field occurs rapidly (<1 d) and, within the vicinity of the front, is accompanied by dramatic temperature and salinity changes (Johnson 1989). Such effects can complicate the interpretation of hydrographic data collected over a large area during a 2- to 3-week cruise and indicate the need to consider the wind history of this region when analyzing such data.

Upwelling

The deeper (~100-m depth) waters offshore of the shelf break of the northern Chukchi Sea provide a potentially important source of nutrient-rich water for this shelf. At still greater depths (~250 m), waters upwelled from the Atlantic layer could exert a substantial change in the temperature–salinity structure of the northern Chukchi Sea as well as provide nutrients. Shelf break upwelling provides a mechanism for the flux of these waters onto the shelf. Episodic upwelling in Barrow Canyon and along the shelf break of the Beaufort Sea is a frequent phenomenon that has been inferred from hydrographic data (Hufford 1974; Bourke and Paquette 1976; Garrison and Paquette 1982) and observed in current-meter records (Mountain et al. 1976; Aagaard and Roach 1990) from the same area. Aagaard and Roach's (1990) results show that upwelling, from depths as great as 300 m, occurred along the Alaskan Beaufort Sea shelf break in the form of eastward propagating events lasting about 5 d. Although these upwellings were not well correlated with local winds, they were most common in fall and early winter, when winds were most variable. The occurrence of these upwellings seem to indicate that these episodes are a response of the shelf break circulation to large-scale, stochastic wind forcing, and theoretical support for this hypothesis is implied by the numerical model results of Philander and Yoon (1982).

Once upwelled, the deeper waters will influence the shelf if they can be advected shoreward. Data from Aagaard and Roach's (1990) mooring anchored on the 150-m isobath within Barrow Canyon suggest that average upcanyon (onshore) velocities during upwelling episodes were about 0.25 m/s. Over the 5-d duration of such an event, water parcels would be displaced 125 km upcanyon or as far inshore as the 60-m isobath. In contrast, their current records from the Beaufort Sea shelf break show that upwelling-related onshore velocities were much smaller than those observed in Barrow Canyon. This is not unexpected given that onshore flow is weaker over regions of large bottom slope (i.e., the Beaufort shelf break) than over regions of gentle bottom slope (Johnson and Rockliff 1986). Hence, the westward decrease in slope along the Chukchi Sea shelf break suggests that the onshore excursion of upwelled water here might be greater than in the Beaufort Sea. If so, then the zonal gradient in bottom slope might also be reflected by zonal gradients in biological productivity. There are no published observations of shelf break upwelling in the Chukchi Sea east of Barrow Canyon, but I believe that this is a reflection of a lack of data rather than a lack of evidence.

Sea Ice

The Chukchi Sea is generally ice covered from November through June. North of Bering Strait, the seasonal retreat of sea ice begins in June and, on average, attains its farthest north position at about 72.5°N in mid-September (NOCD 1986). In comparison to the Beaufort and East Siberian seas, ice retreat begins earlier, and ice advance occurs later on the Chukchi shelf because of warm water advected northward through Bering Strait. However, interannual differences in the seasonal retreat, advance, and position of the ice edge are related to the regional winds (Muench et al. 1991; e.g., anomalous southward winds result in anomalous southward position of the ice edge and vice versa).

The northward retreat of the ice edge during the summer months does not proceed uniformly along its length. Figure 7 (from Paquette and Bourke 1981) shows that the Chukchi Sea ice edge is markedly indented by three "meltback embayments" whose relative positions vary little from year to year. Two of these embayments overlie Herald Sea Valley and Barrow Canyon and the third is observed between Herald and Hanna shoals. All are related to bathymetric steering of the northward flow of warm waters. (The data of Paquette and Bourke [1981] clearly show the northward advection of warm water between Herald and Hanna shoals, and this flow is also suggested by spreading of the bottom isotherms at about 71°N, 165°W in Figures 4C and 5C.) These embayments might be particularly important to the ecosystem of the Chukchi Sea because there are marked differences between the wind-forced circulation along a straight ice edge and a meandering ice edge. The numerical model results of Roed and O'Brien (1983) show that for a straight ice edge, zonally oriented such that open water lies to the south of the ice, westward winds result in upwelling along the ice edge, and eastward winds result in downwelling. Hakkinen's (1986) model results show that if an ice edge has meanders (i.e., embayments), upwelling occurs on the windward side of the meander and downwelling on the lee side. High rates of primary production are usually associated with ice edges (Niebauer 1991), and ice-edge upwelling should further enhance this productivity. In the northern Chukchi Sea, where mean

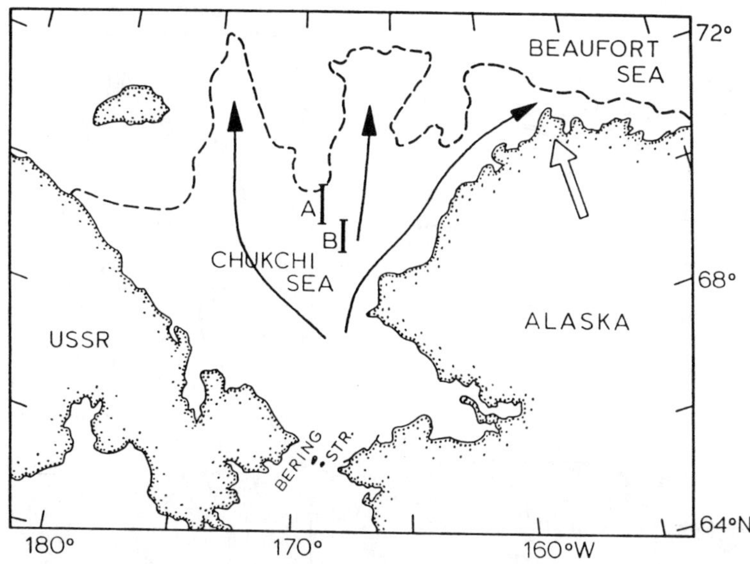

FIGURE 7.—Schematic of the Chukchi Sea ice edge, illustrating the location of "meltwater embayments" (after Paquette and Bourke 1981). Transects A and B correspond to the cross-sectional profiles shown in Figure 8, **A** and **B**, respectively.

winds are westward and vary primarily in the zonal direction, these embayments could be significant sites for carbon fixation. Hakkinen's (1986) model results further imply that biological production along the ice edge of the Chukchi Sea could vary dramatically in both time and space and that field programs must be suitably designed to address this issue.

Ice-edge zones are characterized by large spatial gradients in water (and air) mass properties, currents, winds, ice concentration, and the ocean wave field. All of these variables are interrelated through complicated feedback processes that are rarely in equilibrium. (A nontechnical overview of these processes is given by Muench [1989]. More technical aspects of these processes are found in the June 1987 "Marginal Ice Zone Research" special issue of the *Journal of Geophysical Research*.) An illustration of some of the complexities of these regions in the Chukchi Sea are given in Figure 8 (A and B) from Paquette and Bourke (1981), which shows two cross-sectional profiles of temperature and salinity along the transects A and B shown in Figure 7. Figure 8A (transect A) shows that the ice-edge frontal system consists of an upper-layer front established by meltwater and a lower-layer front established by the temperature contrasts between resident Chukchi water and the northward-flowing Alaska coastal water and Bering shelf water. These upper- and lower-layer fronts do not always coincide, as is evident in other transects shown by Paquette and Bourke (1981). The hydrographic structure suggests a horizontally and vertically sheared, strong westward flow within the surface front. In contrast, Figure 8B (transect B) shows strong vertical salinity gradients, but weak horizontal gradients. In fact, the whole transect is characterized by finescale (1- to 5-m thick) temperature intrusions and weak currents.

Interannual Variability

The relative brevity of research programs seldom affords opportunities to appreciate variability on time scales exceeding the seasonal, and this is especially true for high-latitude oceans where sampling is costly and complicated by harsh environmental conditions. However, accumulating evidence points to large interannual variations in the oceanographic conditions of the Chukchi Sea. A major step forward in understanding this variability was made by Coachman and Aagaard (1988), who related transport variations in Bering Strait to along-strait winds. Their results (Figure 9) show that the mean annual transport varied by a factor of 2 over the period from 1946 to 1985. In addition to the considerable differences from year to year, there occurred a large drop in transport beginning in the late 1960s. In

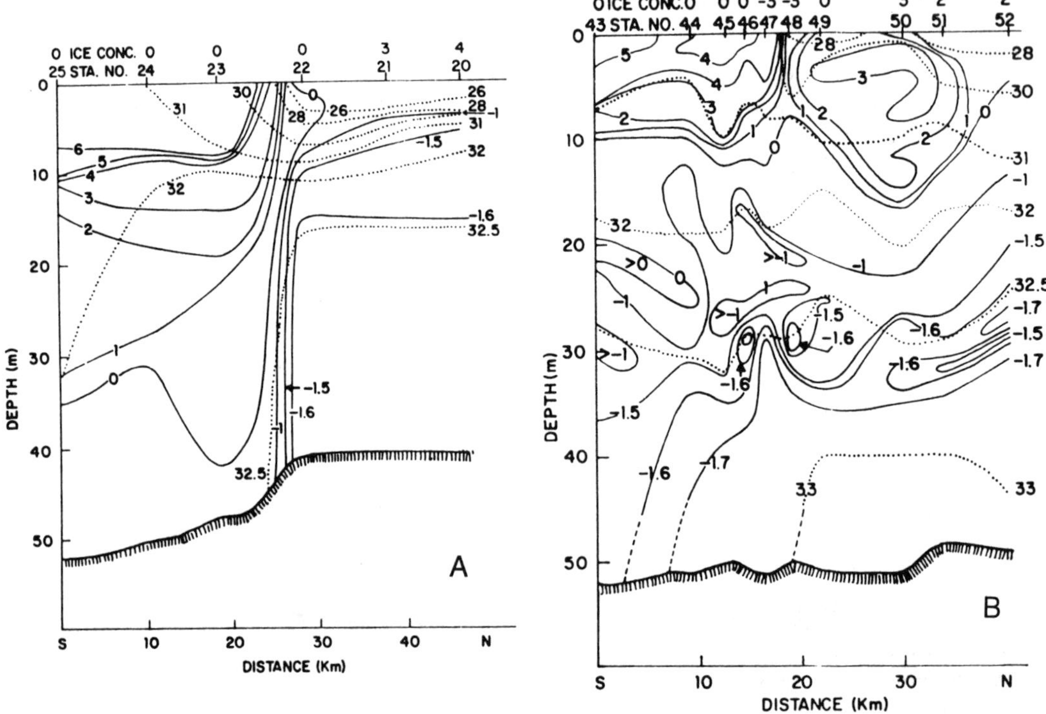

FIGURE 8.—Cross-sectional profiles of temperature (solid lines) and salinity (dotted lines) along two transects normal to the summer ice edge; (**A**) illustrates the case where well-defined upper- and lower-layer fronts coincide. (**B**) shows complex temperature fine-structure associated with strong vertical salinity gradients. Station numbers appear at top of each graph. Temperature is in degrees Celsius and salinity is in practical salinity units.

fact, before 1969, 30% of the estimated mean annual transports were less than the 40-year mean, whereas, from 1969 to 1985, 76% of the transports were less than this mean. These transport variations could have important consequences for the Chukchi Sea ecosystem because of the implied variation in the northward fluxes of heat, salt, nutrients, and plankton. Although the implications of this interannual variability upon the northeastern Chukchi Sea are enormous, little data exists upon which definitive statements regarding effects can be made. Aagaard's (1988) data from September 1981 showed no indication of the bottom front that was argued to be a quasi-permanent feature of the northeastern Chukchi Sea. The absence of this front in 1981 is seen the cross-sectional profiles of temperature and salinity along the Point Lay transect shown in Figure 10 (the corresponding transect for 1982 is shown in Figure 6A). Moreover, no Alaska coastal water is present along this transect, which consists of very cold (~−1°C) resident Chukchi water along the bottom and within 80 km of the coast and meltwater elsewhere. Differences between the 1981 hydrography and the years discussed previously could be associated with anomalies in the regional wind field. Figure 11 shows time series of the anomaly from the mean monthly north–south pseudo-wind stress component at 67.5°N, 167.5°W for the period 1981–1991. Time series of this variable to the south and north of this location are quite similar. The computational procedure for making these estimates is given by Salmon (1992). The data show persistent, anomalously southward wind stress from July through November 1981. In 1986 and 1990 winds were anomalously northward during these months, while in 1982 both northward and southward anomalies were observed. The magnitude and persistence of the 1981 wind stress anomalies would have resulted in (1) anomalously low, northward transport through Bering Strait, thereby increasing the time required to flush winter water from the Chukchi shelf; and (2) westward displacement of Alaska coastal water from the northeastern Chukchi Sea.

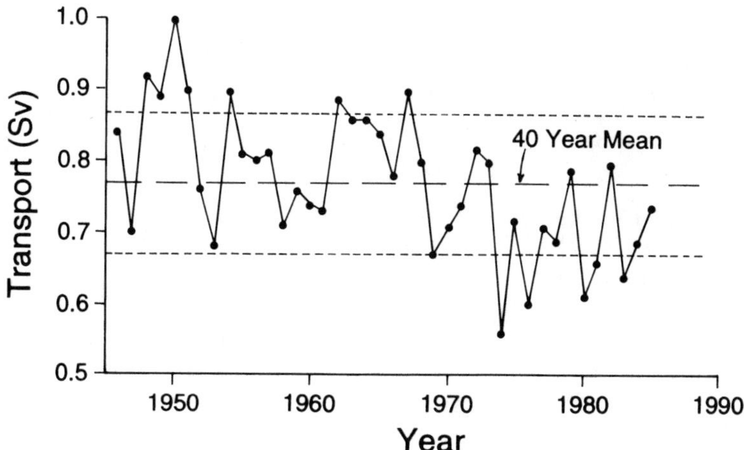

FIGURE 9.—Mean annual northward transport (Sv, Sverdrup) through Bering Strait during 1946–1985 (after Coachman and Aagaard 1988).

Coachman and Shigaev (1992) also documented temperature and salinity (and, by implication, other constituents) differences in the source waters feeding into Bering Strait (Figure 12). For example, in aggregate the salinities for these various water masses were considerably higher in 1967 and 1968 than they were in 1976 and 1977. Temperatures and salinities of southeastern Bering Shelf water also show considerable interannual variations, and these are associated with interannual variability in ice cover over the Bering Sea shelf such that cold, dilute bottom waters in summer follow extensive winter ice cover and vice versa.

Summer–fall sea ice extent in the Chukchi Sea

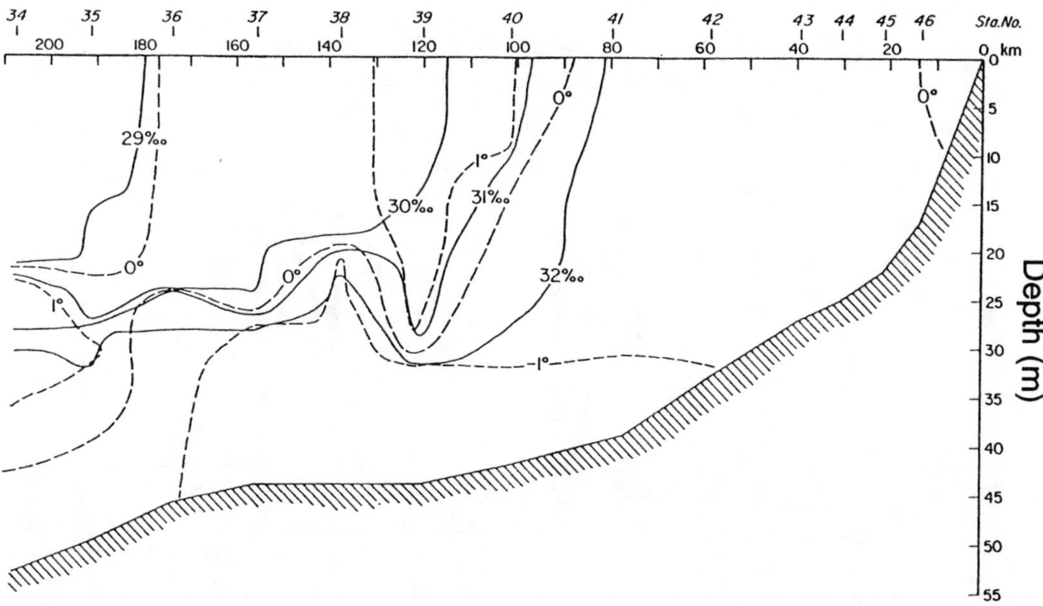

FIGURE 10.—Cross-sectional profiles of temperature (°C, dashed lines) and salinity (practical salinity units, solid lines) along a transect extending northwest from Point Lay in September 1981. The corresponding transect for August 1982 is shown in Figure 6**A** (after Aagaard 1988). Station numbers are along the top of the graph.

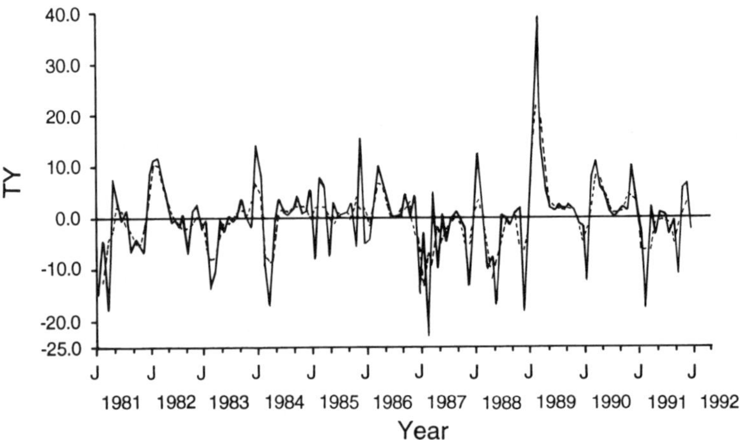

FIGURE 11.—Time series of the monthly (solid line) and 3-month running mean (dashed line) of the north–south pseudowind stress component at 67.5°N, 167.5°W for the period 1981–1991. January of each year is indicated by "J."

also varies enormously from year to year. For example, the Naval Oceanography Command Detachment (NOCD 1986) atlas shows that the mid-September ice-edge position in the northern Chukchi Sea ranges from 70 to 75°N—a distance of more than 500 km. Figure 13, from Mysak et al. (1990), shows smoothed time-series of monthly anomalies of the areal sea-ice extent of the Chukchi and Beaufort seas for the period 1953–1985. Their data show that year-to-year differences in the area covered by sea ice can be as large as 300% and Figure 13 shows that the anomalies persist for 3 to 6 years. These anomalies are related primarily to variations in regional winds and coastal freshwater discharge induced by hemispheric-scale atmospheric fluctuations.

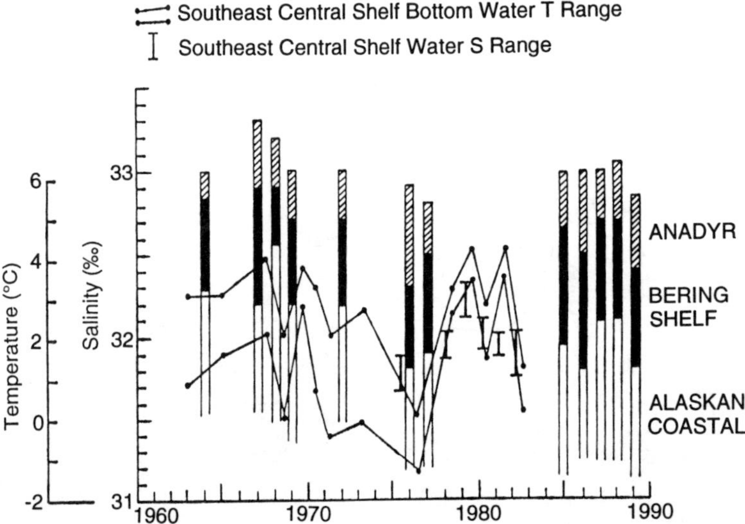

FIGURE 12.—Interannual variations in temperature and salinity of the waters flowing through Bering Strait (after Coachman and Shigaev 1992). The thick vertical bars refer to the salinity ranges of particular water masses observed in Bering Strait in these years (Anadyr—striped, Bering Shelf—solid, and Alaskan Coastal—open). The lines connecting the solid dots denote the temperature range of bottom water observed on the southeast Bering Sea shelf. The thin vertical bars correspond to the salinity range of bottom water observed on the southeast Bering Sea shelf.

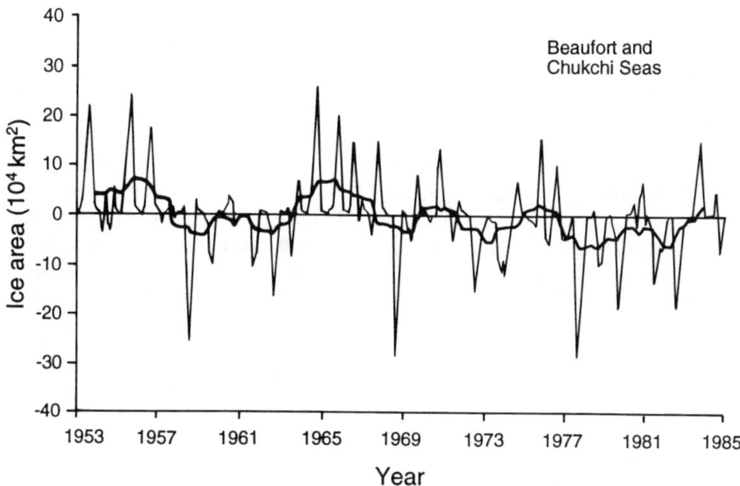

FIGURE 13.—Smoothed time-series of monthly anomalies of the areal sea-ice extent in the Chukchi and Beaufort seas for the period 1953–1985 (after Mysak et al. 1990). The thin curves are the monthly anomalies in sea-ice cover and the thick curves are the 25-month running mean of the monthly anomalies.

Summary

The preceding sections have shown that the bathymetry, meteorology, sea ice distribution, and northward advection of waters through Bering Strait profoundly influence the oceanography of the northeastern Chukchi Sea. One of the more important hydrographic features observed during the open-water season is the existence of a quasi-permanent bottom thermal front that parallels the bathymetry and the core of the ACC. Biologically, the front appears to be an important feeding area for marine mammals preying on benthic organisms and a "boundary" across which benthic and fish community structure change markedly (Feder et al. 1990; Smith et al. 1997a, 1997b; Wyllie-Echeverria et al. 1997, all this volume).

The oceanographic regime of the southern Chukchi Sea supports some of the highest primary production rates observed in the global ocean, and this production is then exported to the northern Chukchi Sea and the Arctic Ocean (Walsh et al. 1989). Although little is known about the northern Chukchi Sea ecosystem, the influence of the ice edge and shelf break upwelling might also lead to high biological production rates here.

This review has also touched upon the enormous, but poorly documented, interannual variability in the physical environment of the Chukchi Sea. This variability must also be reflected in the ecosystem. A better understanding of this variability is required in order to address issues pertaining to the ecological effects of marine industrial development in the Arctic. Resolution of these issues will require an interdisciplinary approach.

Acknowledgments

I thank W. Barber, F. Muter, L. Tuttle, K. Coyle, and S. Okkonen for at-sea collections of conductivity–temperature–depth data. K. Aagaard kindly provided some of the temperature and salinity data used in Figure 2. The preparation of this article benefited from useful discussions with H. Feder, K. Coyle, W. Barber, and M. Johnson. I also thank R. Meyer of the Minerals Management Service (MMS), Department of the Interior, Anchorage, Alaska, for his enthusiastic support. This work was supported by MMS Contract 14-35-0001-30559 to the University of Alaska.

References

Aagaard, K. 1984. The Beaufort undercurrent. Pages 47–71 *in* P. W. Barnes, D. M. Schell, and E. Reimnitz, editors. The Alaskan Beaufort Sea: ecosystems and environment. Academic Press, New York.

Aagaard, K. 1988. Current, CTD, and pressure measurements in possible dispersal regions of the Chukchi Sea. NOAA (National Oceanic and Atmospheric Administration) OCSEAP (Outer Continental Shelf Environmental Assessment Program), Final Report 57: 255–333. U.S. Department of Commerce and U.S. Department of the Interior, Anchorage, Alaska.

Aagaard, K., C. H. Pease, A. T. Roach, and S. T. Salo. 1990. Beaufort Sea mesoscale circulation study—final report. NOAA (National Oceanic and Atmospheric Administration) Technical Memorandum ERL (Environmental Research Lab) PMEL (Pacific Marine Environmental Laboratory) 90, Seattle.

Aagaard, K., and A. T. Roach. 1990. Arctic Ocean-shelf exchange: measurements in Barrow Canyon. Journal of Geophysical Research 95:18163–18175.

Bourke, R. H., and R. G. Paquette. 1976. Atlantic water on the Chukchi Shelf. Geophysical Research Letters 3:3629–3632.

Boyer, D. L., R. Chen, G. C. D'hieres, and H. Didelle. 1987. On the formation and shedding of vortices from side-wall mounted obstacles in rotating systems. Dynamics of Atmospheres and Oceans 11:59–86.

Brower, W. A., Jr., R. G. Baldwin, C. N. Williams, Jr., J. L. Wise, and L. D. Leslie. 1988. Climatic atlas of the outer continental shelf waters and coastal regions of Alaska. Volume 11, Chukchi–Beaufort Sea. National Climatic Data Center, Asheville, North Carolina.

Carmack, E. C. 1990. Large-scale physical oceanography of polar oceans. Pages 171–222 *in* W. O. Smith, Jr., editor. Polar oceanography, part A: physical science. Academic Press, New York.

Coachman, L. K., and K. Aagaard. 1974. Physical oceanography of Arctic and subarctic seas. Pages 1–72 *in* H. Nelson and Y. Herman, editors. Marine geology and oceanography of the Arctic seas. Springer-Verlag, New York.

Coachman, L. K., and K. Aagaard. 1988. Transports through Bering Strait: annual and interannual variability. Journal of Geophysical Research 93:15635–15540.

Coachman, L. K., K. Aagaard, and R. B. Tripp. 1975. Bering Strait: the regional physical oceanography. University of Washington Press, Seattle.

Coachman, L. K., and V. V. Shigaev. 1992. Northern Bering–Chukchi Sea ecosystem: the physical basis. Pages 17–27 *in* P. A. Nagel, editor. Results of the third joint U.S.–U.S.S.R. Berring and Chukchi seas expedition (BERPAC). U.S. Fish and Wildlife Service, Washington, DC.

Colony, R. 1984. An estimate of the mean field of Arctic sea ice. Journal of Geophysical Research 89:10623–10629.

Csanady, G. T. 1982. Circulation in the coastal ocean. Reidel Publishing Co., Boston.

D'Asaro, E. A. 1988. Generation of submesoscale vortices: a new mechanism. Journal of Geophysical Research 93:6685–6693.

Feder, H. M., A. S. Naidu, M. J. Hameedi, S. C. Jewett, and W. R. Johnson. 1990. The Chukchi Sea continental shelf: benthos–environment interactions. NOAA (National Oceanic and Atmospheric Administration) OCSEAP (Outer Continental Shelf Environmental Assessment Program) Final Report 68:25–311, Anchorage, Alaska.

Fleming, R. H., and D. Heggarty. 1966. Oceanography of the southeastern Chukchi Sea. Pages 697–754 *in* N. J. Wilimovsky and J. N. Wolfe, editors. Environment of the Cape Thompson region, Alaska. U.S. Atomic Energy Commission, Washington, DC.

Garrison, G. R., and R. G. Paquette. 1982. Warm water interaction in the Barrow Canyon in winter. Journal of Zephyr Research 87:5853–5859.

Grebmeier, J. M., C. P. McRoy, and H. M. Feder. 1988. Pelagic–benthic coupling on the shelf of the northern Bering and Chukchi seas. I. Food supply source and benthic biomass. Marine Ecology Progress Series 48: 57–67.

Hakkinen, S. 1986. Coupled ice-ocean dynamics in the marginal ice zones: up/downwelling and eddy generation. Journal of Geophysical Research 87:5853–5859.

Hufford, G. L. 1974. On apparent upwelling in the southern Beaufort Sea. Journal of Geophysical Research 79:1305–1306.

Johnson, J. A., and N. Rockliff. 1986. Shelf break circulation processes. Pages 33–62 *in* C. N. K. Mooers, editor. Baroclinic processes on continental shelves. American Geophysical Union, Washington, DC.

Johnson, W. R. 1989. Current response to wind in the Chukchi Sea: a regional coastal upwelling event. Journal of Geophysical Research 94:2057–2064.

Kinder, T. H., D. C. Chapman, and J. A. Whitehead, Jr. 1986. Westward intensification of the mean circulation on the Bering Sea shelf. Journal of Physical Oceanography 16:1217–1229.

Moore, S. E., and J. T. Clarke. 1986. A comparison of gray whale (*Eshrichtus robustus*) and bowhead whale (*Balaena mysticetus*) distribution, abundance and habitat preference and behavior in the northeastern Chukchi Sea, 1982–84. International Whaling Commission Report of the Commission 36:273–279.

Mountain, D. G., L. K. Coachman, and K. Aagaard. 1976. On the flow through Barrow Canyon. Journal of Physical Oceanography 6:461–470.

Muench, R. D. 1989. The sea ice margins: a summary of physical phenomena. NOAA (National Oceanic and Atmospheric Administration) Technical Memorandum ERL (Environmental Research Lab) PMEL (Pacific Marine Environmental Laboratory) 88, Seattle.

Muench, R. D., C. H. Pease, and S. A. Salo. 1991. Oceanographic and meteorological effects on autumn sea-ice distribution in the western Arctic. Annals of Glaciology 15:171–177.

Mysak, L. A., D. K. Manak, and R. F. Marsden. 1990. Sea-ice anomalies observed in the Greenland and Labrador seas during 1901–1984 and their relation to an interdecadal Arctic climate cycle. Climate Dynamics 5:111–133.

Niebauer, H. J. 1991. Bio-physical oceanographic interactions at the edge of the Arctic ice pack. Journal of Marine Systems 2:209–232.

NOCD (Naval Oceanography Command Detachment). 1986. Sea ice climatic atlas: volume III, Arctic west. NOCD, Asheville, North Carolina.

Overland, J. E., and C. H. Pease. 1982. Cyclone climatology of the Bering Sea and its relation to sea-ice extent. Monthly Weather Review 110:5–13.

Overland, J. E., and A. T. Roach. 1987. Northward flow in the Bering and Chukchi seas. Journal of Geophysical Research 92:7097–7105.

Paquette, R. G., and R. H. Bourke. 1981. Ocean circulation and fronts as related to ice melt-back in the

Chukchi Sea. Journal of Geophysical Research 86: 4215–4230.

Pease, C. H. 1987. Meteorology of the Chukchi Sea: an overview. Pages 11–19 *in* D. Hale, editor. Chukchi Sea information update. NOAA (National Oceanic and Atmospheric Administration) OCSEAP (Outer Continental Shelf Environmental Assessment Program), U.S. Department of Commerce and U.S. Department of the Interior, Anchorage, Alaska.

Philander, S. G. H., and J.-H. Yoon. 1982. Eastern boundary currents and coastal upwelling. Journal of Physical Oceanography 12:862–879.

Phillips, R. L., and M. W. Colgan. 1987. Sea-floor feeding traces of gray whales and walrus in the northeast Chukchi Sea. Pages 183–186 *in* J. P. Galloway and T. D. Hamilton, editors. Geologic studies in Alaska by the U. S. Geological Survey during 1987. U. S. Geological Circular 1016, Menlo Park, California.

Roed, L. P., and J. J. O'Brien. 1983. A coupled ice–ocean model of upwelling in the marginal ice zone. Journal of Geophysical Research 88:2863–2872.

Salmon, D. K. 1992. On interannual variability and climate change in the North Pacific. Doctoral dissertation. University of Alaska, Fairbanks.

Springer, A. M., C. P. McRoy, and K. R. Turco. 1989. The paradox of pelagic food webs in the northern Bering Sea: II. Zooplankton communities. Continental Shelf Research 9:359–386.

Smith, R. L., M. Vallarino, E. Barbour, E. Fitzpatrick, and W. E. Barber. 1997a. Population biology of the Bering flounder in the northeastern Chukchi Sea. Pages 127–132 *in* J. Reynolds, editor. Fish ecology in Arctic North America. American Fisheries Society Symposium 19, Bethesda, Maryland.

Smith, R. L., W. E. Barber, M. Vallarino, J. Gillispie, and A. Ritchie. 1997b. Population biology of the Arctic staghorn sculpin in the northeastern Chukchi Sea. Pages 133–139 *in* J. Reynolds, editor. Fish ecology in Arctic North America. American Fisheries Society Symposium 19, Bethesda, Maryland.

Walsh, J. J. 1988. On the nature of continental shelves. Academic Press, New York.

Walsh, J. J., and 20 coauthors. 1989. Carbon and nitrogen cycling within the Bering/Chukchi seas: source regions for organic matter effecting AOU demands of the Arctic Ocean. Progress in Oceanography 22:277–359.

Wyllie-Echeverria, T., W. E. Barber, and S. Wyllie-Echeverria. 1997. Water masses and transport of age-0 Arctic cod and age-0 Bering flounder into the northeastern Chukchi Sea. Pages 60–67 *in* J. Reynolds, editor. Fish ecology in Arctic North America. American Fisheries Society Symposium 19, Bethesda, Maryland.

American Fisheries Society Symposium 19:60–67, 1997

Water Masses and Transport of Age-0 Arctic Cod and Age-0 Bering Flounder into the Northeastern Chukchi Sea

TINA WYLLIE-ECHEVERRIA,[1] WILLARD E. BARBER,
AND SANDY WYLLIE-ECHEVERRIA[2]

University of Alaska, School of Fisheries and Ocean Sciences
Fairbanks, Alaska 99775–1090, USA

Abstract.—It is hypothesized that some fish populations in the northeastern Chukchi Sea are maintained through transport from more southern areas. We examine this hypothesis in relation to the abundance and distribution of age-0 Arctic cod *Boreogadus saida* and Bering flounder *Hippoglossoides robustus*. Ichthyoplankton were sampled with a variety of midwater gear during 1989–1991. Arctic cod dominated the catches and occurred throughout the northeastern Chukchi Sea, with higher concentrations at northern stations during all 3 years. They rarely occurred at southern stations when Bering shelf water was present. Larval Arctic cod (6.3 mm, mean standard length) were caught in mid-July in 1991, indicating reproduction extends into late July. Bering flounder occurred primarily in areas dominated by Alaska coastal water as far north as 71°N. They did not occur when the resident Chukchi water mass was present. We conclude that populations of Bering flounder in the northeastern Chukchi Sea are maintained by the transport of larvae in Alaska coastal water. Although these fish and others may be routinely advected into the northeastern Chukchi Sea by Alaska coastal water, resident Chukchi water may be a critical factor in delimiting their northern distribution.

Many species of fish from the northern Pacific Ocean and the Bering Sea have their northern range bordering the Chukchi Sea, for example, starry flounder *Platichthys stellatus*, Pacific halibut *Hippoglossus stenolepis*, and Pacific cod *Gadus macrocephalus*. Others, however, commonly occur in the region and are considered Arctic, such as, Arctic cod *Boreogadus saida* (Alverson and Wilimovsky 1966; Allen and Smith 1988).

Three species of Gadidae occur in the northeastern Chukchi Sea. Arctic cod, which has a circumpolar distribution, is the dominant gadid species (Pruter and Alverson 1962; Alverson and Wilimovsky 1966; Gillispie et al. 1997, this volume). They are found near the ice edge, migrate, and spawn under the ice (Ponomarenko 1968) and are a key link in the transport of energy from lower to higher trophic levels (Craig et al. 1982). Additionally, Arctic cod is an important prey for 11 species of marine mammals, 20 species of marine birds, and 4 species of fishes (Lowry and Frost 1981).

Six species of flatfishes commonly occur in the Chukchi Sea, where Bering flounder *Hippoglossoides robustus* is the dominant one (Pruter and Alverson 1962; Smith et al. 1997, this volume). Bering flounder range from Tatar Strait in the west to the Chukchi Sea through the Bering Sea to the Aleutian Islands (Andriashev 1955; Pruter and Alverson 1962). Pruter and Alverson (1962) hypothesized that populations of Bering flounder might be maintained by import from the northern Bering Sea to the Chukchi Sea. To investigate this possibility, we sampled planktonic age-0 fish in the late summer and fall of 1989–1991. From these samplings we describe the relationship of age-0 Bering flounder to the physical oceanographic conditions in the northeastern Chukchi Sea and compare this distribution to that of age-0 Arctic cod.

The passive movement of planktonic larvae and postlarvae oceanic fish is affected by currents (Sinclair 1988). The distribution of Atlantic cod *G. morhua* (Elizarov 1965), Atlantic herring *Clupea harengus* (Jakobsson 1969), and capelins *Mallotus villosus* (Stergiou 1991) has been associated with changes in the distribution of water masses in the north Atlantic. Arctic populations of saffron cod *Eleginus gracilis* are known to vary in distribution and abundance with different current regimes (Vasil'kov et al. 1981). Ice cover, as well as other physical factors affected by atmospheric forcing, affects the distribution of walleye pollock *Theragra chalcogramma* (Vasil'kov and Glebova 1984).

Currents in the northeastern Chukchi Sea are dominated by flow from the Bering Sea, which is caused by the Arctic Ocean being approximately 0.5 m lower than the Pacific Ocean (Stigebrandt 1984). Secondarily, the direction and strength of the

---

[1]Present address: University of Washington, Joint Institute for the Study of Atmosphere and Oceans, Seattle, Washington 98105, USA.

[2]Present address: University of Washington, School of Marine Affairs, Seattle, Washington 98105, USA.

wind, especially in summer, influences the rate of northerly flow and periodically imposes southerly flow over the northeastern Chukchi Sea shelf (Coachman and Shigaev 1992). The result is that three water masses predominate in the northeastern Chukchi Sea: Alaska coastal water, resident Chukchi water, and Bering shelf water. Their distribution is influenced by the wind and modified by the freezing and melting of seasonal sea ice (Weingartner 1997, this volume). The flow of currents northward introduces nutrients, phytoplankton, and zooplankton into the Chukchi Sea (Walsh et al. 1989).

Materials and Methods

A Bongo net, Isaacs-Kidd midwater trawl (IKMT), and beam trawl were used to sample age-0 fish during late summer and fall of 1989–1991. The Bongo net was 60 cm in diameter, with 1.0-mm mesh netting and a 0.5-mm mesh cod end, and sampled a 0.28-m^2 cross section of water. The IKMT had a 1.8-m head bar and a net having 5.0-mm mesh, a 1.0-mm mesh cod end, and sampled a 2.65-m^2 cross section of water. Each net was fitted with a calibrated flowmeter suspended in the mouth of the net. In 1989 the IKMT was deployed from the *Alpha Helix*. In 1990 and 1991 the IKMT and the Bongo net were deployed from a 33.5-m trawler. In 1991 a 2.24-m^2 square beam trawl was also used, the net of which graded from 6.5 mm to 4.0 mm with a 1.0-mm lining in the cod end and was deployed from the *Oshoro Maru* (University of Hokkaido, Japan). Also in 1991, the Bongo net and IKMT were deployed from a 9-m skiff operating from an anchored barge. All nets were pulled at a speed of approximately 2 knots and were deployed in double oblique tows from surface to near bottom.

All collections were preserved in 5% formalin in seawater solution and returned to the laboratory. Fish were separated from the plankton samples, identified to the lowest taxonomic level possible following Matarese et al. (1989), and preserved in 80% ethanol. Standard lengths were measured on all undamaged fish to the nearest 0.1 mm; no shrinkage corrections were applied. The densities for each station were calculated as the mean of two replicate tows and reported as number per 1,000 m^3. Total number of fish captured is reported for the beam trawl collections because of the lack of a flowmeter.

General station locations were chosen to increase the probability of sampling different water masses. Latitude and longitude were determined with a global positioning system. At the end of each sampling, vertical profiles of salinity, temperature, and depth were obtained with a Seabird SBE 19 CTD instrument. These data were used to classify each station as to one of three water masses present on the northeastern Chukchi Sea shelf, following the scheme of Coachman and Shigaev (1992) and Weingartner (1997). Two water masses, Alaska coastal water and Bering shelf water, originate in the Bering Sea. The Alaska coastal water flows northward from the Bering Sea along the coast of Alaska and is influenced by freshwater input, primarily the Yukon River (Weingartner 1997); it is characterized by relatively warm temperatures of 2–6° C and salinity less than 31.5 ppt. Bering shelf water is colder (0–3°C) and more saline (>32.5 ppt) than Alaska coastal water and also flows northward to the west of Alaska coastal water. Resident Chukchi water is a combination of Arctic Ocean water and water from the melting and freezing of sea ice. Resident Chukchi water is characterized by low temperatures (<1°C) and high salinity (32–33 ppt).

Wind stress curl at 67.5°N, 167.5°W for 1981–1991 was obtained from Figure 11 in Weingartner (1997). Salmon (1992) calculated the north–south wind stress curl for the Chukchi Sea from atmospheric pressure data for 1981–1990.

Results

Members of 10 families were captured; of these, Cottidae, Gadidae, Pleuronectidae, Agonidae, Stichaeidae, Cyclopteridae, and Ammodytidae were present in all years (Table 1). Zoarcidae, Bathylagidae, and Pholidae occurred occasionally. Arctic cod was dominant in samples of larval and juvenile pelagic fishes in all 3 years. Species and abundance varied greatly between years, with a notable reduction of pleuronectids in 1991.

During 3–9 September 1989, sampling was limited to 21 stations along two lines of longitude (169 and 168°W) in the northeastern Chukchi Sea (Figure 1). Alaska coastal water was present at the surface throughout the study area. Bering shelf water was evident at the most southern station below 10 m. Resident Chukchi water underlay the Alaska coastal water north of 70°N below 20–25 m. Eight families of fishes were caught in the IKMT in 1989, of which age-0 Arctic cod and Bering flounder were most common (Table 1). Arctic cod were present at all stations irrespective of water type. They were most common north of 70°N in Alaska coastal water–resident Chukchi water, where densities at three stations ranged from 130 to 403/1,000 m^3

TABLE 1.—Percent abundance of each species sampled by double oblique tows in the northeastern Chukchi Sea with an Isaacs-Kidd midwater trawl (IKMT), a beam trawl, and a Bongo net during summers of 1989–1991. Data were combined in 1991 for fish caught with the IKMT and beam trawl. Species indicated by a "t" constituted less than 2% of the total catch.

Taxon	1989	1990		1991	
	IKMT	Bongo	IKMT	Beam and IKMT	Bongo
Cottidae					
Myoxocephalus spp.	t	t	t	6	t
M. verrucosus	1	0	t	0	0
Gymnocanthus spp.	0	0	t	2	t
G. tricuspis	3	0	t	0	0
Porocottus sp.	0	0	t	0	0
Gadidae					
Boreogadus saida	79	16	36	83	85
Eleginus gracilis	0	10	19	2	t
Theragra chalcogramma	0	t	t	3	0
Pleuronectidae					
Hippoglossoides robustus	7	23	18	t	0
Pleuronectes spp.	t	t	t	0	0
P. asper	t	t	t	0	0
P. proboscideus	t	t	t	0	0
P. bilineatus	0	t	t	0	0
Platichthys stellatus	0	0	t	0	0
Agonidae					
Aspidophoroides olriki	t	0	t	t	0
A. bartoni	t	0	0	0	0
Podothecus acipenserinus	t	0	0	t	0
Stichaeidae					
Stichaeus punctatus	1	t	7	t	10
Lumpenus sp.	2	t	6	3	t
Cyclopteridae					
Liparis sp.	t	t	5	t	t
Ammodytidae					
Ammodytes hexapterus	3	14	3	0	t
Bathylagidae	0	t	t	0	0
Pholidae	0	0	0	0	t
Zoarcidae					
Lycodes sp.	t	t	t	0	0
Total number of fish	2,068	99	581	379	76
Total number of stations	21	48	48	17	16

(Figure 1). Two stations at the boundary of the Alaska coastal water and Alaska coastal water–resident Chukchi water had densities of 137 and 709/1,000 m^3. South of this area the densities were much lower and ranged from 2 to 31/1,000 m^3. Age-0 Arctic cod were 27.0–52.0 mm long. Bering flounder larvae were present at stations south of 71°N, primarily in Alaska coastal water or Alaska coastal water–Bering shelf water. Densities reached 32/1,000 m^3 at the two stations along 70°N; these fish were 15.0–34.0 mm long.

During 16 August to 16 September 1990, we sampled 48 stations throughout the northeastern Chukchi Sea (Figure 2). Alaska coastal water was evident in surface waters throughout the area. Two additional water masses were present underneath the Alaska coastal water: Bering shelf water occurred in the southwestern stations and resident Chukchi water occurred in the northern stations (Figure 2). Thermoclines and pycnoclines were only evident where Alaska coastal water overlay a colder, more-saline water mass. Nine families of fishes were captured, with Arctic cod, saffron cod, and Bering flounder dominating (Table 1). Age-0 Arctic cod were caught at all stations north of 71°N and made up 36% of the catch with the IKMT. Density exceeded 25/1,000 m^3 at three stations north of Icy Cape (Figure 2). The length of Arctic cod averaged 33 mm and ranged from 12.2 to 51.0 mm. The distribution of larval Arctic cod captured with the Bongo net was similar to that caught with the IKMT. Abundance, however, was much lower, varying from 0.1 to 0.7/1,000 m^3. Bering flounder were also captured in the IKMT but south of 71°N. At stations where these fish occurred, densities ranged from 1.1 to 4.6/1,000 m^3. Bering flounder occurred in Alaska coastal water and where Alaska coastal water overlaid Bering shelf water (Figure 2). Bering

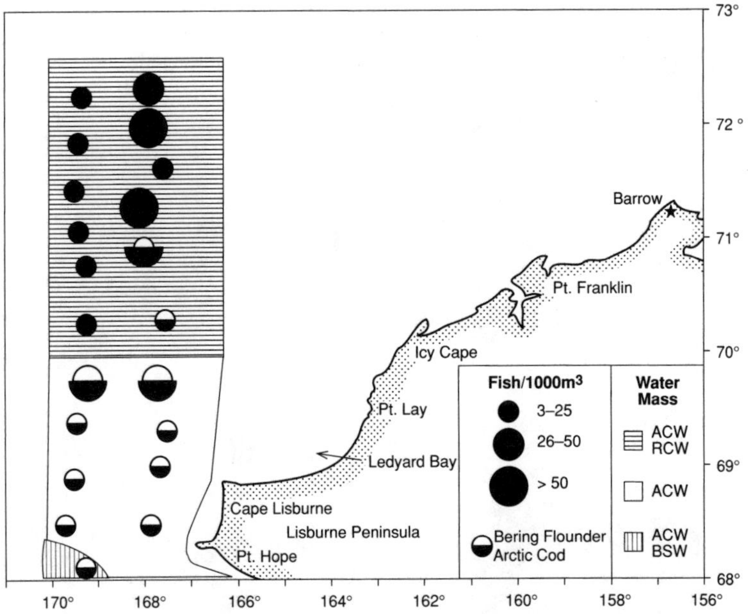

FIGURE 1.—Abundance and distribution of Arctic cod and Bering flounder captured with the Isaacs-Kidd midwater trawl in 1989. Densities are in fish per 1,000 m^3. Water masses are indicated by shading: ACW = Alaska coastal water, RCW = resident Chukchi water, BSW = Bering shelf water. Circles are scaled to reflect fish abundance. Bering flounder are represented by open circles; solid circles represent Arctic cod. Divided circles represent both species at that station.

flounder larvae captured in the Bongo net reflected the same distribution as those captured in the IKMT and also occurred south of 71°N in Alaska coastal water. Densities, however, were much lower, ranging from 0.1 to 0.4/1,000 m^3. The Bongo net captured smaller fish, averaging 15 mm in length and ranging from 7.6 to 25.7 mm; those caught in the IKMT averaged 29 mm and ranged from 8.3 to 51.5 mm.

During 25–31 July 1991, we sampled 17 stations with the beam trawl and IKMT. Alaska coastal water was present south of 70°N and Alaska coastal water–resident Chukchi water and resident Chukchi water occurred north of 69°30′N (Figure 3). Six species of fishes were caught; Arctic cod (83%), *Myoxocephalus* spp., walleye pollock, and *Lumpenus* sp. were the dominant species (Table 1). Arctic cod were caught throughout the study area, with the highest numbers occurring in resident Chukchi water, where up to 65 fish were caught per 0.5-h tow. The average length was 14 mm and the range was 7.0–36.0 mm. One Bering flounder was caught in Alaska coastal water.

Fish were caught in the Bongo net at 5 of 16 stations during 2–25 September 1991. Again, Arctic cod was dominant (85%) and occurred in Alaska coastal water–resident Chukchi water and resident Chukchi water. The average length was 14 mm and the range was 7.0–37.0 mm. No Bering flounder were caught. Arctic cod captured in the Bongo net in samples near the ice edge from the skiff during 16 July averaged 12 mm in length and ranged from 6.3 to 19.5 mm. In samples from 16 August, the average length was 13 mm and ranged from 8.2 to 23 mm.

Discussion

The waters of the northeastern Chukchi Sea contain Bering shelf water and Alaska coastal water, flowing from the Bering Sea, and resident Chukchi water from the Arctic Ocean. Resident Chukchi water is advected onto the shelf and mixes with waters formed from the freezing and melting during the previous winter (Weingartner 1997). Pruter and Alverson (1962) suggested that some of the marine fishes and invertebrates inhabiting the area maintain their populations only by continual recruitment of eggs and larvae transported northward. If so, larvae of those fish with their main distribution primarily in the Bering Sea would tend to be associated with the Alaska coastal water and the Bering shelf water in the northeastern Chukchi Sea. In

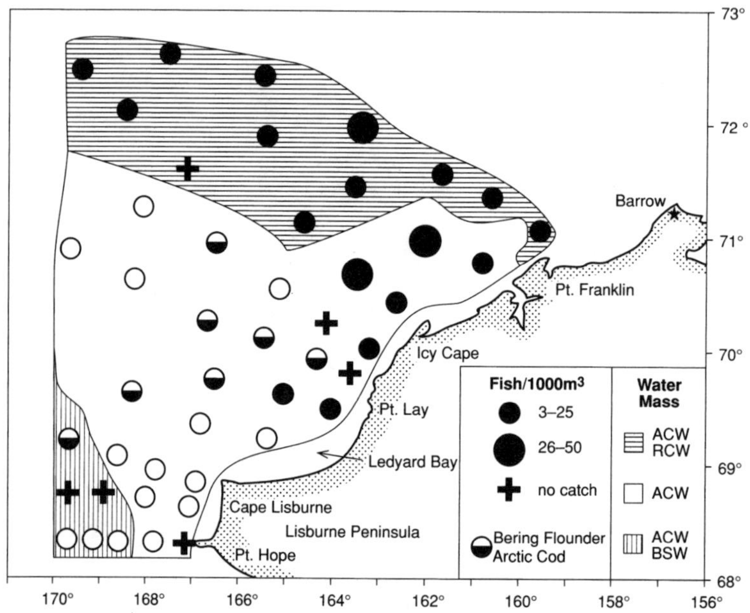

FIGURE 2.—Abundance and distribution of Arctic cod and Bering flounder captured with the Isaacs-Kidd midwater trawl in 1990. Densities are in fish per 1,000 m^3. Water masses are indicated by shading (see Figure 1 caption for definitions and explanation of circles).

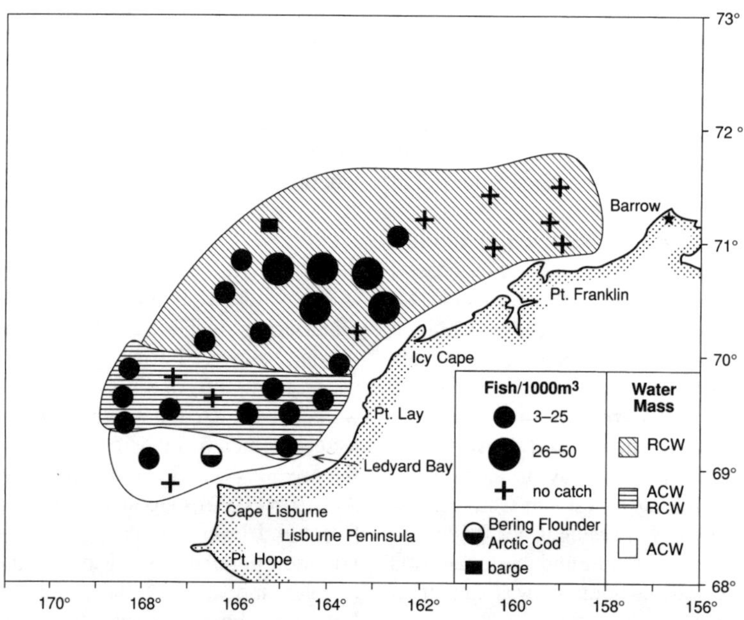

FIGURE 3.—Abundance and distribution of Arctic cod and Bering flounder captured with the Isaacs-Kidd midwater trawl in 1991. Densities are in fish per 1,000 m^3. The small, black rectangle is the barge location. Water masses are indicated by shading (see Figure 1 caption for definitions and explanation of circles).

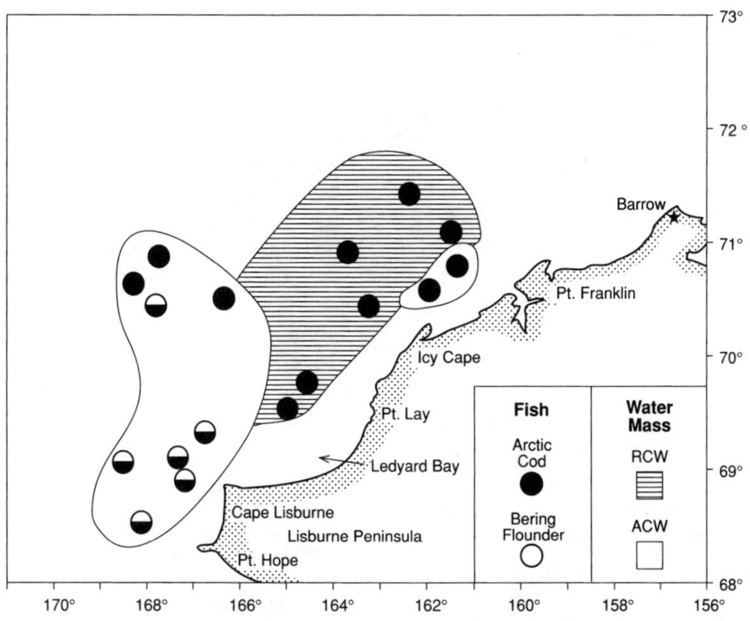

FIGURE 4.—Distribution of Arctic cod and Bering flounder captured with the Isaacs-Kidd midwater trawl in 1970. Water masses are indicated by shading (see Figure 1 caption for definitions). Since no abundance data exists circles are not scaled but only represent presence of each species. (Based on data from Quast 1972 and Ingham and Rutland 1972.)

contrast, those fish with their center of distribution in the Arctic Ocean would tend to be associated with the resident Chukchi water.

In 1989 and 1990, the surface water mass extending throughout the study area was Alaska coastal water (Figures 1 and 2). Resident Chukchi water occurred on the bottom from 70°N northward. In 1991, however, a different distribution was observed. Alaska coastal water extended as far north as 70°N, while resident Chukchi water was prevalent throughout the water column northward. The distribution of age-0 Bering flounder reflects the distribution of these water masses. Bering flounder were present during 1989 and 1990 in Alaska coastal water and Alaska coastal water–Bering shelf water, with only trace numbers in Alaska coastal water–resident Chukchi water. In 1991 only one age-0 Bering flounder was captured; it was in Alaska coastal water (Figure 3). *Pleuronectes* spp., dominated by yellowfin soles (*P. asper*) and Pacific sand lances *Ammodytes hexapterus*, followed the same general pattern. These fishes were primarily distributed in the Bering Sea and extended into the Chukchi Sea (Allen and Smith 1988). In 1989 and 1990, *Pleuronectes* spp. occurred in Alaska coastal water, and sand lances were sampled from Alaska coastal water and Alaska coastal water–resident Chukchi water but were rare in 1991 (Table 1). This suggests that age-0 Bering flounder, and quite possibly yellowfin soles and sand lances, were advected into the northeastern Chukchi Sea with Alaska coastal water.

Change in the distribution and abundance of organisms is coincident with the boundary between Alaska coastal water and resident Chukchi water which forms a semipermanent front along the bottom between 70 and 71°N (Weingartner 1997). The distributions of adult (Smith et al. 1997) and age-0 Bering flounder (Figures 1–3) are closely aligned with the position of resident Chukchi water and the influx of Alaska coastal water along the front which appears to be a boundary. In 1990 adults were most abundant in Alaska coastal water south of 70°N, while few were present in resident Chukchi water (Smith et al. 1997). In 1991 abundance was very low relative to 1990 and restricted to the area south of 70°N. Additionally, Feder et al. (1990) identified two distinct benthic communities that change characteristics near 71°N.

Age-0 Arctic cod were distributed throughout the northeastern Chukchi Sea, with higher concentrations at or near stations where resident Chukchi water was present. In waters off Greenland (between 69 and 72°N), a high abundance of Arctic cod was noted in the mid-1920s and absent since the mid-1930s (Rass 1968). This decline was attributed to the warmer climate that occurred at that time. These observations suggest that the distribution of Arctic cod might change as a result of climatic shifts.

Climatic conditions in the northeastern Chukchi Sea were similar in 1989 and 1990. Winds during July–November were northward, resulting in an eastward displacement of Alaska coastal water and an increased flushing of resident Chukchi water from the shelf (Weingartner 1997). In contrast, during 1991 the winds were predominantly southward, limiting the northern extent of Alaska coastal water and pushing resident Chukchi water southward.

The smallest Arctic cod caught (6.3 mm) in mid-July 1991 were the size of newly hatched larvae (5.5 mm; Rass 1968). Previous studies state that Arctic cod hatch no later than February in these waters (Matarese et al. 1989). Our data suggest that the time of hatching may extend to mid-July.

Our data support Pruter and Alverson's (1962) transport hypothesis. For more evidence, we reexamined an earlier study. In September and October of 1970, a team of scientists conducted a survey of the southeastern Chukchi Sea (Ingham et al. 1972). Using the data reported by Quast (1972), and the temperature and salinity profiles of Ingham and Rutland (1972) reported in this study, we reconstructed the water masses present and age-0 distribution. The Alaska coastal water was present nearshore and to 71°N (Figure 4) while resident Chukchi water was not present south of 69°30′N. Arctic cod were present at all stations but Bering flounder occurred only at stations south of 70°30′N and were associated with Alaska coastal water. Water mass characteristics and distribution of pelagic juvenile fish species indicate that the conditions of 1970 were similar to those in 1989 and 1990.

We conclude that populations of Bering flounder in the northeastern Chukchi Sea are maintained by the transport of larvae in the Alaska coastal water and that the northern limit of Bering flounder is undoubtedly connected to the presence of resident Chukchi water. Hence, although these fishes and others may be routinely advected into the northeastern Chukchi Sea by Alaska coastal water, resident Chukchi water may be a critical factor in limiting their northern distribution.

Acknowledgments

We thank the crews of the *Alpha Helix*, *Ocean Hope III*, *Oshoro Maru*, and *Responder* for their assistance. Shell Oil Company and Chevron Corporation allowed the use of the *Responder* barge and gave considerable assistance. We also appreciated the assistance of the scientific staff, especially Elizabeth Mosenthin, in collecting and processing the samples. Special thanks go to National Marine Fisheries Service personnel, Terry Sample, Richard Bakkala, and Claire Armistead for collecting and processing the samples. The support of Robert M. Meyer throughout the study is especially acknowledged. This study was funded by the Alaska Outer Continental Shelf Region of the Minerals Management Service, U.S. Department of the Interior, Anchorage, Alaska, Contract No. 14-35-0001-3-559.

References

Allen, M. J., and G. B. Smith. 1988. Atlas and zoography of common fishes in the Bering Sea and northeastern Pacific. NOAA (National Oceanic and Atmospheric Administration) Technical Report NMFS (National Marine Fisheries Service)-66.

Alverson, D. L., and N. J. Wilimovsky. 1966. Fishery investigations of the Chukchi Sea. Pages 843–860 *in* N. J. Wilimovsky and J. N. Wolf, editors. Environment of the Cape Thompson Region, Alaska. U.S. Atomic Energy Commission, Washington, DC.

Andriashev, A. P. 1955. Contribution to the knowledge of the fishes from the Bering and Chukchi Seas. U.S. Fish and Wildlife Service Special Scientific Report (Fisheries) 145. (Translated from Russian, Exploration des Mers des L'Russes 25:292–355.)

Coachman, L. K., and V. V. Shigaev. 1992. Northern Bering–Chukchi sea ecosystem: the physical basis. Pages 17–27 *in* P. A. Nagel, editor. Results of the third joint U.S.–U.S.S.R. Bering and Chukchi seas expedition (BERPAC). U.S. Fish and Wildlife Service, Washington, DC.

Craig, P. C., W. B. Griffiths, L. Haldorson, and H. McElderry. 1982. Ecological studies of Arctic cod (*Boreogadus saida*) in Beaufort Sea coastal waters, Alaska. Canadian Journal of Fisheries and Aquatic Sciences 39:395–406.

Elizarov, A. A. 1965. Long-term variations of oceanographic conditions and stocks of cod observed in the areas of west Greenland, Labrador and Newfoundland. International Commission for the Northwest Atlantic Fisheries Special Publication 6:827–831.

Feder, H. M., A. S. Naidu, J. M. Hameedi, S. C. Jewett, and W. R. Johnson. 1990. The Chukchi Sea continental shelf: benthos–environmental interactions. NOAA (National Oceanic and Atmospheric Administration) OCSEAP (Outer Continental Shelf Environmental Assessment Program) Final Report 68:25–311, Anchorage, Alaska.

Gillispie, J. G., R. L. Smith, E. Barbour, and W. E. Barber. 1997. Distribution, abundance, and growth of Arctic cod in the northeastern Chukchi Sea. Pages 81–89 *in* J. Reynolds, editor. Fish ecology in Arctic North

America. American Fisheries Society Symposium 19, Bethesda, Maryland.

Ingham, M. C., and eight coeditors. 1972. An ecological survey in the eastern Chukchi Sea September–October 1970. U.S. Coast Guard, Oceanographic Unit, Oceanographic Report 50 (CG 373-50), Washington, DC.

Ingham, M. C., and B. A. Rutland. 1972. Physical oceanography of the eastern Chukchi Sea off Cape Lisburne–Icy Cape. Pages 1–86 *in* M. C. Ingham and eight coeditors. An ecological survey in the eastern Chukchi Sea September–October 1970. U.S. Coast Guard Oceanographic Unit, Oceanographic Report 50 (CG 373-50), Washington, DC.

Jakobsson, J. 1969. On herring migrations in relation to changes in sea temperature. Jokull 19:134–145.

Lowry, L. F., and K. J. Frost. 1981. Distribution, growth, and foods of Arctic cod (*Boreogadus saida*) in the Bering, Chukchi and Beaufort seas. Canadian Field-Naturalist 95:186–191.

Matarese, A. C., A. W. Kendall, Jr., D. M. Blood, and B. M. Vitner. 1989. Laboratory guide to early life history stages of northeast Pacific fishes. NOAA (National Oceanic and Atmospheric Administration) Technical Report NMFS (National Marine Fisheries Service) 80.

Ponomarenko, V. P. 1968. Some data on the distribution and migrations of polar cod in the seas of the Soviet Arctic. Rapports et Proces-Verbaux des Reunions Conseil International pour l'Exploration de la Mer 158:131–133.

Pruter, A. T., and D. L. Alverson. 1962. Abundance, distribution and growth of flounders in the southeastern Chukchi Sea. Journal du Conseil Conseil International pour l'Exploration de la Mer 27:81–99.

Quast, J. C. 1972. Preliminary report on the fish collected on WEBSCO-70. Pages 203–206 *in* M. C. Ingham and eight coeditors. An ecological survey in the eastern Chukchi Sea September–October 1970. U.S. Coast Guard Oceanographic Unit Oceanographic Report 50 (CG 373-50), Washington, DC.

Rass, T. S. 1968. Spawning and development of Polar cod. Rapports et Proces-Verbaux des Reunions Conseil International pour l'Exploration de la Mer 158: 135–137.

Salmon, D. K. 1992. On interannual variability and climate change in the north Pacific. Doctoral dissertation. University of Alaska Fairbanks, Fairbanks.

Sinclair, M. 1988. Marine populations: an essay on population regulation and speciation. University of Washington, Washington Sea Grant, Seattle.

Smith, R. L., M. Vallarino, E. Barbour, E. Fitzpatrick, and W. E. Barber. 1997. Population biology of the Bering flounder in the northeastern Chukchi Sea. Pages 127–132 *in* J. Reynolds, editor. Fish ecology in Arctic North America. American Fisheries Society Symposium 19, Bethesda, Maryland.

Stergiou, C. I. 1991. Possible implications of climatic variability on the presence of capelin (*Mallotus villosus*) off the Norwegian coast. Climatic Change 3:369–391.

Stigebrandt, A. 1984. The north Pacific: a global-scale estuary. Journal of Physical Oceanography 14:464–470.

Vasil'kov, V. P., N. G. Chupysheva, and N. G. Kolesova. 1981. The possibility of long term forecasting from solar activity cycles of catches of the saffron cod, *Eleginus gracilis*, in the Sea of Japan. Journal of Ichthyology 20:26–33

Vasil'kov, V. P., and S. Y. Glebova. 1984. Factors determining year-class strength of the walleye pollock, *Theragra chalcogramma* (Gadidae), of western Kamchatck. Journal of Ichthyology 24:80–90.

Walsh, J. J., and 15 coauthors. 1989. Carbon and nitrogen cycling within the Bering/Chukchi seas: sources of organic matter affecting AOU demand of the Arctic Ocean. Progress in Oceanography 22:277–359.

Weingartner, T. J. 1997. A review of the physical oceanography of the northeastern Chukchi Sea. Pages 40–59 *in* J. Reynolds, editor. Fish ecology in Arctic North America. American Fisheries Society Symposium 19, Bethesda, Maryland.

American Fisheries Society Symposium 19:68–80, 1997

Population Structure and Feeding Ecology of Arctic Cod Schools in the Canadian High Arctic

Haakon Hop[1]
University of Alberta, Department of Biological Sciences
Edmonton, Alberta, Canada T6G 2E9

Harold E. Welch and Richard E. Crawford[2]
Department of Fisheries and Oceans
501 University Crescent, Winnipeg, Manitoba, Canada R3T 2N6

Abstract.—Large schools of Arctic cod *Boreogadus saida* occur in Canadian High Arctic waters. We observed two types of schools on the south coasts of Cornwallis and Devon islands: juvenile schools occurring in July in association with broken ice cover in relatively shallow water and schools of adults occurring later (August–September), during the open-water season. Juvenile schools, with mean fork length (FL) of 83 mm, were composed mainly (90%) of age-1 fish. Adult schools (mean FL, 165 mm) consisted of age-classes 2–7, with ages 3–4 being dominant, and had gonadosomatic indexes that averaged about 10% of ripe condition. Juvenile schools fed on calanoid copepods, whereas adult schools were feeding only occasionally, primarily on *Themisto* amphipods and *Calanus* copepods. Schooling adults had a much higher frequency of empty stomachs than nonschooling adults (64.6% versus 5.5%); the former also had slightly lower condition factors. Adult Arctic cod exhibited two distinct behavioral strategies: either participate in schools with reduced feeding intensity or disperse individually to feed. The potential benefits of schooling, such as predator protection, apparently outweigh the energetic costs of reduced feeding.

Schools of Arctic cod *Boreogadus saida* occur throughout the circumpolar range of the species and have been observed in Canadian (Bain and Sekerak 1978; Welch et al. 1992, 1993; Crawford and Jorgenson 1993, 1996), Greenlandic (Welch et al. 1993), Alaskan (Craig et al. 1982), and Russian waters (Ponomarenko 1967, 1968; Rass 1968). In addition, nonschooling or scattered individuals are present during both the ice-covered and open-water seasons.

Bain and Sekerak (1978) recorded mass movements of Arctic cod (size range 110–240 mm) into Allen Bay, Cornwallis Island, Northwest Territories, in July–September, and Crawford and Jorgenson (1993) observed that schooling density was negatively related to the amount of drifting ice cover. Craig et al. (1982) sampled both juvenile and adult schools in Simpson Lagoon, Alaska, and such schools have also been sampled in Russian waters (Moskalenko 1964; Andriyashev et al. 1980). However, because of the unpredictable and often ephemeral occurrences of Arctic cod schools, they are not sampled frequently and relatively little is known about their population structure or feeding ecology.

Arctic cod generally feed pelagically on copepods and amphipods, on ice-associated amphipods, and on epibenthic crustacea (Bradstreet et al. 1986; Lønne and Gulliksen 1989). This fish species is a central component of the arctic marine food web (Klumov 1937; Hobson and Welch 1992; Welch et al. 1992), with larger schools of high fish density representing important energy stores in the arctic food web (Welch et al. 1993; Crawford and Jorgenson 1996). Such schools may be subject to intense predation pressure by seabirds and marine mammals (Welch et al. 1993). We investigated ecological aspects of schooling Arctic cod and make comparisons, where possible, between juvenile and adult schools and with nonschooling cod.

Methods

Samples from Arctic cod schools were obtained during July–September 1985–1991 from Resolute and Allen bays on Cornwallis Island, Gascoyne Inlet, Erebus and Radstock bays on the southwest coast of Devon Island, and Grise Fjord on southern Ellesmere Island (Figure 1). We determined the temperature profile daily at one of these locations, Resolute Bay (5 km^2, maximum depth of 29 m with a 10- to 15-m deep sill at the mouth), from mid-July

[1]Present address: Norwegian Polar Institute, Box 399, N-9001 Tromsø, Norway.

[2]Present address: Waquoit Bay National Estuarine Research Reserve, Post Office Box 3092, Waquoit, Massachusetts 02536, USA.

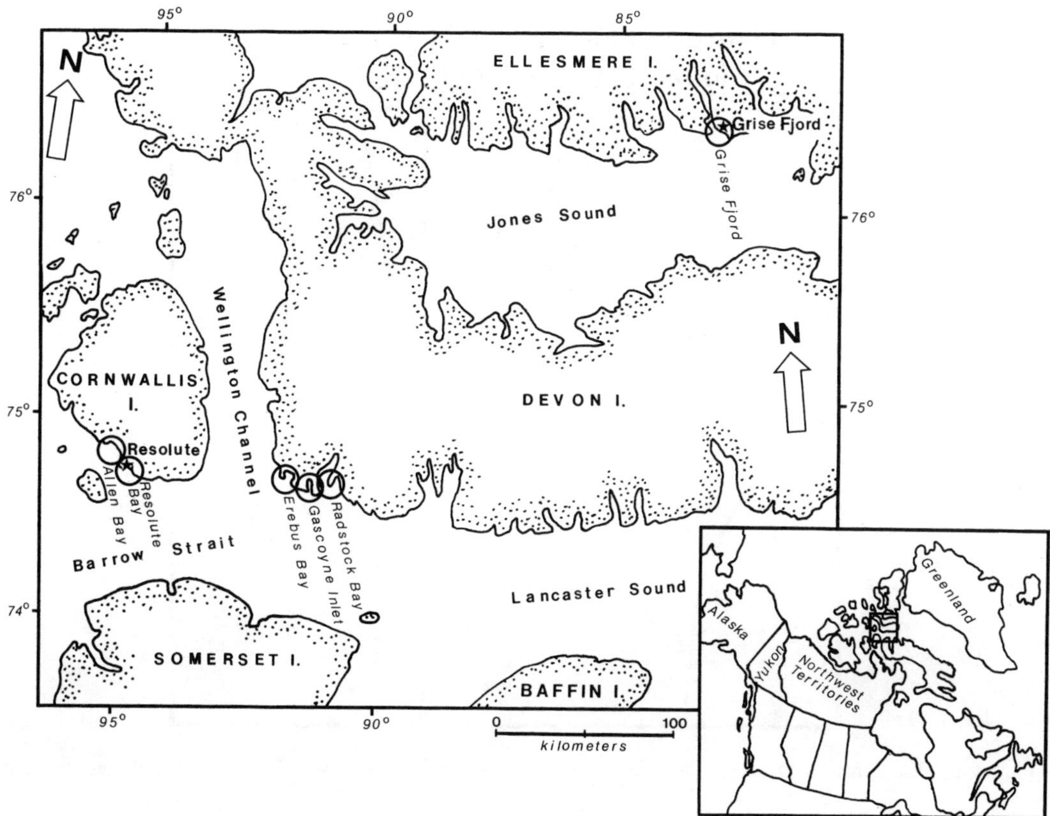

FIGURE 1.—Sampling locations (circles) on the coasts of Cornwallis, Devon, and Ellesmere islands, Northwest Territories, arctic Canada, 1985–1991. Stars indicate inhabited communities.

to mid-September 1989. The other bays and inlets range in size from 15 km^2 (Gascoyne Inlet) to 200 km^2 (Radstock Bay), with the largest bays deepening to 150 m.

Sampling was opportunistic and employed three types of trawls (bottom, Kuichak, and Isaacs-Kidd), trap nets, gill nets, dip nets, and a cast net. The bottom trawl was a small otter trawl with a 4.9-m head rope, 5.1-cm mesh netting, and a 3.8-cm mesh cod end with a 1.3-cm mesh inner liner. The Kuichak midwater trawl had a 7.5-m^2 aperture, with 3.8-cm mesh netting grading down to a 0.6-cm mesh at the cod end, and was rigged with spreader bars connected to a bridle to maintain its mouth opening. Because it could not be towed fast enough to effectively catch adult fish, it was dropped onto a school in relatively shallow water by tossing it from the boat. The Isaacs-Kidd midwater trawl had a 2.2-m^2 aperture with a 6.4-cm mesh body, a 1.3-cm mesh inner liner, and a 0.3-cm mesh cod end. Schools were also caught in trap nets (8-m^3 box, 6-m wings, 20 m-lead), set overnight perpendicular to shore, with the leads in less than 2 m of water and the box at 5- to 8-m depth. The presence of schools was indicated by overnight catches of more than 5,000 individuals, compared with regular catches of less than 500 fish. Other sampling gear included sinking experimental gill nets (1.8-m height, with six 13.7-m long panels of 20-, 37-, 65-, 90-, 110-, 120-mm stretched mesh), dip nets, and a cast net (4.3-m diameter, 0.95-cm mesh).

Fish were weighed whole (TW, g) and reported as fork length (FL, mm). Specimens measured as total length (TL, mm) were converted to fork length: FL = (TL + 2.008)/1.051 (r^2 = 0.998, N = 55, range = 70–210 mm). We determined Fulton's condition factor, CF = (TW · 10^5)/FL^3 (Ricker 1975), and age was determined from cross-sectioned otoliths (Chilton and Beamish 1982). Gonads were dissected, weighed wet (GW, g), and sexed by external morphology (Vladykov 1972; Lear 1980). Gonadosomatic index, GSI = (GW/

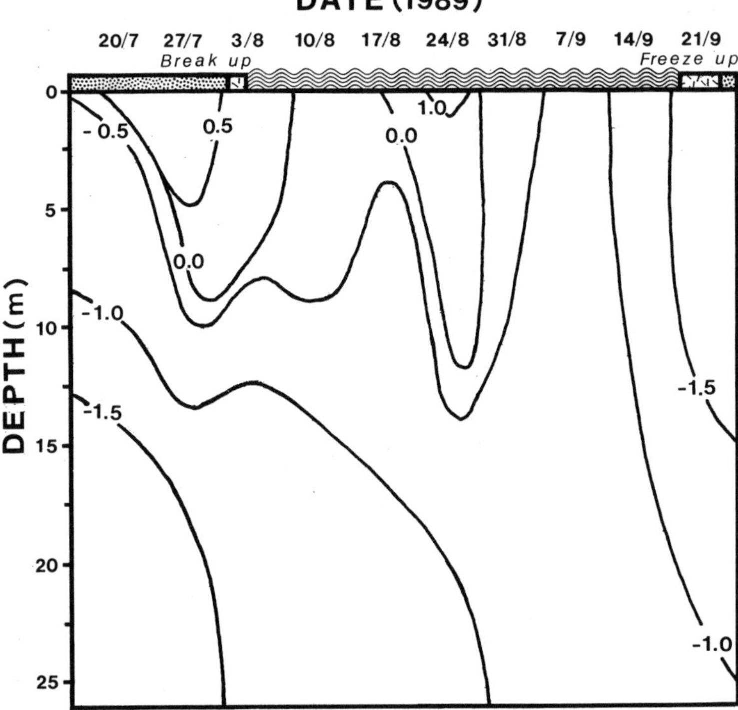

FIGURE 2.—Isotherms for Resolute Bay, July–September 1989, drawn from weekly means of daily temperature records.

TW) · 100, was determined, where GW and TW are gonad weight and whole body weight, respectively.

Stomachs were removed and the apparent degree of stomach fullness was determined on a scale from 1 to 5: (1) empty, (2) less than one-third full, (3) one-third to two-thirds full, (4) greater than two-thirds full, and (5) completely full and distended. Total blotted wet weight of stomach contents was expressed as a percentage of total fish weight.

To assess whether large schools of Arctic cod might deplete zooplankton food sources locally, we sampled zooplankton at several locations in 1990: in Allen Bay near where fish were schooling, at two adjacent sites within Allen Bay, at a neighboring bay (Resolute Bay), and offshore in Resolute Passage. Vertical plankton tows were taken from maximum depth or 30 m, whichever was less, using paired 0.5-m diameter bongo nets of 1,000-μm mesh. Of all zooplankton collected, only *Calanus hyperboreus* and *C. glacialis* were present in sufficient numbers to analyze. They were sorted to stage and grouped as "large" (3.3 to 4.4-mm body length, which included *C. hyperboreus* IV–VI plus *C. glacialis* V–VI) and "small" (1.9 to 2.5-mm body length, which included *C. hyperboreus* III plus *C. glacialis* III–IV).

Results

Environmental Conditions and School Occurrences

Thermal conditions in Resolute Bay were relatively stable, varying about ±1.5°C annually. From October to mid-June, water temperature is less than −1.5°C, the water column being almost isothermal at −1.8°C from March to June (Bergmann 1989). Ice breakup usually occurs in late July, and afterward surface waters warm up slightly and may reach a maximum of 1.0°C (Figure 2). Increasing winds in September mix the water column and subsequent surface cooling lowers the temperature to about −1.5°C before freeze-up.

Juvenile Arctic cod schools were observed in Resolute Bay in July (Table 1), corresponding with the early warming of surface waters. In 1989 these schools occurred in shallow water under the ice, and individuals were frequently observed in ice cracks,

TABLE 1.—Sampling date, gear, and means (±SEs) of fork length (FL), weight, and condition factor (CF) of samples from schools of adult and juvenile Arctic cod *Boreogadus saida* and nonschooling cod in the Canadian High Arctic, 1985–1991. Group means (±SEs) are calculated from the means of individual collections.

Location	Date (d/month/year)	Gear	FL (mm)		Weight (g)		CF (×1,000)
			N	Mean	*N*	Mean	
Adult schools							
Resolute Bay	21 Aug 86	Gill net	93	184.4 ± 2.0	93	48.1 ± 1.4	752 ± 7
Resolute Bay	31 Jul 88	Trap net	511	173.7 ± 1.4	195		794 ± 7
Allen Bay	26–27 Jul 89	Trap net	108	206.8 ± 2.0	42	69.5 ± 4.1	736 ± 1
Allen Bay	2 Sep 89	Bottom trawl[a]	101	146.1 ± 2.8	101	23.0 ± 1.8	655 ± 5
Allen Bay	16 Sep 89	Bottom trawl[a]	99	148.1 ± 2.3	9	24.1 ± 1.3	699 ± 7
Allen Bay	8 Aug 90	Trap net	100	146.8 ± 2.2	100	3.5 ± 1.1	707 ± 7
Allen Bay	15 Aug 90	Bottom trawl[a]	201	151.2 ± 1.8	01	28.9 ± 1.6	730 ± 11
Allen Bay	28 Aug 90	Isaacs-Kidd trawl	400	176.3 ± 1.5	200	43.0 ± 1.7	678 ± 5
Gascoyne Inlet	8 Aug 85	Hand caught	79	179.1 ± 2.4	79	34.2 ± 1.4	576 ± 8
Gascoyne Inlet	23 Aug 85	Kuichak trawl	967	168.6 ± 0.7	180	33.7 ± 0.1	656 ± 0
Gascoyne Inlet	18 Aug 88	Bottom trawl[a]	178	148.8 ± 1.3	178	22.4 ± 0.6	654 ± 5
Erebus Bay	19 Aug 88	Dip net	89	166.5 ± 2.0	89	3.3 ± 1.2	698 ± 6
Radstock Bay	31 Jul 91	Cast net	100	138.2 ± 1.5	97	8.7 ± 0.7	716 ± 7
Grise Fjord	16 Sep 87	Beach stranded	47	178.8 ± 4.4	47	8.2 ± 3.5	783 ± 12
Group mean (±SE)			14	165.2 ± 5.2	14	4.7 ± 4.0	702 ± 15
Juvenile schools							
Resolute Bay	14 and 20 Jul 89	Dip net	50	85.4 ± 0.9	50	4.2 ± 0.2	657 ± 8
Resolute Bay	31 Jul 91	Bottom trawl[b]	100	79.6 ± 0.6	40	3.1 ± 0.1	661 ± 8
Group mean (±SE)			2	82.5 ± 2.9	2	3.7 ± 0.5	659 ± 2
Nonschooling adults							
Resolute Bay	9 and 26 Aug 85	Bottom trawl[b]	34	140.2 ± 5.6	34	27.8 ± 3.8	884 ± 15
Resolute Bay	29–30 Jul 88	Bottom trawl[b]	48	147.8 ± 3.9	48	28.3 ± 2.2	802 ± 15
Resolute Bay	12–31 Aug 88	Bottom trawl[b]	29	156.7 ± 4.5	29	34.0 ± 3.1	840 ± 14
Resolute Bay	1–4 Sep 88	Trap net	40	177.7 ± 3.7	38	40.8 ± 2.1	746 ± 10
Resolute Bay	29–30 Jul 89	Trap net	87	185.6 ± 3.1	87	48.5 ± 2.5	708 ± 9
Resolute Bay	4–9 Sep 89	Trap net	37	173.2 ± 4.9	37	37.7 ± 2.8	687 ± 17
Group mean (±SE)			6	163.5 ± 7.3	6	36.2 ± 3.2	778 ± 32
Nonschooling juveniles							
Resolute Bay	29–30 Jul 89	Trap net	45	89.0 ± 1.5	45	5.0 ± 0.3	674 ± 11
Resolute Bay	9–11 Aug 89	Bottom trawl[b]	98	82.8 ± 1.3	98	4.3 ± 0.2	668 ± 9
Resolute Bay	31 Aug–1 Sep 89	Bottom trawl[b]	116	86.8 ± 1.2	116	4.9 ± 0.2	686 ± 9
Group mean (±SE)			3	86.2 ± 1.8	3	4.7 ± 0.2	676 ± 5

[a]Trawl towed off the bottom, in the water column.
[b]Trawl towed on the bottom.

or even swimming in brackish water on the ice. Elsewhere we observed juvenile Arctic cod in brackish meltwater and frequently caught them near river mouths.

Adult schools were observed from late July to mid-September, most frequently in August (Table 1). Most schools in the bays of Cornwallis and Devon islands occurred in relatively shallow waters during the warm period. For example, at Gascoyne Inlet in August 1985, schools were observed in shallow (<5 m) waters at a temperature of about 2°C (Bergmann 1989). However, schools have also been detected hydroacoustically in deeper (<25 m) water in Allen Bay (Welch et al. 1993). Shallow-water schooling events normally attract feeding seabirds and marine mammals; indeed, marine mammals may have driven some schools into the shallow regions (Welch et al. 1993).

Size Structure of Schools

Mean sizes of fish in adult schools were 138–207 mm FL (Table 1), with the overall size ranges being 110–260 mm (age-1 fish excluded). Length distributions varied among areas and gear (Figure 3, a–d), although some similarities were apparent. Length distributions from gill and trap nets in Resolute Bay were similar when small fish (<100 mm) were excluded from the trap net catch (Smirnov test, $P > 0.05$; Figure 3a). Also, length distributions of three bottom trawl samples from Allen Bay (1989, 1990), with the trawl towed through the school in midwater, did not differ significantly (Smirnov test, $P > 0.05$; Figure 3b). Length distributions may vary between areas; for example, fish caught in Allen Bay were smaller (Kruskal–Wallis [K–W] test, $P < 0.05$) than those caught in Resolute Bay (K–W test, $P <$

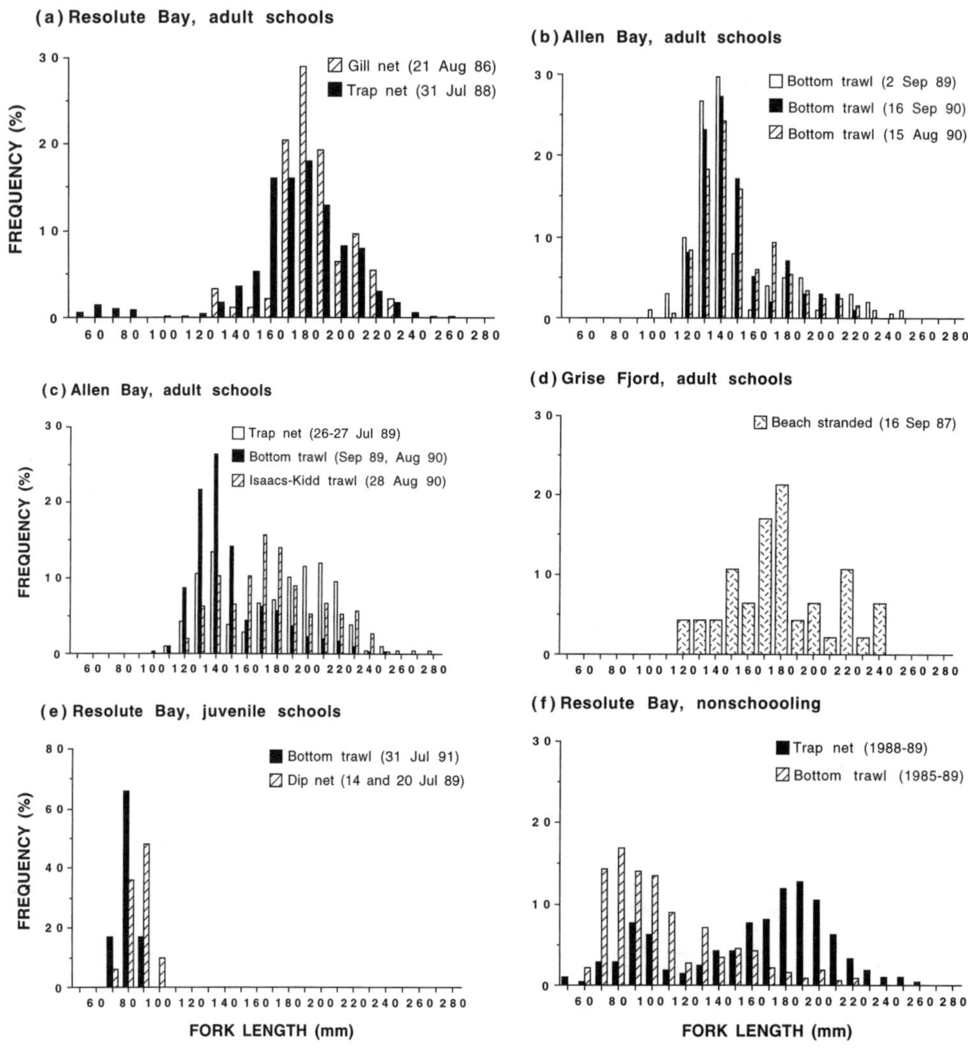

FIGURE 3.—Fork-length distributions of adult schools (**a–d**), juvenile schools (**e**), and nonschooling adults and juveniles (**f**) Arctic cod *Boreogadus saida*. (**a**) Resolute Bay: gill net ($N = 93$) and trap net ($N = 511$). (**b**) Allen Bay: bottom trawl, 2 Sep 89 ($N = 101$), 16 Sep 89 ($N = 99$) and 15 Sep 90 ($N = 201$). (**c**) Allen Bay: trap net ($N = 108$), bottom trawl ($N = 401$), and Isaacs-Kidd trawl ($N = 400$). (**d**) Grise Fjord: beach stranded ($N = 47$). (**e**) Resolute Bay: bottom trawl ($N = 100$) and dip net ($N = 50$). (**f**) Resolute Bay: trap net ($N = 210$) and bottom trawl ($N = 326$).

0.05), and fish caught with the same bottom trawl in Gascoyne Inlet (1988) had a significantly different length distribution (Smirnov test, $P < 0.05$) from those caught in Allen Bay (1989–1990), although their mean lengths were similar (Table 1). When data were pooled among years in Allen Bay, it became apparent that different gear (trap net, bottom trawl, and Isaacs-Kidd trawl) caught different length distributions of fish (Smirnov test, $P < 0.05$; Figure 3c). The beach-stranded fish from Grise Fjord also had a different distribution, skewed toward larger fish, possibly because seabirds had eaten the smaller ones (Figure 3d).

Juvenile schools from Resolute Bay consisted of fish that ranged from 65 to 105 mm (Figure 3e). Nonschooling cod, caught in a trap net and bottom trawl in Resolute Bay, consisted of all sizes >50 mm (Figure 3f).

Age Distribution

Arctic cod are slow growing, reaching a mean length of 224 mm at age 7 (Figure 4a), with the largest schooling individual being 268 mm. Adult schools consisted of age-classes 2–7, with ages 3–4 being dominant (overall mean, 3.7), whereas juvenile schools consisted mostly (>90%) of age-1 fish (Figure 4b). There were generally no significant differences in age distributions among samples (pooled for years) from different areas (K–W, $P >$ 0.05), or between years in Allen Bay ($P > 0.05$). One exception was a trap net sample from Allen Bay (1989) with unusually large fish (Table 1) of mean age 5.1. Age-class 2 was nearly absent from both juvenile and adult schools and was infrequently caught outside schools. Nonschooling fish included both age-1 and older fish (mostly ages 4–5), with age-3 fish being underrepresented.

Feeding Ecology and Condition Factor

Adult schools were composed of fish that were generally not feeding or were only feeding occasionally (Figure 5a). They had a high frequency of empty stomachs, 64.6% on average, compared with 5.5% for nonschooling adults (Table 2; Figure 5b). For example, the 1990 Allen Bay school had empty stomachs on 8 August. On 15 August 55.4% of the fish had evidence of food in their stomachs. Then, on 28 August most of the stomachs (99.0%) were empty again (Table 2; Figure 5a). Juveniles, schooling or not, had low frequencies (<10%) of empty stomachs (Table 2). Both juveniles and nonschooling adults frequently had full to extremely full stomachs (Figure 5b).

Schooling fish had significantly lower condition factors than nonschooling individuals when samples were pooled among years (two-tailed t-test, $P =$ 0.0261) (Table 1). The mean condition factor (CF) for a proportion of the catches, in which the measured weights of stomach contents (Table 2) were subtracted, remained lower for schooling (CF = 0.688) than for nonschooling fish (CF = 0.734). Even though the difference between condition factors was not statistically significant ($P = 0.1064$), this may indicate a 6% weight loss.

The most abundant food items in stomachs of schooling adults were the pelagic hyperiid amphipod *Themisto libellula* and large *Calanus* copepods. Juvenile fish, in schools or not, fed almost exclusively on calanoid copepods but also took cyclopoid and harpacticoid copepods. Nonschooling adults fed on *Calanus* (*C. hyperboreus*, *C. glacialis*), amphipods (genera *Themisto*, *Onisimus*, *Gammarus*, *Gammaracanthus*, *Monoculodes*, *Weyprechtia*, *Paroediceros*, and *Apherusa*), and epibenthic crustacea (*Mysis* spp.) (H. Hop, Norwegian Polar Institute, unpublished data).

Even the occasional feeding of large schools may temporarily deplete planktonic prey populations in their local area. When a large school was present in one of the inner basins of Allen Bay in 1990 (location A, Figure 6), the proportion of large *Calanus* was reduced, as compared with other sites examined (locations B–E, Figure 6). Samples from this school on 15 August indicated feeding (Table 2), and large copepods were depleted at this location on 21 August. Subsequently, the school had empty stomachs on 28 August (Table 2), and plankton sampling on 4–6 and 12–15 September showed no depletion of large *Calanus* relative to Resolute Bay (location D, Figure 6).

Sex Ratio and Gonad Development

The mean sex ratio (males:females) of schooling fish samples was 0.85 (range, 0.55–1.23; Table 3), indicating a slight but significant surplus of females (Binomial test, $P = 0.0163$). Males (mean FL, 155 mm) were somewhat smaller than females (mean FL, 166 mm), although the distributions were not significantly different (Smirnov test, $P = 0.4927$). The overall sex ratio for nonschooling fish (0.95, Table 3) was not significantly different from a 1:1 ratio (Binomial test, $P > 0.05$).

Male gonad development for schooling Arctic cod starts in August (Figure 7). Female gonads start rapid development later (October–November) and reach full development in February, 1–2 months later than males (Hop et al. 1995). Gonads are therefore not well developed during the period reported here, with gonadosomatic indexes (GSI) of 2.4–5.6% for males (mean 3.8%) and 2.1–3.4% for females (mean 2.8%) (Table 3), representing 13% and 6% of the weights of fully mature gonads for males and females, respectively (Hop et al. 1995).

Discussion

Arctic cod typically aggregate annually in the shallows at various locations of the eastern Canadian Arctic during a specific time-window of about 2 months (Welch et al. 1993), although a portion of the population is dispersed at the same time. We identified two types of Arctic cod schools having different age structures and feeding ecologies: nonfeeding schools of mostly adults and actively feeding schools of juveniles.

Observed schooling occurrences may partly reflect seasonal variation in observational efforts as a result of ice conditions. We observed schools from July through September, whereas other researchers have observed schools in winter as well (Moska-

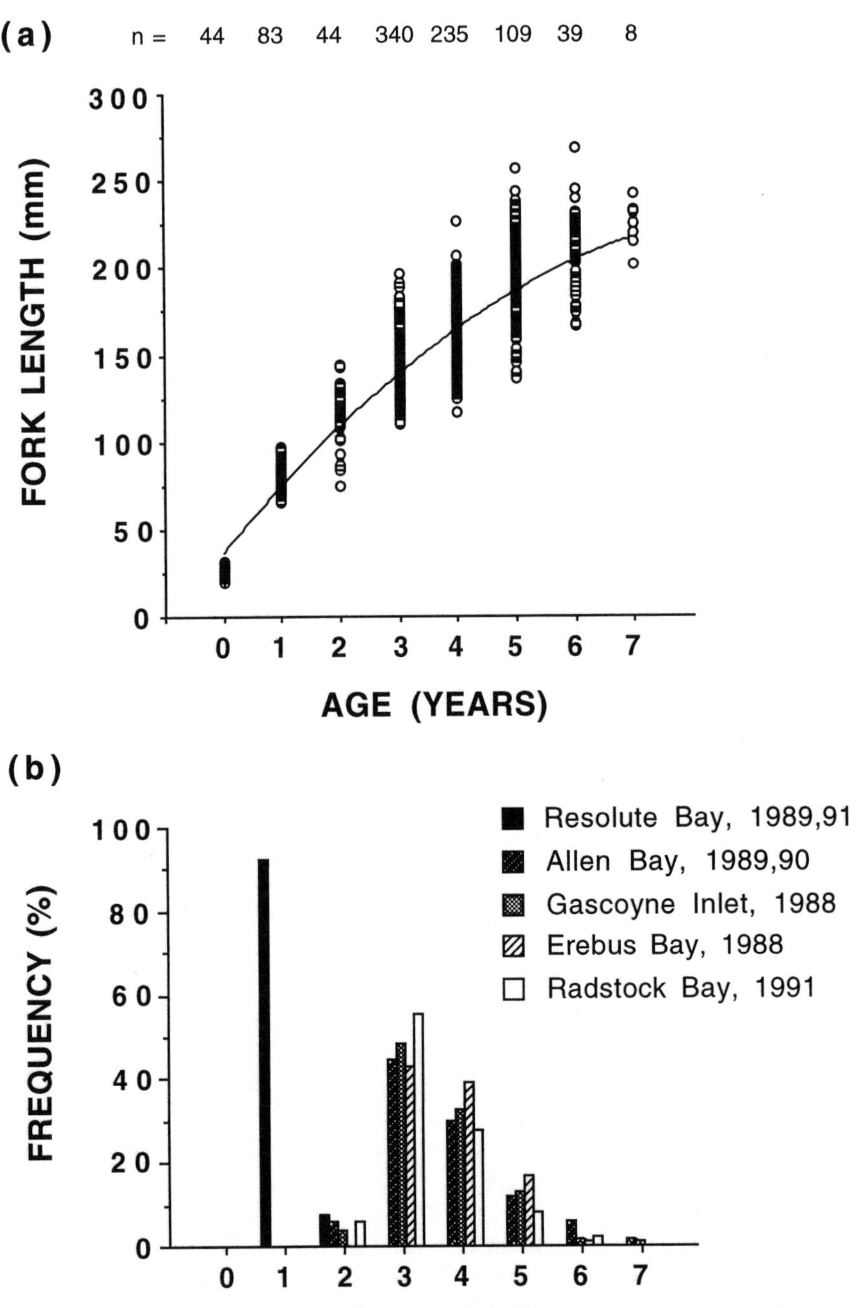

FIGURE 4.—Age relationships for schooling Arctic cod *Boreogadus saida*. (**a**) Length-at-age (age-0 individuals caught in plankton nets are added to include all year-classes). (**b**) Age distributions of juvenile schools from Resolute Bay, 1989 and 1991 ($N = 90$) and adult schools from: Allen Bay, 1989 an 1990 ($N = 438$); Gascoyne Inlet, 1988 ($N = 108$); Erebus Bay, 1988 ($N = 89$); and Radstock Bay, 1991 ($N = 97$).

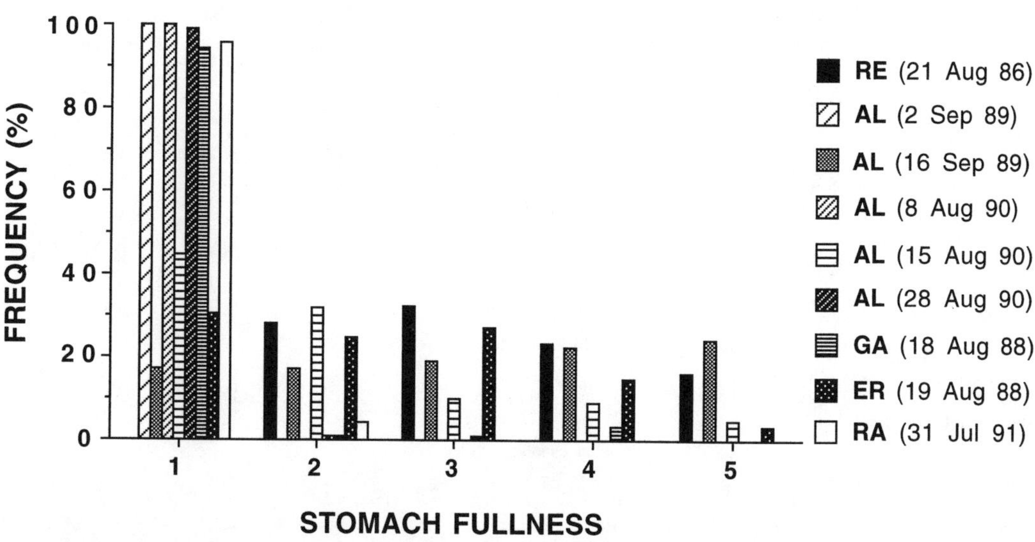

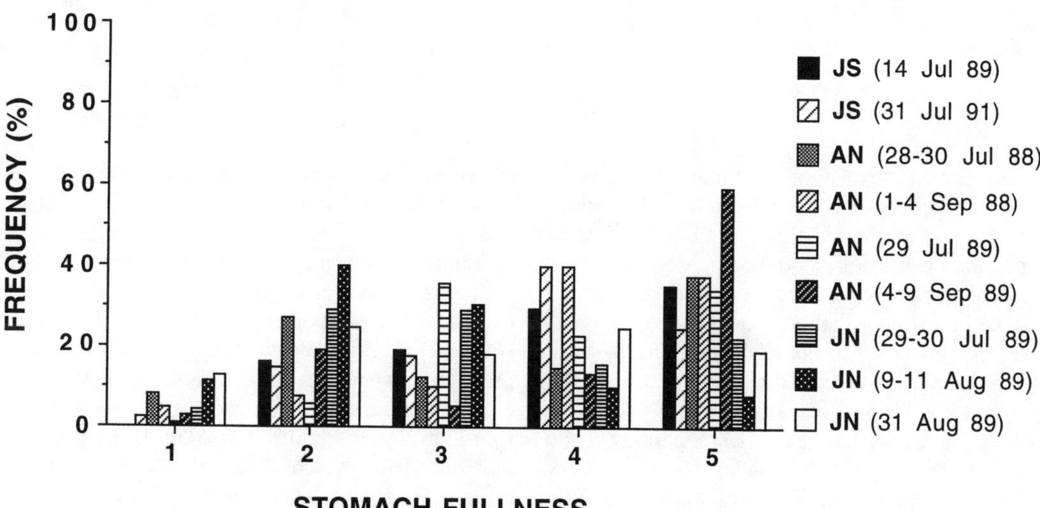

FIGURE 5.—Distribution of stomach fullness for schooling and nonschooling Arctic cod *Boreogadus saida*. 1 = empty, 2 = less than one-third full, 3 = one-third to two-thirds full, 4 = greater than two-thirds full, 5 = completely full and distended. (**a**) Adult schools from: Resolute Bay (RE), Allen Bay (AL), Gascoyne Inlet (GA), Erebus Bay (ER), and Radstock Bay (RA). (**b**) Juvenile schools (JS), nonschooling adults (AN), and nonschooling juveniles (JN) from Resolute Bay.

lenko 1964; Andriyashev et al. 1980); however, it appears that fish disperse after spawning (Moskalenko 1964).

Little is known about the reasons for the inshore schooling behavior of Arctic cod, and environmental or biological factors may affect adult and juvenile schools differently. Responses to temperature (Hognestad 1968; Craig et al. 1982), salinity (Crawford

TABLE 2.—Mean (±SE) stomach fullness (SF, scale 1–5, see Methods section for definitions), percent of empty stomachs, and mean weight (±SE) of stomach contents (as % body weight) of samples from schools of adult and juvenile Arctic cod *Boreogadus saida* and nonschooling cod in the Canadian High Arctic, 1985–1991. Group means (±SEs) are calculated from the means of individual collections.

Location	Date (d/month/year)	*N*	SF	Empty (%)	*N*	Stomach weight (%)
			Adult schools			
Resolute Bay	21 Aug 86	43	3.28 ± 0.16	0.0		
Allen Bay	2 Sep 89	101	1.00 ± 0.00	100.0	101	0.00 ± 0.00
Allen Bay	16 Sep 89	99	3.19 ± 0.14	17.2	36	1.61 ± 0.23
Allen Bay	8 Aug 90	100	1.00 ± 0.00	100.0	100	0.00 ± 0.00
Allen Bay	15 Aug 90	101	1.98 ± 0.12	44.6	101	0.68 ± 0.13
Allen Bay	28 Aug 90	100	1.01 ± 0.01	99.0	100	0.00 ± 0.00
Gascoyne Inlet	18 Aug 88	179	1.13 ± 0.04	94.4	179	0.05 ± 0.02
Erebus Bay	19 Aug 88	89	2.36 ± 0.12	30.3	89	0.82 ± 0.11
Radstock Bay	31 Jul 91	97	1.04 ± 0.02	95.9	97	0.03 ± 0.02
Group mean (±SE)		9	1.78 ± 0.32	64.6 ± 13.7	8	0.40 ± 0.21
			Juvenile schools			
Resolute Bay	14 and 20 Jul 89	37	3.84 ± 0.18	0.0	37	2.30 ± 0.27
Resolute Bay	31 Jul 91	40	3.70 ± 0.17	2.5	40	3.20 ± 0.26
Group mean (±SE)		2	3.77 ± 0.07	1.3 ± 1.3	2	2.75 ± 0.45
			Nonschooling adults			
Resolute Bay	29–30 Jul 88	48	3.46 ± 0.21	8.3	48	2.40 ± 0.31
Resolute Bay	12–30 Aug 88	29	3.30 ± 0.30	10.3	29	2.00 ± 0.38
Resolute Bay	1–4 Sep 88	40	3.98 ± 0.18	5.0	38	3.40 ± 0.45
Resolute Bay	29–30 Jul 89	87	3.84 ± 0.11	1.1	87	3.69 ± 0.28
Resolute Bay	4–9 Sep 89	37	4.08 ± 0.21	2.7	36	3.25 ± 0.43
Group mean (±SE)		5	3.73 ± 0.15	5.5 ± 1.7	5	2.95 ± 0.32
			Nonschooling juveniles			
Resolute Bay	29–30 Jul 89	45	3.22 ± 0.18	4.4	45	1.86 ± 0.26
Resolute Bay	9–11 Aug 89	98	2.64 ± 0.11	11.2	98	1.29 ± 0.13
Resolute Bay	31 Aug–1 Sep 89	116	3.12 ± 0.12	12.9	116	1.79 ± 0.15
Group mean (±SE)		3	2.99 ± 0.18	9.5 ± 2.6	3	1.65 ± 0.18

and Jorgenson 1996), food resources (Craig et al. 1982), and predators (Crawford and Jorgenson 1993, 1996; Welch et al. 1993) have been suggested as potential causes of schooling.

Juvenile schools occurred in July in association with ice in warming shallow waters. These fish may have been attracted to the slightly warmer surface waters during this time of the year but also to the abundant food resources associated with the ice pack, because they fed heavily on copepods at the time.

Adult schools occurred in inshore waters later (August–September). This corresponded to a 1.5°C temperature increase in the upper 10 m of the water column, but schools were also observed in deeper water where the temperature was less than 0°C. It has been suggested that schooling behavior may be a response to temperature before spawning (Sameoto 1984), implying that increased temperature would influence gonad development. Our observations indicate that male gonads start developing at the time of schooling in August, while female gonads show little development at this time, and spawning does not occur until late winter (Barenkova et al. 1966; Ponomarenko 1967; Craig et al. 1982).

Adults schools are often observed in association with aggregations of marine mammals and seabirds exploiting this temporarily abundant food resource (Finley et al. 1990; Welch et al. 1993). Even though predators may not cause schools to form (Welch et al. 1993), mass strandings, such as the one in Grise Fjord and others reported in Allen Bay (Bain and Sekerak 1978; Welch et al. 1993), are probably often induced by predators forcing fish into shallow water (Welch et al. 1993). However, mass strandings have also been observed after severe storms in Alaskan and Russian waters (Klumov 1937; Shibanoff 1958; Moskalenko 1964; Craig et al. 1982).

Ages 3–4 were the dominant year-classes in adult schools; these represent adult fish since most Arctic cod are mature at age 2 to 3 (Craig et al. 1982). Age distributions will depend somewhat on gear selectivity, but dominant year-classes probably were correctly represented because these same year-classes were also detected in other studies (e.g.,

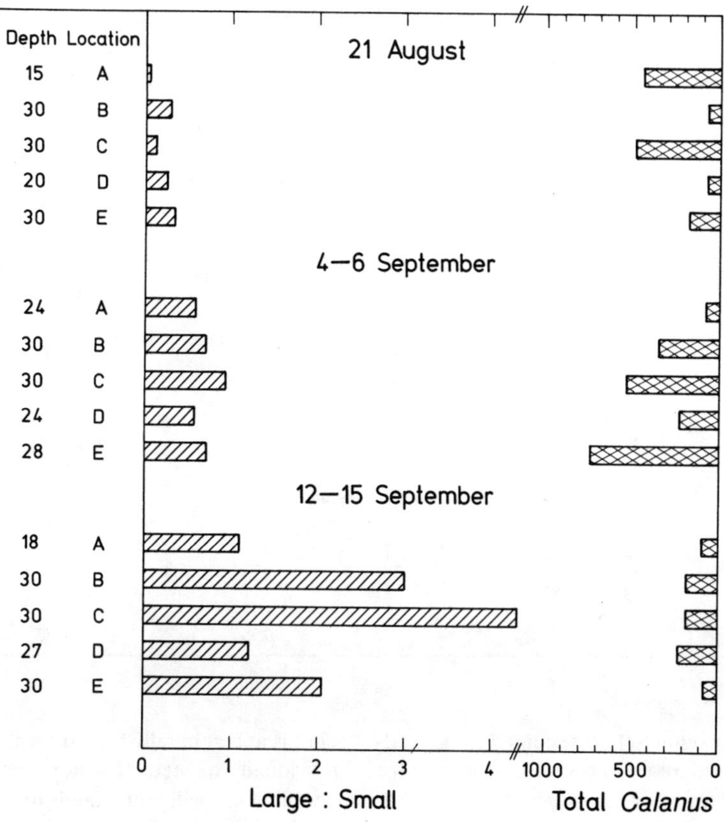

FIGURE 6.—Calanoid copepods (*Calanus hyperboreus* and *C. glacialis*) captured in vertical plankton tows at different locations, 1990. On the left are ratios of large to small *Calanus*; the right half shows the total number of *Calanus*. Depth indicates the depth of the tow at each of the following locations: A: a small basin between May Island and Cornwallis Island in Allen Bay, where schools of Arctic cod *Boreogadus saida* were observed from 8 August to 4 September 1990 (Welch et al. 1993); B: about 1 km south of location A, where there were no schools; C: middle of Allen Bay, several km from location A; D: Resolute Bay; E: offshore at a depth of 100 m, about 4 km outside Resolute Bay.

Moskalenko 1964). Age distributions varied little among locations and dates, and there was no evidence of a cohort dominating from 1 year to the next in Allen Bay. Age-3 fish were underrepresented in nonschooling samples, probably because this year-class was abundant in schools. Length distributions of schools were also similar over time within an area, although they varied between areas. It is not known whether Arctic cod in the study areas belong to the same population, but assuming that they do, some of the variation between areas may have resulted from gear selectivity.

Age-2 fish did not appear to any extent in either juvenile or adult schools. Fish schools, most likely consisting of Arctic cod, have been detected hydroacoustically in Canadian arctic waters deeper than 100 m (R. E. Crawford, Waquoit Bay National Estuarine Research Reserve, unpublished data), but it was uncertain whether these aggregations included this year-class. In offshore catches from the western coast of Alaska, the most abundant year-class was age-2 (Wolotira et al. 1979). Fish in the age-0 year-class, common in the waters of our study area, are planktonic (Barenkova et al. 1966) and not vulnerable to the gear we used.

The overall sex ratio in adult schools indicated a surplus of females. A higher proportion of the small fish were males, whereas the largest fish were predominately females (also observed by Moskalenko 1964; Wolotira et al. 1979; Craig et al. 1982). Possible explanations are faster growth rates of females than males (J. S. Christiansen, Norwegian Institute of Fisheries and Aquaculture, unpublished) differential mortality rates.

TABLE 3.—Mean sex ratio (number of males:number of females) and mean (±SE) gonadosomatic index (= gonad weight as % body weight) of male and female schooling and nonschooling Arctic cod *Boreogadus saida* 1985–1991. Group means (±SEs) are calculated from the means of individual collections.

Location	Date (d/month/year)	N	Sex ratio	Gonadosomatic index (%) Male	Female
		Adult schools			
Resolute Bay	21 Aug 86	93	0.55	5.33 ± 0.38	2.86 ± 0.05
Allen Bay	26–27 Jul 89	35	0.75		
Allen Bay	2 Sep 89	100	0.92		
Allen Bay	16 Sep 89	98	1.04		
Allen Bay	8 Aug 90	100	0.82	2.98 ± 0.22	2.07 ± 0.09
Allen Bay	15 Aug 90	100	0.89	3.27 ± 0.11	2.33 ± 0.11
Allen Bay	28 Aug 90	100	0.69	5.60 ± 0.35	2.91 ± 0.10
Gascoyne Inlet	8 Aug 85	77	0.64	2.44 ± 0.05	3.37 ± 0.13
Gascoyne Inlet	23 Aug 85	80	0.82	4.05 ± 0.20	3.29 ± 0.09
Gascoyne Inlet	18 Aug 88	179	1.08	3.81 ± 0.16	2.99 ± 0.08
Erebus Bay	19 Aug 88	89	0.71		
Radstock Bay	31 Jul 91	67	1.23	2.70 ± 0.16	2.41 ± 0.22
Group mean (±SE)		12	0.85	3.77 ± 0.42	2.78 ± 0.16
		Nonschooling adults			
Resolute Bay	9 and 26 Aug 85	34	0.62	3.83 ± 0.54	2.03 ± 0.13
Resolute Bay	Jul and Aug 1988	77	0.93	4.26 ± 0.35	2.38 ± 0.11
Resolute Bay	1–4 Sep 88	38	0.73		
Resolute Bay	29–30 Jul 89	89	1.02		
Resolute Bay	4–9 Sep 89	37	1.47		
Group mean (±SE)		5	0.95	4.05 ± 0.22	2.21 ± 0.18

Although the juvenile schools seemed to be feeding aggregations, there was no similar evidence for adult schools. Instead our data suggested that adult Arctic cod exhibit two distinct behavioral strategies during this time of the year: (1) participate in schools and generally forego feeding or (2) disperse individually to feed. It is not known whether nonschooling fish will join schools for periods of time or whether schooling fish will leave temporarily to feed (Bradstreet et al. 1986). Schools have been ob-

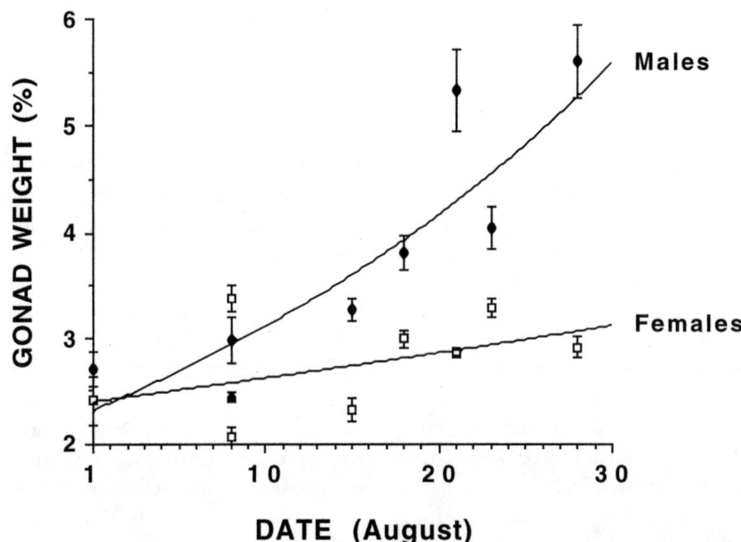

FIGURE 7.—Mean gonad weight (±SE) as percent body weight (= gonadosomatic index) for schools of Arctic cod *Boreogadus saida* during August 1985–1991 in the Canadian High Arctic. Exponential curves are fitted to show gonad development over time (males are solid diamonds, r^2 = 0.796; females are open squares, r^2 = 0.214).

served periodically at the same general location in Allen Bay for up to 2 months (Welch et al. 1993). High frequencies of empty stomachs (90–100%) indicate that individuals are not returning to the schools after feeding, and individuals with empty stomachs must have been starving for at least 2 weeks because of slow gastric evacuation rates (Hop 1994). However, the occasional feeding by fish in large schools may be sufficient to deplete zooplankton locally, as exemplified by the low ratio of large to small copepods in the vicinity of the Allen Bay school in August 1990. It is unlikely that fish school as a result of low food supply, because there was no evidence of resource depletion in Allen Bay in September 1990.

In contrast, nonschooling individuals fed extensively on copepods, amphipods, and mysids, and a high frequency of full to extremely full stomachs suggested that they fed almost continuously during July–September.

There appeared to be an energetic cost of schooling for adults. The absence of feeding was reflected in the lower condition factors of schooling fish, although the difference was not significant when adjusted for stomach weight, possibly because of low sample size of nonschooling fish. However, any energetic cost, as measured by condition factor, would be minor because fasting Arctic cod lose less than 10% of their body weight per month at low ambient temperatures (Hop and Graham 1995); with a high assimilation efficiency (about 0.80), this loss may be reduced substantially if they feed only occasionally. The small reduction in condition factor suggested weight loss of about 6%, but most samples had been collected within only 3 weeks of the first recorded schooling events (end of July).

Schools occurred at a time when food resources were generally abundant, but schools did not move to nearby resource-rich areas after their local area became depleted. Energetic costs of schooling are apparently outweighed by benefits, such as protection from predators.

Acknowledgments

We thank personnel at the Department of Fisheries and Oceans, Resolute Bay Marine Laboratory, Resolute Bay, Northwest Territories, for assistance in the field. The sample from Grise Fjord was collected by R. E. A. Stewart, Freshwater Institute, Winnipeg. Support for this study was received from the Department of Fisheries and Oceans, Polar Continental Shelf Project, Northern Heritage Society and Science Institute of N.W.T., Norwegian Council of Fisheries Research, a Circumpolar Boreal Alberta Research Grant, and a Roger Tory Peterson Institute Research Grant. We thank W. M. Tonn and three anonymous reviewers for comments. This article is contribution No. 19 from the Resolute Marine Laboratory.

References

Andriyashev, A. P., B. F. Mukhomediayarov, and E. A. Pavshtiks. 1980. On mass congregations of the cryopelagic cod fishes (*Boreogadus saida* and *Arctogadus glacialis*) in circumpolar regions of the Arctic. Pages 196–211 *in* M. E. Vinogradov and I. A. Mel'nikov, editors. Biology of the central Arctic basin. Shirov Institute of Oceanology, Academy of Sciences, U.S.S.R. (In Russian.) English translation, 1981: U.S. National Marine Fisheries Service, Northwest and Alaska Fisheries Center, Seattle.

Bain, H., and A. D. Sekerak. 1978. Aspects of biology of Arctic cod, *Boreogadus saida*, in the central Canadian Arctic. LGL Ltd., Toronto, Ontario.

Barenkova, A. S., V. P. Ponomarenko, and N. S. Khokhlina. 1966. Distribution, size and growth of larvae and fry of *Boreogadus saida* (Lep.) in the Barents Sea. Voprosy Ikhtiologii 6:498–518. (In Russian.) English translation, 1977: Canadian Fisheries and Marine Service Translation Series 4025.

Bergmann, M. A. 1989. CTD (conductivity, temperature, depth) profiles: Barrow Strait, N.W.T., 1984–1986. Canadian Data Report of Fisheries and Aquatic Sciences 752.

Bradstreet, M. S. W., and six coauthors. 1986. Aspects of the feeding biology of Arctic cod (*Boreogadus saida*) and its importance in Arctic marine food chains. Canadian Technical Report of Fisheries and Aquatic Sciences 1491.

Chilton, D., and R. Beamish. 1982. Age determination methods for fishes studied by the groundfish program at the Pacific Biological Station. Canadian Special Publication of Fisheries and Aquatic Sciences 60.

Craig, P. C., W. B. Griffiths, L. Haldorson, and H. McElderry. 1982. Ecological studies of Arctic cod (*Boreogadus saida*) in Beaufort Sea coastal waters. Canadian Journal of Fisheries and Aquatic Sciences 39:395–406.

Crawford, R. E., and J. Jorgenson. 1993. Schooling behaviour of arctic cod, *Boreogadus saida*, in relation to drifting pack ice. Environmental Biology of Fishes 36:345–357.

Crawford, R. E., and J. Jorgenson. 1996. Quantitative studies of Arctic cod, *Boreogadus saida*: important energy stores in the Arctic food web. Arctic 49:181–193.

Finley, K. J., M. S. W. Bradstreet, and G. W. Miller. 1990. Summer feeding ecology of harp seals (*Phoca groenlandica*) in relation to arctic cod (*Boreogadus saida*) in the Canadian high Arctic. Polar Biology 10:609–618.

Hobson, K. A., and H. E. Welch. 1992. Determination of trophic relationships within a high-arctic marine food web using $\delta^{13}C$ and δ^{15} N analysis. Marine Ecology Progress Series 84:9–18.

Hognestad, P. T. 1968. Polar cod, *Boreogadus saida* Lep. in Norwegian waters. Astarte 31:1–3.

Hop, H. 1994. Bioenergetics and trophic relationships of Arctic cod (*Boreogadus saida*) in the Canadian high Arctic. Doctoral dissertation. University of Alberta, Edmonton.

Hop, H., and M. Graham. 1995. Respiration of juvenile Arctic cod (*Boreogadus saida*): effects of acclimation, temperature, and food intake. Polar Biology 15:359–367.

Hop, H., M. Graham, and V. L. Trudeau. 1995. Spawning energetics of Arctic cod (*Boreogadus saida*) in relation to seasonal development of the ovary and plasma sex steroid levels. Canadian Journal of Fisheries and Aquatic Sciences 52:541–550.

Klumov, S. K. 1937. Polar cod (*Boreogadus saida*) and its importance for certain life processes in the Arctic. Izvestiya Akademii Nauk SSSR Seriya Biologicheskaya 1937(1):175–188. (In Russian.) English translation: Alaska Department of Fish and Game, Anchorage.

Lear, W. H. 1980. Morphometrics, meristics and gonadal development of Arctic cod from northern Labrador during September, 1978. Canadian Atlantic Fisheries Scientific Advisory Commission Research Document 80/7. St. John's, Newfoundland.

Lønne, O. J., and B. Gulliksen. 1989. Size, age and diet of polar cod, *Boreogadus saida* (Lepechin 1773), in ice covered waters. Polar Biology 9:187–191.

Moskalenko, B. F. 1964. On the biology of polar cod *Boreogadus saida* (Lepechin). Voprosy Ikhtiologii 4:32, 433–443. (In Russian.) English translation: Alaska Department of Fish and Game, Anchorage.

Ponomarenko, V. P. 1967. Areas of autumn–winter distribution of polar cod in southern Barents Sea as an indicator of the inshore summer distribution of capelin and cod feeding on capelin. Voprosy Ikhtiologii 7:1073–1079. (In Russian.) English translation, 1979: Fisheries and Marine Service Canada, St. John's, Newfoundland.

Ponomarenko, V. P. 1968. Some data on the distribution and migrations of polar cod in the seas of the Soviet Arctic. Rapports et Procès-Verbaux des Réunions Conseil International pour l'Exploration de la Mer 158:131–135.

Rass, T. S. 1968. Spawning and development of polar cod. Rapports et Procès-Verbaux des Réunions Conseil International pour l'Exploration de la Mer 158: 135–137.

Ricker, W. E. 1975. Computation and interpretation of biological statistics of fish populations. Bulletin of the Fisheries Research Board of Canada 191.

Sameoto, D. 1984. Review of current information on Arctic cod (*Boreogadus saida* Lepechin) and bibliography. Bedford Institute of Oceanography, Dartmouth, Nova Scotia.

Shibanoff, S. V. 1958. Dynamics of Arctic fox numbers in relation to breeding, food and migration conditions. Translations of Russian Game Reports, volume 3.

Vladykov, V. D. 1972. Morphological differences in male gonads among nine genera of Gadidae (Pisces). Journal of the Fisheries Research Board of Canada 29: 1709–1716.

Welch, H. E., R. E. Crawford, and H. Hop. 1993. Occurrence of Arctic cod (*Boreogadus saida*) schools and their vulnerability to predation in the Canadian high Arctic. Arctic 46:331–339.

Welch, H. E., and seven coauthors. 1992. Energy flow through the marine ecosystem of the Lancaster Sound region, arctic Canada. Arctic 45:343–357.

Wolotira, R. J., Jr., T. M. Sample, and M. Morin, Jr. 1979. Baseline studies of fish and shellfish resources of Norton Sound and the southeastern Chukchi Sea. Pages 258–572 *in* Environmental assessment of the Alaskan continental shelf. Final report of the principal investigators, volume 6, biological studies. BLM (U.S. Bureau of Land Management)/NOAA (National Oceanic and Atmospheric Administration) OCSEAP (Outer Continental Shelf Environmental Assessment Program), Boulder, Colorado.

American Fisheries Society Symposium 19:81–89, 1997

Distribution, Abundance, and Growth of Arctic Cod in the Northeastern Chukchi Sea

J. G. Gillispie, R. L. Smith, E. Barbour, and W. E. Barber
University of Alaska, Institute of Marine Science, Fairbanks, Alaska 99775, USA

Abstract.—Distribution, abundance, and growth of Arctic cod *Boreogadus saida* from the northeastern Chukchi Sea were examined. Arctic cod was the fish species most frequently caught by benthic trawl in 1990 and 1991. In 1990 highest abundance and biomass occurred in the southern portion of the study area in Bering shelf water, followed by Alaska coastal water, and resident Chukchi water. In 1991 none of the stations sampled occurred in Bering shelf water; however, the stations with higher abundance and biomass still occurred toward the south occupied by resident Chukchi water. Warmer (>1°C), more productive water masses (Bering shelf water and Alaska coastal water) from the south were more prevalent in 1990 than in 1991. Differences in age distribution between the 2 years suggest that survival may have been reduced in 1991. Warmer water masses may also increase growth: fish younger than 4 years old were larger in 1990 compared to 1991.

Arctic cod *Boreogadus saida* are one of the most abundant and widely distributed circumpolar fishes in the Arctic. Along the Alaskan coast, they can be found from the northern Bering Sea through the Chukchi Sea and east to the Canadian border. In the Bering Sea, they were found as far south as 60°N latitude by Frost and Lowry (1981), with the more commonly reported southern limit just south of Norton Sound (Pereyra and Wolotira 1977; Lowry and Frost 1981). Arctic cod have been found in the Chukchi Sea (Alverson and Wilimovsky 1966; Wolotira et al. 1977); however, before this study little information was available regarding Arctic cod in the northeastern Chukchi Sea (Walters 1955; Frost and Lowry 1983; Fechhelm et al. 1985). Along the northern Alaska coast, most studies have been conducted in the Beaufort Sea and Prudhoe Bay (Cannon et al. 1991). Kleinenberg et al. (1969) reported finding Arctic cod as far north as 88°N off the Russian coast. Andriyashev (1964) observed Arctic cod near the North Pole and hypothesized that they would also occur under the polar ice cap.

In general, Arctic cod abundance and biomass are higher in arctic waters than in the more southerly Bering Sea (Wolotira et al. 1977). Lowry and Frost (1981) hypothesized that fish moved northward every spring and summer with the receding ice edge from the northern Bering Sea and southward in the fall with the advancing ice edge. This annual pattern is consistent with their known tolerances for temperatures ranging from −1.8°C (Andriyashev 1964; Alverson and Wilimovsky 1966) to 6°C (Craig et al. 1982). However, Arctic cod have been found inhabiting waters of up to 13.5°C (Craig et al. 1982).

Arctic cod contribute significantly to the diets of marine mammals (Frost and Lowry 1980; Lowry et al. 1981; Bradstreet 1982; Finley and Evans 1983) and seabirds (Springer et al. 1987). In many areas this fish is the only abundant food source for these animals and appears to be an important link from lower to upper trophic levels (Bradstreet and Cross 1982). Abundance of ringed seal *Phoca hispida* has been related to Arctic cod populations and densities by Chapskii (1971). It has been proposed that the primary prey for beluga (white whales) *Delphinapterus leucas* (Kleinenberg et al. 1969); during their summer stay in the Chukchi Sea is Arctic cod. Kleinenberg et al. (1969) also related beluga whale migrations to movement of Arctic cod. Other marine mammals that consume Arctic cod include sei whales *Balaenoptera borealis* (Tomilin 1957), killer whales *Orcinus orca* (Tomilin 1957), ribbon seals *P. fasciata* (Frost and Lowry 1980), and fin whales *B. physalus* (Klumov 1937). Arctic cod was an important prey item for nesting seabirds on Saint Lawrence Island, eastern Chukchi and western Beaufort seas (Springer and Byrd 1988).

Physical Oceanographic Features

Weingartner's (1997, this volume) review of the hydrography of the northeastern Chukchi Sea identifies the water masses and their characteristics and origins. Three primary water masses, Alaska coastal water, Bering shelf water, and resident Chukchi water, are found in the Chukchi Sea. Both Bering shelf water and Alaska coastal water flow north from the Bering Sea. The Bering shelf water is a mixture of Bering Sea and Gulf of Anadyr waters, with temperatures ranging from 0 to 3°C and salinity from 32.5 to 33 practical salinity units (psu). Alaska coastal water is formed in coastal areas and mixed with fresher water from Kotzebue Sound.

Temperatures of Alaska coastal water range from 2 to 13°C and salinities are less than 32.2 psu. Resident Chukchi water, found offshore in the northern Chukchi Sea, is derived from the upper layers of the Arctic Ocean or shelf water left from the previous winter (Weingartner 1997). Resident Chukchi water has a temperature less than 1°C and salinity ranges from 32 to 33 psu.

Two other factors that affect hydrography in the Chukchi Sea are ice cover and melting ice. Ice formation begins in September and October (Lowry et al. 1981) and covers the Chukchi Sea from about November to July (Aagaard 1988). Freezing seawater results in the formation of a highly saline brine that can contribute to the highly saline water of resident Chukchi water (Weingartner 1997). Ice cover, especially landfast ice, reduces the impact of wind-driven surface currents (Aagaard 1988). Melting ice contributes to the freshwater input and may produce an additional water mass (Weingartner 1997). In addition, the ice edge does not melt at a uniform rate, resulting in meltwater embayments (Paquette and Bourke 1981) where upwelling may occur (Hakkinen 1986).

Study Purpose

The purpose of this study was to expand our knowledge of Arctic cod abundance, distribution, and age in the northeastern Chukchi Sea. We attempted to assess these aspects of Arctic cod biology within the context of biological and physical oceanographic features of the Chukchi Sea.

Methods

Stations in the northeastern Chukchi Sea were sampled for fish from 16 August to 16 September 1990 and from 14 September to 23 September 1991. Forty-eight stations were sampled in 1990 and 17 stations in 1991. Sampling was done using a National Marine Fisheries Service (NMFS) 83-112 survey otter trawl. The net had a 25.2-m headrope and a 34.1-m footrope that was set back 7.1 cm from a tickler chain. The cod end was 90-mm stretched mesh with a 33-mm stretched mesh liner inserted in it. Two 30-min hauls at a trawling speed of approximately 2 knots were taken from each sampling station. The area sampled (m^2) was calculated by multiplying the width of the net opening (measured electronically by a NMFS Scanmar) by the distance trawled. The distance was determined from the ship's position (latitude and longitude) at the beginning and end of each haul.

Biomass (kg/km^2 trawled) and abundance (number of fish/km^2 trawled) were determined as the average of the two hauls from each station. A subset of captured fish was frozen before transport to the laboratory; 287 individual fish were examined from 1990 and 60 from 1991. In addition to biological sampling, conductivity–temperature–depth information was collected with a Seabird SBE 19.

Fish were measured (total and fork lengths) to the nearest 1 mm and weighed to the nearest 0.1 g. The fish weight used in all calculations was the total body weight minus the stomach content weight. Otoliths were removed for aging and stored in 50% glycerin.

The annulus on otoliths viewed with transmitted light was defined as the translucent zone, whereas the annulus in burnt sections of otoliths was defined as the dark zone. Age was determined by counting the number of annuli. Otoliths were measured with a calibrated ocular micrometer.

Statistical analyses included t-tests for differences between means (Freund 1979) and F-statistics (Neter et al. 1990) for slopes and elevations. A Mann–Whitney U-test was conducted to detect differences in biomass and abundance between years. A Kruskal–Wallis test (Zar 1984) was used to determine differences in biomass and abundance between water masses within each year. When only two water masses were present, a Mann–Whitney U-test was conducted.

Results

In 1990 bottom temperature ranged from 12.7°C nearshore by Point Hope, decreasing offshore and northward to as low as −1.2°C (maps of temperature and salinity can be found in Weingartner [1997]). Salinity was lowest, 29.7 psu, nearshore between Point Hope and Point Lay. It increased both offshore and northward to 33.3 psu. In 1991 bottom temperature ranged from 7°C nearshore by Point Hope, decreasing offshore and northward to as low as −1.7°C. Salinity was lowest nearshore at Point Hope and increased offshore and northward to as high as 33.5 psu. In 1990 the sea ice edge was positioned at approximately 74°N, whereas in 1991 it was around 71°30′N.

Distribution and Abundance

Arctic cod was the most abundant of the fishes caught in this study: 76% of the total numeric catch in 1990, 66% in 1991. In 1990 Arctic cod were present at all 48 stations sampled and ranged in numbers from 10 to 120,000 fish/km^2 (Figure 1A). They tended to be most abundant in the southern

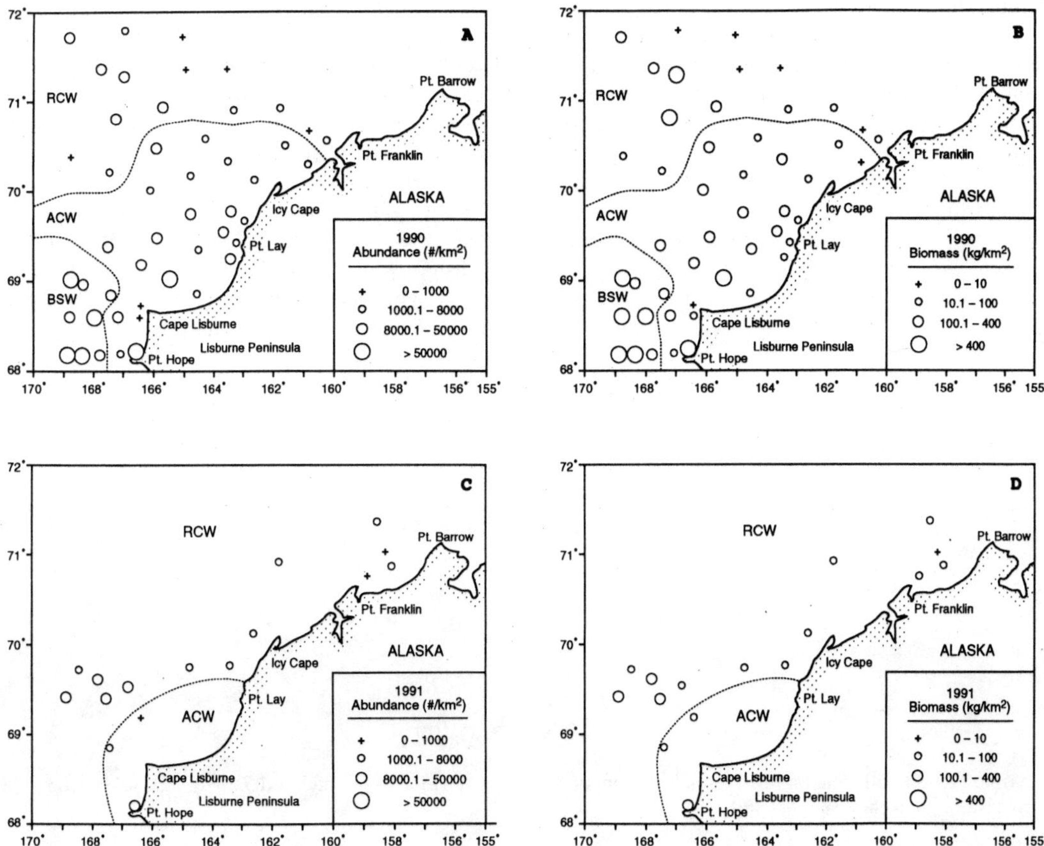

FIGURE 1.—Abundance (fish/km^2) and biomass (kg/km^2) of Arctic cod *Boreogadus saida* from the northeastern Chukchi Sea: (**A**) abundance in 1990, (**B**) biomass in 1990, (**C**) abundance in 1991, and (**D**) biomass in 1991. Abbreviations are RCW = resident Chukchi water, ACW = Alaska coastal water, and BSW = Bering shelf water.

part of the sampling area, off Point Hope, Alaska. Of the six stations where abundance was greater than 50,000 fish/km^2, four occurred in Bering shelf water and two in Alaska coastal water. Bering shelf water had the highest average abundance (59,700 fish/km^2), followed by Alaska coastal water (13,400 fish/km^2), and resident Chukchi water (8,340 fish/km^2). Abundance in Bering shelf water was significantly different from that in Alaska coastal water ($P < 0.005$) and in resident Chukchi water ($P < 0.001$). There was no significant difference between abundance in Alaska coastal water and that in resident Chukchi water ($P > 0.50$).

Arctic cod biomass from 1990 was 61% of the total fish biomass. Biomass ranged from 0.45 to 1,830 kg/km^2 (Figure 1B). Of the nine stations where biomass was greater than 400 kg/km^2, five were in Bering shelf water, two in Alaska coastal water, and two in resident Chukchi water. Bering shelf water had the highest average biomass (864 kg/km^2), followed by resident Chukchi water (196 kg/km^2) and Alaska coastal water (188 kg/km^2). Biomass in Bering shelf water was significantly different from that in resident Chukchi water ($P < 0.002$) and in Alaska coastal water ($P < 0.010$). There was no significant difference between biomass in Alaska coastal water and that in resident Chukchi water ($P > 0.500$)

In 1991 Arctic cod were present at 16 of the 17 stations. Their abundance ranged from 394 to 15,700 fish/km^2 (Figure 1C). As in 1990, they tended to be most abundant off Point Hope. However, fish were generally present in fewer numbers at each station in the sampling area; there were no stations in which abundance was greater than 50,000 fish/km^2. Resident Chukchi water had the highest average abundance, 6,110 fish/km^2, while

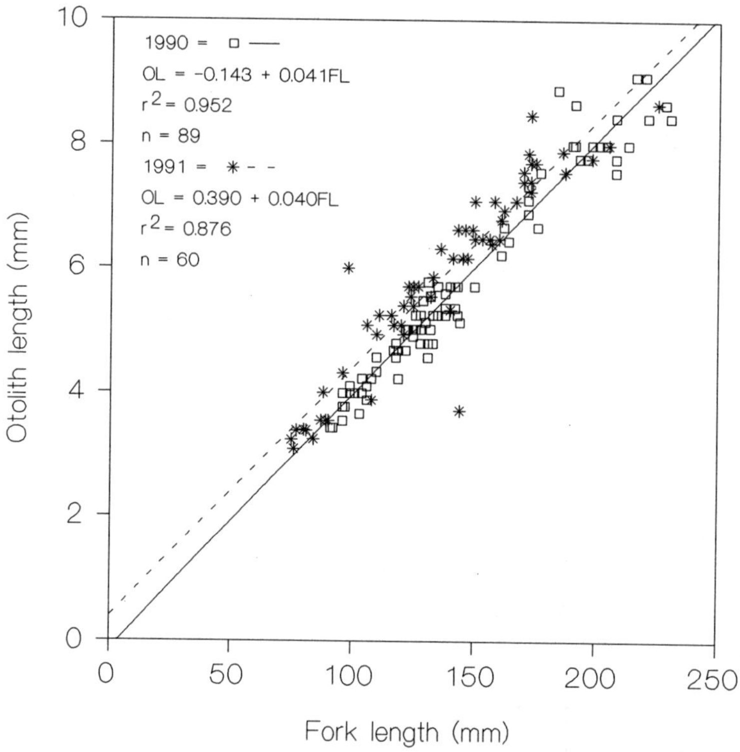

FIGURE 2.—Otolith length (OL) as a function of fork length (FL) for Arctic cod *Boreogadus saida* from the northeastern Chukchi Sea, 1990 and 1991.

Alaska coastal water had an average of 5,200 fish/km^2. Bering shelf water was not present at any of the stations sampled. The difference in abundance between Alaska coastal water and resident Chukchi water was not significant (P = 0.946). Arctic cod biomass was 47% of the total fish biomass. Biomass ranged from 0 to 220 kg/km^2 (Figure 1D). Unlike in 1990, none of the stations had a biomass greater than 400kg/km^2. Alaska coastal water had an average biomass of 88.4 kg/km^2 and resident Chukchi water had an average of 53.1 kg/km^2. The difference in biomass between the two water masses was not significant (P = 0.946).

The average Arctic cod biomass for the northeastern Chukchi Sea was 304 kg/km^2 in 1990 and 60 kg/km^2 in 1991. The difference in biomass between years was significant (P = 0.001). The average abundance for the area was 19,540 fish/km^2 in 1990 and 5,370 fish/km^2 in 1991. The difference in abundance between years was significant (P = 0.001).

Length–Weight Relationships

Fork length (FL) was a linear function of total length (TL), FL = 2.223 + 0.952 TL; r^2 = 0.998. For 1990 fish, otolith length (OL) as a function of fork length was OL = −0.143 + 0.041 FL; r^2 = 0.953; for 1991 fish, and OL = 0.390 + 0.040 FL, r^2 = 0.876 (Figure 2). The equations for OL and FL were significantly different between the 2 years ($P <$ 0.001).

Weight (W) was a curvilinear function of FL: $W = (4.989 \times 10^{-6}) \times FL^{3.072}$, r^2 = 0.990 for 1990 fish; $W = (9.233 \times 10^{-6}) \times FL^{2.948}$, r^2 = 0.995 for 1991 fish. These two regression equations were not significantly different (P = 0.536), therefore the data were pooled. The equation from the pooled data was $W = (5.398 \times 10^{-6}) \times FL^{3.056}$; r^2 = 0.991 (Figure 3).

Age

Of the fish examined from 1990, the maximum age recorded was one 8-year-old female, FL = 228 mm. In 1991 the oldest fish was a 5-year-old male, FL = 187 mm. There were no age-7 fish for either year and no age-6 fish in 1991. In 1990 61% of the total sample were age-1 fish, whereas in 1991 only 32% were age 1 (Figure 4). In 1990 78% were less than 3 years old, whereas in 1991 70% were less

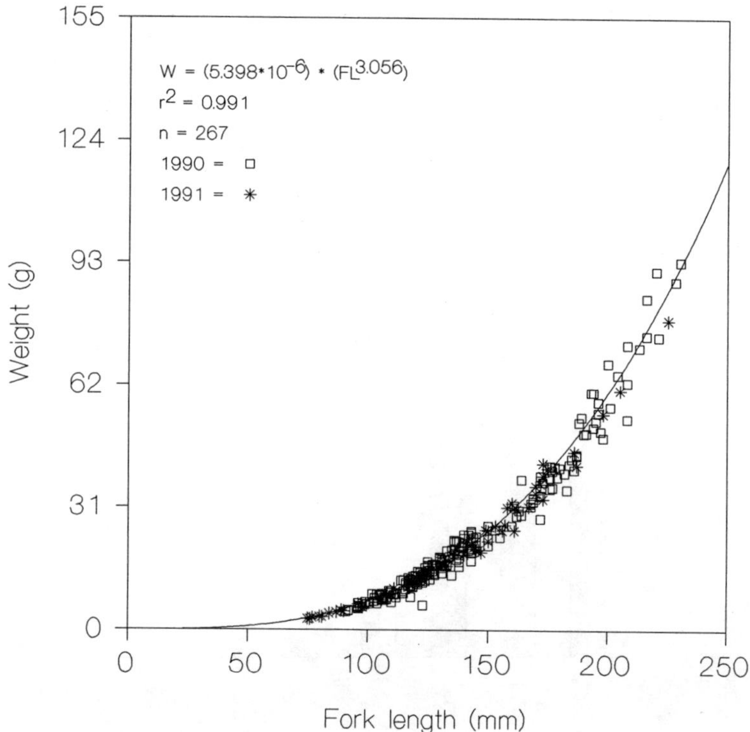

FIGURE 3.—Weight (*W*) as a function of fork length (FL) for Arctic cod *Boreogadus saida* from the northeastern Chukchi Sea. Data were pooled for the 2 years of the study.

than 3. Females and males were about equally represented in age-classes 1 and 2 in 1990 and 1991, whereas fish in age-classes 3 and 4 were primarily females in both years (Table 1).

The average FL at age (Table 2) was significantly larger in 1990 than 1991 for age groups 1 ($P <$ 0.001) and 3 ($P = 0.007$). In 1991 the average FL for age group 2 was significantly ($P = 0.005$) greater than in 1990. There was no significant difference between mean FLs for the 4-year-old fish in 1990 versus 1991 ($P = 0.979$), and insufficient data were available for comparing age groups greater than 4 years.

Fish captured in the northern part of the study area did not have significantly different FLs from those captured in the southern part for fish less than 3 years old ($P = 0.465$) and fish greater than or equal to 3 years old ($P = 0.944$). However, when inshore fish were compared with offshore fish, offshore fish greater than or equal to 3 years old were significantly larger ($P < 0.001$). Fish less than 3 years old did not differ significantly ($P = 0.246$) in size between the two areas.

Discussion

Abundance and Distribution

Studies relating Arctic cod to physical parameters have been done only in the inshore environment. No relationships have been determined for offshore habitats. Moulton and Tarbox (1987) found that Arctic cod in the inshore Beaufort Sea were more concentrated along the transition of cold (<−1°C), high-salinity (28–32 psu) marine bottom water and warm (2–9°C), low-salinity (6–27 psu) coastal surface water. Our data (not restricted to inshore) did not follow a similar trend because the highest concentrations were in the southern offshore part of the study area by Point Hope in 2 to 3°C (bottom temperatures) Bering shelf water. The −1°C bottom isotherm was farther north off of Point Franklin (Weingartner 1997). Bering shelf water is known to have a higher abundance of zooplankton compared with Alaska coastal water and resident Chukchi water (Springer et al. 1989), therefore Arctic cod may be more attracted to it for food. Moulton and Tarbox (1987) reported high concentrations of

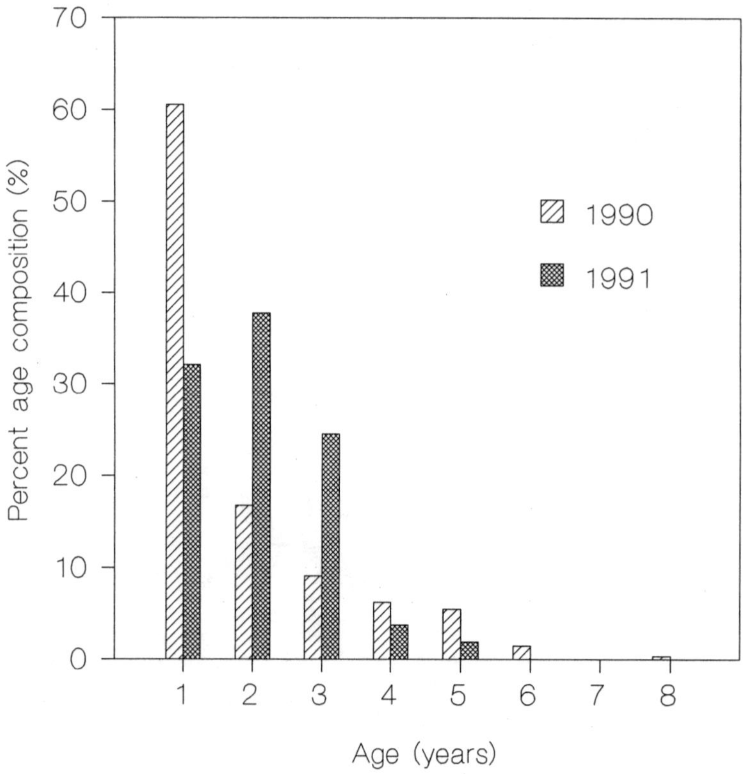

FIGURE 4.—Age composition of Arctic cod *Boreogadus saida* from the northeastern Chukchi Sea, 1990 and 1991.

copepods and mysids in the transition layer, along with Arctic char *Salvelinus alpinus* and Arctic ciscoes *Coregonus autumnalis*.

Wolotira et al. (1977) conducted a study in the southeastern Chukchi Sea in 1976 using the same net type as ours. Biomass in their northernmost stratum, occupied by Alaska coastal water (determined by examining their temperature and salinity data), was 12.3 kg/km^2. This biomass was considerably lower than that found in the northeastern Chukchi Sea, where the average biomass was 25 times greater in 1990 and 5 times greater in 1991. Average biomass for Alaska coastal water was 15 times greater in 1990 and 7 times greater in 1991. The differences between these two studies, separated by 13 years, may be the result of extreme interannual differences. Interannual differences in abundance and biomass were evident in our study between 1990 and 1991. In addition, time of year may have influenced the abundance and biomass, particularly if Lowry and Frost's (1981) hypothesis of annual migration is correct. Wolotira et al. (1977) conducted their study during September–October, whereas ours was in August–September.

TABLE 1.—Percentage of Arctic cod *Boreogadus saida* that were females in age-classes 1–8 in the northeastern Chukchi Sea, 1990–1991.

Age	1990		1991	
(years)	*N*	Females (%)	*N*	Females (%)
1	149	54	15	47
2	41	37	20	55
3	23	70	13	92
4	16	69	2	100
5	14	57	1	0
6	3	33	0	0
8	1	100	0	0

Environmental Conditions and Growth

Environmental conditions determine the growth and survival of fish. Warmer years in the Chukchi Sea tend to be more favorable than cold years to the growth and abundance of some fish (Springer et al. 1984). Arctic cod may fall into this category as evidenced by the differences in length of age-1, -2,

TABLE 2.—Comparison of mean ($\bar{x}$) and range of fork length at age for Arctic cod *Boreogadus saida* from the northeastern Chukchi Sea, 1990 and 1991.

	This study (1990)			This study (1991)			Frost and Lowry (1983)		Craig et al. (1982)		
Age	*N*	$\bar{x}$	Range	*N*	$\bar{x}$	Range	$\bar{x}$	Range	*N*	$\bar{x}$	Range
1	127	122	91–161	15	91	75–120	72	45–117	225	84	54–110
2	30	129	99–193	20	147	111–173	116	97–144	137	128	88–177
3	25	173	98–221	13	147	121–198	141	129–160	94	159	120–196
4	17	199	184–220	2	199	172–225	171	161–180	19	180	129–203
5	15	189	160–230	1	187	187			9	209	153–250
6	1	216	216						4	240	198–257
8	1	228	228								

and -3 fish between 1990 and 1991. In 1990 the sea ice edge in our study area was farther north than in 1991. The prevailing wind direction in 1990 was from the south, whereas in 1991 the dominant wind direction was from the north. These conditions resulted in 1990 being warmer than 1991 and presumably more conducive to growth and survival. In addition, Weingartner (1997) linked warm surface temperatures with increased northward flow from the northern Bering Sea, which may bring a greater abundance of pelagic prey than would otherwise be resident in the Chukchi Sea. This situation is reflected in the recruitment of Arctic cod to the age-1 year-class. In 1990 the percentage of age-1 fish in the samples was nearly twice that of 1991. Our mean length-at-age data (Table 2) also support the contention that 1990 was more favorable for growth than 1991.

Mean length at age in this study was greater than that reported in previous studies (Craig et al. 1982; Frost and Lowry 1983), which suggests that growth conditions in the study area were better during the years preceding this study than in the years preceding the previous studies. In addition, the significant difference in length at age of fish less than 4 years old in this study indicates that 1990 was better for growth than 1991. The difference in mean length at age for older fish between offshore and inshore may be related to water masses. Bering shelf water has higher concentrations of dissolved nutrients and chlorophyll than Alaska coastal water (Walsh et al. 1989) and is a major source of zooplankton for the Chukchi Sea (Springer et al. 1989). In 1990 offshore stations were located in Bering shelf water, where higher concentration of prey items, contributing to greater growth, may have occurred, compared to inshore stations located in Alaska coastal water.

Age

Of the northern gadids, Arctic cod appear to have the shortest life span. Walleye pollock *Theragra chalcogramma* live to 28 years (McFarlane and Beamish 1990), Atlantic cod *Gadus morhua* to 16 years (Fleming 1960), Pacific cod *G. macrocephalus* to 12 years (Craig et al. 1982), Greenland cod *G. ogac* to 11 years (Mikhail and Welch 1989), and saffron cod *Eleginus gracilis* to 9 years (Wolotira et al. 1977); the maximum observed age of Arctic cod in previous studies was 7 years (Bradstreet et al. 1986). Our study suggests that Arctic cod may live to age 8. Only one fish was determined to be this age and the possibility of observer errors in aging individual otoliths may have been present. However, two readers independently read the otolith, and both agreed to the assessment of age 8. Even with this unusually old individual, Arctic cod exhibit the shortest longevity of northern gadids studied.

Arctic cod spawning has been reported to occur under the ice from late November to early February (Craig et al. 1982) and in some instances as late as March (Rass 1968). Aronovich et al. (1975) found the incubation period of the eggs could be prolonged by extended subzero winter water temperatures. An extended spawning and temperature-dependent development could lengthen the time period in which larval fish appear. Wyllie-Echeverria et al. (1997, this volume) observed newly hatched larvae as late as mid-July. These events, either singularly or together, may result in a wide range of lengths at age, which would explain the large variability in lengths at age found in this and other studies. Another source of variability was collection date. Arctic cod in 1990 were collected nearly 1 month later than those in 1991.

Life History Strategy

Species employing *r*-selection reproductive strategies generally occur in environments that are unstable or unpredictable, with the possibility of high mortality rates. In these environments it is advantageous to invest resources in the production of as many offspring as early in the life cycle as possible

(Adams 1980). Craig et al. (1982) concluded that Arctic cod exhibited *r* strategy traits: small body size, relatively short life span, early maturity, rapid growth, and large numbers of offspring. Our data support the *r*-selected life strategy traits of relatively short life span and small size.

Due to the unpredictable population size of *r*-selected species, Arctic cod may exert a regulatory effect on its predators (Craig et al. 1982). In years where environmental conditions were not conducive to population growth of Arctic cod, there would subsequently be fewer fish (or lower biomass) available for consumption by marine mammals and seabirds, possibly resulting in lower reproductive success of these predators.

References

Aagaard, K. 1988. Current, CTD, and pressure measurements in possible dispersal regions of the Chukchi Sea. Pages 255–333 *in* Final reports of principal investigators. NOAA (National Oceanic and Atmospheric Administration) OCSEAP (Outer Continental Shelf Environmental Assessment Program) 57, Anchorage, Alaska.

Adams, P. B. 1980. Life history patterns in marine fishes and their consequences for fisheries management. U.S. National Marine Fisheries Service Fishery Bulletin 78:1–12.

Alverson, D. L., and N. J. Wilimovsky. 1966. Fishery investigations of the southeastern Chukchi Sea. Pages 843–860 *in* N. J. Wilimovsky and J. N. Wolfe, editors. Environment of the Cape Thompson region, Alaska. U.S. Atomic Energy Commission, Oak Ridge, Tennessee.

Andriyashev, A. P. 1964. Fishes of the northern seas of the U.S.S.R. Translated from Russian: Israel Program for Scientific Translations, Catalogue Number 836, Jerusalem.

Aronovich, T. M., S. I. Doroshev, L. V. Spectorova, and V. M. Makhotin. 1975. Egg incubation and larval rearing of navaga (*Eleginus navaga* Pall.), polar cod (*Boreogadus saida* Lepechin) and arctic flounder (*Liopsetta glacialis* Pall.) in the laboratory. Aquaculture 6:233–242.

Bradstreet, M. S. W. 1982. Occurrence, habitat use, and behavior of seabirds, marine mammals, and Arctic cod at the pond inlet ice edge. Arctic 35:28–40.

Bradstreet, M. S. W., and W. E. Cross. 1982. Trophic relationships at high arctic ice edges. Arctic 35:1–12.

Bradstreet, M. S. W., and six coauthors. 1986. Aspects of the biology of Arctic cod (*Boreogadus saida*) and its importance in arctic marine food chains. Canadian Technical Report of Fisheries and Aquatic Sciences 1491.

Cannon, T. C., D. R. Glass, and C. M. Prewitt. 1991. Habitat use patterns of juvenile Arctic cod in the coastal Beaufort Sea near Prudhoe Bay, Alaska. Pages 157–162 *in* C. S. Benner and R. W. Middleton, editors. Fisheries and oil development on the continental shelf. American Fisheries Society Symposium 11, Bethesda, Maryland.

Chapskii, K. K. 1971. The ringed seal of western seas of the Soviet Arctic. Proceedings of the Arctic Sciences Research Institute, Leningrad, USSR 145:1–72. (In Russian.) English translation, 1971: Fisheries Research Board of Canada Translation Series 1665.

Craig, P. C., W. B. Griffiths, L. Haldorson, and H. McElderry. 1982. Ecological studies of Arctic cod (*Boreogadus saida*) in Beaufort Sea coastal waters, Alaska. Canadian Journal of Fisheries and Aquatic Sciences 39:395–406.

Fechhelm, R. G., P. C. Craig, J. S. Baker, and B. J. Gallaway. 1985. Fish distribution and use of nearshore waters in the northeastern Chukchi Sea. Pages 121–297 *in* Final reports of principal investigators. NOAA (National Oceanic and Atmospheric Administration) OCSEAP (Outer Continental Shelf Environmental Assessment Program) 32, Anchorage, Alaska.

Finley, K. J., and C. R. Evans. 1983. Summer diet of the bearded seal (*Erignathus barbatus*) in the Canadian high arctic. Arctic 36:82–89.

Fleming, A. M. 1960. Age, growth and sexual maturity of cod (*Gadus morhua* L.) in the Newfoundland area, 1947–1950. Journal of the Fisheries Research Board of Canada 17:775–809.

Freund, J. E. 1979. Modern elementary statistics. Prentice-Hall, Englewood Cliffs, New Jersey.

Frost, K. J., and L. F. Lowry. 1980. Feeding of ribbon seals (*Phoca fasciata*) in the Bering Sea in spring. Canadian Journal of Zoology 58:1601–1607.

Frost, K. J., and L. F. Lowry. 1981. Trophic importance of some marine gadids in northern Alaska and their body–otolith size relationships. U.S. National Marine Fisheries Service Fishery Bulletin 79:187–192

Frost, K. J., and L. F. Lowry. 1983. Demersal fishes and invertebrates trawled in the northeastern Chukchi and western Beaufort seas, 1976–77. NOAA (National Oceanic and Atmospheric Administration) Technical Report NMFS (National Marine Fisheries Service) SSRF (Special Scientific Report Fisheries)-764.

Hakkinen, S. 1986. Coupled ice–ocean dynamics in the marginal ice zones: up/downwelling and eddy generation. Journal of Geophysical Research 87:5853–5859.

Kleinenberg, S. E., A. V. Yablokov, B. M. Bee'konvich, and M. N. Tarasevich. 1969. Beluga (*Delphinapterus leucas*) investigation of the species. Publication by the Israel Program for Scientific Translations for the Smithsonian Institute and the National Science Foundation, Washington, DC.

Klumov, S. K. 1937. Polar cod and their importance for certain life processes in the Arctic. Istitutua Akademiya Nauk SSSR (Biology) 1:175.

Lowry, L. F., and K. J. Frost. 1981. Distribution, growth, and foods of Arctic cod (*Boreogadus saida*) in the Bering, Chukchi, and Beaufort seas. Canadian Field-Naturalist 95:186–191.

Lowry, L. F., K. J. Frost, and J. J. Burns. 1981. Trophic relationships among ice-inhabiting phocid seals and functionally related marine mammals in the Chukchi Sea. *In* Environmental assessment of the Alaskan continental shelf. Final reports of principal investigators, volume 11. Biological studies. NOAA (National Oceanic and Atmospheric Administration) OCSEAP (Outer Continental Shelf Environmental Assessment Program), Boulder, Colorado.

McFarlane, G. A., and R. J. Beamish. 1990. An examination of age determination structures of walleye pollock (*Theragra chalcogramma*) from five stocks in the northeast Pacific Ocean. Pages 37–56 *in* L. Low, editor. Proceedings of the symposium on application of stock assessment techniques to gadids. International North Pacific Fisheries Commission Bulletin 50.

Mikhail, M. Y., and H. E. Welch. 1989. Biology of Greenland cod, *Gadus ogac*, at Saqvaqjuac, northwest coast of Hudson Bay. Environmental Biology of Fishes 26:49–62.

Moulton, L. L., and K. E. Tarbox. 1987. Analysis of arctic cod movements in the Beaufort sea nearshore region, 1978–79. Arctic 40:43–49.

Neter, J., W. Wasserman, and M. H. Kutner. 1990. Applied linear statistical models, third edition. Richard D. Irwin, Inc., Boston.

Paquette, R. G., and R. H. Bourke. 1981. Ocean circulation and fronts as related to ice melt-back in the Chukchi Sea. Journal of Geophysical Research 86: 4215–4230.

Pereyra, W. T., and R. J. Wolotira. 1977. Baseline study of fish and shellfish resources of Norton Sound and southeastern Chukchi Sea. Pages 288–319 *in* Environmental Assessment Alaskan Continental Shelf, volume 8. Receptors–fish. NOAA (National Oceanic and Atmospheric Administration)/BLM (Bureau of Land Management), OCSEAP (Outer Continental Shelf Environmental Assessment Program), Boulder, Colorado.

Rass, T. S. 1968. Spawning and development of polar cod. Pages 135–137 *in* R. W. Blacker, editor. Symposium on the ecology of pelagic fish species in arctic waters and adjacent seas. Rapports et Proces-Verbaux des Reunions Conseil International pour l'Exploration de la Mer 158.

Springer, A. M., and G. V. Byrd. 1988. Seabird dependence on walleye pollock in the southeastern Bering Sea. Pages 667–677 *in* Proceedings of the international symposium on the biology and management of walleye pollock. Alaskan Sea Grant, Anchorage.

Springer, A. M., C. P. McRoy, and K. R. Turco. 1989. The paradox of pelagic food webs in the northern Bering Sea: II. Zooplankton communities. Continental Shelf Research 9:359–386.

Springer, A. M., E. C. Murphy, D. G. Roseneau, C. P. McRoy, and B. A. Cooper. 1987. The paradox of pelagic food webs in the northern Bearing Sea—I. Seabird food habits. Continental Shelf Research 7:895–911.

Springer, A. M., D. G. Roseneau, E. C. Murphy, and M. I. Springer. 1984. Environmental controls of marine food webs: food habits of seabirds in the eastern Chukchi Sea. Canadian Journal of Fisheries and Aquatic Sciences 41:1202–1215.

Tomilin, A. G. 1957. Cetacea. *In* V. G. Heptner, editor. Mammals of the USSR and adjacent countries. Translated from Russian 1967: Israel Program for Scientific Translations, volume 9, Jerusalem.

Walsh, J. J., and 20 coauthors. 1989. Carbon and nitrogen cycling within the Bering/Chukchi seas: source regions for organic matter effecting AOU demands of the Arctic Ocean. Progress in Oceanography 22:277–359.

Walters, V. 1955. Fishes of western arctic America and eastern arctic Siberia. Bulletin of the American Museum of Natural History 106:264–354.

Weingartner, T. J. 1997. A review of the physical oceanography of the northeastern Chukchi Sea. Pages 40–59 *in* J. Reynolds, editor. Fish ecology in Arctic North America. American Fisheries Society Symposium 19, Bethesda, Maryland.

Wolotira, R. J., T. M. Sample, and M. Morin. 1977. Demersal fish and shellfish resources of Norton Sound, the southeastern Chukchi Sea, and adjacent waters in the baseline year 1976. NOAA (National Oceanic and Atmospheric Administration) NMFS (National Marine Fisheries Service) Northwest and Alaska Fisheries Center Processed Report, Seattle.

Wyllie-Echeverria, T., W. E. Barber, and S. Wyllie-Echeverria. 1997. Water masses and transport of age-0 Arctic cod and age-0 Bering flounder into the northeastern Chukchi Sea. Pages 60–67 *in* J. Reynolds, editor. Fish ecology in Arctic North America. American Fisheries Society Symposium 19, Bethesda, Maryland.

Zar, J. H. 1984. Biostatistical analysis, 2nd edition. Prentice-Hall, Englewood Cliffs, New Jersey.

American Fisheries Society Symposium 19:90–103, 1997

Wind-Driven Transport and Dispersion of Age-0 Arctic Ciscoes along the Alaska Beaufort Coast

JOSEPH M. COLONELL

Woodward-Clyde Consultants
3501 Denali Street, Suite 101, Anchorage, Alaska 99503, USA

BENNY J. GALLAWAY

LGL Ecological Research Associates, Inc.
1410 Cavitt Street, Bryan, Texas 77801, USA

Abstract.—Movements of age-0 Arctic ciscoes *Coregonus autumnalis* along the Beaufort Sea coast from Canada to Alaska are considered herein to be analogous to the transport of particles in the turbulent flow of a stream of fluid. By representing their movements as a one-dimensional advection and turbulent dispersion process, which is driven by coastal winds, the "concentration" of fish per unit length of coast is computed as a function of summer wind history. Recruitment strengths at selected locations along the Alaska Beaufort coast are computed for comparison with observations obtained from fish-monitoring programs around the causeways near Prudhoe Bay and with records from the commercial and subsistence fisheries in the Colville River delta. Given the simplicity of the model and its underlying assumptions, the agreement between observed and predicted recruitment strengths is remarkably good.

The anadromous Arctic cisco *Coregonus autumnalis* is found in the coastal waters of the Beaufort Sea from Point Barrow, Alaska, to Cambridge Bay, Canada (Scott and Crossman 1973). Despite this broad coastal distribution, Gallaway et al. (1983) hypothesized that most of the Arctic ciscoes found in Canada and Alaska originate from spawning stocks in Canada's Mackenzie River system and that some portion of the age-0 fish are transported westward into Alaskan waters by wind-driven coastal currents. These fish take up residence in the deltas of the larger North Slope rivers (e.g., Colville and Sagavanirktok rivers) and remain in Alaska until the onset of sexual maturity, about age 7, when they migrate back to the Mackenzie River to spawn. Extensive fish-monitoring studies in Alaska and Canada have supported this life cycle scenario for Arctic ciscoes (e.g., Bond and Erickson 1987; Cannon et al. 1987; Moulton 1989; Reub et al. 1991). Genetic studies also support the Mackenzie River origin hypothesis (Bickham et al. 1989).

Support for the hypothesis that wind-driven coastal currents are the primary mechanism for the westward transport of age-0 Arctic ciscoes was provided by Fechhelm and Griffiths (1990), who correlated catch trends with the strength and persistence of east winds. Results of their study suggested that (1) strong east winds are necessary for good recruitment of age-0 Arctic ciscoes to the central Alaska Beaufort Sea, and (2) weak east winds may be insufficient for transporting large numbers of fish into this area. These results are consistent with observed correlations between coastal wind patterns during the year of recruitment and subsequent commercial fishery catches 5–7 years later in the Colville River delta (Fechhelm and Fissel 1988).

Fechhelm and Griffiths (1990) characterized wind conditions for each recruitment season as the net easterly wind components for the period 1 July to 15 August, which they considered to be the critical interval for recruitment to the Prudhoe Bay region. Because this approach yielded such a clear distinction between good and poor recruitment years, the results provided encouragement for the prospect of predicting both time of arrival and strength of recruitment by considering only the physics of the wind-driven motions of the coastal ocean.

If age-0 Arctic ciscoes are totally passive travelers, then their movements along the Beaufort Sea coast are governed only by advection and turbulent diffusion, which serve to transport and disperse them along the coast in response to wind. Wind-induced surface currents are approximately 3% of the wind speed, such that even moderate breezes of 5–10 m/s (10–20 knots) produce surface currents of 15–30 cm/s, the speed at which age-0 Arctic ciscoes would move along the coast.

The presumption that the age-0 Arctic ciscoes are passive travelers is based on our understanding of swimming abilities of small fish. Bernatchez and Dodson (1987) report the swimming speed for runs of James Bay ciscoes *C. artedi* to be about one body length per second (actual value = 0.86 body length per second). Age-0 Arctic ciscoes are typically about 10 cm in length upon arrival in the Prudhoe Bay region. If we assume that the age-0 Arctic ciscoes have an average length of 5–6 cm during their passage from the Mackenzie River delta to Prudhoe Bay, we conclude that their average swimming speed for the trip is about 5–6 cm/s, a substantially lower speed compared to that caused by wind-driven transport.

If wind-driven currents are primarily responsible for movement of age-0 Arctic ciscoes, wind-based predictions of the appearances (timing and abundance) of these fish along the Beaufort Sea coast should agree well with the observed appearances. Our purpose was to evaluate a model of particle diffusion in coastal currents from the Mackenzie River delta to Prudhoe Bay by comparing the predicted and observed appearances of age-0 Arctic ciscoes in various years.

Methods

Movements of age-0 Arctic ciscoes along the Beaufort Sea coast are assumed to be analogous to the transport of particles in the turbulent flow of a narrow stream of fluid. Gross movements are the result of wind-induced advection of the coastal waters, whereas localized dispersal is the result of turbulent diffusion processes that are related to the length and time scales of the flow and the distribution of the fish. The following analysis represents the flow as one-dimensional and aligned with the alongshore direction; that is, cross-shore variations in either the flow or in the concentrations of fish are not considered. Many studies of this recruitment support the notion that most of the age-0 Arctic ciscoes are migrating westward within a narrow band of water adjacent to the Alaska coast (e.g., Fechhelm and Griffiths 1990).

The flow described above is typical of the situation found in rivers and narrow estuaries. Advection and turbulent dispersion of particles in such one-dimensional flows can be described mathematically (Ippen 1966) by

$$\partial C/\partial t + u(\partial C/\partial x) = D(\partial^2 C/\partial x^2); \qquad (1)$$

C = "concentration" (number of fish per unit length of coastline);
t = time;
u = wind-induced current speed;
x = alongshore coordinate, measured west–northwest (288° true azimuth) along the Beaufort coast from the Mackenzie River delta (see Figure 1);
D = turbulent diffusion coefficient.

Equation (1) was programmed to compute daily concentrations of fish as a function of date and location along the Beaufort coast. The wind-induced current speed, u, was estimated as 3% of the daily average of the alongshore component of the wind velocity, an approximate relationship that is supported by observational evidence (e.g., Wiegel 1964). The diffusion coefficient, D, was determined as a function of the diffusion length scale which, in turn, depends on both the size of the fish "cloud" and the fluctuations in the current speed.

The diffusion coefficient was computed according to

$$\log_{10} D = 1.0 + (4/3 \times [\log_{10} L - 3.0]), \qquad (2)$$

in which L is the smaller of two length scales, one being a measure of the size of the fish cloud and the other representing the excursion amplitude of the current caused by fluctuations in its speed (Tennekes and Lumley 1972).

The fish cloud length scale (L_f) is defined as the standard deviation of the instantaneous alongshore concentration profile of fish. The other length scale (L_c) is due to the current and is obtained by determining the distance between two particles (i.e., fish), one traveling at the mean current speed and the other traveling at the instantaneous current speed. In the present version of the model, L_c was calculated using the maximum excursion length that occurred within a given year. There are many alternatives for representation of L_c, but experience shows that little difference in the results is obtained by more elaborate schemes for its computation (Fischer et al. 1979).

As already noted, the choice of length scale for computation of the diffusion coefficient is according to

$$L = \min [L_f, L_c], \qquad (3)$$

and we note that L is a function of time because L_f varies with time. This is consistent with our intuitive understanding of turbulent diffusion—that is, when the fish cloud is small, it is subject to dispersion only by current fluctuations that are similar to its own size or smaller; as the fish cloud grows in size, it is subject to dispersion by ever larger current fluctuations, which are themselves ultimately limited by the largest fluctuations that actually occur.

The computational model of equation (1) was driven by daily averages of the alongshore component of wind velocity, with calculation of the advective current speed and the diffusion coefficient as outlined above. The distribution of fish along the

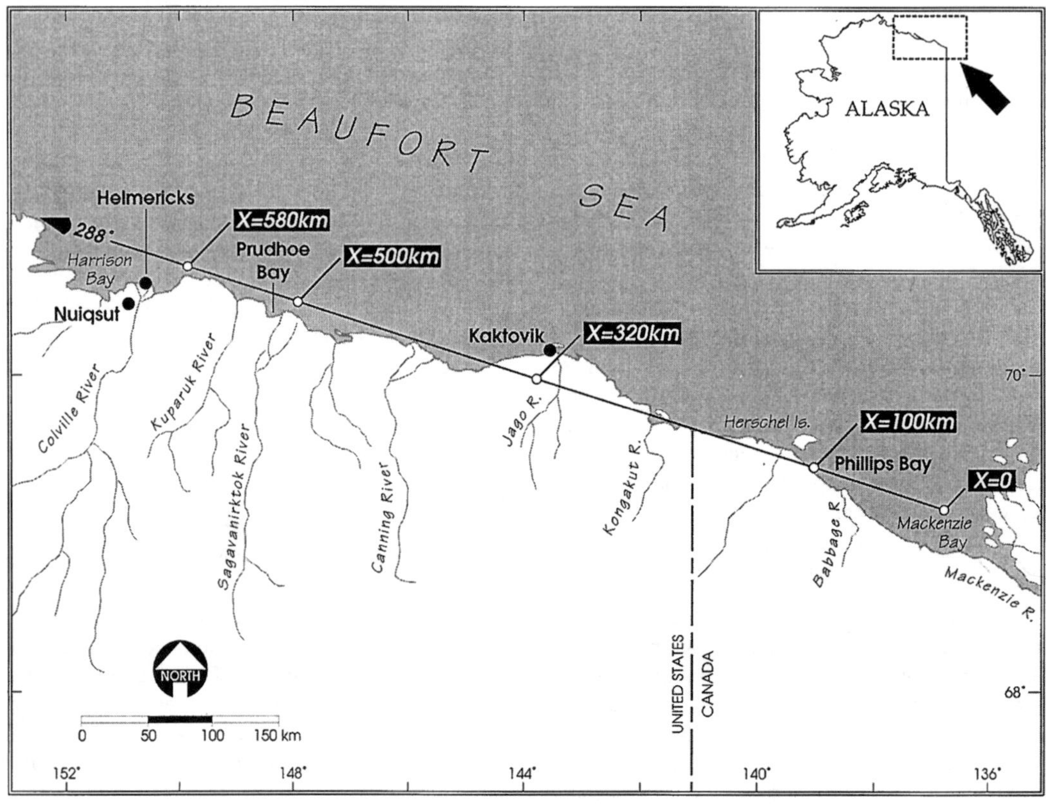

FIGURE 1.—Study area, with distance coordinate system (X = alongshore coordinate).

Beaufort Sea coast was represented by numerical values of fish concentration as a function of distance and time, $C(x, t)$. The initial concentration profile of fish for all computations was 1,000,000 fish distributed uniformly along the 40-km width of the Mackenzie River delta. The "pulse," or rate of discharge, of fish from the delta was taken for most computations to be 20 d, but we also experimented with pulses as short as 5 d. Experience with dispersion computations of this type shows that the precise shape of the initial concentration profile is relatively unimportant, because dispersion soon obliterates the original geometrical characteristics of the profile.

The start date for the fish pulses from the Mackenzie River delta can be important because early-season wind variations can have a significant influence on arrival dates at downstream locations. The exodus of age-0 Arctic ciscoes from the Mackenzie River delta is believed to coincide with the river ice breakup, the annual dates of which were not available for this study. However, the best estimates of start date and pulse length are that the age-0 Arctic ciscoes depart from the Mackenzie River delta between late May and late June during the breakup of winter ice cover, with peak "discharge" occurring about mid-June (W. Bond, Canada Department of Fisheries and Oceans, personal communication).

For most of the computations reported herein, the pulse was assumed to begin on 1 June; however, when arrival time information was available (e.g., as a result of monitoring activities in the Prudhoe Bay region), we adjusted the Mackenzie departure date to ensure arrival at Prudhoe in accordance with observations. Playing this computational wild card neither enhances nor detracts from evaluation of the model's predictive capabilities; rather, it simply facilitates comparison of the predicted and actual recruitment rates of age-0 Arctic ciscoes at Prudhoe Bay.

Results

Using wind data from Deadhorse (Alaska) airport for the years 1970–1991, the westward migration of age-0 Arctic ciscoes along the Alaska Beaufort coast was estimated in terms of (1) pulse characteristics and arrival times at selected locations from Phillips Bay (Yukon Territory, Canada) to the Colville River, (2) corresponding catch rates at Prudhoe Bay, and (3) representation of individual year-classes in the commercial fishery in the Colville River delta. Several examples of these computations serve to illustrate the role of wind-driven currents in supporting either weak or strong recruitment to the Prudhoe Bay region and, ultimately, to the Colville River fisheries.

Movement of age-0 Arctic ciscoes along the Alaska Beaufort coast in 1990, as predicted by the model, is shown in Figure 2. The 1990 recruitment of these fish to the Prudhoe Bay region was the largest that had been observed in a decade of monitoring. For the 1990 computation, the departure of fish from the Mackenzie River delta was adjusted to 20 June to cause the predicted arrival date at Prudhoe Bay to coincide with the observed arrival date. Predicted daily catch rates, shown as a normalized catch per unit effort (CPUE), are shown in Figure 3 with actual data (displayed as dots) pooled for the Prudhoe Bay region. Another presentation of this comparison is provided in Figure 4, in which the actual and predicted cumulative CPUEs are shown. The cumulative CPUE presentation (Figure 4) provides a better basis for evaluating model predictions than the daily catch rate (Figure 3) because localized short-term fluctuations in catch rates are effectively smoothed out of the actual data by the cumulative catch presentation.

Figures 5–7 depict comparisons of actual and predicted cumulative CPUE for 1985, 1986, and 1987, respectively. For all of these years, the departure dates from the Mackenzie River delta were adjusted by trial-and-error to obtain coincidence of predicted and observed arrival dates in the Prudhoe Bay region.

The usefulness of a model such as this can be evaluated in terms of its ability to predict catch rates at locations where such data exist. Two sources of data were available for this purpose: (1) annual fish-monitoring data from the Prudhoe Bay region (1982–1991) for comparison of catch rates in terms of CPUE, yielding information on current-year recruitment; and (2) commercial and subsistence fisheries data from the Colville River delta (1970–1992), providing information on year-class representation in the fisheries.

Table 1 summarizes results for 10 years, 1982–1991, in which Prudhoe Bay fish-monitoring data could be compared to model predictions. Recruitment strengths for this period were based upon fyke net catches of age-0 Arctic ciscoes as number of fish caught per net per 24 h of operation. During this period the strongest recruitment occurred in 1990, when the CPUE was 340 fish · net^{-1} · d^{-1}. In four other years (1983, 1985, 1986, and 1987) there were moderate-to-strong recruitments, with CPUEs ranging from 24 to 154 fish · net^{-1} · d^{-1}. The remaining 5 years (1982, 1984, 1988, 1989, and 1991) had virtually zero recruitments of age-0 Arctic ciscoes. For model evaluation purposes the latter years, except 1991, were termed "weak" recruitments, whereas all of the former plus 1991 were termed "strong" recruitments. The 1991 recruitment is a special case that is discussed below.

Predicted recruitments for the 1982–1991 period are listed in Table 1 as percent of total arriving in the Prudhoe Bay region at biweekly intervals from 30 July to 30 September. Predicted recruitment was considered strong if 40% of the year-class arrived by 1 September and weak otherwise. Using this criterion, predictions were deemed to agree with observations for 8 of the 10 years, although the case for 1991 deserves additional explanation.

In 1991 fish-monitoring activities around Prudhoe Bay were curtailed on 1 September, without any apparent recruitment of age-0 Arctic ciscoes. When monitoring resumed in June 1992, early-season, large catches of age-1 Arctic ciscoes showed that there probably had been "a modest recruitment after 1 September 1991" (LGL 1993). Because the model predicted arrival of 59% by 30 September, this was considered sufficient to judge the model in agreement with observation.

Predicted and observed recruitment strengths of age-0 Arctic ciscoes in the Colville River commercial fisheries are listed in Table 2. Records maintained by J. W. Helmericks (Fairbanks, Alaska, personal communication), the operator of the commercial fishery, include catch data by species for the period 1969 to 1992. Since 1985 the Nuiqsut village subsistence fishery has been monitored in conjunction with extensive monitoring of causeways in the Prudhoe Bay region.

Contributions of Arctic ciscoes to each of these fisheries were calculated from CPUEs and age data reported by Craig and Haldorson (1981) for the 1970–1972 year-classes and Moulton et al. (1993) for 1978–1987 year-classes. Consequently, the contributions of 13 year-classes are available to evaluate the capability of the fish dispersion model to predict recruitment to the Colville River fisheries. Each CPUE is the representation of that year-class in all the years where data are available for at least

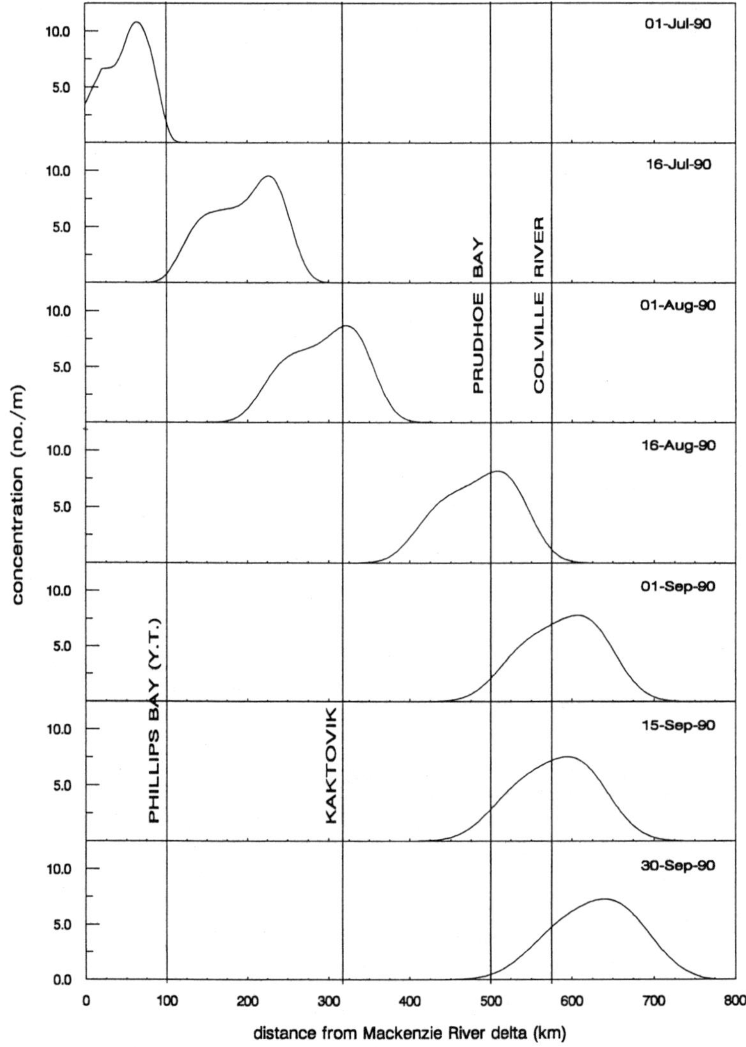

FIGURE 2.—Movement of age-0 Arctic ciscoes *Coregonus autumnalis* along Alaska Beaufort Sea coast for summer 1990 wind conditions. Abbreviations are no. = number and Y.T. = Yukon Territory.

2 of the 3 years from age 5 to 7, which are the ages of maximum contribution to the fishery.

Total CPUEs for year-classes 1970–1972 and 1978–1987 fell into two distinct groups: for 4 of those years, the total CPUE ranged from 0.3 to 9.3 fish $\cdot$ net^{-1} $\cdot$ d^{-1}, with all but one of them being less than or equal to 1.4; for the other 9 years, the total CPUE ranged from 35 to 97 (Table 2). For the purpose of judging agreement between observed and predicted year-class recruitments, years of the former group (CPUE $<$ 10 fish $\cdot$ net^{-1} $\cdot$ d^{-1}) were deemed weak recruitment years, and those of the latter group (CPUE $>$ 35 fish $\cdot$ net^{-1} $\cdot$ d^{-1}) were deemed strong recruitment years.

Recruitments predicted by the wind transport model also fell into two distinct groups (Table 2). In 8 of the 13 years, the model predicted at least 50% recruitment to the Colville River delta by the end of September. For the other 5 years, the predicted recruitment was zero. Accordingly, predicted recruitments in the former category were deemed strong and those in the latter category were deemed weak.

As indicated in Table 2, for 10 of the 13 year-classes there is judged to be agreement between observed and predicted recruitments; for 7 of those years both actual and predicted recruitments were strong, whereas for the other 3 years,

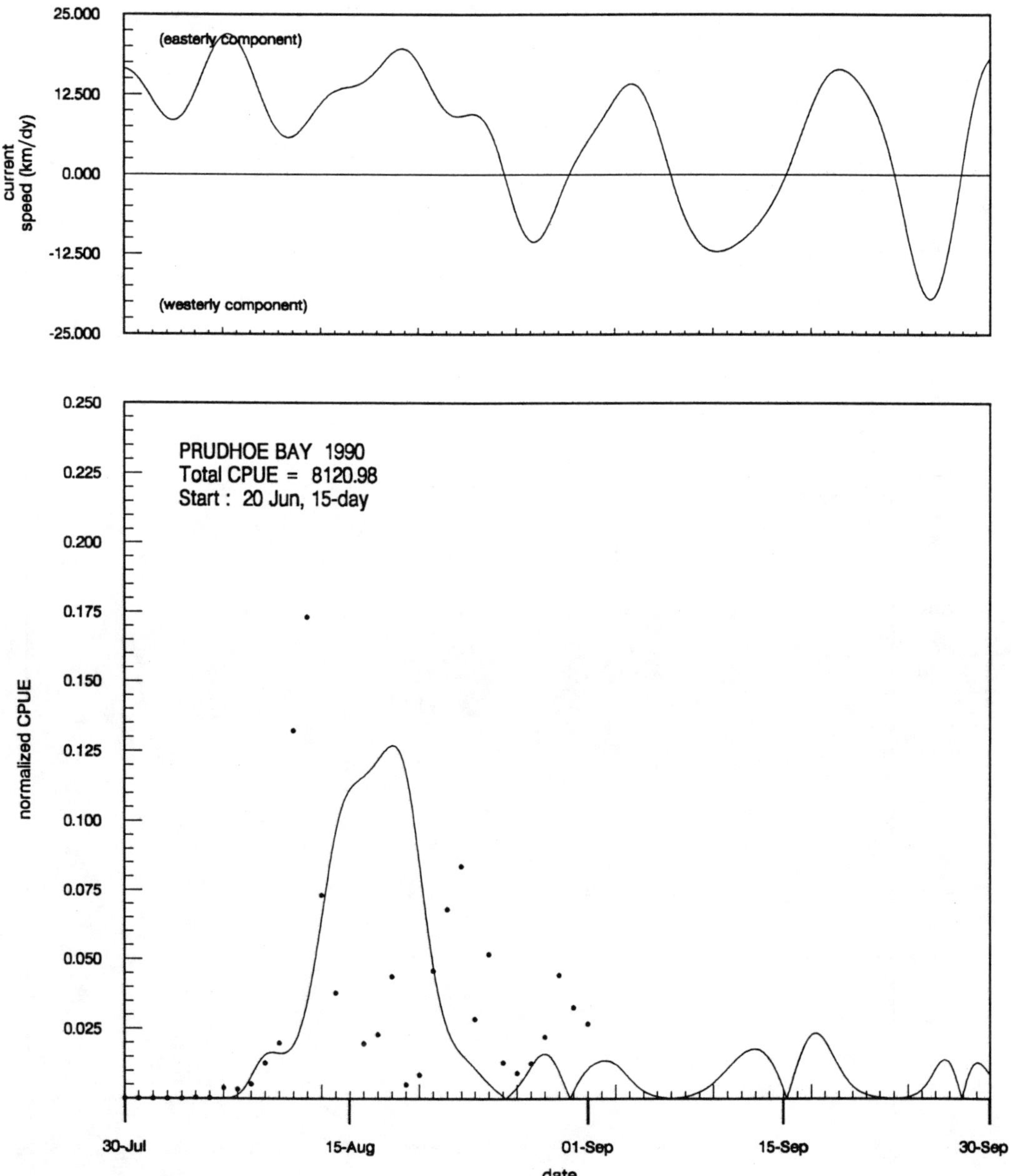

FIGURE 3.—Wind-driven current speed and Prudhoe Bay catch per unit effort (CPUE; dots = actual, line = model) for 1990.

both observed and predicted recruitments were weak. For two of the three nonagreements (1970 and 1983), the observed recruitments were strong, but the predicted recruitments were weak. The third nonagreement was the prediction of a strong 1982 year-class recruitment, which failed to occur.

Discussion

The fish transport model described herein relies only on the daily alongshore wind speed component for computing the movement and dispersal of age-0 Arctic ciscoes along the Alaska Beaufort Sea coast from their origin in the Mackenzie River delta.

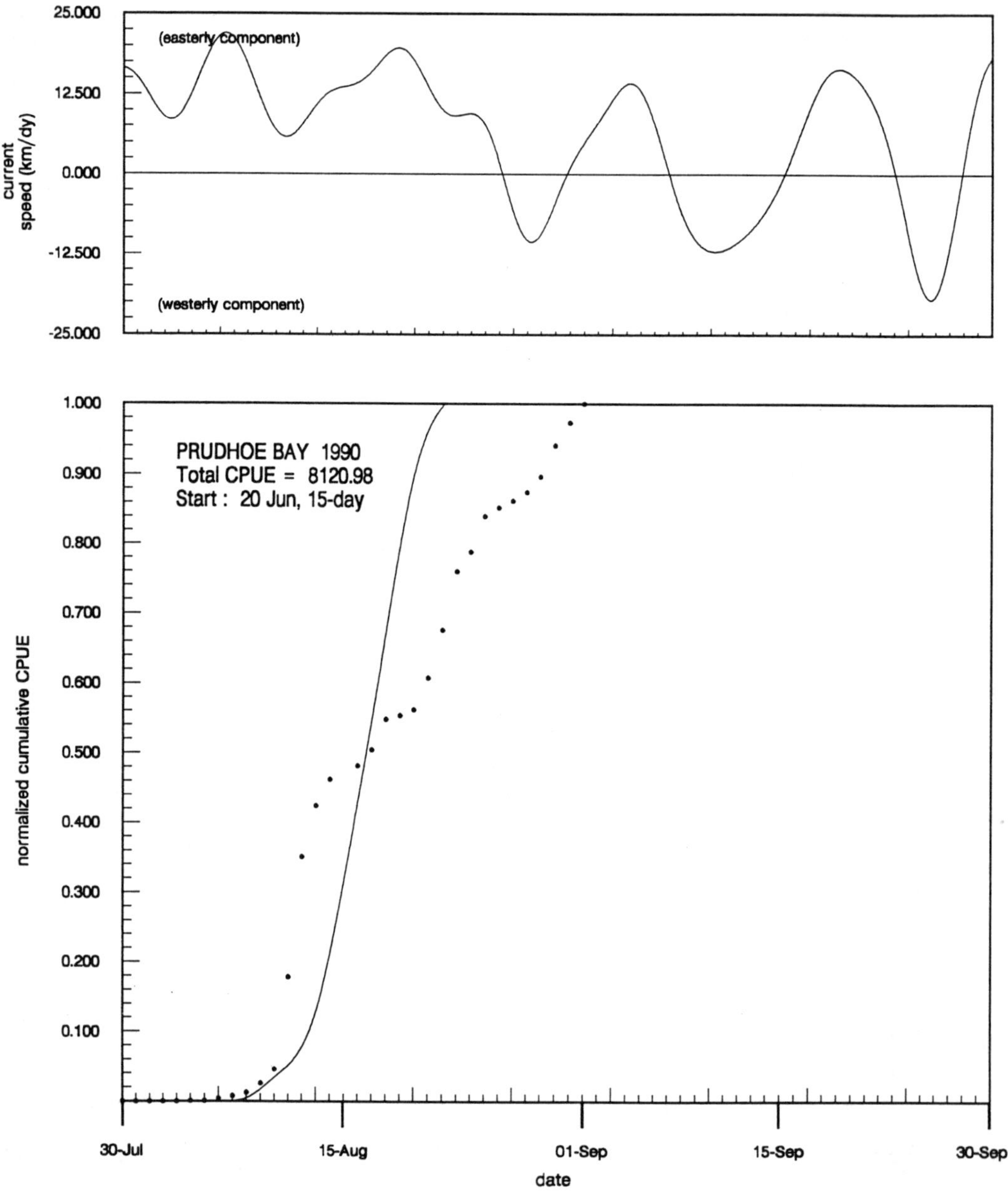

FIGURE 4.—Wind-driven current speed and Prudhoe Bay cumulative catch per unit effort (CPUE; dots = actual, line = model) for 1990.

Given the success of earlier efforts to correlate Arctic cisco recruitment with net winds on a seasonal basis, the general success of this effort should not be a surprise. However, in evaluating the performance of this model, it should be recognized that there are four distinct conditions of agreement or nonagreement between observations (which are regarded as "truth") and predictions (which are necessarily "suspect"). Tables 3 and 4 are a type of correlation matrix in which all observations and predictions of year-class recruitment strengths are categorized according to these conditions of agreement and nonagreement.

The most satisfying condition for agreement is

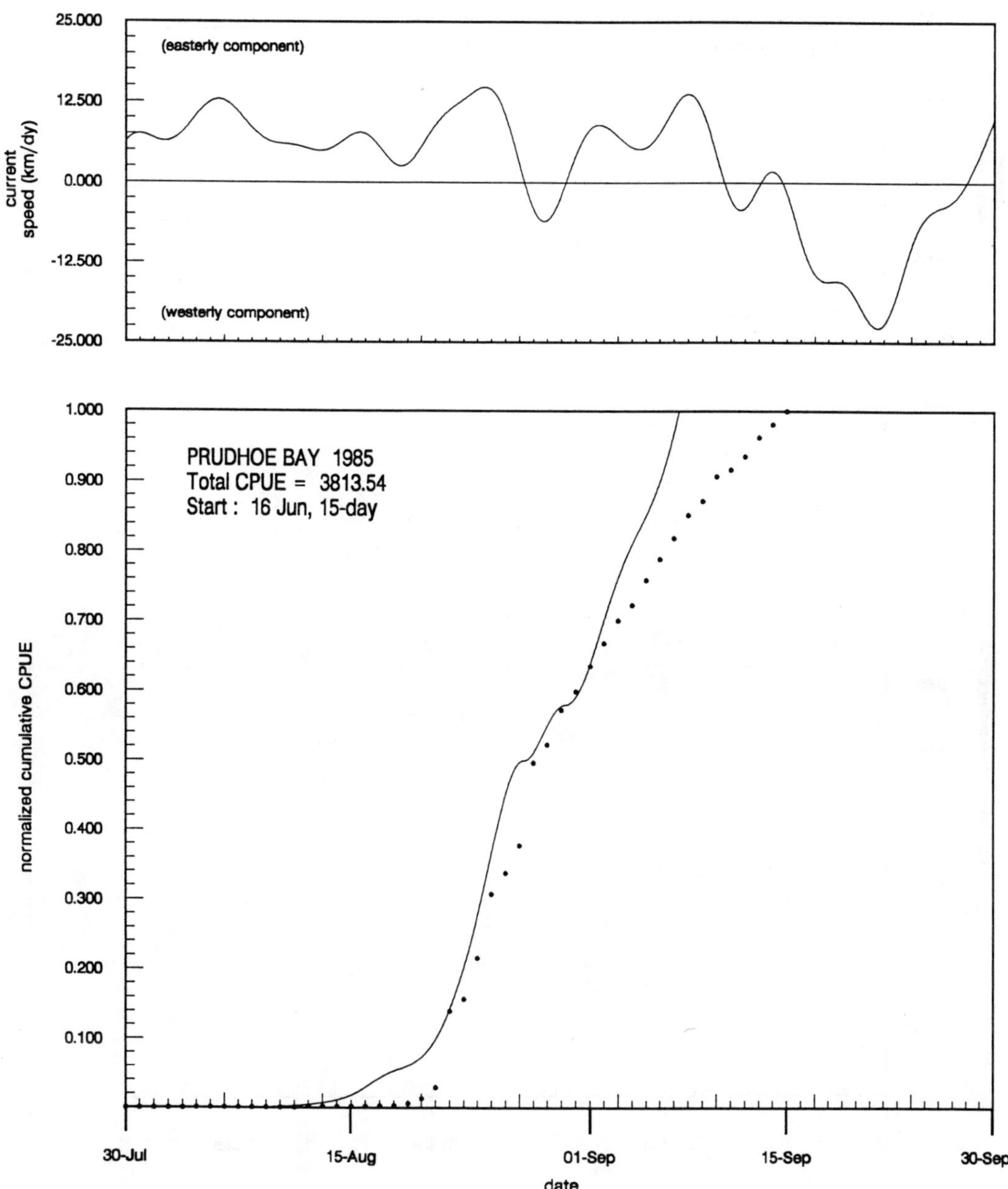

FIGURE 5.—Wind-driven current speed and Prudhoe Bay cumulative CPUE for 1985. Abbreviation and symbols same as in Figure 4.

when a prediction of strong recruitment is supported by observations of strong recruitment; this condition is subsequently referred to as the "strong–strong" condition. When both prediction and observation indicate weak recruitment, the "weak–weak" condition, some doubt remains as to whether the agreement is for the correct reason or if it is just fortuitous.

Of the two conditions for nonagreement, the easier one to reconcile is the "strong–weak" condition, which occurs when the prediction for strong recruitment is not supported by observations; in this case,

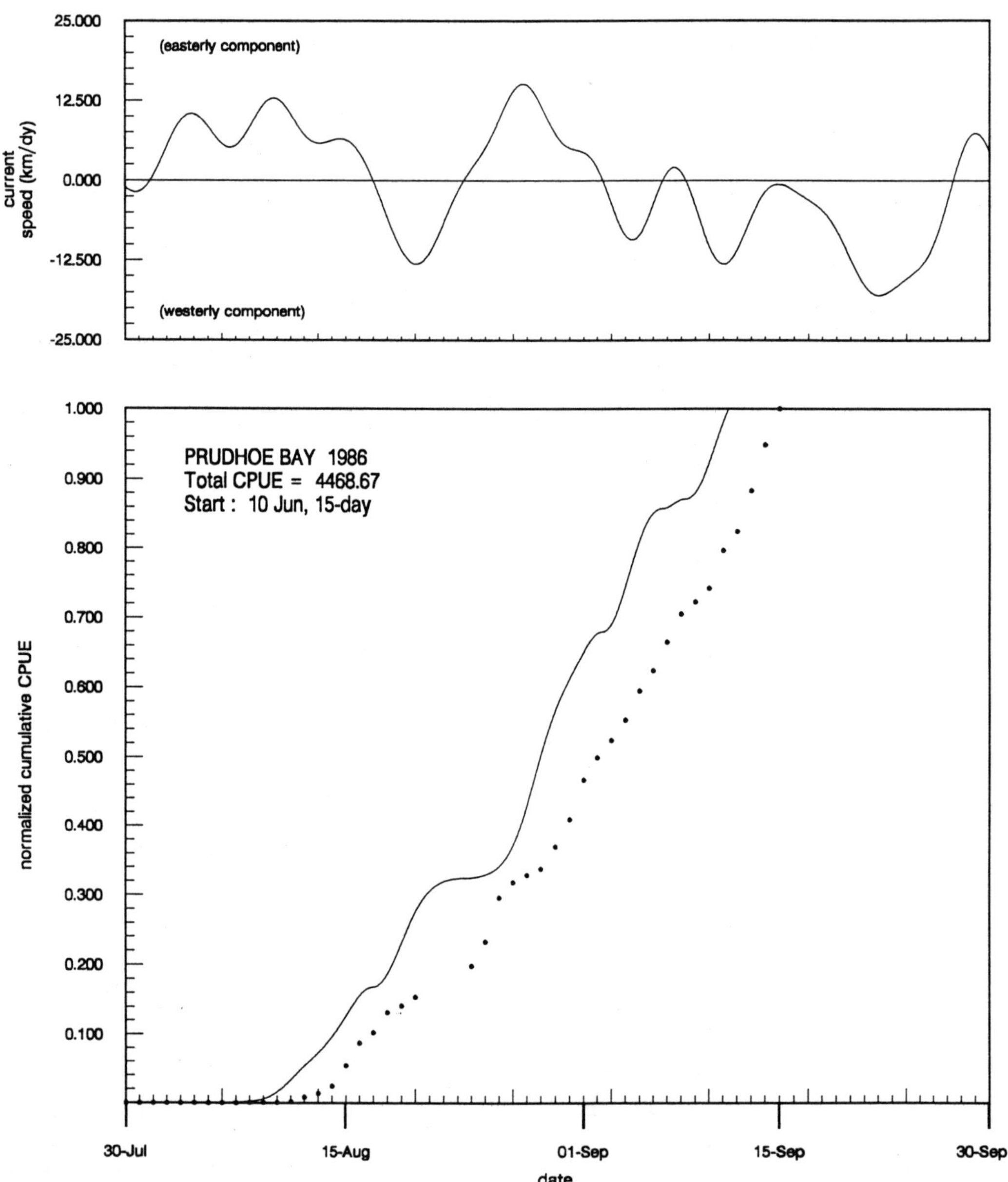

FIGURE 6.—Wind-driven current speed and Prudhoe Bay cumulative CPUE for 1986. Abbreviation and symbols same as in Figure 4.

the nonagreement is probably attributable to factors that are not incorporated in the model. For example, a late breakup of sea ice would inhibit the capability of winds to induce favorable surface currents, or the year-class is not as large as usual for any number of physical or biological reasons. One year, 1982, definitely exhibits this condition (Tables 3 and 4) and it could be argued that 1991 also belongs in this category. The observed weak recruitment of age-0 Arctic ciscoes to Prudhoe Bay in 1982 is further corroborated by the very weak representation of that year-class in the subsequent Colville River delta fisheries.

The remaining case for nonagreement is the

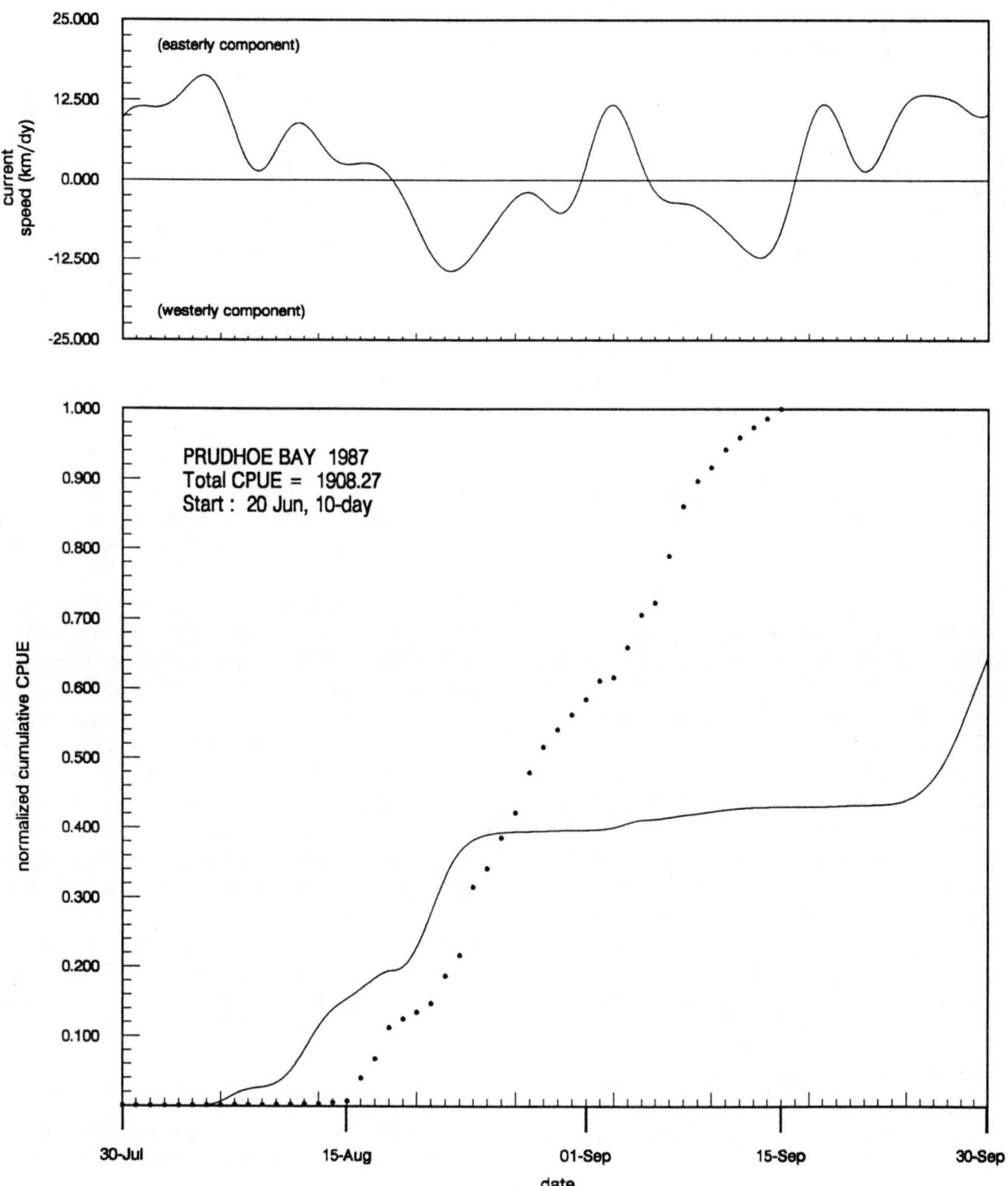

FIGURE 7.—Wind-driven current speed and Prudhoe Bay cumulative CPUE for 1987. Abbreviation and symbols same as in Figure 4.

"weak–strong" condition, which is the occasion of an observed strong recruitment that apparently occurs without benefit of favorable wind-driven currents to support a migration all the way from the Mackenzie delta. Two year-classes, 1970 and 1983, exhibit the "weak–strong" condition (Tables 3 and 4). Both had moderately strong showings in the Colville fishery (Table 2), but winds during the summers of 1970 and 1983 were not sufficient to transport the young fish much beyond Kaktovik (Figures 8 and 9), which is almost 200 km east of Prudhoe Bay.

Given the demonstrated importance of wind-driven currents to westward migrations of age-0

TABLE 1.—Comparison of predicted and observed recruitment of age-0 Arctic ciscoes *Coregonus autumnalis* to the Prudhoe Bay region, 1982–1991.

Year	Observed		Predicted percent arrival in Prudhoe Bay						Agreement?
	Arrival date	CPUE[a]	Start date	30 Jul	15 Aug	1 Sep	15 Sep	30 Sep	
1982	n/a	0	1 Jun	28	40	50	81	100	No
1983	27 Aug	24.2	1 Jun	0	0	0	0	0	No
1984	n/a	0	1 Jun	0	0	0	0	0	Yes
1985	16 Aug	144.0	16 Jun	0	2	68	100	00	Yes
1986	10 Aug	153.6	10 Jun	0	13	66	100	00	Yes
1987	13 Aug	67.3	20 Jun	0	16	40	44	64	Yes
1988	3 Sep	1.1	1 Jun	1	1	2	6	26	Yes
1989	30 Aug	1.9	1 Jun	0	0	0	0	0	Yes
1990	5 Aug	340.0	20 Jun	0	34	100	100	100	Yes
1991	n/a	0[b]	1 Jun	0	0	0	20	9	Yes[b]

[a]CPUE = catch per unit effort, which is the number of fish caught per net per day, upon arrival of age-0 Arctic ciscoes.
[b]Monitoring ceased on 1 Sep 1991, but resumption of monitoring in Jun 1992 showed that a modest recruitment of age-0 Arctic ciscoes had occurred after 1 Sep 1991.

Arctic ciscoes from the Mackenzie River, we cannot suddenly discount the importance of wind in those years that appear to favor an eastward transport. Indeed, one must wonder if the 1970 and 1983 recruitments originated from some place west of the Colville River. Although direct evidence for this notion is nonexistent, there is some information to suggest it.

First, some long-term residents of Barrow believe that Arctic ciscoes spawn in the area and, in fact, subsistence fishermen in Barrow sometimes take large, mature Arctic ciscoes in their fall fishery. Second, Lockwood and Bickham (1991, 1992) and Morales et al. (1993) all provide genetic evidence that a small population of Arctic ciscoes occurs in western Alaska. These fish are differentiated at the stock level from fish that originate in the Mackenzie River of Canada. These observations argue for a spawning Arctic cisco population west of the Colville River.

The notion that the 1970 and 1983 year-classes originated west of the Colville River presents a new set of questions. For example, why are these fish not seen more frequently in years that favor eastward transport? Do these fish provide a regular but heretofore unrecognized contribution to the Colville River fisheries? What is their geographic origin? The available information is insufficient to answer these questions, but the results of this modeling exercise, along with the cited genetic research results, provide the foundation for a hypothesis that there is a spawning population of Arctic ciscoes west of the Colville River. The issue is worthy of attention because of its potential ramifications for impact assessments of coastal and offshore developments in the central Alaskan Beaufort Sea.

TABLE 2.—Comparison of predicted and observed recruitment strengths of age-0 Arctic ciscoes *Coregonus autumnalis* year-classes 1970–1972 and 1978–1987 in the Colville River delta commercial and subsistence fisheries.

Year-class	Total CPUE[a]	Predicted recruitment (%) by			Agreement
		1 Sep	15 Sep	30 Sep	
1970	35.6	0	0	0	No
1971	9.3	0	0	0	Yes
1972	51.1	10	49	51	Yes
1978	96.9	30	64	100	Yes
1979	59.6	0	1	78	Yes
1980	58.0	100	100	100	Yes
1981	1.2	0	0	0	Yes
1982	0.3	3	19	63	No
1983	40.0	0	0	0	No
1984	1.4	0	0	0	Yes
1985	44.8	34	75	76	Yes
1986	39.1	71	77	87	Yes
1987	36.8	97	99	100	Yes

[a]Total CPUE (catch per unit effort, which is the number of fish caught per net per day) for all years in which year-class was identified: Craig and Haldorson (1981) for 1970–1972 and Moulton and Field (1994) for 1978–1987.

Conclusions

The advective–dispersion model developed for this analysis provides valuable insight into the role of wind-driven currents in moving and dispersing age-0 Arctic ciscoes along the Alaska Beaufort Sea

TABLE 3.—Correlation of observations and model predictions of recruitment strengths of age-0 Arctic ciscoes *Coregonus autumnalis* in Prudhoe Bay region monitoring by year-class.

Observations of recruitment strength	Model predictions of recruitment strength	
	Strong	Weak
Strong	1985, 1986, 1987, 1990	1983
Weak	1982	1984, 1988, 1989, 1991

TABLE 4.—Correlation of observations and model predictions of recruitment strengths of age-0 Arctic ciscoes *Coregonus autumnalis* in Colville River delta commercial and subsistence fisheries by year-class.

Observations of recruitment strength	Model predictions of recruitment strength	
	Strong	Weak
Strong	1972, 1978, 1979, 1980, 1985, 1986, 1987	1970, 1983
Weak	1982	1971, 1981, 1984

coast. Although evaluation of the model's capability to predict arrival times and recruitment strengths is only semiquantitative, the results provide additional support for the proposition that age-0 Arctic ciscoes are recruited westward under the usually prevailing easterly winds from the Mackenzie River delta region into Alaska as passive travelers, with their primary mode of transport being only wind-driven surface currents. Results for those years when east-

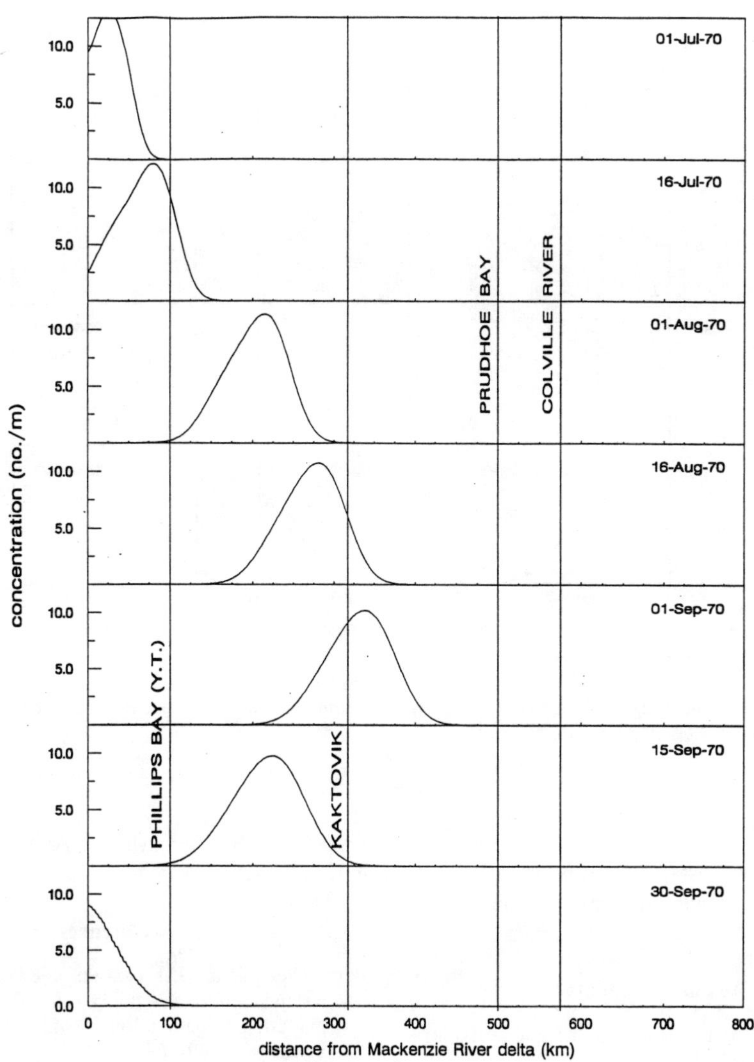

FIGURE 8.—Movement of age-0 Arctic ciscoes *Coregonus autumnalis* along Alaska Beaufort Sea coast for summer 1970 wind conditions.

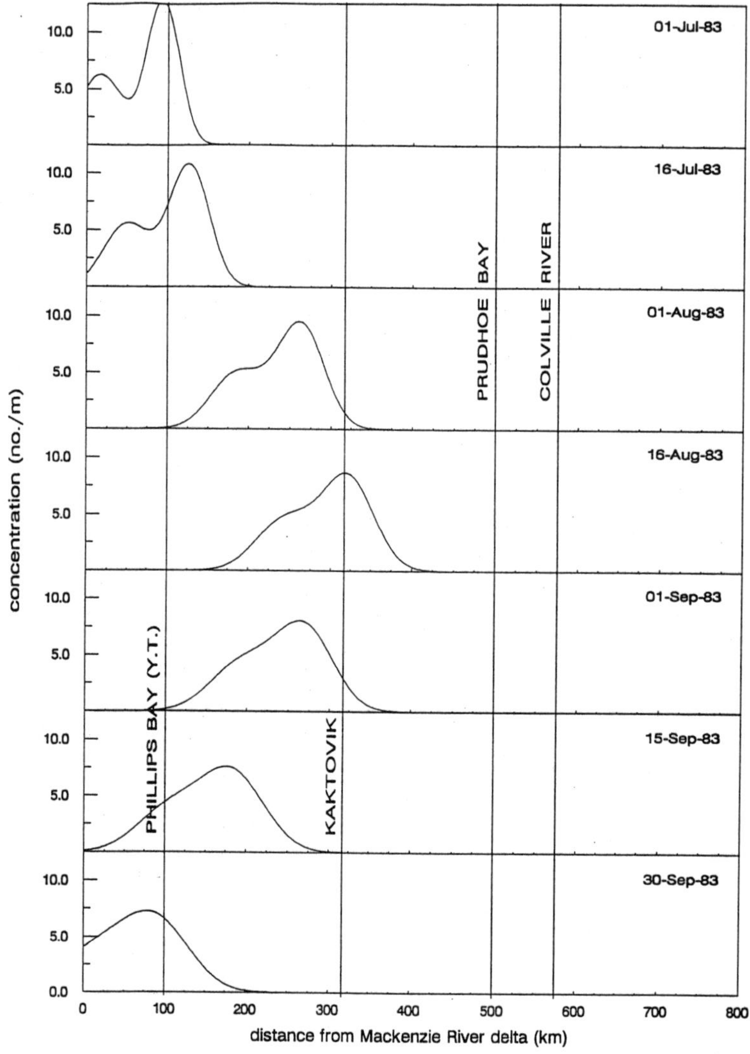

FIGURE 9.—Movement of age-0 Arctic ciscoes *Coregonus autumnalis* along Alaska Beaufort Sea coast for summer 1983 wind conditions. Abbreviations are as in Figure 2.

ward transport prevailed (i.e., under west winds) support a hitherto unstated hypothesis of an Arctic cisco spawning population west of the Colville River.

Acknowledgments

We are grateful to Alan Niedoroda and Christopher Reed of Woodward-Clyde Consultants (Tallahassee, Florida) for their contributions to the mathematical formulation of this analysis. The continuing support of BP Exploration (Alaska), Inc. for research on the behavior of Arctic ciscoes is also gratefully acknowledged.

References

Bernatchez, L., and J. J. Dodson. 1987. Relationship between bioenergetics and behavior in anadromous fish migration. Canadian Journal of Fisheries and Aquatic Sciences 44:349–407.

Bickham, J. W., S. M. Carr, B. G. Hanks, D. W. Burton, and B. J. Gallaway. 1989. Genetic analysis of population variation in the Arctic cisco (*Coregonus autumnalis*) using electrophoretic, flow cytometric, and mitochondrial DNA restriction analyses. Biological Papers of the University of Alaska 23:112–122.

Bond, W. A., and R. N. Erickson. 1987. Fishery data from Philips Bay, Yukon, 1985. Canadian Data Report Fisheries and Aquatic Sciences 635.

Cannon, T. C., B. Adams, D. Glass, and T. Nelson. 1987. Fish distribution and abundance. Chapter 5 *in* 1985 Endicott environmental monitoring program final report, volume 6. Prepared by Envirosphere Company for U.S. Army Corps of Engineers Alaska District, Anchorage.

Craig, P. C., and L. Haldorson. 1981. Beaufort Sea barrier island ecological process studies: Simpson Lagoon, part 4. Pages 384–678 *in* Environmental assessment of Alaskan continental shelf, volume 7. NOAA (National Oceanic and Atmospheric Administration) OCSEAP (Outer Continental Shelf Environmental Assessment Program) BLM (Bureau of Land Management), Boulder, Colorado.

Fechhelm, R. G., and D. B. Fissel. 1988. Wind-aided recruitment of Canadian Arctic cisco (*Coregonus autumnalis*) into Alaskan waters. Canadian Journal of Fisheries and Aquatic Sciences 45:906–910.

Fechhelm, R. G., and W. B. Griffiths. 1990. Effect of wind on the recruitment of Canadian Arctic cisco-(*Coregonus autumnalis*) into the central Alaskan Beaufort Sea. Canadian Journal of Fisheries and Aquatic Sciences 47:2164–2171.

Fischer, H. B., E. J. List, R. C. Y. Koh, J. Imberger, and N. H. Brooks. 1979. Mixing in inland and coastal waters. Academic Press, New York.

Gallaway, B. J., W. B. Griffiths, P. C. Craig, W. J. Gazey, and J. W. Helmericks. 1983. An assessment of the Colville River delta stock of Arctic cisco—migrants from Canada? Biological Papers of the University of Alaska 21:4–23.

Ippen, A. T., editor. 1966. Estuary and coastline hydrodynamics. McGraw-Hill, New York.

LGL (LGL Alaska Research Associates, Inc.). 1993. The 1991 Endicott development fish monitoring program, volume I: analysis of fyke net data. Prepared by LGL for BP Exploration (Alaska), Inc. and North Slope Borough, Anchorage, Alaska.

Lockwood, S. F., and J. W. Bickham. 1991. Genetic stock assessment of spawning Arctic cisco (*Coregonus autumnalis*) populations by flow cytometric determination of DNA content. Cytometry 12:260–267.

Lockwood, S. F., and J. W. Bickham. 1992. Genome size in Beaufort Sea coastal assemblages of Arctic ciscoes. Transactions of the American Fisheries Society 121: 13–20.

Morales, J. C., B. G. Hanks, J. W. Bickham, and B. J. Gallaway. 1993. Allozyme analysis of population structure in Arctic cisco (*Coregonus autumnalis*) from the Beaufort Sea. Copeia 1993:863–867.

Moulton, L. L. 1989. Recruitment of Arctic cisco (*Coregonus autumnalis*) into the Colville delta, Alaska. Biological Papers of the University of Alaska 24:107–111

Moulton, L. L., and L. J. Field. 1994. The 1992 Colville River fishery. *In* The 1992 Endicott development fish monitoring program, volume II. Prepared by MJM Research for BP Exploration (Alaska), Inc. and North Slope Borough, Anchorage, Alaska.

Moulton, L. L., L. C. Lestelle, and L. J. Field. 1993. The 1991 Colville River fishery. *In* The 1991 Endicott development fish monitoring program, volume II. Prepared by MJM Research for BP Exploration (Alaska), Inc. and North Slope Borough, Anchorage, Alaska.

Reub, G. S., J. D. Durst, and D. Glass. 1991. Fish distribution and abundance. *In* The 1987 Endicott environmental monitoring program, final report, volume 6. Prepared by Envirosphere Company for U.S. Army Corps of Engineers Alaska District, Anchorage.

Scott W. B., and E. J. Crossman. 1973. Freshwater fishes of Canada. Bulletin of the Fisheries Research Board of Canada 184.

Tennekes, H., and J. L. Lumley. 1972. A first course in turbulence. The MIT Press, Cambridge, Massachusetts.

Wiegel, R. L. 1964. Oceanographical engineering. Prentice-Hall, Englewood Cliffs, New Jersey.

American Fisheries Society Symposium 19:104–108, 1997

Relative Abundance and Maturity of Lake Trout from Walker Lake, Gates of the Arctic National Park and Preserve, Alaska, 1987–1988

F. Jeffrey Adams
U.S. Fish and Wildlife Service, King Salmon Fishery Resources Office
Post Office Box 277, King Salmon, Alaska 99613, USA

Abstract.—To provide information for management, gill nets and angling were used to investigate the population of lake trout *Salvelinus namaycush* in Walker Lake, south-central Brooks Range, Alaska, during summers 1987–1988. Significantly more lake trout were captured during spring than fall. Significantly more lake trout were captured at stream mouths and straight shorelines than at the outlet of the lake. Large lake trout were more abundant along shorelines, whereas small lake trout were more abundant in offshore areas. Earliest maturity occurred at age 12 for both sexes, with minimum lengths of 422 mm for females and 435 mm for males. Females spawned every other year. This population was similar to other arctic populations of lake trout in habitat preferences, maturity, and spawning frequency. The population of lake trout in Walker Lake appears to be healthy but is susceptible to overharvest.

Lake trout *Salvelinus namaycush* can attain great size, are notably admired as a trophy fish, and are valued as a food fish (Martin and Olver 1980). The species is confined to the northern temperate and subarctic-arctic freshwater systems of North America (Ryder 1972). Slow growth, late maturity, and intermittent spawning make this species vulnerable to overexploitation (Martin and Olver 1980).

Most information about lake trout comes from southern Canada and the Great Lakes region, where lake trout populations experience drastically different climatic conditions than Alaskan stocks (Burr 1987a). Few reports are available concerning lake trout in Alaska, and these lack substantial information about the fish's habitat preferences and reproductive characteristics.

The goal of this study was to provide information for management and conservation of lake trout in Walker Lake. To meet this goal, the study objectives were to (1) estimate relative abundance as an indicator of seasonal and habitat preferences, and (2) describe maturity and spawning frequency.

Study Area

Walker Lake (67°08′N, 154°20′W) is located in Gates of the Arctic National Park and Preserve in the south-central Brooks Range of northern Alaska. The lake is 60 km north of the Arctic Circle and lies in the subarctic climate zone of the boreal forest (taiga) (R. G. Ring and coworkers, Gates of the Arctic National Park and Reserve, unpublished). It is an oligotrophic lake (Jones et al. 1990) that is 21 km long with a surface area of 3,800 ha (R. E. Reanier and P. M. Anderson, University of Washington, unpublished). The lake is contained in a narrow, glacially carved valley and has maximum depths of 120 m in the north basin and 122 m in the south basin. The middle area of the lake is relatively shallow and contains several islands. The lake is drained by a tributary of the Kobuk River.

Methods

Relative Abundance

Sampling in 1987 was not random and could not be used for relative abundance estimation. However, sampling in 1988 was strictly designed to estimate relative abundance by stratifying the lake into shoreline and offshore (pelagic) areas.

Shoreline sampling consisted of two 3-week seasons: an early season from 13 June to 4 July and a late season from 24 August to 15 September. Double (end to end) monofilament experimental gill nets (a 45.7- × 4.9-m net with 15-m panels of 1.3-, 1.9-, and 2.5-cm bar mesh; and, on the deeper end, a 45.7- × 4.9-m net with 3.8-cm bar mesh) were attached to shore and set on the bottom, perpendicular to the contour. Nets were fished for 4 h per set and were checked every 20 min to minimize fish mortality.

The shoreline of the lake was stratified into six types: (1) straight—areas with no abrupt change in shoreline, (2) peninsulas—areas of increased shoreline development, (3) stream mouths—areas where permanent creeks (except the inlet) entered the lake, (4) islands—areas where islands or shoals existed, (5) inlet—area where Kaluloktok Creek enters the lake, and (6) outlet—area where an unnamed creek drains the lake. Sample sites were chosen in proportion to the amount of each shoreline type in the lake with a minimum of two sample sites per type. Thirty-four sample sites were randomly chosen. Each site was fished twice for a total of 8 h of effort per sampling season. The difference in catch per unit effort (CPUE) between sampling seasons was tested with the Wilcoxon paired sample test ($\alpha = 0.05$) (Zar 1984). The differences between CPUEs for shoreline features were tested with the nonparametric Tukey multiple comparison test ($\alpha = 0.05$) (Zar 1984). Significant differences in CPUEs were interpreted as indications of seasonal or habitat preferences.

Offshore sampling was performed from 6 July to 2 August 1988 with two experimental gill nets (45.7 × 4.9 m with 15-m panels of 1.3-, 1.9-, and 2.5-cm bar mesh). One net was set on the bottom of the lake in depths ranging from 40 to 115 m. The second net was concurrently set approximately halfway between the surface of the lake and the bottom in depths from 20 to 50 m. Twenty-two sample sites were randomly chosen; these were divided between the two basins in proportion to the pelagic area of the two basins. Nets were initially set for 1 h to assess the number of mortalities, then each site was fished an additional 24 h. The difference between CPUEs for bottom and midwater sets was tested with the Mann–Whitney U-test ($\alpha = 0.05$) (Zar 1984). A significant difference in CPUEs was interpreted as an indication of a habitat preference.

Maturity

Sex was determined by inspection of the fresh gonads from lake trout sacrificed for a related study during 1987 and 1988; age of these fish was estimated by counting annuli on otoliths. Sex was also noted for fish that released gametes during handling. Fish were captured by gill nets and angling.

Lake trout that would have spawned in the season of capture or that had spawned in the past were considered mature. Females with egg diameters larger than 1 mm and mesovaria that were stretched and fragmented were considered mature (Martin and Olver 1980). A female with an ovary containing retained eggs from a previous spawning was also considered mature. Males with flattened testes and a maximum testis width greater than 3 mm were classified as mature (Martin and Olver 1980). Fish whose sex could be determined but were not mature were classified as immature. Fish whose sex could not be determined were classified as juvenile.

TABLE 1.—Estimates of relative abundance for lake trout *Salvelinus namaycush* from Walker Lake, 1988. Catch per unit effort (CPUE) cannot be compared between shoreline and offshore areas because of differences in gill net deployment.

Factor or feature	Number of sets	Number captured	Effort (h)	CPUE (fish/h)	
				Mean	Range
		Shoreline sites			
Factor					
Early season	68	179	272	0.66	0–2.25
Late season	68	144	272	0.53	0–2.00
Feature					
Stream mouth	24	79	96	0.82	0–2.00
Straight	48	123	192	0.64	0–2.25
Peninsula	40	85	160	0.53	0–1.75
Island	8	17	32	0.49	0–1.75
Inlet	8	15	32	0.47	0–1.50
Outlet	8	4	32	0.13	0–0.25
Total per factor	136	323	544		
		Offshore sites			
Factor					
Midwater	22	1	550	0.002	0–0.04
Bottom	22	11	550	0.02	0–0.16
Total per factor	44	12	1,100		

Results

Relative Abundance

Three hundred twenty-three lake trout were captured in 136 gill net sets (544 h of effort) during shoreline sampling in 1988 (Table 1). Of these, 179 fish were captured during the early sampling season and 144 during the late season. Fork lengths ranged from 312 to 850 mm (Figure 1). The CPUE for the early season was significantly greater than that during the late season ($P < 0.05$) (Table 1). For the combined sampling seasons in 1988, only the CPUEs of the stream mouth and straight shoreline areas were significantly greater than the CPUE from the outlet ($P < 0.05$).

At offshore sites, 44 gill net sets (1,100 h of effort) captured 12 lake trout (Table 1). One lake trout (309 mm) was captured in the midwater sets, whereas 10 small lake trout (203–323 mm) and one large lake trout (667 mm) were captured from the bottom sets (Figure 1). However, only one of these fish (323 mm) was captured at a depth greater than 50 m. The 11 small lake trout were juveniles. There was no significant difference in CPUEs between bottom and midwater sets ($P > 0.05$).

Maturity

Of the 80 lake trout whose gonads were inspected, 15% were juveniles, 15% immature females, 25% mature females, 17% immature males,

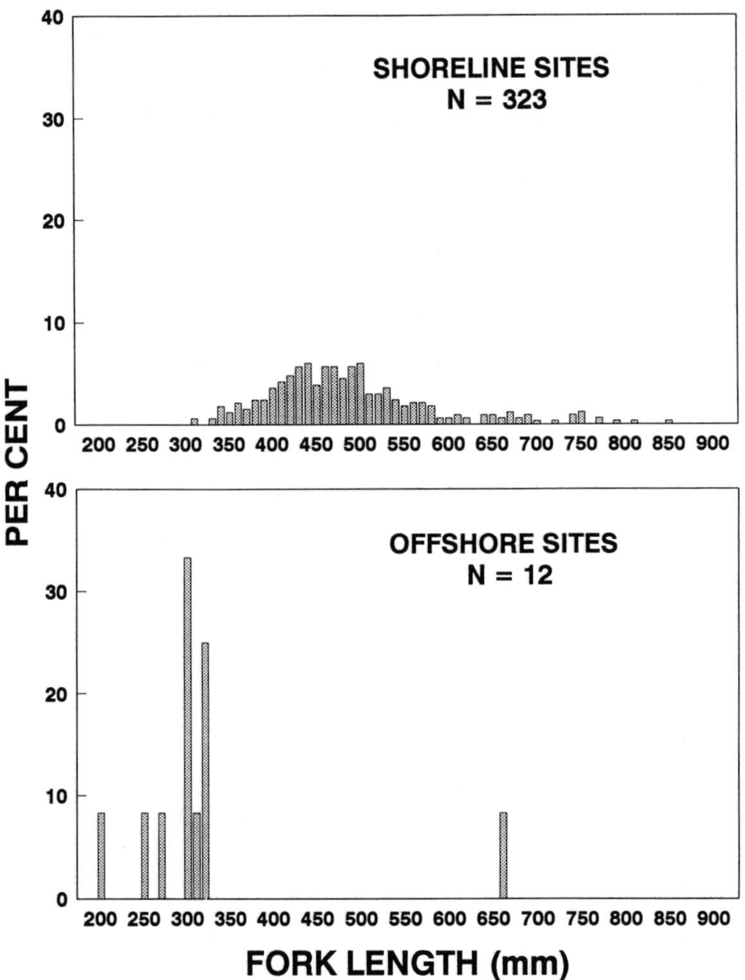

FIGURE 1.—Length frequencies of lake trout *Salvelinus namaycush* captured with gill nets in Walker Lake, 1988. *N* is number of fish captured.

and 28% mature males. The ranges of fork lengths and ages were, respectively, 203–426 mm and 5–14 years for juveniles; 425–494 mm and 11–15 years for immature females; 422–795 mm and 12–25 years for mature females; 428–456 mm and 11–14 years for immature males; and 435–805 mm and 12–26 years for mature males. All lake trout larger than 525 mm were mature. All females older than age 15 were mature; all males older than age 14 were mature.

No females released eggs during handling, but three fish contained residual eggs. Lengths of these fish ranged from 695 to 795 mm, with ages from 18 to 25 years. The females that contained residual eggs in their body cavities also contained small (1–2 mm) developing eggs in their ovaries. These fish apparently had spawned during the previous fall but would not spawn during the upcoming fall. The other mature females did not contain residual eggs but did contain large (3–5 mm) developing eggs. These fish apparently had not spawned the previous fall but would spawn during the upcoming fall. Therefore, it appears that female lake trout in Walker Lake do not spawn consecutively. The most frequently that spawning could occur would be every other year.

Thirteen males were confirmed as spawners when they released milt during handling. Fork lengths ranged from 454 to 818 mm, but these fish were not sacrificed for aging. Spawning frequency for males could not be determined.

Discussion

Relative Abundance

In subarctic–arctic areas, lake trout concentrate on their spawning grounds along shorelines from mid-August through October (Johnson 1972; Martin and Olver 1980). This suggested that gill net sets made in the late season would capture significantly more lake trout than sets made earlier in the season. However, the CPUE during the early season was significantly greater than that during the late season. This observation agrees with the work of Martin (1952), who found that the number of lake trout captured was higher after ice-out than before freeze-up. He attributed this difference to greater feeding activity after ice-out. The lower CPUE for the late season sets suggests that lake trout may have moved to feeding, staging, or spawning grounds in areas that were not sampled. Therefore, CPUE may not be a good indicator of spawning abundance or behavior.

Preferences for shoreline or offshore areas of the lake were related to the size of lake trout. Johnson (1972) reported that in northern Canada, large lake trout (>400 mm) were captured along shorelines, whereas small lake trout (<400 mm) were captured in offshore areas. These findings were similar to results from Walker Lake where small lake trout constituted 92% of the fish captured at offshore sites, but only 14% of the fish captured at shoreline sites. Large lake trout enter the shallow portions of a lake to feed (Martin and Olver 1980), and their presence may cause small lake trout to flee and seek refuge in the offshore areas.

The capture of few small lake trout is common for lake trout studies (Martin and Olver 1980; Burr 1987a) because these fish are sparsely distributed throughout the offshore areas of a lake (Peck 1982). Also, the sampling gear fished only a small fraction of the water column. Therefore, the combination of low abundance and inefficient sampling resulted in few captures.

Small lake trout feed on planktonic and benthic invertebrates (Martin and Olver 1980). Ten of the 11 small lake trout captured in offshore nets were captured in bottom sets in water less than 50 m deep. Apparently, small lake trout in Walker Lake were also benthic feeders but did not frequent the deepest parts of the lake.

Catch rates at particular shoreline features illustrated behavioral characteristics of lake trout. The high CPUE at stream mouths probably resulted from availability of drift from the stream. The drift would not only provide food for lake trout, but it would also attract other species that are forage for lake trout. Straight shorelines had high CPUEs probably because they were the predominant feature of the lake shoreline. A lake trout moving between areas would pass through this shoreline type frequently. The outlet area had the lowest CPUE because it was shallow and provided no cover for lake trout.

Maturity

Length and age of first maturity for lake trout in Walker Lake were similar to the findings for other Alaska populations. Lake trout from subarctic Alaska mature at lengths from 308 to 526 mm, and the youngest age at maturity ranges from 5 to 20 years (Burr 1987b, 1989). Lake trout from Arctic lakes in the Brooks Range reached first maturity between 340 and 500 mm and between 10 and 15 years of age (Craig and Wells 1975; P. McCart and H. Bain, University of Calgary, and P. Craig, University of California, unpublished).

Nonconsecutive spawning in lake trout is common in northern populations. Females in Great Slave Lake spawned every other year (Kennedy 1954), and in Great Bear Lake they spawned every third year (Miller and Kennedy 1948). A population from an alpine lake in Alberta contained females that might spawn every 4 years (Donald and Alger 1986), whereas females in Keller Lake, Northwest Territories, may spawn only three times in a lifetime (Johnson 1972). Female lake trout from populations in the Brooks Range were suspected to spawn every other year (Craig and Wells 1975; McCart, Bain, and Craig, unpublished). Although spawning frequency for male lake trout in Walker Lake could not be determined, males from an alpine lake in Alberta spawned every year (Donald and Alger 1986).

The population of lake trout from Walker Lake exhibited characteristics that were typical of an arctic lake trout population. Their slow growth, late maturity, and intermittent spawning make them a fragile resource that could be easily overexploited (Martin and Olver 1980), especially during spring when large fish are available along shorelines. The current status of the population appears to be healthy, and the remoteness of the lake should allow the population to remain stable; however, conservative management is recommended.

Acknowledgments

I thank the biologists, technicians, and volunteers who assisted with the project. Special thanks are extended to the U.S. Fish and Wildlife Service's Fairbanks and King Salmon Fishery Resources offices and Gates of the Arctic National Park and Preserve.

References

Burr, J. M. 1987a. Synopsis and bibliography of lake trout in Alaska. Alaska Department of Fish and Game Fishery Manuscript 5, Juneau.

Burr, J. M. 1987b. Stock assessment and biological characteristics of lake trout populations in interior Alaska, 1986. Alaska Department of Fish and Game Fishery Data Series 35, Juneau.

Burr, J. M. 1989. Stock assessment and biological characteristics of lake trout populations in interior Alaska, 1988. Alaska Department of Fish and Game Fishery Data Series 99, Juneau.

Craig, P. C., and J. Wells. 1975. Fisheries investigations in Chandalar River region, northeast Alaska. Pages 67–75 *in* P. C. Craig, editor. Fisheries investigations in a coastal region of the Beaufort Sea. Arctic Gas Biological Report Series 34, Calgary, Alberta.

Donald, D. B., and D. J. Alger. 1986. Stunted lake trout from the Rocky Mountains. Canadian Journal of Fisheries and Aquatic Sciences 43:608–612.

Johnson, L. 1972. Keller Lake: characteristics of a culturally stressed salmonid community. Journal of the Fisheries Research Board of Canada 29:731–740.

Jones, J. R., J. D. LaPerriere, and B. D. Perkins. 1990. Limnology of Walker Lake and comparisons with other lakes in the Brooks Range, Alaska. Verhandlungen Internationale Vereinigung fur Theoritische Ange Wandte Limnologie 24:302–308.

Kennedy, W. A. 1954. Growth, maturity and mortality in the relatively unexploited lake trout of Great Slave Lake. Journal of the Fisheries Research Board of Canada 11:827–852.

Martin, N. V. 1952. A study of the lake trout in two Algonquin Park, Ontario, lakes. Transactions of the American Fisheries Society 81:111–137.

Martin, N. V., and C. H. Olver. 1980. The lake charr. Pages 205–277 *in* E. K. Balon, editor. Charrs: salmonid fishes of the genus *Salvelinus*. Dr. W. Junk, The Hague, Netherlands.

Miller, R. B., and W. A. Kennedy. 1948. Observations on the lake trout of Great Bear Lake. Journal of the Fisheries Research Board of Canada 74:176–189.

Peck, J. W. 1982. Extended residence of young-of-the-year lake trout in shallow water. Transactions of the American Fisheries Society 111:775–778.

Ryder, R. A. 1972. The limnology and fishes of oligotrophic glacial lakes in North America (about 1800 A.D.). Journal of the Fisheries Research Board of Canada 29:617–628.

Zar, J. H. 1984. Biostatistical analysis. Prentice-Hall, Englewood Cliffs, New Jersey.

American Fisheries Society Symposium 19:109–118, 1997

Growth, Density, and Biomass of Lake Trout in Arctic and Subarctic Alaska

JOHN M. BURR

Alaska Department of Fish and Game, Sport Fish Division
1300 College Road, Fairbanks, Alaska 99701–1599, USA

Abstract.—Growth of lake trout *Salvelinus namaycush* in 11 Alaskan lakes was characterized and compared with that of other North American populations. Asymptotic length varied from 517 to 984 mm and was positively correlated with length at 50% maturity (LM_{50}), lake surface area, and fish species richness. Incremental growth of subadults was correlated with LM_{50} only. Estimates of density of adult lake trout from 7 of the 11 lakes varied from 3.1 to 31.9 fish/ha. Adult density was inversely related to lake surface area and to fish species richness. The complexity of the fish community, which is strongly related to lake surface area, is an important factor influencing maximum size, maturity schedules, and population density in these lake trout populations.

Populations of lake trout *Salvelinus namaycush* are characterized by slow growth, late maturity, and low density. The species is primarily limited to oligotrophic environments, which, together with its life history traits, results in low annual productivity. Growth rate, size at maturity, age at maturity, and population density of lake trout are variable and reflect the productivity of the environment and the food habits of the population (Martin and Olver 1980).

Lake trout in Alaska represent the westernmost extent of the native range of the species. Information is available from lake trout populations in high-latitude lakes in Canada (e.g., Falk et al. 1974; Bond 1975; Johnson 1976; Yaremchuk 1986), but most of these lakes are very large. The population dynamics of lake trout in Alaskan lakes, which are relatively smaller, are likely to be different.

Lake surface area has a significant effect on the growth, mortality, and productivity of lake trout populations in Ontario (Carl et al. 1990; Payne et al. 1990). The number of predators and competitors of lake trout increases with lake size. Mean size, maximum size, and age at maturity of lake trout also increase with lake size. Hence, lake size, or surface area, can be viewed as a key surrogate factor representing a complexity of factors affecting lake trout populations.

Herein, I summarize the results of stock assessments of lake trout in Alaska and examine some aspects of the relationships of lake area and fish species richness to growth, maturity, and density of lake trout.

Methods

Collection of Samples

Data on length, age, and maturity were collected from lake trout in nine lakes in the Alaska Range in subarctic Alaska and from two lakes in the Brooks Range of arctic Alaska. All lakes except Paxson Lake and Walker Lake are above tree line (Table 1).

Sinking gill nets (3 × 46 m) with panels of 26-, 38-, 51-, and 76-mm (stretch measure) mesh were primarily used to capture lake trout. Lake trout were generally entangled in these nets rather than gilled. Gill nets were generally checked at 0.5-h intervals to minimize netting mortality. Other capture gear included 2.4-m × 0.6-m hoop nets, baited with cut fish, and fyke nets with 1.2-m × 1.2-m square frames and 50-m center leads. Hoop and fyke nets were checked on a daily basis. Spawning fish at Paxson Lake were captured with a 3-m × 100-m beach seine (25-mm mesh) at night.

Samples were collected during mid-June through September 1987–1990. Fork length of all lake trout was recorded to the nearest 1 mm. Lake trout greater than 250 mm that were captured in good condition were marked with individually numbered Floy T-bar tags. The adipose fin was removed from all tagged fish as a second mark. Dead lake trout were weighed and then dissected to obtain otoliths (saggitae) for age determination. Sex and state of maturity of dead fish were determined by field inspection of gonads.

Age, Growth, and Maturity

Ages were determined from whole otoliths using techniques described by Sharp and Bernard (1988) and Burr (1993). Parameters of the von Bertalanffy growth equation were estimated from length and age data for each population using an iterative method to produce least-squares estimates of the parameters (PROC NLIN, SAS 1988). The mean length of the 10 largest lake trout captured from

TABLE 1.—Characteristics of 11 Alaskan lakes ordered by increasing mean age to 50% maturity of lake trout *Salvelinus namaycush*. Abbreviations of fish names are AC = Arctic char *S. alpinus*, BB = burbot *Lota lota*, CS = chum salmon *Oncorhynchus keta*, GR = Arctic grayling *Thymallus arcticus*, HW = humpback whitefish *Coregonus pidschian*, LC = least cisco *C. sardinella*, LS = longnose sucker *Catostomus catostomus*, NP = northern pike *Esox lucius*, RW = round whitefish *Prosopium cylindraceum*, SC = slimy sculpin *Cottus cognatus*, and SS = kokanee *O. nerka*.

Lake	Location	Elevation (m)	Area (ha)	Maximum depth (m)	Additional fish species present
Sevenmile	63°06′N, 145°37′W	925	33	12	BB, SC
Paxson	62°55′N, 145°33′W	778	1,575	26	BB, GR, HW, RW, SC, SS
Summit	63°08′N, 145°33′W	978	1,650	70	BB, GR, RW, SC, SS
Fielding	63°10′N, 145°40′W	906	538	24	BB, GR, RW, SC
Upper Tangle	63°02′N, 146°03′W	868	150	27	BB, GR, LS,RW, SC
Butte	63°10′N, 147°50′W	1,022	318	24	GR, RW, SC
Glacier	63°07′N, 146°15′W	1,124	177	27	BB, GR, RW, SC
Landlocked Tangle	63°00′N, 146°06′W	872	241	37	BB, GR, RW, SC
Twobit	63°08′N, 145°38′W	1,006	109	24	SC
Chandler	68°14′N, 152°44′W	914	1,300	22	AC, BB, GR, RW, SC
Walker	67°08′N, 154°20′W	206	3,800	122	AC, BB, CS, GR, LC, NP, RW, SC

each population was used to set bounds for estimating the asymptotic length (L_∞).

The ages and lengths at maturity were estimated for the 11 lake trout populations by Burr (1993). Lake trout were classified as mature if they would have spawned in the season of capture or if they had spawned in the past. Male lake trout in these populations spawned at a younger age and at a similar or smaller size than did females (Burr 1993), hence only female maturity is considered in the present study. The age (AM_{50}) and length (LM_{50}) at 50% maturity were selected for comparison between populations.

Average change in mean length at age for the 2 years preceding AM_{50} was used as an estimate of annual incremental growth of subadult lake trout. Annual growth of subadult fish was calculated because the age–length data for mature lake trout in some of the populations was extremely variable.

Abundance and Density

At Glacier, Sevenmile, Landlocked Tangle, and Upper Tangle lakes, a modified Petersen mark–recapture estimator (Chapman 1951) was used to estimate population abundance of lake trout (both sexes) larger than 250 mm (Burr 1992). Recruitment was assumed to be minimal when marking and recapture events were performed within a season, as was done in Summit, Fielding, Upper Tangle, Landlocked Tangle, Chandler, and Walker lakes.

Abundance of lake trout in Twobit, Butte, Sevenmile, and Glacier lakes was also estimated with the Petersen estimator but with marking events and recapture events being performed in separate years (Seber 1982). To evaluate recruitment through growth between the marking period and the recapture period (1 year later), a nonparametric method for detecting and culling recruitment was used (Robson and Flick 1965). The abundance of spawning lake trout in Paxson Lake was estimated using the Jolly–Seber model (Seber 1982). The goodness-of-fit method (Seber 1982) was used to detect differences in estimated abundance between years.

To facilitate comparison of abundance between different years and between populations, abundance estimates of lake trout of mature size (LM_{50} and larger) were calculated. The abundance of mature (adult) lake trout in the populations was calculated by reducing the abundance estimate for fish 250 mm and larger by the proportion of fish in the samples that were between 250 mm and the estimated length at 50% maturity (LM_{50}).

The surface area of each lake was estimated from scale 1:63,000 aerial photographs with a computerized digitizing table. Density of adult lake trout (fish/ha) was calculated directly from the abundance estimates and the estimated surface area. Adult biomass (kg/ha) was calculated from the estimated total weight of lake trout of mature size and the lake surface area.

Selected lake trout population characteristics (AM_{50}, LM_{50}, L_∞, and incremental growth) were investigated along with physical characteristics of the lakes (elevation, latitude, surface area) and fish species richness (number of fish species other than lake trout) using simple correlation analysis to examine potential relationships between the population parameters and community and habitat attributes.

TABLE 2.—Maturity and growth of lake trout *Salvelinus namaycush* in 11 Alaskan lakes. N is the sample size. Estimates of AM_{50} and LM_{50} are from Burr (1993). Incremental growth is the mean annual growth for 2 years preceding AM_{50}.

Lake	N	AM_{50} (years)	LM_{50} (mm)	von Bertalanffy parameters t_0	K	L_∞ (mm)	Incremental growth (mm)
Sevenmile	159	5	386	0.45	0.41	519	43
Paxson	392	6	426	−2.2	0.15	619	47
Summit	132	7	452	0.2	0.22	574	43
Fielding	99	8	481	0	0.10	824	31
Upper Tangle	71	8	400	0	0.09	785	37
Butte	113	8	370	−1.1	0.11	603	26
Glacier	406	9	375	0	0.18	454	24
Landlocked Tangle	129	10	350	−2.9	0.10	468	14
Twobit	178	11	343	0.95	0.15	482	16
Chandler	70	12	410	0	0.05	900	26
Walker	98	14	476	2.2	0.05	1,032	32

Results

Growth

Growth of lake trout was variable both within and between populations, particularly for adult fish (Table 2). Slow growth (subadult incremental growth < 20 mm) was observed in Landlocked Tangle and Twobit lakes. Growth was moderate (24–26 mm/year) in Butte, Glacier, and Chandler lakes. Annual growth was relatively rapid (31–47 mm/year) for the remaining six populations.

Incremental growth of immature lake trout was not correlated with age at maturity or any of the habitat attributes (Table 3). A significant positive correlation was detected between incremental growth and length at maturity.

Asymptotic length, estimated by L_∞, was also positively correlated with length at maturity. Positive correlations were also found between L_∞ and lake surface area and fish species richness (Table 3). Asymptotic length was not correlated with age at maturity, elevation, or latitude.

Abundance and Density

Estimates of the abundance of mature lake trout from seven of the study lakes varied from 96 lake trout in Upper Tangle Lake to 5,066 lake trout in Paxson Lake (Table 4). For most lakes, only one estimate of abundance was available. Multiple estimates were calculated for three of the populations: Differences were not significant between estimates of abundance for Glacier Lake ($P = 0.3$) or for Paxson Lake ($P = 0.4$). Differences in estimates of abundance for Sevenmile Lake were grouped by method of estimation; estimates from the single-season Petersen method (1987 and 1991) were less ($P < 0.03$) than were estimates from data across seasons (1988, 1989, 1990; Table 4).

Estimates of the density (and biomass) of lake trout of mature size varied from 0.6 fish/ha (1.2 kg/ha) to 32.8 fish/ha (33.5 kg/ha), with the highest densities in the smallest lakes (Table 4, Figure 1). The estimated density of lake trout in Upper Tangle Lake (0.6 fish/ha) was very low relative to estimates

TABLE 3.—Correlation of maturity and growth of lake trout *Salvelinus namaycush* with characteristics of 11 Alaskan lakes. Pearson correlation coefficients are listed with P values in parenthesis; significant correlations are indicated by asterisks.

	AM_{50}	LM_{50}	Elevation	Latitude	Area	Species richness
AM_{50}		0.20	−0.50	0.71*	0.47	0.33
		(0.56)	(0.11)	(0.02)	(0.14)	(0.32)
LM_{50}			−0.54	0.38	0.70*	0.67*
			(0.09)	(0.24)	(0.02)	(0.03)
Species richness			−0.73*	0.51	0.83*	
			(0.01)	(0.11)	(<0.01)	
Incremental growth	−0.38	0.68*	−0.29	0.06	0.45	0.31
	(0.26)	(<0.02)	(0.39)	(0.86)	(0.05)	(0.35)
Asymptotic length (L_∞)	0.27	0.78*	−0.49	0.52	0.70*	0.90*
	(0.42)	(<0.01)	(0.13)	(0.10)	(0.02)	(<0.01)

TABLE 4.—Estimated abundance, density, and biomass of lake trout *Salvelinus namaycush* of mature size (>LM_{50}) for seven populations in Alaska.

Lake	Year estimated	Estimated Abundance	Estimated SE	Density (fish/ha)	Biomass (kg/ha)
Sevenmile	1987	459	85	13.9	14.2
	1988	792	158	23.9	24.5
	1989	1,054	138	31.9	32.6
	1990	1,084	175	32.8	33.5
	1991	505	73	15.3	15.6
Twobit	1987	1,112	171	10.2	5.2
Upper Tangle	1988	96	17	0.6	1.2
Glacier	1986	1,724	403	9.7	9.9
	1989	1,474	324	8.3	8.5
Landlocked Tangle	1987	1,645	359	6.8	4.3
Butte	1988	2,124	347	6.7	6.7
Paxson	1988	4,895	955	3.1	5.0
	1989	5,066	318	3.2	5.2

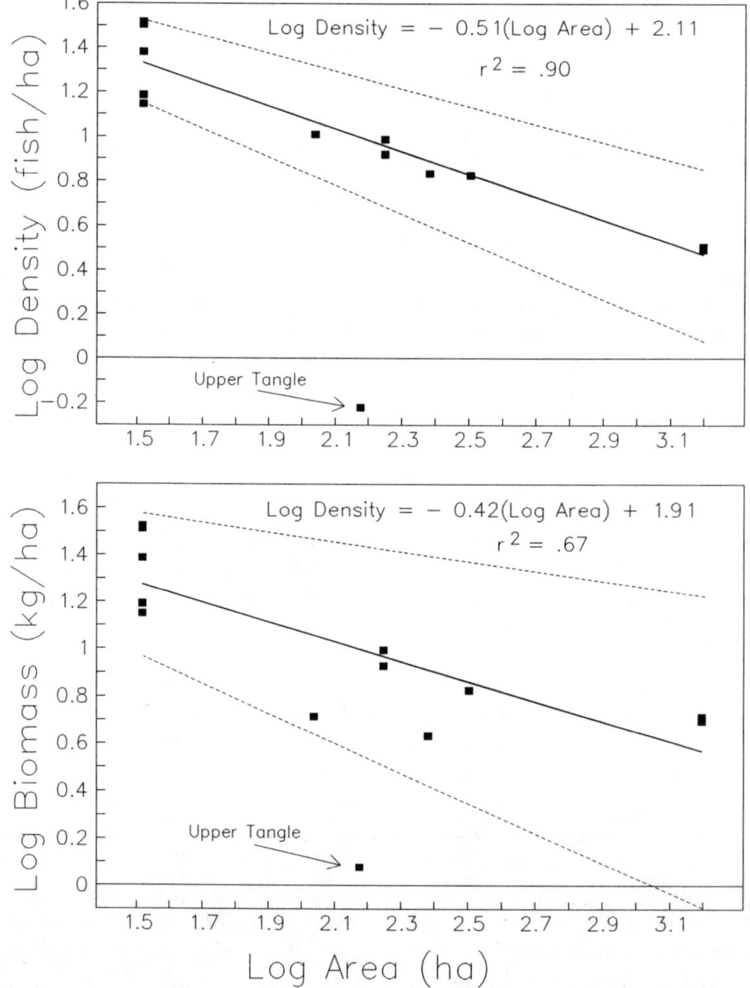

FIGURE 1.—Relationship between lake surface area and lake trout density (**top**) and biomass (**bottom**). Confidence limits (95%) are shown as broken lines.

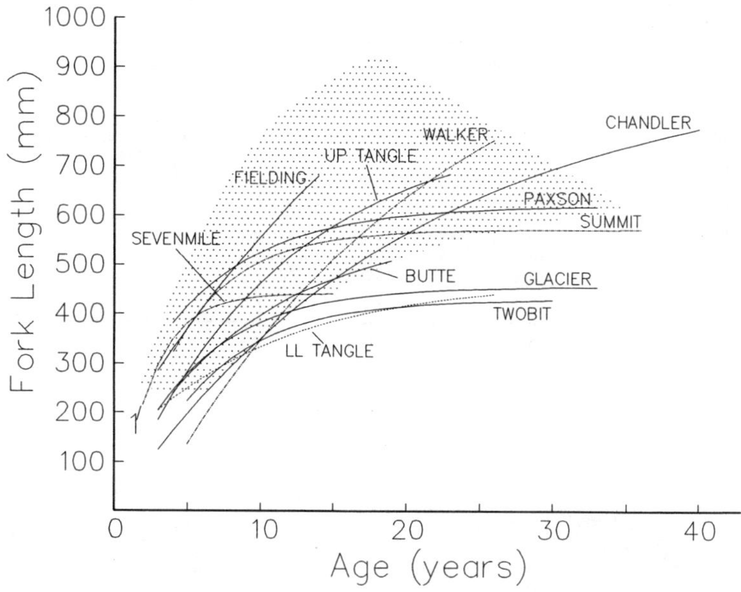

FIGURE 2.—Fitted growth curves for 11 populations in Alaska. Shaded area indicates range of observed growth of lake trout as reported by Healey (1978).

from other populations. A significant inverse relationship between lake surface area and adult density and biomass was evident with the Upper Tangle Lake estimate removed (Figure 1). The relationship between surface area and biomass ($r^2 = 0.67$; $P < 0.01$) was not as strong as that between area and density ($r^2 = 0.90$; $P < 0.01$).

Discussion

Growth and Maturity

Growth rates observed in the Alaskan populations (Figure 2) are within the range reported for the species (Healey 1978; Martin and Olver 1980; Payne et al. 1990). Growth of lake trout from Twobit and Landlocked Tangle lakes was slower than that reported from most other areas. In many populations, growth was rapid for immature fish and slowed markedly for mature fish. However, as shown in Figure 3, growth did not slow with the onset of maturity in the two arctic lakes (Chandler and Walker), in Fielding and Upper Tangle lakes, and for a portion of the fish in Paxson, Summit, and Glacier lakes. The faster growth observed in a portion of the adult lake trout in some of the populations was seen in populations where forage fish were available and where some of the lake trout are presumably piscivorous. The continued fast growth for mature fish may represent a shift in feeding habits from a largely invertebrate diet to a mainly piscivorous one. Fast-growing and slow-growing adults may coexist in most populations but may not have been detected from the samples collected in this study. The slow growth of older fish observed in some of the populations may be associated with the lack of sufficient appropriately sized forage, as suggested by Kerr (1971). The absence or unavailability of suitably sized forage fish for lake trout resulted in a diet of benthic invertebrates, smaller fish (including lake trout), and plankton in many northern populations (Falk et al. 1973; Ball 1988). Such populations may become dominated by small, slow-growing or "stunted" individuals (Donald and Alger 1986; Carl et al. 1990). Small lakes in Alaska are inhabited by few fish species (Table 1). Alaskan lake trout populations in smaller lakes are generally planktivorous and benthivorous; fish species richness and piscivory increases with lake size.

Estimates of AM_{50}, (5–14 years) for lake trout from these 11 Alaskan populations were within the range of values reported for this species from other areas (Burr 1993). Lengths at maturity for the lake trout populations (335–481 mm) were generally less than those observed for populations farther south but were within the range reported.

Incremental growth of subadult fish was posi-

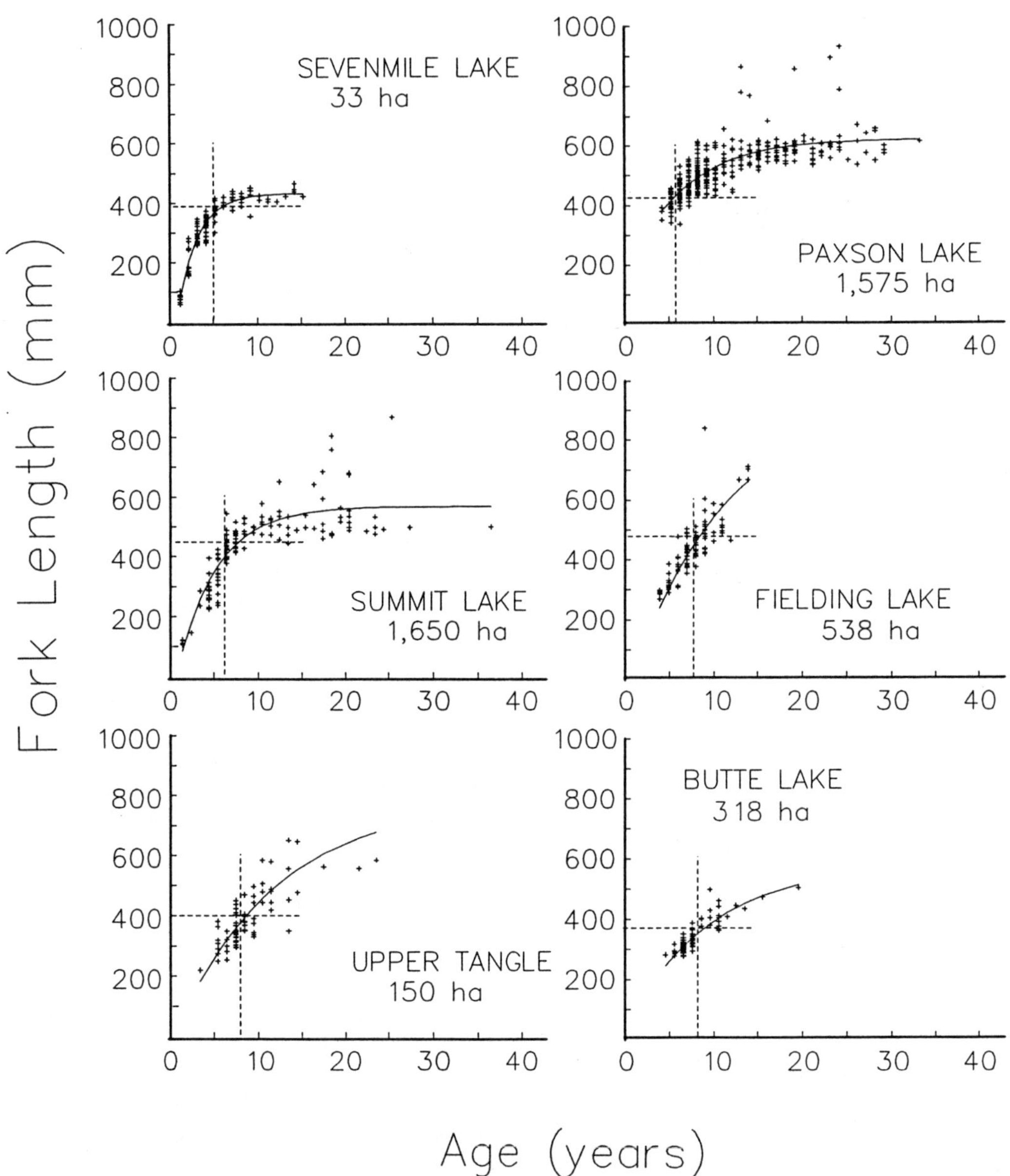

FIGURE 3.—Growth of the lake trout *Salvelinus namaycush* from 11 Alaskan populations. Fitted von Bertalanffy curves are shown with length and age data. Age and length at 50% maturity are shown by dashed lines. Lake area (ha) is shown on each graph.

tively related to LM_{50} (Table 3). Differences in age at maturity among populations have been attributed to differences in growth rates such that age at maturity and size at maturity are inversely related (Alm 1959; Hanson and Wickwire 1967; Paterson 1968). In contrast, Healey (1978) found that size at maturity and age at maturity were not correlated. In some populations (primarily planktivorous), small fish mature at a young age (Martin 1966; Donald and Alger 1986). An inverse relationship between age at maturity and length at maturity between populations was not observed in the results from the 11 Alaskan populations (Tables 2 and 3).

In addition to growth, diet may affect the age at

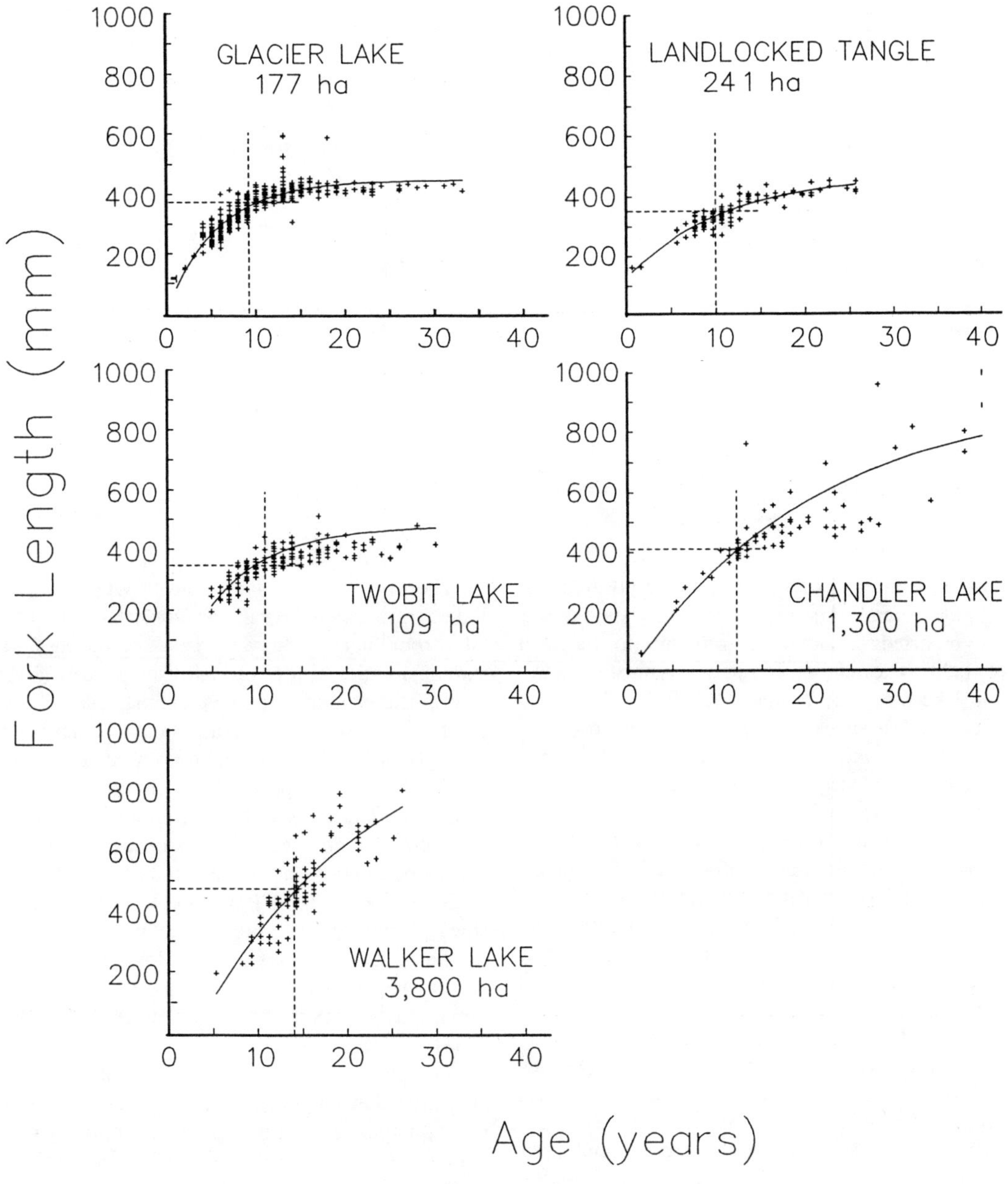

FIGURE 3—Continued.

maturity and size at maturity for lake trout. Available forage, represented by fish species richness, was positively correlated with LM_{50} and with L_∞ but not with incremental growth (Table 3). Slow-growing lake trout in Ontario that fed on plankton matured earlier than fast-growing piscivorous lake trout, and a change in diet from plankton to fish resulted in an increase in size at maturity and age at maturity (Martin 1966). In contrast, lake trout in Alaskan lakes containing only small invertebrate forage matured at smaller size and older age than did lake trout inhabiting lakes with forage fish species (Burr 1993).

Asymptotic length, L_∞, of lake trout was positively related to lake surface area. Lake surface area was also positively correlated with LM_{50} and fish

TABLE 5.—Estimates of density and biomass of mature lake trout *Salvelinus namaycush* from various water bodies.

Location	Surface area (ha)	Density (fish/ha)	Biomass (kg/ha)	Reference
East Blue Lake				
Manitoba	97	2.93	2.24	B. H. Gibson, Ontario Ministry of Natural Resources, unpublished in Martin and Olver 1980
Swan Lake				
Alberta	200	1.13		Paterson 1968
Squeers Lake				
Ontario	384	18	7.4	Ball 1988
Alexie Lake				
Northwest Territories	547	1.37		Healey 1978
		1.65		
Goldstream Pond				
Maine	1,468	1.38		DeRoche and Bond 1957
		0.87		
Thompson Lake				
Maine	1,791	10.74[a]		S. E. DeRoche, Maine Department of Inland Fisheries and Game, unpublished in Martin and Olver 1980
Lake Opeongo				
Ontario	5,860	0.41		Martin and Fry 1973

[a]For lake trout 356 mm fork length and larger, size at maturity for this population is not documented.

species richness (Table 3). Significant positive correlation between lake area and LM_{50} and L_∞ was also reported for lake trout from Ontario (Payne et al. 1990). The relationship between maximum size and lake area may be the result of the lake area effect on fish species diversity. Carl et al. (1990) showed that lakes with forage fish (especially coregonines) produced larger fish than did lakes without suitably sized prey.

Because of lower energy availability to populations in high-latitude and -altitude lakes (Henderson et al. 1973), growth and maturity is likely to be slower (Healey 1978). A positive correlation between AM_{50} and latitude was observed in the study lakes (Table 3). An increase in age at maturity with increase in latitude was reported by Falk et al. (1973) and by Healey (1978). Length at maturity may also decrease with increasing latitude (Healey 1978). Mean lengths at maturity for the Alaskan lake trout populations were smaller than estimates from lakes of comparable size from populations outside of Alaska (Burr 1993). However, within the data set, LM_{50} was not inversely related to latitude (Table 3). The effect of elevation on growth or maturation was not evident in the Alaskan lakes, but the range of elevations is very limited and the lowest-altitude lake was at high latitude (Table 1).

Density and Biomass

The density of mature lake trout from studies outside Alaska is generally less than 3 fish per ha (Healey 1978, Table 5), whereas six of seven Alaskan populations had greater than 3 fish/ha (Table 4). The low density of lake trout in Upper Tangle Lake is probably the result of suboptimal habitat for lake trout (deep, cold water is scarce during midsummer) and a history of high fishing effort (Burr 1989). The estimates from Sevenmile Lake (14–33 fish/ha and 14–34 kg/ha) are much higher than most. However, most populations studied outside Alaska have been primarily piscivorous or have come from larger lakes than the populations in this data set, or both. Where planktivorous and benthivorous populations have been studied, higher densities have been found. For example, reported densities of mature lake trout from Ontario were 12 fish/ha in Lake Louisa (B. P. Monroe and F. J. Hicks, Ontario Ministry of Natural Resources, unpublished) and 18 fish/ha from Squeers Lake (Ball 1988).

The inverse relationship between density and lake area that has been observed in Alaskan lake trout populations is consistent with reports by Carlander (1977), Goddard et al. (1987) and Payne et al. (1990), which implies that smaller lakes produce more fish than larger lakes on a per unit area basis. However, density does not always correlate well with biomass. Populations without forage fish (e.g., Twobit Lake) and primarily piscivorous populations (e.g., Paxson Lake) of similar biomass per area (5 kg/ha, Table 4) may differ widely in density (10.2 versus 3.2 fish/ha) because of the small size typical of planktivorous and benthivorous fish. Lake trout from six of the seven study lakes in Table 4 (excluding Paxson) averaged less than 420 mm fork length (Burr 1992). The lengths at 50% maturity for these lake trout populations are very similar (343–400 mm, Table 2) and are typically less than what was reported from other geographical areas (Burr 1993). Martin and Olver (1980) found that the highest densities of lake trout (4.1–9.8 fish/ha) generally occur in those lakes where fish mature at a small size, are planktivorous or benthivorous or both, and where the average size of fish is between 300 and 400 mm. Carl et al. (1990) demonstrated the important negative effect of additional fish species (particularly predators) on lake trout production. The relatively high densities of lake trout found in the seven lakes in Alaska is probably attributable to the small surface area of the lakes studied, the small average size of lake trout, the small size at maturity, and the simple fish communities in these populations.

Estimates of rates of growth and maturation and estimates of density reported here are within ranges reported from more southerly populations, particularly when the effects of lake area are considered. In this set of Alaskan lakes, the availability of appropriate-sized prey items appears to be of primary importance. However, better quantitative information about lake trout food habits in Alaskan populations is needed. Until additional data are available, lake surface area can be used to indicate the likely ranges of lake trout size, size at maturity, population density, and type of prey species present.

References

Alm, G. 1959. Connection between maturity, size, and age in fishes. Institute of Freshwater Research Drottningholm Report 40.

Ball, H. E. 1988. The dynamics of a polyphagous lake trout, *Salvelinus namaycush* (Walbaum) population in a northwestern Ontario lake. Master's thesis. Lakehead University, Thunder Bay, Ontario.

Bond, W. A. 1975. Data on the biology of lake whitefish and lake trout from Kaminuriak Lake, District of Keewatin, N.W.T. Canada Department Environment Fisheries Marine Service, Data Report Series CEN/D-75-4, Winnipeg.

Burr, J. M. 1989. Stock assessment and biological characteristics of lake trout populations in interior Alaska, 1988. Alaska Department of Fish and Game Fisheries Data Series 99, Juneau.

Burr, J. M. 1992. A summary of abundance and density estimates for interior Alaska's lake trout populations and an examination of trends in yield. Alaska Department of Fish and Game Fisheries Manuscript Series 92-1, Juneau.

Burr, J. M. 1993. Maturity of lake trout from eleven lakes in Alaska. Northwest Science 67:78–87.

Carl, L., and eight coauthors. 1990. Fish community and environmental effects on lake trout. Lake trout synthesis. Ontario Ministry of Natural Resources, Toronto.

Carlander, K. D. 1977. Biomass, production, and yields of walleye (*Stizostedion vitreum*) and yellow perch (*Perca flavescens*) in North American lakes. Journal of the Fisheries Research Board of Canada 34:1602–1612.

Chapman, D. G. 1951. Some properties of the hypergeometric distribution with applications to zoological censuses. University of California Publications in Statistics 1:131–160.

DeRoche, S. E., and L. H. Bond. 1957. The lake trout of Cold Stream Pond, Enfield, Maine. Transactions of the American Fisheries Society 85:257–270.

Donald, D. B., and D. J. Alger. 1986. Stunted lake trout from the Rocky Mountains. Canadian Journal of Fisheries and Aquatic Sciences 43:608–612.

Falk, M. R., D. V. Gilman, and L. W. Dahkle. 1973. The 1972 sports fisheries of Great Bear and Great Slave lakes, N.W.T. Canada Department Environment Fisheries Marine Service, Data Report Series CEN/T-73-8, Winnipeg.

Falk, M. R., D. V. Gilman, and L. W. Dahkle. 1974. Data on the biology of lake trout from Great Bear and Great Slave lakes, N.W.T., 1973. Canada Department Environment Fisheries Marine Service, Data Report Series CEN/D-74-4, Winnipeg.

Goddard, C. I., D. H. Loftus, J. A. MacLean, C. H. Olver, and B. J. Shuter. 1987. Evaluation of the effects of fish community structure on observed yields of lake trout (*Salvelinus namaycush*). Canadian Journal of Fisheries and Aquatic Sciences 44(Supplement 2): 239–248.

Hanson, J. A., and R. H. Wickwire. 1967. Fecundity and age at maturity of lake trout, *Salvelinus namaycush*, in Lake Tahoe. California Fish and Game 53:154–164.

Healey, M. C. 1978. Dynamics of exploited lake trout populations and implications for management. Journal of Wildlife Management 42:307–328.

Henderson, H. F., R. A. Ryder, and A. W. Kudhongnia. 1973. Assessing fishery potential of lakes and reservoirs. Journal of the Fisheries Research Board of Canada 30:2000–2009.

Johnson, L. 1976. Ecology of Arctic populations of lake trout, *Salvelinus namaycush*, lake whitefish, *Coregonus clupeaformis*, Arctic char, *S. alpinus*, and associated species in unexploited lakes of the Canadian Northwest Territories. Journal of the Fisheries Research Board of Canada 33:2459–2488.

Kerr, S. R. 1971. A simulation model of lake trout growth. Journal of the Fisheries Research Board of Canada 28:815–819.

Martin, N. V. 1966. The significance of food habits in the biology, exploitation, and management of Algonquin Park, Ontario, lake trout. Transactions of the American Fisheries Society 95:415–422.

Martin, N. V., and F. E. Fry. 1973. Lake Opeongo: effects of exploitation and introductions on the salmonid community. Great Lakes Fishery Commission Technical Report 24:1–34.

Martin, N. V., and C. H. Olver. 1980. The lake charr, *Salvelinus namaycush*. Pages 205–277 *in* E. K. Balon, editor. Charrs: salmonid fishes of the genus Salvelinus. Dr. W. Junk, The Hague, Netherlands.

Paterson, R. J. 1968. The lake trout (*Salvelinus namaycush*) of Swan Lake, Alberta. Alberta Department of Lands and Forests, Fish and Wildlife Division Research Report 2:1–149.

Payne, N. R., and six coauthors. 1990. The harvest potential and dynamics of lake trout populations in Ontario. Lake trout synthesis, Ontario Ministry Natural Resources, Toronto.

Robson, D. S., and W. A. Flick. 1965. A non-parametric statistical method for culling recruits from a mark–recapture experiment. Biometrics 21:936–947.

SAS. 1988. SAS/STAT users guide, release 6.03 edition. SAS Institute, Inc., Cary, North Carolina.

Seber, G. A. F. 1982. The estimation of animal abundance and related parameters, 2nd edition. Charles Griffin & Company, Ltd., London.

Sharp, D., and D. R. Bernard. 1988. Precision of estimated ages of lake trout (*Salvelinus namaycush*) from five structures. North American Journal of Fisheries Management 8:367–372.

Yaremchuk, G. C. B. 1986. Results of a nine year study (1972–1980) of the sport fishing exploitation of lake trout (*Salvelinus namaycush*) on Great Bear and Great Slave lakes, N.W.T.: the nature of the resource and management options. Canadian Technical Report of Fisheries and Aquatic Sciences 1436.

American Fisheries Society Symposium 19:119–126, 1997

Some Reproductive Characteristics of Least Ciscoes and Humpback Whitefish in Dease Inlet, Alaska

LAWRENCE L. MOULTON
MJM Research
5460 NE Tolo Road, Bainbridge Island, Washington 98110, USA

LEE MICHAEL PHILO AND JOHN C. GEORGE
Department of Wildlife Management, North Slope Borough
Post Office Box 69, Barrow, Alaska 99723, USA

Abstract.—Least ciscoes *Coregonus sardinella* and humpback whitefish *C. pidschian* were collected from Dease Inlet, Alaska, during 1988–1990 for baseline data (length, weight, sex, maturity, fecundity, and age) on the reproductive cycle of these two lightly exploited species. Gonadosomatic indexes ranged from 4.1 to 13.8% of body weight in least ciscoes (mean = 8.3%) and 2.9 to 8.7% in humpback whitefish (mean = 5.7%). Fecundity was highly correlated with body weight in both species; it ranged from 12,206 to 100,939 in least ciscoes, with an average of 95 eggs/g somatic weight and from 10,824 to 44,387 in humpback whitefish, with an average of 45 eggs/g somatic weight. Fecundities were higher than those reported for populations of the same species in the Chatanika River, Alaska, and in the Canadian Arctic. Approximately half of the least ciscoes and humpback whitefish spawned each year. The mature portion of both populations was dominated by older fish (>10 years).

The Dease Inlet region is a major area for subsistence harvest of coregonines by residents of Barrow, Alaska. Most of the fish-harvesting sites are along five major rivers that drain into the southern portion of the inlet (Figure 1). Despite the importance of this region as a harvest area, few investigations have been conducted on the utilized fish populations. Investigations into Dease Inlet fish populations began in 1988 to develop a database for fishery management.

The objective of our study was to obtain information on the maturity and fecundity of least ciscoes *Coregonus sardinella* and humpback whitefish *C. pidschian* in the Dease Inlet region. Least cisco is the most abundant coregonine species in the coastal region of Dease Inlet and probably originates from many of the rivers and lakes connected to the inlet. Humpback whitefish are less abundant, but are frequently caught by subsistence fishers. Few studies of fecundity in either species have been reported; most of this information comes from more southerly locations (Morrow 1980).

We also compared the results of our study with those of other arctic and subarctic regions.

Methods

Sampling was conducted along the east coast of Dease Inlet, Alaska, during 30 July–11 August 1988, and 27–31 July 1990 (Figure 1). Fish were collected with fyke nets and variable-mesh gill nets, and samples representing the length range in the catch were retained for further analysis. Fork length was measured to the nearest 1 mm, weight to the nearest 0.1 g. Sex and stage of maturity were determined, and otoliths were removed from each specimen. Ages were estimated from otoliths using the break-and-burn technique (Chilton and Beamish 1982).

Stage of maturity followed that described by Nikolsky (1963). Ovaries were removed and weighed to the nearest 0.1 g. Ovaries were preserved in Gilson's fluid. The stomach between the esophagus and pyloric sphincter was removed and the remaining body tissue was weighed to obtain somatic weight. Somatic weight data were free of weight measurement error caused by weight of stomach contents.

Fecundity was estimated using dry weights of ovaries and three subsamples of ova from each fish (Bagenal and Braum 1968). The gonadosomatic index (GSI) was calculated by dividing the ovary weight by somatic weight. Mean individual egg weight was estimated by dividing the wet ovary weight by the estimated fecundity. Egg density was calculated by dividing fecundity by somatic weight. Age, fork length, and somatic weight were evaluated as factors of fecundity. Analysis of covariance (Snedecor and Cochran 1967) was used to test for differences in fecundity relationships between species and locations.

The GSI samples were collected in late July to early August, approximately 5 weeks before the onset of spawning in this region (Bendock and Burr 1985). During summer in the Mackenzie River delta (Bond and Erickson 1985), the GSI increases

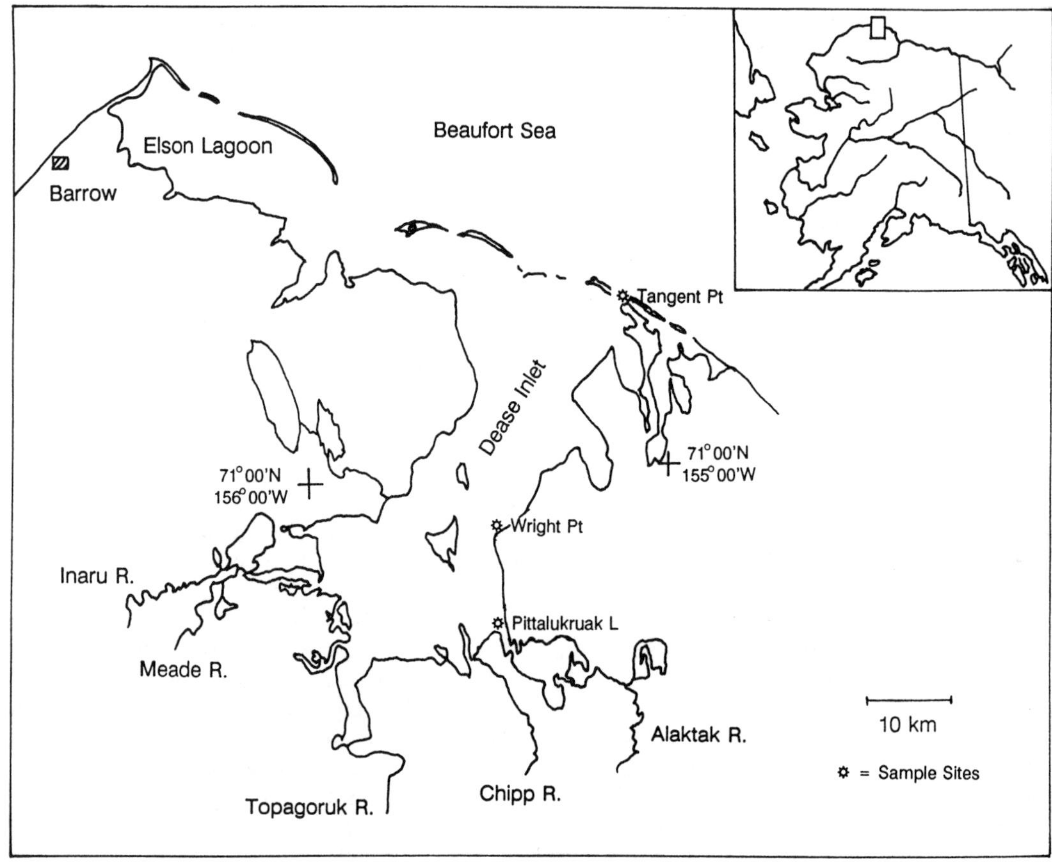

FIGURE 1.—Dease Inlet study area in Alaska, showing major drainages in southern portion.

by 12.5% every week in least ciscoes and by 20% in lake whitefish *C. clupeaformis*. In order to correct for our time of sampling, the GSI values were adjusted based on the above rates indicated by the Bond and Erickson (1985) data.

Results

Age Composition

The sampled least ciscoes were older, on average, than the sampled humpback whitefish, but ages 10 to 12 were most abundant in both species (Figure 2). We do not know if the sample distributions reflect actual age distribution in the two populations.

Maturity

Least cisco females reached 50% maturity by age 7, males by age 8. Male humpback whitefish also appeared to reach maturity at a later age than females (Figure 3). For those least ciscoes judged to be mature, 49% of the females and 53% of the males were classified as prespawning fish that would spawn that fall. Among humpback whitefish, 38% of the females and 70% of the males were judged to be prespawning fish. A χ^2 test indicated that none of the prespawning percentages differed significantly from 50% ($P > 0.05$). It is therefore likely that spawning occurs every 2 years, as has often been reported for coregonine populations in northern latitudes (Morin et al. 1982).

The GSI was greater in least ciscoes than in humpback whitefish (Table 1, Figure 4). If the GSI in Dease Inlet fish increases at the rate reported by Bond and Erickson (1985), the expected values by early September would be 14.6% for least ciscoes and 13.4% for humpback whitefish. The GSI was positively correlated with fecundity ($P < 0.01$), whereas individual egg weight was not significantly correlated with fecundity ($P > 0.05$), thereby indi-

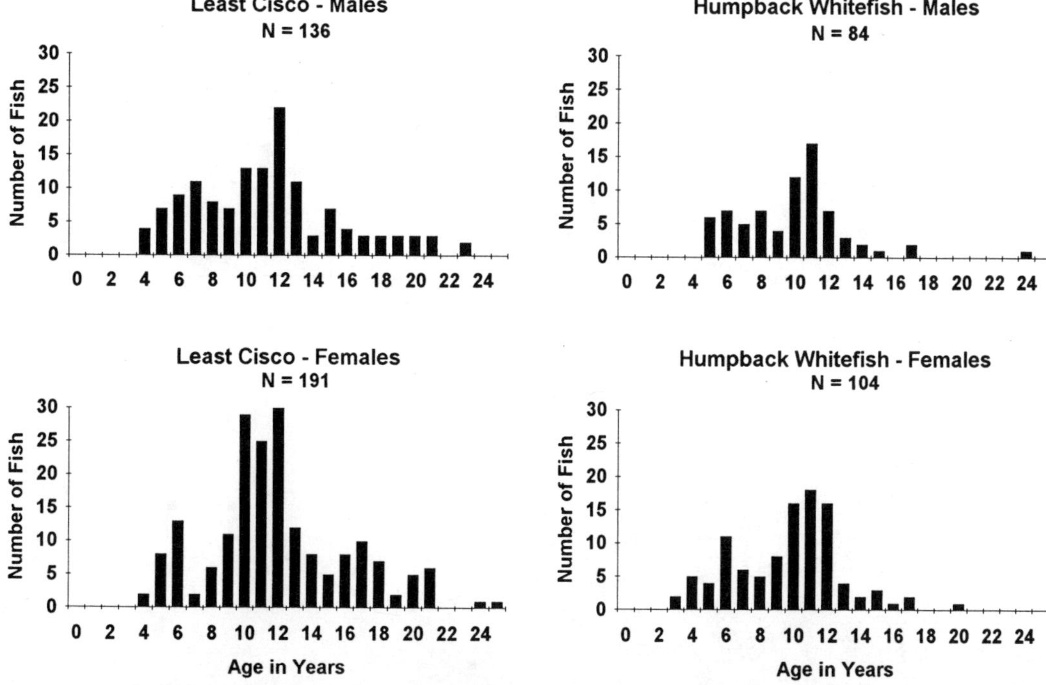

FIGURE 2.—Age distribution of least ciscoes *Coregonus sardinella* and humpback whitefish *C. pidschian* captured by fyke nets in the coastal region of Dease Inlet, 1988–1990.

cating that egg size did not change with size of fish. As females grow, increase in ovary size is the result of additional eggs, not larger eggs.

Fecundity

Fecundity, in both species, was most strongly correlated with somatic weight, although the relationships with length and age were also significant (Table 2). Analysis of covariance indicated that the slopes of the fecundity–somatic weight relationship for the two species were not different, but there was a significant difference in the intercept ($P < 0.01$). At equivalent somatic weights, fecundity of least ciscoes exceeded that of humpback whitefish over the entire weight range sampled (Figure 5).

Fecundity (eggs/female) in least ciscoes averaged 33,791 compared to 22,723 for humpback whitefish, although the latter was the larger of the two species (Table 1). When standardized to the somatic weight, least cisco produced more than twice as many eggs per unit weight.

The fecundity measurements of Dease Inlet least ciscoes are among the highest recorded for this species (Figure 6). Individuals from the populations reported by Mann (1974) appear to be freshwater resident forms that do not reach the size of the anadromous fish from Dease Inlet (Table 3). Least cisco from the Chatanika River, a tributary to the Yukon River (Clark and Bernard 1988), are similar in size to the Dease Inlet fish, but the fecundity–weight relationship indicates that the latter have higher fecundity.

Few estimates of humpback whitefish fecundity have been reported, the Chatanika River population (Clark and Bernard 1988) providing the most suitable information for comparison (Figure 7). Dease Inlet humpback whitefish had a significantly greater fecundity than Chatanika River fish (Table 3).

Discussion

As expected, age at maturity in the Dease Inlet populations is greater than that observed farther south. Least ciscoes from Dease Inlet reached 50% maturity at ages 7–8 and humpback whitefish at ages 11–14. In the Chatanika River, least ciscoes were mostly mature at age 4 and humpback whitefish were mostly mature at age 5 (Clark and Bernard 1988). The Chatanika populations have been heavily exploited (Clark and Bernard 1988) com-

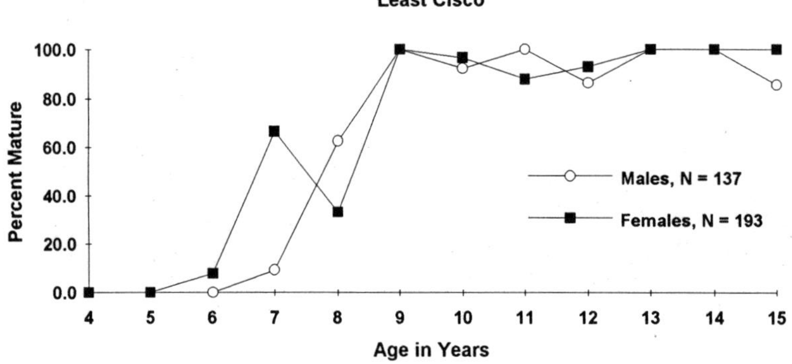

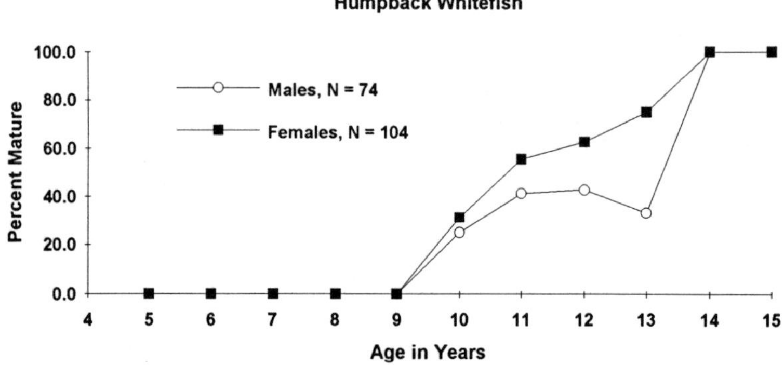

FIGURE 3.—Percent mature least ciscoes *Coregonus sardinella* and humpback whitefish *C. pidschian* by sex and age group from samples obtained in the coastal region of Dease Inlet, 1988–1990.

pared to those in Dease Inlet, so the older fish are more scarce in the former. The maturity data from Dease Inlet suggest that both species are on a 2-year spawning cycle.

TABLE 1.—Life history variables measured in prespawning female least ciscoes *Coregonus sardinella* (N = 36) and humpback whitefish *C. pidschian* (N = 13) from Dease Inlet, 1988–1990. Data shown are means with minima and maxima in parentheses.

Variable	Least ciscoes	Humpback whitefish
Length (mm)	324 (288–385)	344 (313–389)
Weight (g)	341 (191–683)	493 (334–730)
Age (years)	12.4 (7–20)	12.8 (11–16)
Fecundity[a]	33.8 (12.2–100.9)	22.7 (10.8–44.4)
Egg density/g[b]	94.6 (52.9–147.9)	45.3 (30.2–73.7)
GSI[c]	8.3 (4.1–13.8)	5.7 (2.9–8.7)

[a]In thousands of eggs per female.
[b]Number of eggs/g somatic body weight.
[c]GSI = gonadosomatic index, which is the ovary weight divided by the somatic weight.

Morin et al. (1982) examined the latitudinal trend in fecundity for lake whitefish, which is closely related to humpback whitefish, and cisco *Coregonus artedi*. They predicted that fecundity should decrease, and age at maturity should increase, with increasing latitude. Although fecundity data were limited, there was a trend toward this prediction for lake whitefish (Table 4).

The data obtained during our study indicate an opposite trend for least ciscoes and humpback whitefish when compared with the limited information available from other studies (Table 4). The Dease Inlet fish had the highest fecundity reported to date for both species. This finding is paradoxical when the general trend is for reduced reproductive potential with increasing latitude. Morin et al. (1982) reported that their data on ciscoes in James and Hudson bays did not conform to this general trend. They attribute this anomaly to physiological consequences of harsh climatic conditions in north-

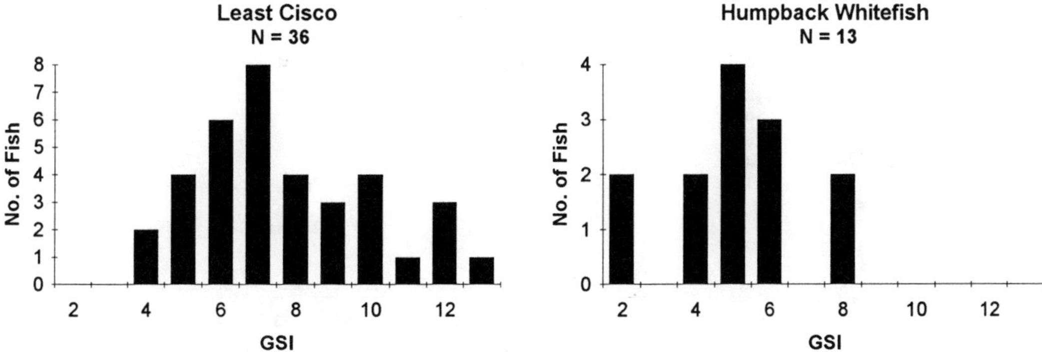

FIGURE 4.—Frequency of gonadosomatic index (GSI) values observed in least ciscoes *Coregonus sardinella* and humpback whitefish *C. pidschian* from the coastal region of Dease Inlet, 1988–1990.

TABLE 2.—Fecundity relationships in least ciscoes *Coregonus sardinella* and humpback whitefish *C. pidschian* from Dease Inlet, 1988–1990. In all regression equations, the dependent variable is $\log_{10}$ (fecundity).

Independent variable	Regression statistics				
	Slope	Intercept	*N*	*r*	*P*
Least ciscoes					
Age	0.032	4.083	35	0.504	<0.01
Log_{10} (length)	5.468	−9.238	36	0.885	<0.01
Log_{10} (weight)	1.549	0.584	36	0.941	<0.01
Humpback whitefish					
Age	0.060	3.551	13	0.646	<0.05
Log_{10} (length)	1.253	−6.460	13	0.740	<0.01
Log_{10} (weight)	1.313	0.802	13	0.802	<0.01

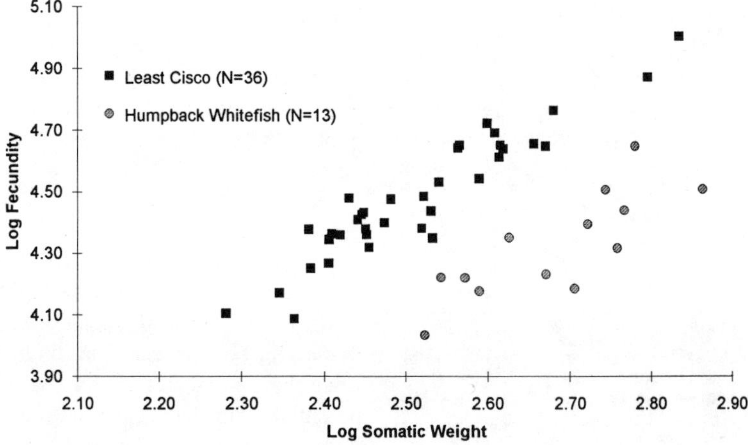

FIGURE 5.—Relationship between fecundity ($\log_{10}$) and somatic weight ($\log_{10}$) for least ciscoes *Coregonus sardinella* and humpback whitefish *C. pidschian* in Dease Inlet, 1988–1990 (least cisco: $r = 0.941$, $P < 0.01$; humpback whitefish: $r = 0.802$, $P < 0.01$).

TABLE 3.—Comparison of $\log_{10}$(fecundity)–$\log_{10}$(body weight) regressions and egg density (number of eggs/g somatic body weight) in least ciscoes *Coregonus sardinella* and humpback whitefish *C. pidschian* from Dease Inlet.

Water body	Regression statistics					Egg density	Source
	Slope	Intercept	*N*	*r*	*P*		
				Least ciscoes			
Dease Inlet	1.549	0.584	34	0.941	<.01	94.6	This study
Chatanika	1.215	1.193	38	0.743	<.01	57.3	Clark and Bernard (1988)
Trout Lake	0.351	3.209	13	0.217	NS	48.0	Mann (1974)
Lake 105	1.065	1.347	22	0.739	<.01	36.3	Mann (1974)
Peter Lake	1.077	1.376	28	0.531	<.01	39.9	Mann (1974)
				Humpback whitefish			
Dease Inlet	1.313	0.802	13	0.802	<.01	45.3	This study
Chatanika	0.518	2.786	72	0.384	<.01	27.0	Clark and Bernard (1988)

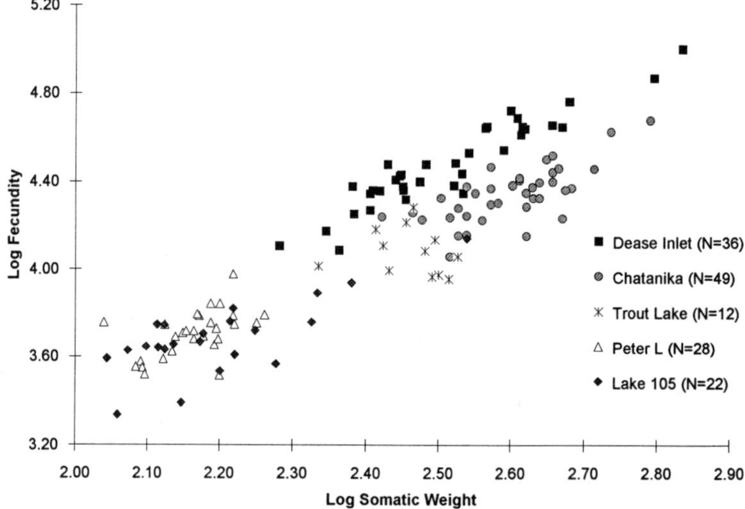

FIGURE 6.—Comparison of fecundity–weight relationships for five populations of least ciscoes *Coregonus sardinella* in Alaska and Yukon Territory (Chatanika from Clark and Bernard 1988; Trout Lake, Peter Lake, and Lake 105 from Mann 1974).

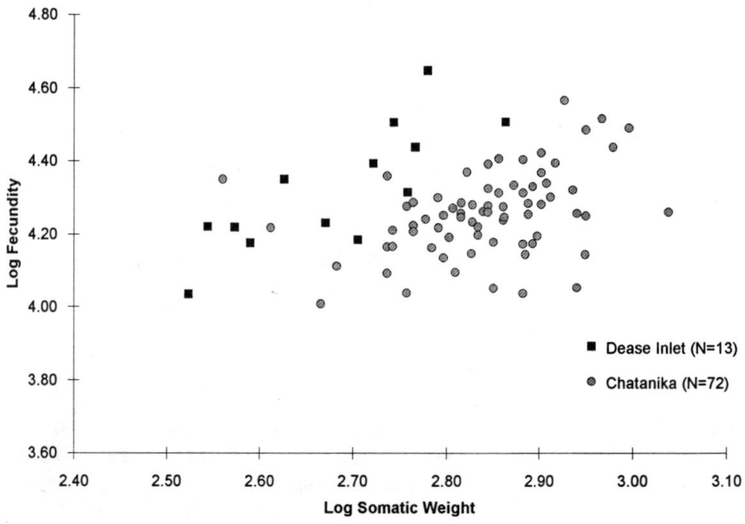

FIGURE 7.—Comparison of fecundity–weight relationships for two populations of humpback whitefish *C. pidschian* in Alaska (Chatanika from Clark and Bernard 1988).

TABLE 4.—Relationship between egg density and latitude in selected populations of least ciscoes *Coregonus Sardinella* and whitefish. Egg density data for Dease Inlet fishes were recalculated using total body weight instead of somatic body weight to permit comparison with egg density data from other studies.

Water body	Latitude (°N)	Egg density (number of eggs/g total body weight)	Source
		Least ciscoes	
Dease Inlet	71	84.9	This study
Trout Lake	69	43.2	Mann (1974)
Lake 105	69	32.0	Mann (1974)
Peter Lake	69	35.7	Mann (1974)
Chatanika River	65	51.5	Clark and Bernard (1988)
		Humpback whitefish	
Dease Inlet River	71	41.6	This study
Chatanika River	65	23.9	Clark and Bernard (1988)
		Lake whitefish	
Great Slave Lake	62	14.9	In Morin et al. (1982)
Lake La Ronge	55	17.6	In Healey (1975)
La Grande River	53	20.2	Morin et al. (1982)
Pigeon Lake	53	18.0	Bidgood (1974)
Buck Lake	53	27.7	Bidgood (1974)
Georgian Bay	45	18.1	In Morin et al. (1982)
Lake Ontario	44	21.8	In Morin et al. (1982)
Lake Erie	42	35.9	In Morin et al. (1982)

ern James and Hudson bays. Lambert and Dodson (1990) also describe a similar anomaly for the Eastmain River, James Bay, and hypothesized that longer migration distances and a set of rapids in the Eastmain River, in the southern part of their study area, imposed a greater somatic cost of reproduction as compared with the northern area, where migration distances were shorter and rapids were nonexistent. Similar phenomena could be operating for the Dease Inlet–Chatanika situation. The migration distances in Dease Inlet, though not well known, probably are not great. The water currents within the inlet and rivers are very low. The energetic cost of migrations may be low relative to that in the Chatanika system, which would allow for a greater allocation of energy to reproduction. It is likely, therefore, that the influence of local environmental factors on energy allocation in these two systems overrides the general latitudinal effect. However, the number of observations in our study is low for both species. Further analysis of trends can be made when data from more systems become available.

Acknowledgments

This study was funded by the Department of Wildlife Management, North Slope Borough, under the direction of B. Nageak, with scientific review by T. Albert. Assistance with fieldwork was provided by C. Brower, M. Leimberger, K. Bagne, B. Adams, and R. Suydam. We thank K. Chang-Kue and W. Griffiths for reviewing the manuscript.

References

Bagenal, T. B., and E. Braum. 1968. Eggs and early life history. Pages 159–181 *in* W. E. Ricker, editor. Methods for assessment of fish production in fresh waters. IBP (International Biological Programme) Handbook 3, IBP, Blackwell Scientific Publications, London.

Bendock, T. N., and J. Burr. 1985. Freshwater fish distributions in the central Arctic Coastal Plain (Topagoruk River to Ikpikpuk River). Alaska Department of Fish and Game, Sport Fish Division, Fairbanks.

Bidgood, B. F. 1974. Reproductive potential of two lake whitefish (*Coregonus clupeaformis*) populations. Journal of the Fisheries Research Board of Canada 31: 1631–1639.

Bond, W. A., and R. N. Erickson. 1985. Life history studies of anadromous coregonid fishes in two freshwater lake systems on the Tuktoyaktuk Peninsula, Northwest Territories. Canadian Technical Report of Fisheries and Aquatic Sciences 1336.

Chilton, D. E., and R. J. Beamish. 1982. Age determination methods for fishes studied by the groundfish program at the Pacific Biological Station. Canadian Special Publication of Fisheries and Aquatic Sciences 60.

Clark, J. H., and D. R. Bernard. 1988. Fecundity of humpback whitefish and least cisco, Chatanika River,

Alaska. Alaska Department of Fish and Game, Division of Sport Fish, Fishery Data Series 77, Juneau.

Healey, M. C. 1975. Dynamics of exploited whitefish populations and their management with special reference to the Northwest Territories. Journal of the Fisheries Research Board of Canada 32:427–448.

Lambert, Y., and J. J. Dodson. 1990. Influence of freshwater migration on the reproductive patterns of anadromous populations of cisco (*Coregonus artedii*) and lake whitefish (*C. clupeaformis*). Canadian Journal of Fisheries and Aquatic Sciences 47:335–345.

Mann, G. J. 1974. Life history types of the least cisco (*Coregonus sardinella*, Valenciennes) in the Yukon Territory north slope and eastern Mackenzie River delta drainages. Arctic Gas Biological Report Series 18:1–132.

Morin, R., J. J. Dodson, and G. Power. 1982. Life history variations of anadromous cisco (*Coregonus artedii*), lake whitefish (*C. clupeaformis*), and round whitefish (*Prosopium cylindraceum*) populations of eastern James–Hudson Bay. Canadian Journal of Fisheries and Aquatic Sciences 39:958–967.

Morrow, J. E. 1980. The freshwater fishes of Alaska. Alaska Northwest Publishing Company, Anchorage.

Nikolsky, G. V. 1963. The ecology of fishes. Academic Press, New York.

Snedecor, G. W., and W. G. Cochran. 1967. Statistical methods, 6th edition. Iowa State University Press, Ames.

American Fisheries Society Symposium 19:127–132, 1997

Population Biology of the Bering Flounder in the Northeastern Chukchi Sea

R. L. Smith, M. Vallarino, E. Barbour, E. Fitzpatrick, and W. E. Barber

University of Alaska, Institute of Marine Science
Fairbanks, Alaska 99775, USA

Abstract.—Population characteristics of Bering flounder *Hippoglossoides robustus* in the northeastern Chukchi Sea were estimated from benthic trawl samples in 1990 and 1991. Weight–length and age–length relationships indicated that age-3 females grew faster and to a larger size than males. Asymptotic total lengths of males and females were 211 mm and 241 mm, respectively; maximum age was 11 years. Biomass and abundance differed dramatically in the 2 years of this study. Mean biomass declined significantly, from 17.2 kg/km^2 in 1990 to 0.79 kg/km^2 in 1991. Our data, plus historical information, suggest that this species is subject to an unstable physical environment that may cause either mass mortalities, recruitment failures, or both. The possibility of environmental effects is consistent with the difference in growth rates between the 1950s and the 1980s and is consistent with recent data on interannual variability in wind and current directions and larval abundance.

The Bering flounder *Hippoglossoides robustus* differs from its close relative the flathead sole *H. elassodon* in several subtle anatomical characteristics. In addition to having a more markedly curved lateral line and a wider interorbital space, the Bering flounder has lower dorsal (D) and anal (A) fin ray counts (D: 67–79; A: 51–60) than the flathead sole (D: 76–86; A: 60–69) (Shmidt 1950; Andriyashev 1964). The geographic ranges of these two species overlap considerably. The flathead sole occurs from northern California, through the Gulf of Alaska, across the Bering Sea, and southward to Japan (Hart 1973). The Bering flounder is much more restricted in its distribution, occurring from Tatar Strait to the Chukchi Sea and extending through the Bering Sea to the Aleutian Islands (Quast and Hall 1972). In the area of overlap it is unclear whether interbreeding occurs or to what extent niche specialization has led to reduced competition between these two similar forms.

The Bering flounder falls prey to Arctic cod *Boreogadus saida* (Coyle et al. 1997, this volume) and to several marine mammals including white whales *Delphinapterus leucas* (Frost and Lowry 1981) and bearded seals *Erignathus barbatus* (Lowry and Frost 1981). Pruter and Alverson (1962) found the Bering flounder to be the most abundant flatfish in the southeastern Chukchi Sea, most frequently occurring at depths ≥44 m. Moiseev (1953) found that the Bering flounder occurs in greater abundance at subzero temperatures than at temperatures above 0°C. Bering flounder occurred at extremely low population densities in the Chukchi Sea during the late 1950s (Pruter and Alverson 1962).

The purpose of our study was to expand the knowledge about the population biology of Bering flounder in the northeastern Chukchi Sea. We report on the distribution, abundance, biomass, and weight–length and age–length relationships and make inferences about the life history of this species.

Methods

Fish were captured from 16 August through 16 September 1990 and from 14 through 23 September 1991. A National Marine Fisheries Service (NMFS) 83–112 survey otter trawl was towed at approximately 2 knots. The net had a 25.2-m headrope and a 34.1-m footrope set back 7.1 cm from a tickler chain. The cod end was 90-mm stretched mesh into which a liner of 33-mm stretched mesh was inserted. Each trawl station consisted of two 30-min hauls. Area sampled (m^2) was calculated by multiplying the width of the net opening by the distance trawled. In 1990 the opening of the net was verified by use of a Scanmar mensuration unit attached to the wings of the net. Distance trawled was determined from the ship's position (latitude and longitude) at the start and end of each haul. Biomass (kg/km^2 trawled) and abundance (number of fish/km^2 trawled) were estimated by averaging the two 30-min hauls at each station. Mean biomass and abundance values for the 2 years were compared using a Mann–Whitney *U*-test (Zar 1984).

A total of 232 individual fish were returned to the laboratory and examined. Initial laboratory data collection included measurement of fish total and standard lengths (mm), fish weight (g), stomach content weight (g), gonad weight (g), and the preservation of otoliths, stomach contents, and ovaries.

The value for fish weight used in subsequent calculations and ratios was determined by subtracting stomach content weight from total weight. Gonadosomatic index (GSI) is defined as follows: (gonad weight/fish weight) × 100.

The annulus on the otolith viewed with transmitted light was defined as the thin, translucent zone. In fish age 4 or younger, this zone could be traced entirely around the otolith; in older fish, the zone was incomplete but most easily visible on the rostrum (anterior end). The annulus observed in burnt sections of otoliths was defined as the dark zone. The surface pattern of annuli on otoliths of 125 fish was read using a dissecting microscope. Otoliths were measured with an ocular micrometer calibrated with a stage micrometer.

Results

Distribution and Abundance

Bering flounder occurred at 32 of 48 stations sampled in 1990 and at 8 of 16 stations sampled in 1991. Nineteen of the 24 stations at which the Bering flounder was missing were north of 70°N. The northernmost point of distribution in this study was above 72°N (1990 Station 30) and the easternmost point was 159°W (1991 Station 90–32). Where present, Bering flounder abundance ranged from 11 to 6,436 fish/km^2; biomass ranged from 0.1 to 223 kg/km^2. The 1990 data (Figure 1, A and B) show that, in general, the highest abundance and biomass of Bering flounder occurred in the southernmost part of the study area (south of 69°30′N and west of 167°W). In 1991, far fewer Bering flounder were caught; where present (8 of 17 stations), abundance was 12–100 fish/km^2 and biomass was 0.1–4.6 kg/km^2 (Figure 1, C and D).

Considerable variability in abundance was observed among stations and also between hauls at the same station. For instance, at several stations one of the two hauls yielded no Bering flounder while the other one did. Replicate hauls at the same station differed up to 15-fold in biomass and up to threefold in abundance. Mean abundance estimates for all stations sampled in 1990 and 1991 (995 and 429 fish/km^2, respectively) differed significantly (U = 785; $P < 0.001$). Similarly, 1990 and 1991 mean biomass estimates (17.2 and 0.79 kg/km^2, respectively) also differed significantly (U = 825; $P < 0.001$). Eight stations were sampled in both field seasons. Mean abundance and biomass at these stations in 1990 (207 fish/km^2; 7.6 kg/km^2) were significantly higher (U = 85; $P < 0.001$) than the estimates for 1991 (19.7 fish/km^2; 0.66 kg/km^2). Reduced abundance and biomass in 1991 were associated with significantly lower temperatures in 1991. Comparing the eight stations common to both years, we found mean bottom temperatures of 5.4 and 0.9°C (U = 54; $P < 0.05$).

Age–Size Relationships

Total length (TL) and otolith length (OL) were linear functions of standard length (SL) in Bering flounder (TL = 1.55 SL + 4.24, N = 126, r^2 = 0.99; OL = 0.026 SL + 0.383, N = 135, r^2 = 0.92). Rearranging the equation of otolith length versus standard length allowed the prediction of SL based on OL: SL = (OL − 0.383)/0.026. Weight (W) was a power function of standard length ($W = 4.89 \cdot 10^{-6} \cdot SL^{3.25}$, N = 135, r^2 = 0.99).

Maximum longevity among the 133 individuals examined in this study was 11 years for females and 8 years for males (Figure 2). About 75% of the population consisted of fish that were at least 5 years old. Mean length at age indicates that in the first 3 years males and females grow at the same rate. By the end of the 4th year, however, females appear to be significantly larger (Figure 3). Von Bertalanffy growth parameters were calculated from the mean length-at-age (SL) data according to

$$L_t = L_\infty [1 - e^{-K(t - t_0)}] \quad (1)$$

where t = age (years), L_t = length (mm) at age, L_∞ = asymptotic length, K = instantaneous growth coefficient, and t_0 = theoretical age at L_t = 0. The von Bertalanffy equation for males is

$$L_t = 180 [1 - e^{-0.230(t + 0.185)}], \quad (2)$$

and for females

$$L_t = 206 [1 - e^{-0.215(t - 0.009)}]. \quad (3)$$

Gonadosomatic index was plotted as a function of standard length for males and females collected in 1990. For males, the result was a scatter of points with wide variation in gonad development for a particular size interval of fish. Male GSIs ranged from 0 to about 2%. Female GSIs were related to SL with a power curve ($GSI = 2.37 \cdot 10^{-7} \cdot SL^{3.20}$, N = 69, r^2 = 0.84). Ovary weights ranged from near 0 to about 7% of body weight.

Discussion

Distribution and Abundance

Distribution of Bering flounder over the study area was not uniform. Replicate trawl hauls at the same location did not necessarily match with re-

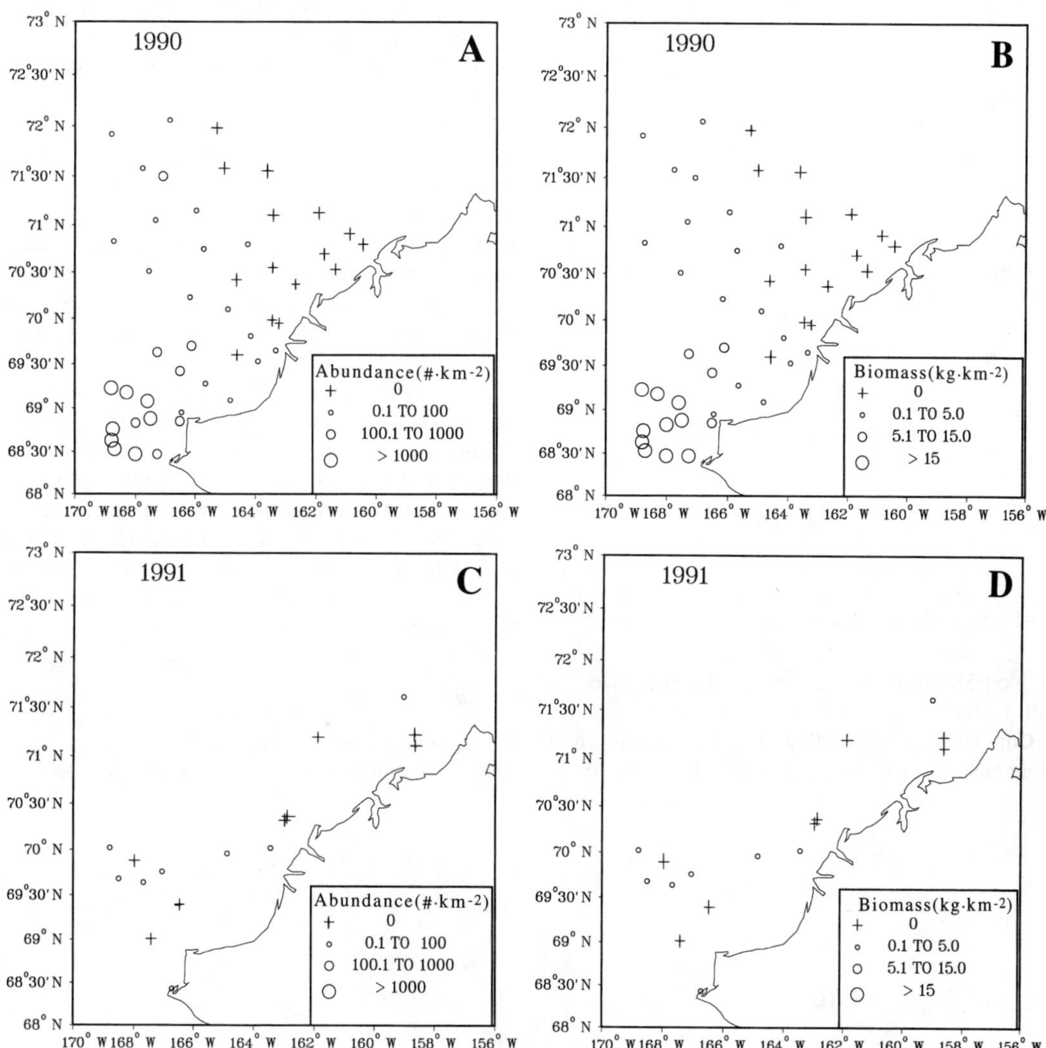

FIGURE 1.—Abundance ($\# \cdot km^{-2}$) and biomass ($kg \cdot km^{-2}$) of Bering flounder *Hippoglossoides robustus* in the northeastern Chukchi Sea. (**A**) Abundance in 1990. (**B**) Biomass in 1990. (**C**) Abundance in 1991. (**D**) Biomass in 1991.

spect to the abundance or even the presence of this species. We were unable to discern bottom type directly at trawl stations and therefore could not correlate abundance or presence of this species with bottom type. We plotted the station locations on the map of sediment types for the northeastern Chukchi Sea (see Figure 3 in Feder et al. 1989) and found that gravel was the only sediment type absent at stations where this species was captured. Other than this apparent avoidance of gravel, sediment type in the study area did not seem to have an obvious influence on the spatial distribution of Bering flounder. In this study, the Bering flounder occurred over a salinity range of 29.4 to 33.5‰ and was missing at stations exhibiting a salinity range of 29.7 to 33.4‰, suggesting that this salinity range had little or no effect on distribution.

Although present in the Chukchi Sea (southern part of our study area) in August 1959, the Bering flounder occurred at extremely low population densities; Pruter and Alverson (1962) caught 289 individuals (<23 kg total) in 59 hauls of 30-min duration each. The highest catch rate Pruter and Alverson reported for Bering flounder was 30 individuals in a 30-min tow using an "eastern otter trawl." In contrast, one of our 1990 stations

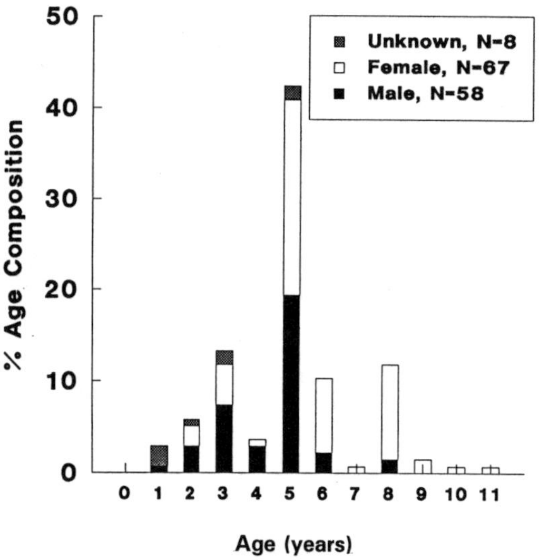

FIGURE 2.—Population age structure of Bering flounder *Hippoglossoides robustus* in the northeastern Chukchi Sea, 1990 data combined. N = 133.

yielded 587 Bering flounder (20.4 kg) in two 30-min hauls.

Comparisons of 1990 and 1991 biomass and abundance values suggest considerable interannual variation in these parameters. Using the same NMFS 83–112 benthic trawl and the same liner, Wolotira et al. (1977) reported a biomass of 4.1 kg/km^2 for Bering flounder in the southeastern Chukchi Sea, 2.4 kg/km^2 in Kotzebue Sound, and 0.59 kg/km^2 in the northern Bering Sea. In their study the starry flounder *Platichthys stellatus* was found in far higher biomasses than Bering flounder in both the southeastern Chukchi Sea and in Kotzebue Sound. Additional evidence of temporal variability in distribution and abundance of Bering flounder is provided by Andriyashev's (1964) contention that Bering flounder did not occur in the Chukchi Sea before 1933. Taken together, these observations on abundance suggest that Bering flounder may experience periodic population increases and also periodic mass mortalities resulting from either direct mortality, recruitment failure, or both. Data on age distribution (below) and larval abundance (Wyllie-Echeverria et al. 1997, this volume) are consistent with this scenario of population fluctuation.

Age and Growth

Our data on length and age (Figure 3) indicate that maximum size of Bering flounder in 1990 ap-

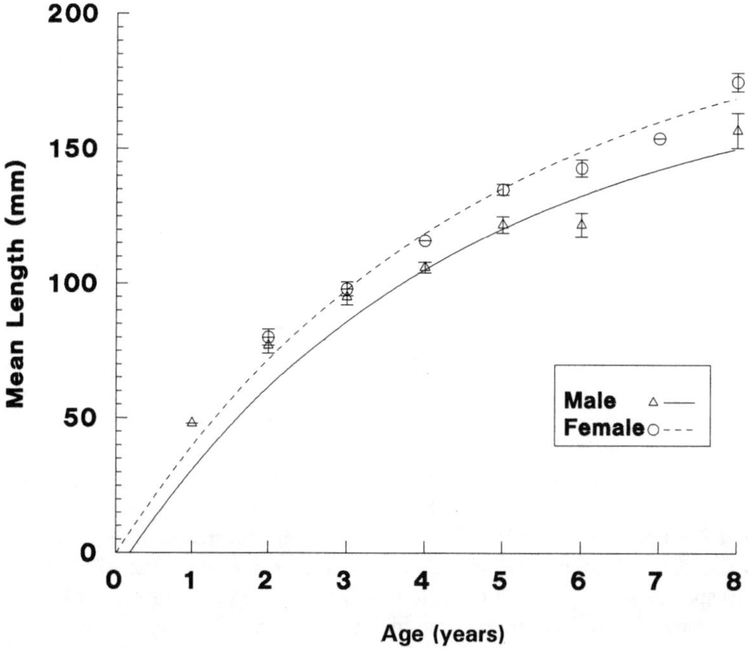

FIGURE 3.—Mean length (= standard length, see Results) at age for male and female Bering flounder *Hippoglossoides robustus* from the northeastern Chukchi Sea. Fish were sampled in 1990. Vertical bars = 2 SD.

proached 200 mm SL. These sizes correspond to the older fish in the study, ages 9–11. Andriyashev (1964) reported the length of the largest specimen from the Chukchi Sea at 150 mm TL. In Pruter and Alverson's (1962) collections, Bering flounder were 140–260 mm, averaging 199 mm TL. Small length at age and low biomass in the 1930s and 1950s led Alverson and Wilimovsky (1966) to conclude that the physical climate of the Chukchi Sea limited population size and depressed normal growth patterns in the Bering flounder.

In our sampling, all ages from 1 to 11 were represented, with age 5 dominating (Figure 2); 75% were at least 5 years old. In contrast, the ages of 89 Bering flounder as determined by Pruter and Alverson (1962) ranged from 6 to 13 years with 7-, 8- and 9-year-olds constituting 90% of the population. Apparently, there are dramatic shifts in population age structure over time, as well as variability in abundance.

Based on the lack of juvenile fish in their samples, Pruter and Alverson (1962) suggested that this species may not successfully reproduce in the Chukchi Sea. Drift may carry the larvae to conditions even harsher than those experienced by the adult spawners. Alternatively, the Chukchi Sea population may rely on the drift of pelagic larvae northward on the prevailing surface currents from more southerly locations to maintain population levels. Wyllie-Echeverria et al. (1997) presented evidence that recruitment of Bering flounder larvae to the study area varies interannually. Larvae were relatively abundant in the southern part of the study area in 1990 but were absent in 1991. Weingartner (1997, this volume) described temporal variability in the northward flow of the Alaska Coastal Water from the southern Chukchi Sea caused by changes in wind directions. In 1990 winds were consistently northeastward, increasing the northward flow of the Alaska coastal water and the advection of larvae from the south. In contrast, 1991 was characterized by more southerly winds that would decrease or reverse the typical northward flow of the Alaska coastal water. This reversal would prevent the recruitment of Bering flounder larvae to the northeastern Chukchi Sea.

Acknowledgments

We thank the crew of the *Ocean Hope III* and T. Sample, R. G. Bakkala, and T. Dark for their assistance. We particularly thank R. M. Meyer for his enthusiastic support and encouraging comments throughout the study. This study was funded by the Alaska Outer Continental Shelf Region of the Minerals Management Service, U.S. Department of the Interior, Anchorage, Alaska, under contract 14-35-0001-3-559.

References

Alverson, D. L., and N. J. Wilimovsky. 1966. Fishery investigations of the southeastern Chukchi Sea. Pages 843–860 *in* N. J. Wilimovsky and J. N. Wolfe, editors. Environment of the Cape Thompson region, Alaska. U.S. Atomic Energy Commission, Washington, DC.

Andriyashev, A. P. 1964. Fishes of the northern seas of the U.S.S.R. Translated from Russian: Israel Program for Scientific Translations, Catalog Number 836, Jerusalem.

Coyle, K. O., J. A. Gillispie, R. L. Smith, and W. E. Barber. 1997. Food habits of four demersal Chukchi Sea fishes. Pages 310–318 *in* J. Reynolds, editor. Fish ecology in Arctic North America. American Fisheries Society Symposium 19, Bethesda, Maryland.

Feder, H. M., A. S. Naidu, M. J. Hameedi, S. C. Jewett, and W. R. Johnson. 1989. The Chukchi Sea continental shelf: benthos–environmental interactions. NOAA (National Oceanic and Atmospheric Administration), OCSEAP (Outer Continental Shelf Environmental Assessment Program) Final Report 68:25–311, Anchorage, Alaska.

Frost, K. J., and L. F. Lowry. 1981. Foods and trophic relationships of cetaceans in the Bering Sea. Pages 825–836 *in* D. W. Hood and J. A. Calder, editors. The eastern Bering Sea shelf: oceanography and resources. NOAA (National Oceanic and Atmospheric Administration) Office of Marine Pollution Assessment, Boulder, Colorado.

Hart, J. L. 1973. Pacific fishes of Canada. Fisheries Research Board of Canada Bulletin 180.

Lowry, L. L., and K. J. Frost. 1981. Feeding and trophic relationships of phocid seals and walruses in the eastern Bering Sea. Pages 813–824 *in* D. W. Hood and J. A. Calder, editors. The eastern Bering Sea shelf: oceanography and resources. NOAA (National Oceanic and Atmospheric Administration) Office of Marine Pollution Assessment, Boulder, Colorado.

Moiseev, P. A. 1953. Cod and flounders of far-eastern waters. Fisheries Research Board of Canada Translation Series 119, Nanaimo, British Columbia.

Pruter, A. T., and D. L. Alverson. 1962. Abundance, distribution and growth of flounders in the southeastern Chukchi Sea. Journal du Conseil Conseil International pour l'Exploration de la Mer 27:81–99.

Quast, J. C., and E. L. Hall. 1972. List of fishes of Alaska and adjacent waters with a guide to some of their literature. NOAA (National Oceanic and Atmospheric Administration) Technical Report NMFS (National Marine Fisheries Service) SSRF (Special Scientific Report Fisheries) 658.

Shmidt, P. Y. 1950. Fishes of the Sea of Okhotsk. Translated from Russian: Israel Program for Scientific Translations, Catalog Number 1263, Jerusalem.

Weingartner, T. J. 1997. A review of the physical oceanography of the northeastern Chukchi Sea. Pages 40–59 *in* J. Reynolds, editor. Fish ecology in Arctic North America. American Fisheries Society Symposium 19, Bethesda, Maryland.

Wolotira, R. J., Jr., T. M. Sample, and M. Morin, Jr. 1977. Demersal fish and shellfish resources of Norton Sound, the southeastern Chukchi Sea, and adjacent waters in the baseline year 1976. NOAA (National Oceanic and Atmospheric Administration) NMFS (National Marine Fisheries Service), Northwest and Alaska Fisheries Center, Processed Report, Seattle.

Wyllie-Echeverria, T., W. E. Barber, and S. Wyllie-Echeverria. 1997. Water masses and transport of age-0 Arctic cod and age-0 Bering flounder into the northeastern Chukchi Sea. Pages 60–67 *in* J. Reynolds, editor. Fish ecology in Arctic North America. American Fisheries Society Symposium 19, Bethesda, Maryland.

Zar, J. H. 1984. Biostatistical analysis, 2nd edition. Prentice-Hall, Englewood Cliffs, New Jersey.

American Fisheries Society Symposium 19:133–139, 1997

Population Biology of the Arctic Staghorn Sculpin in the Northeastern Chukchi Sea

R. L. Smith, W. E. Barber, M. Vallarino, J. Gillispie, and A. Ritchie
University of Alaska, Institute of Marine Science, Fairbanks, Alaska 99775, USA

Abstract.—Distribution, abundance, age, growth, and reproduction were examined for the Arctic staghorn sculpin *Gymnocanthus tricuspis* captured by trawl in the northeastern Chukchi Sea, Alaska. High biomass and numbers generally occurred inshore and south of Icy Cape (70°15′N). Biomass means for 1990 and 1991 were 8.4 and 4.7 kg/km^2, respectively. Mean biomass and abundance were significantly higher in 1990 than in 1991. The oldest female observed was 9 years old; the oldest male was 8. The age structure changed dramatically from 1990 to 1991. In 1990, 42% of the population was greater than or equal to 4 years old; in 1991 that was true of only 9%. After 3 years of age, females grew faster and reached larger size than males. Males began to mature at 60–70 mm standard length, females at about 90 mm. The Arctic staghorn sculpin exhibited interannual variability in distribution, abundance, and age structure. This variability suggests that the species is existing in an unpredictable and dynamic habitat that may result in recruitment failures, mass mortalities, or dispersal of individuals.

The Arctic staghorn sculpin *Gymnocanthus tricuspis* is a circumpolar member of the family Cottidae inhabiting continental shelves of the Arctic and subarctic oceans (Andriyashev 1964). In the subarctic, it is found in the northern Atlantic south to the Gulf of St. Lawrence (Leim and Scott 1966), the Chukchi Sea, and the Bering Sea to the Gulf of Anadyr and Nataliya Bay at 61°N (Andriyashev 1964). It is common in the Chukchi and Beaufort seas (Frost and Lowry 1981). The purpose of our study was to expand the knowledge about the population biology of this species.

Methods

Fish were captured with a National Marine Fisheries Service (NMFS) 83-112 survey otter trawl towed at approximately 2 knots. The trawl had a 25.2-m headrope and a 34.1-m footrope set back 7.1 cm from a tickler chain. The cod end was 90-mm stretched mesh into which a liner of 33-mm stretched mesh was inserted. Each trawl station consisted of two 30-min hauls. Width of the net opening was determined electronically with a Scanmar system incorporating sensor-transmitters attached to the outer wings of the net opening. Area sampled (m^2) was determined by multiplying the width of the net opening by the distance trawled. Distance trawled was determined from the ship's position (latitude and longitude as determined with a Global Positioning System) at the start and end of each haul. Biomass (kg/km^2) and abundance (number of fish/km^2) were calculated by averaging the two 30-min hauls at each station. Mean biomass and abundance values for the 2 years were compared with a Mann–Whitney *U*-test (Zar 1984).

A total of 255 (in 1990) and 315 (1991) fish were returned to the laboratory for examination. We measured total and standard lengths (mm), total fish weight (g), stomach content weight (g), gonad weight (g), and pelvic fin length (mm). Body weight was determined by subtracting stomach content weight from total fish weight. Gonadosomatic index (GSI) is defined as follows: (gonad weight/fish weight) × 100. We removed and preserved otoliths, ovaries, and stomach contents. Analysis of stomach contents is reported by Coyle et al. (1997, this volume). Mean lengths and other indexes were compared using a *t*-test; slopes of regressions were compared using a *t*-statistic (Zar 1984).

Two techniques were used to evaluate age, surface aging, and break-and-burn. In the surface technique, the annulus on the otolith viewed with transmitted light was defined as the translucent zone. In fish younger than or equal to age 4, this zone could be traced entirely around the otolith; in older fish, the zone was incomplete but most easily visible on the rostrum (anterior end). The annulus observed in burnt otolith sections was defined as the dark zone. The surface pattern of annuli on otoliths of 417 fish was read using a dissecting microscope. The ages determined from both techniques were compared for 178 fish—only two did not agree; both differed by 1 year. Otoliths were measured along their longest axis with an ocular micrometer.

Fecundity estimates were determined volumetrically for six of the most mature females. Eggs were separated from ovarian tissue by gently rolling the

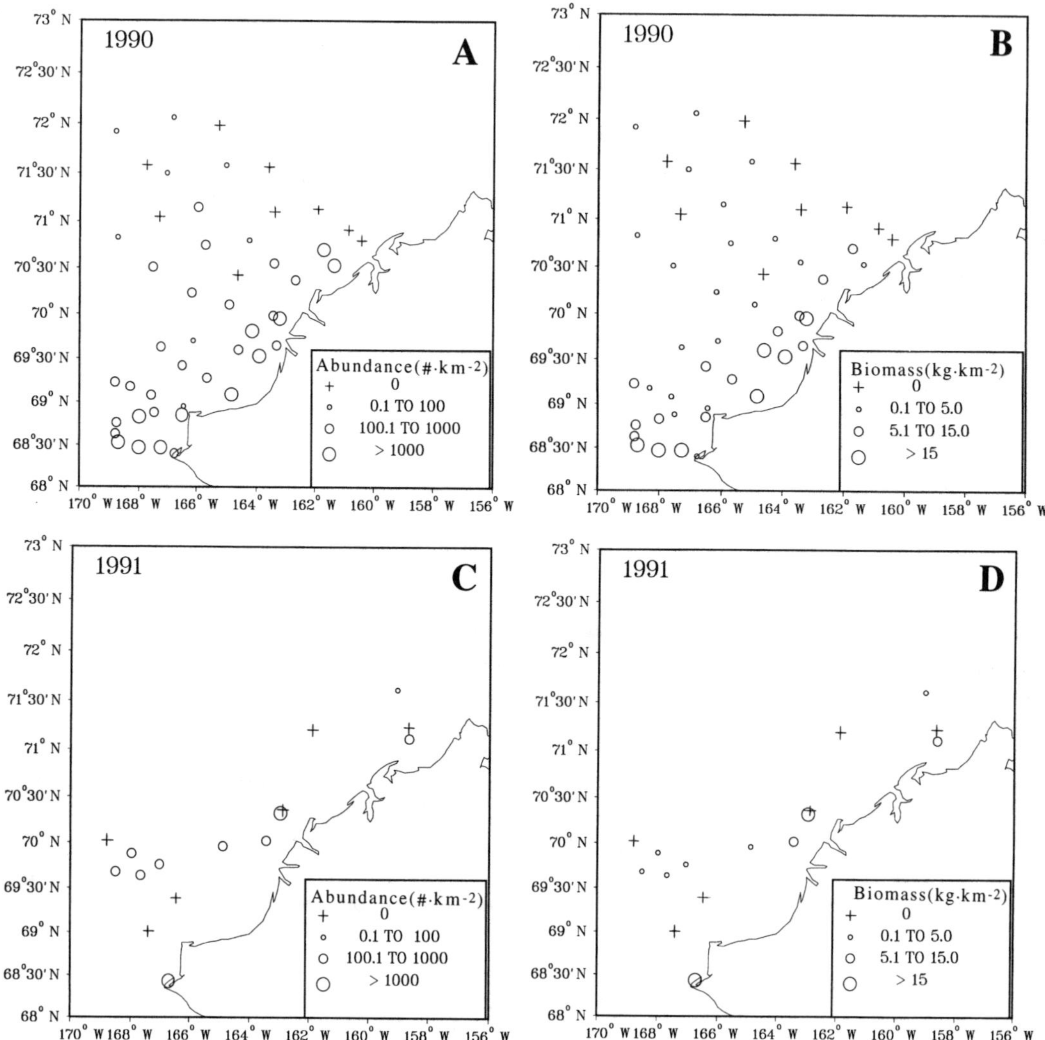

FIGURE 1.—Abundance (numbers of fish/km^2) and biomass (kg/km^2) of Arctic staghorn sculpin *Gymnocanthus tricuspis* in the northeastern Chukchi Sea. (**A**) Abundance in 1990. (**B**) Biomass in 1990. (**C**) Abundance in 1991. (**D**) Biomass in 1991.

ovary between thumb and index finger and then carefully rinsing away the extraneous ovarian tissue. The volume of ova was measured in a graduated cylinder; a sample of 0.5 mL was removed and counted. From this sample 100 ova were randomly selected and measured along their longest axis.

Results

Distribution and Abundance

Arctic staghorn sculpin were present at 39 of 48 stations sampled in 1990. Where present, Arctic staghorn sculpin occurred in numbers ranging from 13 to 8,050 individuals/km^2 (Figure 1A), and biomass varied from 0.45 to 66.6 kg/km^2 (Figure 1B). Generally, high biomass and numbers occurred inshore and south of Icy Cape (70°15′N). Considerable variability in abundance was observed among stations and also between hauls at the same station. For instance, at several stations one of the two hauls yielded no Arctic staghorn sculpin while the other one did. Mean abundance for all 48 stations was 716 (SD = 1,345) fish/km^2; mean biomass for all 48 stations was 8.4 (SD = 13.5) kg/km^2.

TABLE 1.—Body relationships for Arctic staghorn sculpin *Gymnocanthus tricuspis* from the northeastern Chukchi Sea, 1990. Body weight is total fish weight less weight of stomach contents. Carcass weight is body weight less gonad weight.

Dependent variable (Y)	Independent variable (X)	Equation	r^2	N
Total length (mm)	Standard length (mm)	$Y = 1.14X + 2.64$	0.98	274
Otolith length (mm)	Standard length (mm)	$Y = 0.043X + 0.444$	0.92	102
Body weight (g)	Standard length (mm)	$Y = 5.73 \cdot 10^{-6} X^{3.29}$	0.98	274
Eggs/carcass weight (g)	Body weight (g)	$Y = -2.02X + 215$	0.85	6
Ovary weight (g), Aug	Body weight (g)	$Y = 0.109X - 1.09$	0.87	27
Testes weight (g), Aug	Body weight (g)	$Y = 0.023X - 0.017$	0.37	15
Ovary weight (g), Sep	Body weight (g)	$Y = 0.145X - 0.281$	0.91	50
Testes weight (g), Sep	Body weight (g)	$Y = 0.0553X - 0.091$	0.84	40
Pelvic fin length (mm), females	Standard length (mm)	$Y = 0.224X - 4.92$	0.82	40
Pelvic fin length (mm), males	Standard length (mm)	$Y = 0.265X - 5.15$	0.68	39

In 1991 Arctic staghorn sculpin were present at 10 of 17 stations, ranging from 32 to 2,720 individuals/km^2 (Figure 1C) and from 0.7 to 27.5 kg/km^2 (Figure 1D). Again, highest abundances and biomasses were from inshore stations south of Icy Cape. Mean abundance for the 16 stations for which we have data was 429 (SD = 776) fish/km^2; mean biomass was 4.7 (SD = 7.8) kg/km^2. Comparison of abundance and biomass values for the 2 years indicated significantly higher values in 1990 ($P < 0.001$).

Age and Growth

Total length was a linear function of standard length for 1990 fish (Table 1). Otolith length also was a linear function of standard length (Table 1). A standard length–weight regression calculated for all fish sampled in 1990 conformed to a power curve (Table 1).

Maximum longevity among the 417 individuals examined from 1990 and 1991 was 9 years for females and 8 years for males. The age structure of the samples in the northeastern Chukchi Sea differed markedly in 1990 and 1991 (Figure 2). In 1990, 41.6% of the samples consisted of fish greater than or equal to 4 years old; 4.4% were greater than or equal to 6 years old. In contrast, in 1991 only 8.9% of the samples were greater than or equal to 4 years old; only 1.4% were greater than or equal to 6 years old. Among the 270 individuals aged from 1990 only 5 were age 3. Of 147 individuals aged from 1991 only 3 were age 4.

Males and females grew at the same rate in the first 3 years. By the end of the fourth year, however, females were significantly larger (Figure 3). Von Bertalanffy growth curves were calculated and fitted to the combined mean length-at-age data according to

$$L_t = L_\infty [1 - e^{-K(t - t_0)}] \quad (1)$$

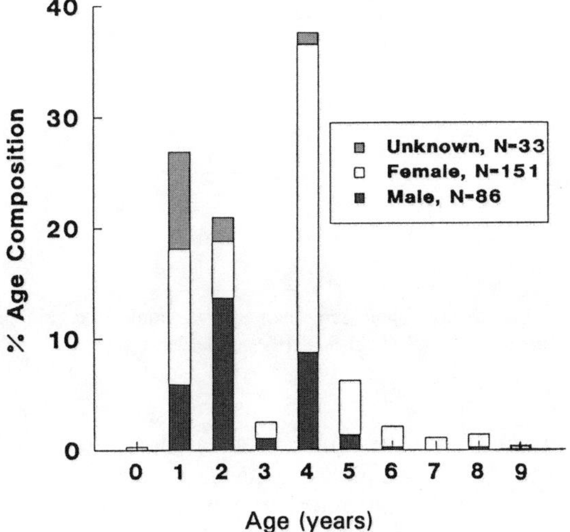

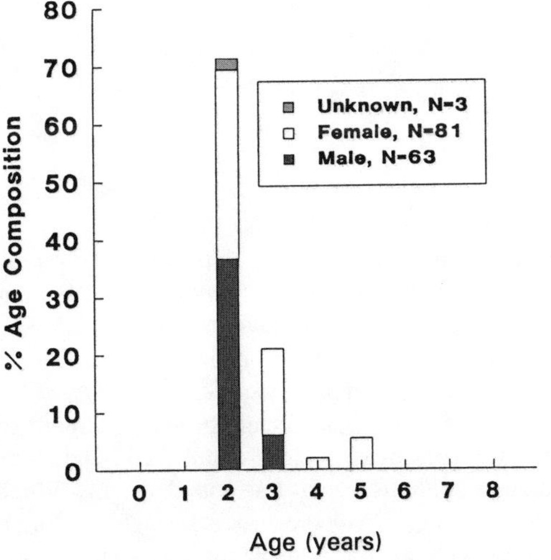

FIGURE 2.—Population age structure of Arctic staghorn sculpin *Gymnocanthus tricuspis* in the northeastern Chukchi Sea (upper panel, 1990 data combined; lower panel, 1991 data combined).

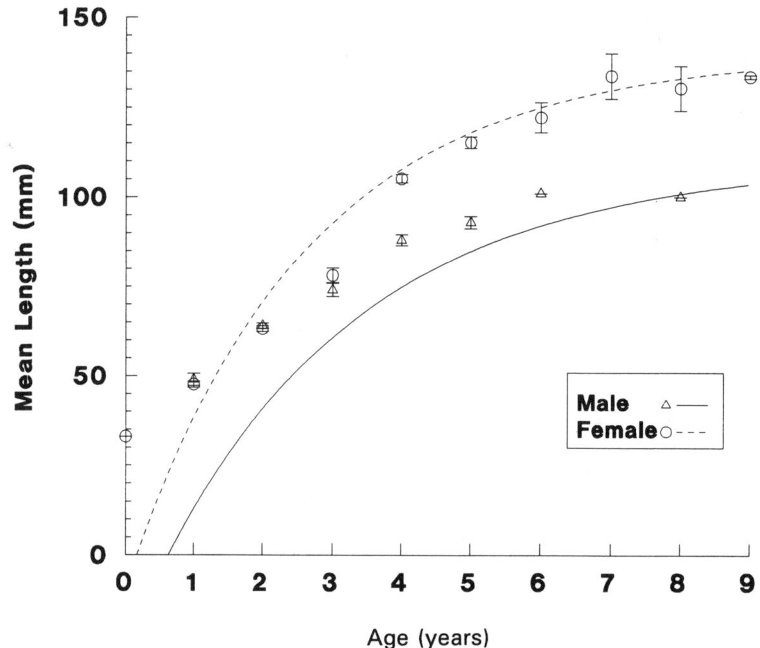

FIGURE 3.—Mean length at age for male and female Arctic staghorn sculpin *Gymnocanthus tricuspis* from the northeastern Chukchi Sea, 1990 and 1991 data.

where t = age (years), L_t = standard length (mm) at age, L_∞ = asymptotic length, K = instantaneous growth coefficient, and t_0 = theoretical age at L_t = 0. For females, the equation was

$$L_t = 140\,[1 - e^{-0.383(t - 0.165)}], \quad (2)$$

with N = 235; for males, the equation was

$$L_t = 110\,[1 - e^{-0.338(t - 0.631)}], \quad (3)$$

with N = 150.

Reproductive Biology

Fecundity, based on egg counts of only six females, ranged from 3,030 to 5,414 eggs in fish ranging from 112 to 134 mm standard length (25.6–57.4 g body weight). Fecundity as a linear function of body weight exhibited a poor fit of the data (F = 4.5; r^2 = 0.53). These fecundity values equate to 154–91 eggs/carcass weight (total fish weight less stomach contents weight less gonad weight). When number of eggs divided by carcass weight is plotted against body weight, a linear regression with negative slope results (Table 1).

Among 1990 Arctic staghorn sculpin, there was wide variation in gonadosomatic index (GSI) at each interval of standard length. Among males, GSI ranged from 0 to 7%; among females, from 0 to 19%. Body weight–gonad weight relationships were developed from fish sampled at station 10 (68°39′N, 168°47′W) on 22 August 1990 and at Station 35 (69°58′N, 163°15′W) on 10 September (Table 1). The slopes of the relationships from the two sample dates were significantly different for females (t = 5.32; $P < 0.001$) and males (t = 6.97; $P < 0.001$).

Males and females of this species are divergent not only in growth rate and longevity but also in pelvic fin development (Figure 4). Males have proportionally longer pelvic fins, and in mature fish, these fins are banded in coloration. Pelvic fin length was a linear function of standard length (SL) for each sex (Table 1). These regressions differed significantly in slope (t = 5.94; $P < 0.001$). The regression lines diverged at about 60 mm SL and were quite distinct at SL greater than 70 mm; sexual maturity in males probably occurs at lengths of 60–70 mm SL. Some males as small as 60–70 mm have testes as large as 6–7% of the body weight.

Discussion

Distribution and Abundance

Distribution of Arctic staghorn sculpin over the study area was not uniform. Replicate trawl hauls at

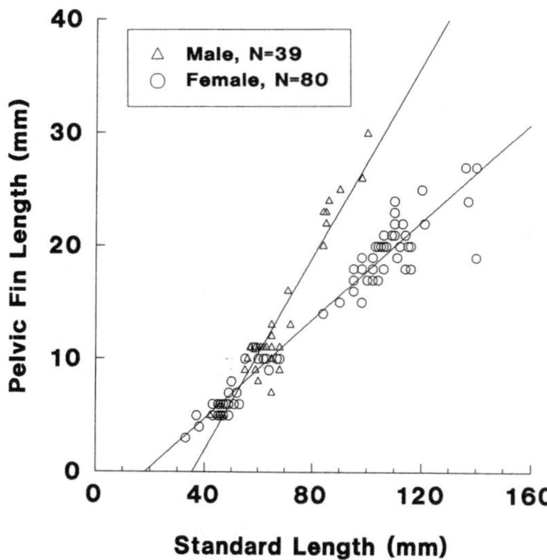

FIGURE 4.—Pelvic fin length as a function of standard length in male and female Arctic staghorn sculpin *Gymnocanthus tricuspis* from the northeastern Chukchi Sea, 1990 data.

the same location did not necessarily agree with respect to the abundance or even the presence of this species. We were unable to discern bottom type directly at trawl stations and therefore could not correlate abundance or presence with bottom type. However, using sediment maps (see Figure 3 in Feder et al. 1989), we concluded that Arctic staghorn sculpin were absent at stations with sediment consisting of mud, gravelly mud, or gravelly sand. Andriyashev (1964) reported that this species burrowed in sandy or sandy–muddy bottoms, was often present on pebbly bottoms, and was rarely encountered on mud or clay bottoms. These observations suggest that much of the variation in abundance of this species may be explicable based on the broad scale and local distribution of sediment types in the Chukchi Sea. Over its entire geographic range, the Arctic staghorn sculpin tolerates considerable ranges in temperature (−1.7 to +12.5°C) and salinity (16–35‰), but it is primarily a species inhabiting temperatures below or close to 0°C and salinities of 32–35‰ (Andriyashev 1964).

Wolotira et al. (1977) used the same gear as that in our study (NMFS 83-112 trawl with the same cod end liner, towed at approximately the same speeds); they reported biomasses of 3.5 kg/km^2 in the southeastern Chukchi Sea, 2.9 kg/km^2 in Kotzebue Sound, and 8.2 kg/km^2 in the northern Bering Sea north of St. Lawrence Island. Our biomasses of 8.4 and 4.7 kg/km^2 for the northeastern Chukchi Sea, in 1990 and 1991 are similar in magnitude.

Age and Growth

Age-3 fish were very scarce in 1990 and age-4 fish were scarce in 1991, suggesting that the 1987 class experienced poor recruitment. This apparent recruitment failure was widespread—we sampled over a wide latitude within the northeastern Chukchi Sea (68°23′N to 71°55′N)—and could have resulted from a large-scale perturbation in the environment.

Males and females grow at similar rates for the first 3 years. Beyond 3 years, females grow faster and ultimately reach a much larger size. Mean length at age in males plateaus at about 110 mm SL, females at about 140 mm SL. Data for Arctic staghorn sculpin from the Laptev Sea (Andriyashev 1964) support our observed disparity in growth rates between sexes in fish older than 3 years.

The largest specimens of Arctic staghorn sculpin, 300 mm TL, came from the western coast of Greenland (Andriyashev 1964). The population off Greenland must find conditions for growth much more favorable than populations in the Chukchi and Laptev seas. The Arctic staghorn sculpin, with a maximum age of 9 years, appears to be shorter lived than its congeners. Tokranov (1988) found maximum ages of threaded sculpin *Gymnocanthus pistilliger* and armorhead sculpin *G. galeatus* from the coast of Kamchatka to be 11 and 13 years for males and females, respectively.

Reproductive Biology

The variation in GSIs in our specimens probably resulted from pooling fish caught over a 1-month period during which gonads were maturing. In comparison to those from station 10, the ovaries of females at Station 35 appeared to be more mature and contained well-developed ova. Based on the slopes of these regressions, one could infer that gonad weight increased from about 2 to 5% of body weight in males and from about 11 to 15% of body weight in females during August. We conclude that the sampling date at Station 35 was nearer the spawning period for this population. Our conclusion is based on an assumption that station location had no effect on maturation schedule.

Based on GSIs and sexual dimorphism in the pelvic fin of Arctic staghorn sculpin, it appears that males become sexually mature at lengths of 60–70 mm SL, corresponding to ages 2 and 3. Andriyashev (1964) indicated that sexual maturity appears to be attained toward the 4th year of life. Dimorphism in pelvic fin length in Arctic staghorn sculpin was also reported by Backus (1957); based on his data (see his Figure 2) only males and females with SL greater than or equal to 120 had different lengths of pelvic fins.

The onset of maturity in females, as indicated by GSI values greater than or equal to 10%, occurs at ages 3 and 4 or at about 90 mm SL. These values agree with those of Andriyashev (1964) and are similar to those reported by Tokranov (1981) for *G. detrisus* in Kamchatka waters. Tokranov (1981) reported first maturity at ages 3 and 4 and at lengths of 180–220 mm. All males were mature at age 5 (220–240 mm), all females at age 6 (240–260 mm).

G. detrisus spawns in December and January in Kamchatka waters when GSI is 17–24%. The September ovary indexes exhibited by Arctic staghorn sculpin (15% of body weight) are approaching the spawning values for *G. detrisus*. Furthermore, Andriyashev (1964) reported female Arctic staghorn sculpin from the Kara and White seas with ripe eggs in the second half of September. Absolute fecundities in our study (3,030–5,414) are similar to those reported for Arctic staghorn sculpin by Andriyashev (1964) (2,060–3,512 for females of 117–158 mm) but lower than the 6,100–72,000 per female reported for *G. detrisus* (Tokranov 1981). The lower fecundity values are undoubtedly a result of the size difference between Arctic staghorn sculpin and *G. detrisus*. Relative fecundity in Arctic staghorn sculpin was 91–154 eggs/g (our study); *G. detrisus* produces 40–225 eggs/g (Tokranov 1981).

Food Webs

The Arctic staghorn sculpin was found as prey in Arctic cod *Boreogadus saida* and Bering flounder *Hippoglossoides robustus* in this research project (Coyle et al. 1997). It also is prey of Atlantic cod *Gadus morhua* in eastern arctic Canada (Dunbar and Hildebrand 1952) and of polar eelpout *Lycodes turneri* at Point Barrow (Walters 1955). Based on the occurrence of pelagic larvae of the Arctic staghorn sculpin in spring and early summer (Pertseva, cited by Andriyashev 1964), this species is susceptible to planktotrophic predators during this season. Age-0 fish recruit to the benthic habitat in late summer (Andriyashev 1964) and, at that time, become available as prey to benthophages.

Sculpins, perhaps including Arctic staghorn sculpin, are an occasional prey of ringed seals *Phoca hispida* in the Chukchi Sea (Lowry et al. 1980). Bearded seals *Erignathus barbatus* also prey on sculpins incidentally but primarily feed on benthic crustaceans and clams (Lowry and Frost 1981), including some of the same genera as are found in the diet of Arctic staghorn sculpin. Gray whales *Eschrichtius robustus* and, to a lesser extent, bowhead whales *Balaena mysticetus* feed on benthic amphipods in the Chukchi Sea (Frost and Lowry 1981) and therefore share a food resource with the Arctic staghorn sculpin.

Acknowledgments

We thank the crew of the *Ocean Hope III* and T. Sample, R. G. Bakkala, C. Armistead, and T. Dark for their assistance. We particularly thank R. M. Meyer for his enthusiastic support and encouragement throughout the study. This study was funded by the Alaska Outer Continental Shelf Region of the Minerals Management Service, U.S. Department of the Interior, Anchorage, Alaska, under contract 14-35-0001-3-559.

References

Andriyashev, A. P. 1964. Fishes of the northern seas of the U.S.S.R. Translated from Russian: Israel Program for Scientific Translations, Catalog Number 836, Jerusalem.

Backus, R. H. 1957. The fishes of Labrador. Bulletin of the American Museum of Natural History 113:273–338.

Coyle, K. O., J. A. Gillispie, R. L. Smith, and W. E. Barber. 1997. Food habits of four demersal Chukchi Sea fishes. Pages 310–318 *in* J. Reynolds, editor. Fish ecology in Arctic North America. American Fisheries Society Symposium 19, Bethesda, Maryland.

Dunbar, M. J., and H. H. Hildebrand. 1952. Contribution to the study of the fishes of Ungava Bay. Journal of the Fisheries Research Board of Canada 9:83–128.

Feder, H. M., A. S. Naidu, M. J. Hameedi, S. C. Jewett, and W. R. Johnson. 1989. The Chukchi Sea continental shelf: benthos–environmental interactions. NOAA (National Oceanic and Atmospheric Administration), OCSEAP (Outer Continental Shelf Environmental Assessment Program) Final Report 68:25–311, Anchorage, Alaska.

Frost, K. J., and L. L. Lowry. 1981. Foods and trophic relationships of cetaceans in the Bering Sea. Pages 825–836 *in* D. W. Hood and J. A. Calder, editors. The eastern Bering Sea shelf: oceanography and resources. NOAA (National Oceanic and Atmospheric Administration), Office of Marine Pollution Assessment, Boulder, Colorado.

Leim, A. H., and W. B. Scott. 1966. Fishes of the Atlantic coast of Canada. Bulletin of the Fisheries Research Board of Canada 155.

Lowry, L. L., and K. J. Frost. 1981. Feeding and trophic relationships of phocid seals and walruses in the eastern Bering Sea. Pages 813–824 *in* D. W. Hood and J. A. Calder, editors. The eastern Bering Sea shelf: oceanography and resources. NOAA (National Oceanic and Atmospheric Administration), Office of Marine Pollution Assessment, Boulder, Colorado.

Lowry, L. L., K. J. Frost, and J. J. Burns. 1980. Variability in the diet of ringed seals, *Phoca hispida*, in Alaska. Canadian Journal of Fisheries and Aquatic Sciences 37:2254–2261.

Tokranov, A. M. 1981. Rate of sexual maturation and fecundity of the staghorn sculpin, *Gymnocanthus detrisus* (Cottidae), off the east coast of Kamchatka. Journal of Ichthyology 21(1):76–82.

Tokranov, A. M. 1988. Reproduction of sculpins of the genus *Gymnocanthus* (Cottidae) in the coastal waters of Kamchatka. Journal of Ichthyology 28(3):124–128.

Walters, V. 1955. Fishes of western Arctic America and eastern Arctic Siberia: taxonomy and zoogeography. Bulletin of the American Museum of Natural History 106:255–274.

Wolotira, R. J., Jr., T. M. Sample, and M. Morin, Jr. 1977. Demersal fish and shellfish resources of Norton Sound, the southeastern Chukchi Sea, and adjacent waters in the baseline year 1976. NOAA (National Oceanic and Atmospheric Administration) NMFS (National Marine Fisheries Service), Northwest and Alaska Fisheries Center, Processed Report, Seattle.

Zar, J. H. 1984. Biostatistical analysis, 2nd edition. Prentice-Hall, Englewood Cliffs, New Jersey.

American Fisheries Society Symposium 19:140–147, 1997

Biology of the Fish Doctor, an Eelpout, from Cornwallis Island, Northwest Territories, Canada

JOHN M. GREEN AND LORRI R. MITCHELL
Memorial University of Newfoundland, Department of Biology
St. John's, Newfoundland, Canada A1B 3X9

Abstract.—Samples of the fish doctor, an eelpout, *Gymnelus viridis* were collected with the aid of scuba from sites 3–30 m deep at Cornwallis Island, Northwest Territories. Specimens were 3–20 cm long and from less than 1 to 86 g total weight. Body coloration was highly variable. Although there were significant differences in the frequency of specific color patterns between males and females, these differences were not reliable predictors of sex. Stomach content analysis revealed that amphipods and polychaetes were the most important prey. Other prey included copepods, mysids, and the siphons of bivalves. Males consumed significantly more amphipods than females; females consumed significantly more polychaetes. Females with mature eggs were 9.7–15.9 cm long. There were 20–82 mature eggs per female; mean egg diameter per female was 1.9–4.6 mm. Only 10.1% of females larger than 9.7 cm had mature eggs. Both egg diameter and egg number were positively correlated with female length. Growth data from length frequency distributions and otoliths fitted to the von Bertalanffy growth equation gave $K = 0.13$ and $L_\infty = 22$. Based on otoliths, the oldest fish were greater than or equal to 12 years.

The fish doctor *Gymnelus viridis* is a shallow-water, benthic zoarcid of arctic and boreal seas. Its habitat ranges from rocky intertidal kelp beds to mud, sand, and gravel substrates. Most specimens have been caught at depths less than 50 m, but some have been taken from at least 256 m (Anderson 1982). The fish doctor has a circumpolar distribution ranging from the Aleutian Islands, Alaska, through the Chukchi Sea eastward discontinuously to the Canadian Arctic Archipelago (Anderson 1982). The distribution records of specimens in the National Museums of Canada have been summarized by Hunter et al. (1984). Despite its widespread distribution and its numerical importance as a component of arctic demersal fish assemblages (Green 1983), little is known of its biology. Anderson (1982) summarized the available life history information and Green (1983) provided data on the food of specimens caught by divers in December and May at Resolute Bay, Cornwallis Island, Northwest Territories.

As part of a continuing study of demersal fish assemblages in arctic Canada, collections of fish doctors were made at Cornwallis Island, Northwest Territories, in August 1990. Herein, we describe aspects of the biology of fish doctors based on underwater observations and an analysis of 321 specimens. In particular, we provide data on body coloration, diet, reproduction, and growth.

Methods

Collecting Methods and Sites

Specimens were collected by divers using scuba at several sites near Resolute Bay, Cornwallis Island, Northwest Territories (74°40′N, 95°52′W) in August 1990. The principal sites were in east Resolute Bay near the mouth of the Mecham River and in Allen Bay at the mouth of the McMaster River (Figure 1). At both sites a relatively steep, cobble and gravel slope, starting at a depth of 2–3 m, terminated in a relatively level, soft substrate. In Resolute Bay the slope was 5–8 m in height, whereas at the McMaster River it began closer to shore and terminated at a depth of 12–15 m. The slopes were devoid of macroalgae, presumably because of their instability and the relatively small size of the cobbles and gravel. At the bottom of the slope at both sites the substrate consisted of sediments rich in organic matter and highly reduced except for a thin (0.5–2 mm) surface layer of oxidized material. At the Mecham River site, scattered clumps of *Desmerestia* and individual *Laminaria* and *Agarum* plants grew where suitable substrates (e.g., stones or debris of human origin) occurred. No macrophytes were present within 5 m of the bottom of the slope at the McMaster River site.

Fish were collected with small (20-cm diameter) fine-mesh dip nets. Starting at the bottom of the slope and working upward, a diver disrupted the substrate, causing the slope material to cascade downward. This disturbance caused fish to swim to the surface of the substrate, where they were easily caught.

Fish were also collected at several other sites, but their densities were much lower at these locations. Other sites consisted of both rocky substrates with dense *Laminaria* stands swept by tidal currents (e.g., off Cape Martyr) and soft substrates with low

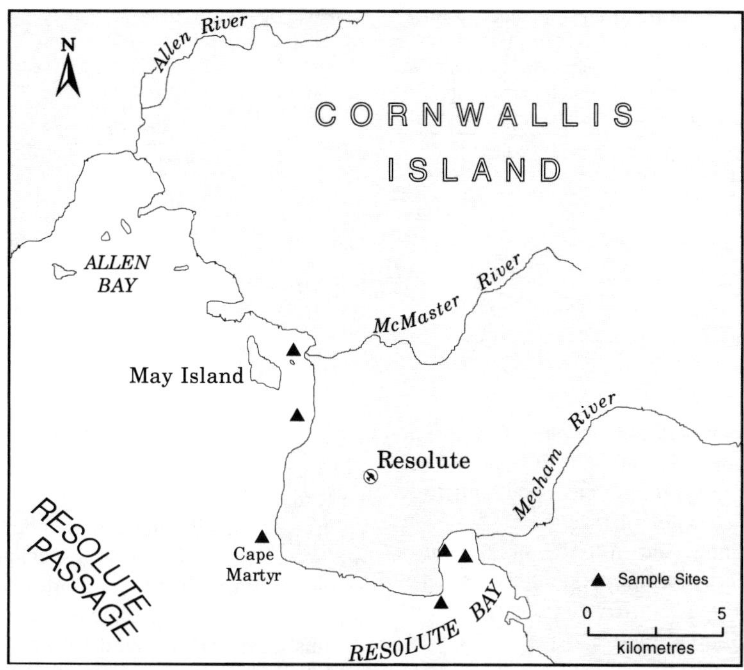

FIGURE 1.—General location of sample sites (black triangles) at Cornwallis Island, Northwest Territories.

densities of attached macrophytes, such as *Agarum* and *Laminaria* (e.g., in west Resolute Bay, Figure 1). At these sites fish were captured in the same way as described above except that individual rocks or bottom debris had to be moved to expose them. During dives it was rare at all sites to see a fish doctor out in the open.

Laboratory Methods

Specimens were initially preserved in 10% formalin and later transferred to 40% isopropyl alcohol. Total length was measured to the nearest 0.5 mm using a measuring board or dial calipers, and total weight was measured to the nearest 0.01 g on a digital balance. Body coloration and the presence and number of ocelli on the dorsal fin were noted at the time of measurement. Fish were sexed by direct observation of the gonads or, if necessary, with the aid of a stereomicroscope. Gonads were removed, blotted dry, and weighed on a digital balance, after which they were stored in 40% isopropyl alcohol. All mature eggs in the ovaries of ripe females were counted and their diameters measured to the nearest 0.01 mm using dial calipers.

Stomachs were removed, blotted dry, and weighed. Stomach contents were removed by making a slit in the stomach wall and flushing the stomach with 40% isopropyl alcohol. Emptied stomachs were blotted dry and reweighed.

The presence or absence and number of prey items of each type were recorded. These data were used to calculate both a percent frequency of occurrence and a percent number. The volume of each food type was measured by the "squash" method (Hyslop 1980). Stomach contents were placed in a petri dish with the bottom marked in square centimeters and pressed down to a uniform depth. The area covered by each prey type was recorded, along with the total area covered. Percent volume for prey items was then calculated.

Three percentage measures (number, frequency, and volume) of each food type for all fish (141) were combined and an index of relative importance (IRI) was calculated as described by Cailliet et al. (1986): IRI = (volume + number) × frequency, all in percent.

Specimens with food in their stomachs were divided into 4-cm length groups and the frequency of occurrence of prey was determined for each size-class. In addition, the frequency of occurrence of food types in 130 fish of known sex were compared using a z-proportion test (Devore 1987) to determine if there were differences in diet between the sexes.

TABLE 1.—Frequency of occurrence, number, volume, and index of relative importance (IRI) of prey items for fish doctors *Gymnelus viridis* from Cornwallis Island, 1990. See text for explanation of IRI. Miscellaneous refers to plant, crustacean, and unidentified tissue remains.

Taxon	Frequency (%)	Number (%)	Volume (%)	IRI
Amphipoda	30	16	19	1,050
Copepoda	25	29	3	800
Mysidacea	6	6	10	96
Polychaeta	30	10	23	990
Bivalvia	13	40	7	611
Miscellaneous	47		37	

TABLE 3.—A comparison of percent frequency of occurrence of prey items in male and female fish doctors *Gymnelus viridis* from Cornwallis Island, 1990. Asterisk indicates statistical significance ($P = 0.05$).

	Frequency of occurrence		
Taxon	Female ($N = 73$)	Male ($N = 57$)	z-value
Amphipoda	29	54	2.89*
Copepoda	26	25	0.13
Mysidacea	4	9	1.17
Polychaeta	52	23	3.36*
Bivalvia	12	16	0.66

Ages were estimated both from otoliths and length frequency data. Otoliths (both right and left sagittae) were removed through a dorsal, longitudinal incision in the center of the head. They were examined while immersed in 100% ethanol on a depression slide using a compound microscope. Total length data for 321 fish were combined into 2-mm increments and plotted as a length frequency histogram.

Results

Body Coloration

The three patterns of body coloration described by Anderson (1982) were observed: a vertical banding pattern with alternating light and dark, usually yellow and green-brown, bands; a mottled pattern with patches of yellow and green-brown pigmentation; and a gray-brown monotone pattern. More females (73%) than males (48%) were banded, whereas roughly the same proportion of females (26%) and males (21%) were mottled. Significantly more males (31%) than females (1%) were monotone. Some males showed distinct melanization of the anal fin, a condition never observed in females. Melanization of the anal fin was always observed in males with a monotone body pattern. Site location did not appear to be associated with body color pattern.

Dorsal fin ocelli ranged in number from one to six and were observed in 36% of specimens. The smallest specimen with one or more ocelli was 4.5 cm long. Of the 321 specimens, 65% had only one ocellus, 22% had two, and 10% had three; one fish each had four, five, and six ocelli. Larger fish tended to have greater numbers of ocelli, but the specimen with five ocelli was only 6.8 cm long; 54% of females had ocelli, compared with 46% of males.

Stomach Analysis

Of the 309 stomachs examined, 63% contained food. Amphipods and polychaetes were the most common prey items and were present in 30% of the stomachs. Copepods were the next most common prey, occurring in 25% of the stomachs. Mysids and bivalves were present in 6% and 13% of the stomachs, respectively (Table 1). Unlike other prey, which had been ingested whole, the bivalves consisted only of nipped-off siphons. The relative value of each prey type based on its abundance, frequency, and volume are presented in Table 1. Based on IRI values, the prey ranked in the following order of importance: Amphipoda, Polychaeta, Copepoda, Bivalvia, and Mysidacea (Table 1).

Of the copepods in stomachs, 54% were harpacticoids and 17% were calanoids. The dominant amphipods were *Gammarus* spp., *Weyprechtia pinguis*, and *Ischryocerus anquipes*. The smallest fish doctors may have a narrower diet than the larger size-classes (Table 2), feeding primarily upon amphipods and copepods. Percent frequency of occurrence data indicated that males consumed significantly more amphipods but fewer polychaetes than did females (Table 3), but both sexes ate the same kinds of prey.

TABLE 2.—Percent frequency of occurrence of prey items for different size ranges of 184 fish doctors *Gymnelus viridis* from Cornwallis Island, 1990. Sample sizes are in parentheses.

	Size range (cm)				
Taxon	2.0–5.9 (11)	6.0–9.9 (59)	10.0–13.9 (58)	14.0–17.9 (22)	18.0–21.9 (4)
Amphipoda	18	27	48	45	50
Copepoda	55	36	22	14	25
Mysidacea	0	0	9	9	25
Polychaeta	9	47	43	45	0
Bivalvia	0	12	17	23	25

Reproduction

Ova size was useful for distinguishing between females that would spawn in the year of capture and those that would not. The smallest female with mature eggs was 9.7 cm long; of females this size or larger, only 8 of 79 (10.1%) would have spawned in the year of capture. The range of number of mature eggs per female was 20–82 and the mean diameter of the eggs in a female was 1.9–4.6 mm (Table 4). Egg number, mean egg diameter, and gonad weight as a percent of total body weight increased with female length (Table 4 and Figure 2).

TABLE 4.—Average number and diameter of ripe eggs and gonad weight for eight female fish doctors *Gymnelus viridis* from Cornwallis Island, 1990.

Total length (cm)	Ripe eggs (*N*)	Mean egg diameter (mm)	Range in diameter (mm)	Ovary weight (g)	100 × Ovary weight/body weight
9.7	20	2.1	1.8–2.4	0.17	4.5
9.8	22	1.9	1.6–2.1	0.17	4.6
10.2	27	2.1	1.8–2.4	0.27	5.3
10.9	33	2.3	1.9–2.5	0.35	5.7
12.7	35	2.5	2.1–2.8	0.41	6.1
13.0	37	3.0	2.7–3.1	0.68	7.7
13.7	25	3.8	3.5–4.3	1.18	8.5
15.9	82	4.6	4.0–5.2	5.94	21

Growth

Modal lengths at age were based both on visual inspection of the length frequency data (Figure 3) and an examination of otoliths from fish representative of the total range of lengths. Although some otoliths were difficult to interpret, what were considered to be annuli could be easily discerned on others. The otolith from a 13.9-cm specimen judged to be at least 6 years old is shown in Figure 4. Based on otoliths, the oldest fish in the sample were at least 12 years old. Fitting the modal length data (Figure 3) to the von Bertalanffy growth equation gave $K = 0.13$ and $L_\infty = 22$ cm. Males were heavier than females of equal length and grew to a larger size than females (Figure 5).

Discussion

Distribution

Fish doctors were observed and collected at all dive sites, including both hard and soft substrates, with and without macrophytic cover, throughout the depth range surveyed (<30 m). The primary factor affecting their distribution and density was the availability of cover. Fish doctors were rarely seen in the open and virtually all specimens were collected after disturbance of the substrate. Suitable cover included rocks, debris of terrestrial origin such as pieces of metal and wood, and prostrate blades of *Laminaria* and *Agarum*. Although it was not uncommon to find more than one fish doctor under the same object, each individual appeared to be physically separated from others.

The only other fishes commonly found in the same under-shelter habitat were several species of eelpout of the genus *Lycodes*. Like fish doctors, *Lycodes* spp. were most abundant in the gravel and cobble slope habitats, but their densities were much lower than those of fish doctors. These eelpouts were also rarely seen in the open.

The highly variable coloration of fish doctors has been noted by other workers. Anderson (1982)

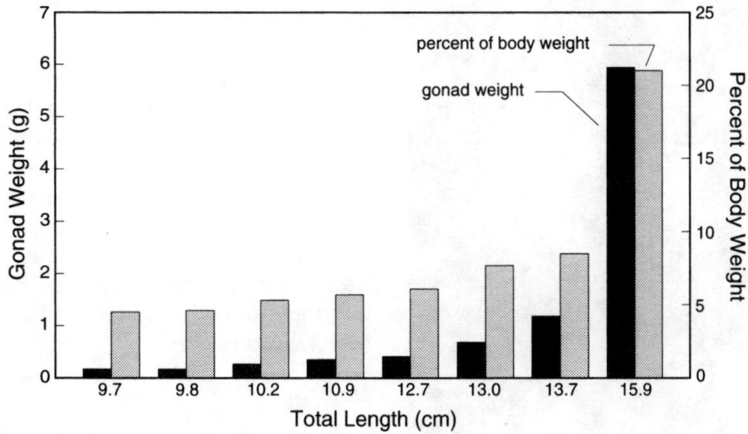

FIGURE 2.—Gonad weight (black bars) and its percentage of body weight (shaded bars) for eight female fish doctors *Gymnelus viridis* from Cornwallis Island, 1990.

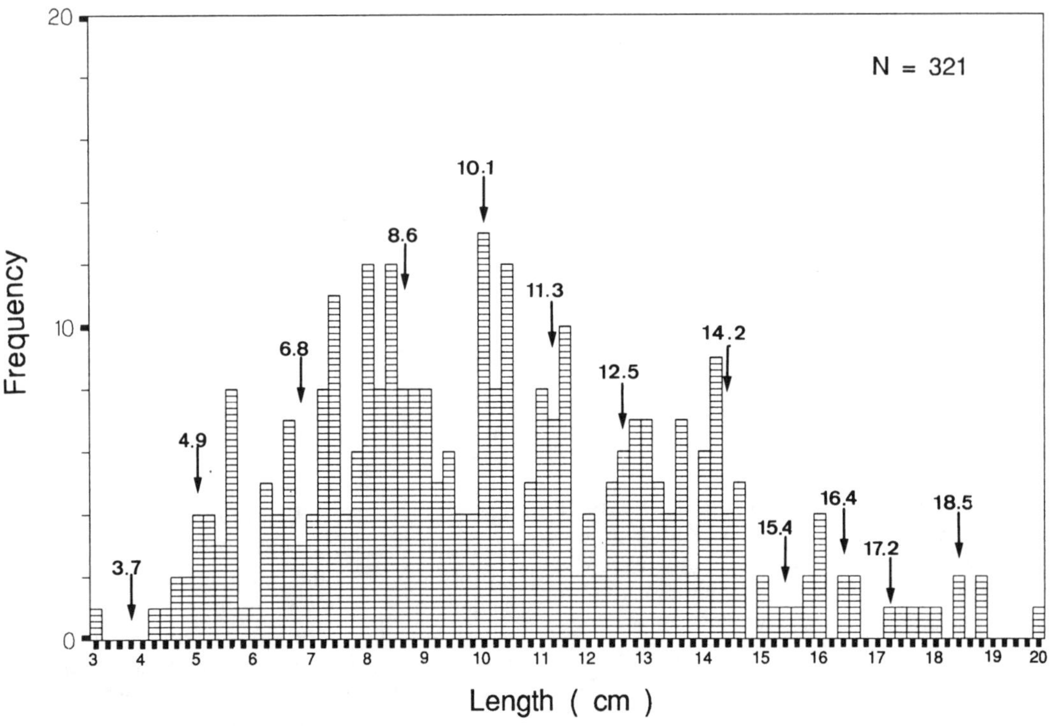

FIGURE 3.—Length frequency data for fish doctors *Gymnelus viridis* (N = 321) from Cornwallis Island, 1990.

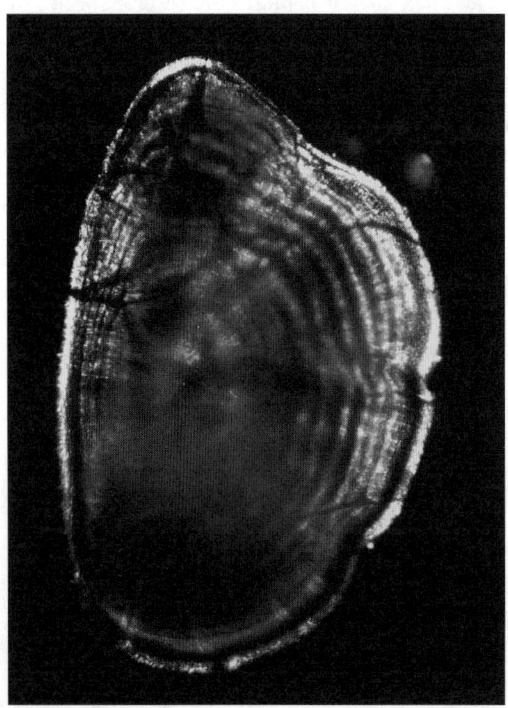

FIGURE 4.—Otolith from a 13.9-cm fish doctor *Gymnelus viridis* from Cornwallis Island, 1990.

summarized the patterns he observed on museum specimens collected throughout the species geographical range. The Cornwallis Island specimens encompassed the range of variation he described. Whereas he noted that males had more ocelli than did females, the opposite was true for our sample.

Our field observations did not indicate that color pattern was related to microhabitat. Although only males were observed to have melanized anal fins, this pattern was not associated with the reproductive state of a male as judged by the size of its gonads. No fish with the bright orange and blue coloration were observed during dives in August at arctic sites (Emery 1973, cited by Anderson 1984). In late August one of us (JMG) observed a single pair of brightly colored fish doctors (deep orange, blue-black, and yellow pigmentation) at a depth of 3 m in Brantford Bay, Boothia Peninsula, Northwest Territories. Subsequent examination showed that these fish were a male and an 11-cm female with 42 mature eggs.

Diet

Because most specimens were collected near midday at a time of year with 24 h of daylight, the percentage of individuals with empty stomachs

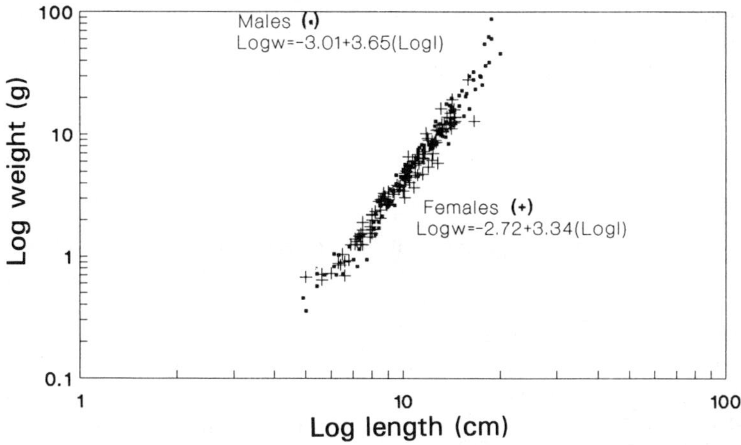

FIGURE 5.—Length–weight relationships of male and female fish doctors *Gymnelus viridis* from Cornwallis Island, 1990. In the equations, l = length and w = weight.

(37%) seems particularly high. There was no indication that fish regurgitated ingested food at the time of capture. In addition, there was no food in the lower digestive tracts of many specimens. Because prey, particularly small crustaceans, seemed to be abundant at the collecting sites, the low incidence of prey in specimens may be indicative of the general sluggishness and inactivity of this species.

Prey items were in general the same as those described by Green (1983) for a much smaller sample from a soft-substrate site in Resolute Bay. The shift from copepods to larger prey as fish doctors increase in length is not unexpected. However, reasons for the differences in prey between males and females of the same size are not readily apparent. Both sexes were taken in the same microhabitats, where they presumably had access to the same array of prey. The relatively larger head, maxillary bone, and adductor mandibulae muscles of males compared with females (Anderson 1982) may account for this difference in prey. The siphons of bivalves were numerous in some individuals and numerically represented an important food item. Their absence from smaller fish is anomalous in view of the small size (2–4 mm × 1 mm) of the siphons. This may be indicative of the strength or experience, or both, required to forage on this prey. With the exception of mysids, prey items were benthic. Mysids were found only in larger fish and were associated with a site in west Resolute Bay where a dense school of mysids was present during the sampling period.

Reproduction

Little is known about the reproductive habits of fish doctors or of other zoarcids. Most information is anecdotal. The fish doctor is reported to spawn in late August or early September, depositing relatively few, large eggs that adhere to the bottom (Anderson 1982, 1984). Our inability to find spawned eggs or to observe pairs of prespawning fish indicate that spawning does not commence before mid-August at Cornwallis Island. Also, none of the females in our collections showed any indication of recently having spawned.

It is both surprising and significant that only 10% of females longer than 9.7 cm (the minimum size of females in which mature eggs were found) were going to spawn in the year of capture. It appears that not only are few eggs produced per female, but that few females mature in any year. As females increase in size, their fecundity increases, as does egg size. As a consequence, the investment in eggs as a percent of total weight increases from about 5.0% for females that are 10 cm long to more than 20% for females greater than 15 cm. Delaying maturity in favor of an increase in body size would have a significant effect on both fecundity and egg size. Evidence is provided by at least one zoarcid that larger larvae have higher survival rates than smaller larvae (Methven and Brown 1991).

Nonannual reproduction is not uncommon in arctic freshwater fishes. Burton (1991) has experimentally demonstrated that the level of energy intake determines whether winter flounder *Pleu-*

ronectes americanus, a north-temperate marine teleost, spawns in a given year. If energy intake is low, which the percentage of specimens with empty stomachs may indicate, it may not be surprising that a low proportion of female fish doctors mature in any given year. Delaying maturity in favor of somatic growth would significantly increase fecundity, especially if females are semelparous. The proportion of males maturing in a particular year is not known, but it also may be low. Competition for those females that do mature may be intense, and this may favor somatic growth over gonadal development in males. The larger heads of males and the bright coloration of spawning pairs may be indications of strong sexual selection.

The time of hatching in fish doctors is unknown. Methven and Brown (1991) found that eggs of ocean pout *Macrozoarces americanus* spawned in the field before October hatched in 3–4 months when held under laboratory conditions. Larvae were well developed at hatching and strongly resembled the adults. Methven and Brown (1991) suggested that a larval stage was absent and that ocean pout hatch as juveniles. This may also be true for fish doctors. If hatching occurs in winter, a benthic rather than a pelagic larval or juvenile stage may have a higher chance of initiating exogenous feeding under arctic winter conditions. One of us (JMG) collected two benthic specimens that were 24–25 mm in late May 1975 in Resolute Bay, neither of which showed any evidence of a yolk sac.

Age and Growth

There are no published data on age and growth of fish doctors. The largest specimens in our collections were approximately 20 cm long, less than the maximum recorded size of 25.6 cm standard length (Scott and Scott 1988). There was good agreement between length and age as determined from analyses of otoliths and length frequency distributions. The oldest fish in our collection were greater than or equal to 12 years old, indicating that these fish are relatively long lived but slow growing.

The length frequency distribution data indicated that fish of ages 0, 1, and 2 were underrepresented in our collections. The reason for this is not clear. It is unlikely that the sampling was biased in favor of the age-3 and older fish. Small fish were as easy to catch as larger fish, and the same sampling techniques have produced different size-frequency distributions at other sites (Green, personal observation). It also seems unlikely that these smaller year-classes were occupying habitats different from those sampled. Although most specimens were taken from the gravel and cobble slope habitats, our collecting efforts and observations covered the range of benthic habitats available to this species. It is possible that our sampling did not reach the depths where the smallest fish were buried. It is also possible that the length-frequency distribution accurately reflects the age structure of a population in which recent recruitment has been low. More fieldwork will be necessary to answer this question.

Acknowledgments

We thank D. Barker and T. P. Birt for help with the collection of samples, D. H. Steele for identifications of stomach contents, and W. Chiasson for assistance with the reading of otoliths. H. Welch, Department of Fisheries and Oceans, Winnipeg, Manitoba, and the Polar Continental Shelf Project provided important logistical support at Resolute Bay. Financial support was from an operating grant from the Natural Sciences and Engineering Research Council and a Northern Science Training Grant (Department of Indian Affairs and Northern Development) to J. M. G.

References

Anderson, M. E. 1982. Revision of the fish genera *Gymnelus* Reinhart and *Gymnelopsis* Soldatov (Zoarcidae) with two new species and comparative osteology of *Gymnelus viridis*. National Museum of Natural Sciences (Ottawa), Publications in Zoology 17:1–76.

Anderson, M. E. 1984. Zoarcidae: development and relationships. Pages 578–582 *in* W. G. Moser and five coeditors. Ontogeny and systematics of fishes, Ahlstrom symposium. American Society of Ichthyologists and Herpetologists, Special Publication 1, Lawrence, Kansas.

Burton, M. P. M. 1991. Induction and reversal of the non-reproductive state in winter flounder, *Pseudopleuronectes americanus* Walbaum, by manipulating food availability. Journal of Fish Biology 39:909–910.

Cailliet, G. M., M. S. Love, and A. W. Ebeling. 1986. Fishes: a field and laboratory manual on their structure, identification and natural history. Wadsworth, Belmont, California.

Devore, J. L. 1987. Probability and statistics for engineering and the sciences, 2nd edition, Brooks/Cole Publishing, Monterey, California.

Emery, A. 1973. Biology survey—summer expedition. *In* Arctic diving. Advisory Committee, Northern Development Department of Indian and Northern Affairs 60, Ottawa. (Not seen; cited in Anderson 1984.)

Green, J. M. 1983. Observations on the food of marine fishes from Resolute Bay, Cornwallis Island, Northwest Territories. Astarte 12:63–67.

Hunter, J. G., S. T. Leach, D. E. McAllister, and M. B. Steigerwald. 1984. A distributional atlas of records of the marine fishes of Arctic Canada in the National Museums of Canada and Arctic Biological Station. National Museum of Canada Syllogeus Series 52.

Hyslop, E. J. 1980. Stomach contents analysis—a review of methods and their application. Journal of Fish Biology 17:411–429.

Methven, D. A., and J. A. Brown. 1991. Time of hatching affects development, size, yolk volume and mortality of newly hatched *Macrozoarces americanus* (Pisces: Zoarcidae). Canadian Journal of Zoology 69:2161–2167.

Scott, W. B., and M. G. Scott. 1988. Atlantic fishes of Canada. Canadian Bulletin of Fisheries and Aquatic Sciences 219.

American Fisheries Society Symposium 19:148–154, 1997

Effects of Low Temperatures and Starvation on Resistance to Stress in Presmolt Coho Salmon

ADAM MOLES, SID KORN, AND STANLEY RICE

National Marine Fisheries Service
Alaska Fisheries Science Center, Auke Bay Laboratory
11305 Glacier Highway, Juneau, Alaska 99801-8626, USA

Abstract.—Arctic winter conditions in Alaska typically cause a decline in juvenile salmonid populations. We reared presmolt coho salmon *Oncorhynchus kisutch* for 5 months in the laboratory at arctic (0.2°C) winter temperatures with and without food. Survival, swimming stamina, salinity tolerance, and disease resistance of these fish were compared with those of control fish reared at subarctic (2°C) and temperate (4°C) winter temperatures. After 150 d of rearing, only starved fish at 4°C had reduced survival: 15% of the fish at 4°C died in the last month of rearing. Although growth at 0.2°C was unexpected, fed fish grew at all three temperatures; fish at 0.2°C grew the least and only after the spring equinox. Swimming stamina was proportional to size in fed fish and was equal in all fed groups when corrected for size. Unfed fish deteriorated and had reduced salinity tolerance, swimming stamina, and disease resistance. Although food limitation is a major factor in overwinter survival of temperate fishes, food limitation is less stressful for fish at arctic temperatures because of their comparatively lower metabolic rates.

Coho salmon *Oncorhynchus kisutch* are found in coastal streams ranging from Monterey, California, north to Point Hope, Alaska, and well inland in the Yukon and Kuskokwim river drainages. Thus, they may experience temperatures ranging from near their tolerance maxima to freezing, sometimes in a single year. Research on salmonid temperature physiology has used fish from streams in the southern part of their range that are not particularly adapted to near-freezing temperatures (Brett et al. 1969). No work has been done on the impacts on salmonids of temperatures below 3°C, common winter temperatures in arctic and subarctic streams.

Abundance of juvenile coho salmon typically declines in the winter (Crone and Bond 1976; Tschaplinski and Hartman 1983; Koski et al. 1984; Murphy et al. 1984). Research examining this decline has focused on temperate streams in the Pacific Northwest and has been attributed to the decline to mortality from freshets, starvation, predation, or disease (Murphy et al. 1984). In Alaska, low winter temperatures may cause greater stress to coho salmon than to fish in areas with warmer winter conditions. Stream temperatures in Alaska fall below 4°C for 5 months or more and are often near 0°C for several months (K. V. Koski, Auke Bay Laboratory, personal communication). Instream winter distribution of juvenile coho salmon is quite different from summer distribution (Heifetz et al. 1986); however, the effects of near-freezing temperatures for several months on fish stocks is largely unknown. Do fish feed at these lower temperatures? Is food availability at these lower temperatures a factor in overwinter survival?

There is a plethora of data on the response of salmonids at the upper reaches of their temperature range but virtually no empirical data at the lower range. It is this lower range that is of particular interest to arctic and subarctic researchers. Based on extrapolation of the temperature–growth curves of Brett et al. (1969), salmon would not be expected to feed or grow at near-freezing temperatures. As such, survival, growth, and the ability to cope with stress should not depend on overwinter feeding.

The hypothesis being tested was that food availability has less of an impact on overwinter survival and stress response than does rearing temperature. Juvenile coho salmon were reared overwinter at simulated arctic (0.2°C), subarctic (2°C), and temperate (4°C) temperatures with and without food. Subarctic fish from southeastern Alaska were used because they are exposed to both freezing and fairly warm temperatures in the winter and because this location is midpoint in the species range distribution. Resistance to stress was measured in tests of salinity tolerance, swimming stamina, and disease resistance. Salinity tolerance, although not an overwinter mortality factor, is often used as a general measure of stress response in salmonids.

Methods

Presmolt coho salmon were obtained in November 1987 from the Snettisham Hatchery located 35 km southeast of Juneau, Alaska. All tests were conducted at the Auke Bay Laboratory at Juneau. Fish (mean weight: 5.6 g; mean fork length: 8.2 cm) were transferred to six fiberglass tanks (208 × 55 × 56 cm high) and acclimated to the rearing temperatures of 0.2, 2, and 4°C (two tanks per temperature). Each tank held 450 fish. Loading was adjusted every 30 d by removing fish for growth determinations. Excess fish were to ensure that loading never exceeded 2,500 g per tank. To assess the role of food availability in temperature effects, three tanks of fish were fed Oregon Moist Pellets[1] ad libitum at each of 0.2, 2, and 4°C. The remaining three tanks of fish were reared at the same three temperatures but were not fed. Dead fish were removed from the tanks daily. Condition factors, fork lengths, and body, liver, and gut weights were determined every 30 d.

Temperature control was maintained with chilled ethylene glycol circulated through stainless steel heat exchangers beneath false bottoms of the experimental tanks. Thermoregulators signaled a pump to push chilled glycol through the heat exchanger on demand. This was balanced against the heating effects of the air and incoming water. Temperature was monitored continuously using electronic thermometers calibrated to National Bureau of Standards thermometers. The oxygen content of the tanks was always at full saturation because of the constant agitation of the water around the chillers. Temperature was maintained within 0.01°C in the 0.2°C tanks and within 0.04°C in the other tanks. New water entered the tanks at the rate of 1 L/min. Each tank was monitored for changes in ammonia, pH, and temperature over time. Ammonia remained at undetectable amounts (<0.001 mg/L) throughout the tests; pH varied between 6.4 and 6.6. Laboratory photoperiod was adjusted using artificial light to match natural photoperiod.

On 14 May (after 150 d of cold-temperature rearing), fish in each tank were gradually (1°C/d) acclimated to an ambient temperature of 7°C for immediate salinity, stamina, and disease testing. The new temperature simulated warming after the spring ice breakup. The period of cold water rearing simulated a typical winter for each of the habitats and used natural photoperiod. All stress tests occurred in May, before the coho parr would undergo smoltification.

[1]Reference to trade names does not imply endorsement by the National Marine Fisheries Service, National Oceanic and Atmospheric Administration.

Salinity Tests

Salinity bioassays (96 h) were run in December 1987 at the start of the rearing period (baseline), at 60 d, and at the end of the testing period (150 d) to determine the tolerance response curve. After preliminary exposures to determine the dose range, four 600-L tanks were filled with varying proportions of freshwater and salt water to yield final salinities of 22, 25, 27, and 29‰. Sodium chloride was added to four similar tanks filled with salt water to attain salinities of 31, 33, 35, and 38‰. Temperature, oxygen, and circulation were maintained in the static tanks by means of Frigid Unit water chillers, which kept the external bath temperature at 7°C. Each tank was divided into six compartments where 10–12 fish from each of the three test and control groups were placed. Mortality, oxygen concentration, ammonia, and pH were recorded at 12-h intervals over a 96-h period.

Median salinity tolerance at which one-half of the fish died (LC_{50}) at each 12-h interval was calculated by logit analyses (Silverstone 1957).

Stamina Tests

Fish used for stamina tests were held without food at 4°C for 24 h before testing. The fish were previously maintained for 14 d at 7°C in rectangular living-stream tanks with a current speed of 5 cm/s—a mild form of exercise to prepare the fish for stamina tests.

We tested the swimming ability of the fish in a stamina tunnel, which has a 30.5-cm diameter, 182-cm long plexiglass tube with a screen at the forward end and an electrical grid at the rear. The tube is designed for uniform velocity across the cross-sectional swimming area. The tube has a large tank at each end and is plumbed in a loop with a variable speed pump. Twelve fish from each of the 6 treatments (72 total) were placed in the tunnel; fish were coded-wire tagged to identify them as to treatment group. The stamina tests were replicated six times. Fish were forced to swim against progressively higher water velocities (19.8, 31.1, 39.6, 49.4, and 59.8 cm/s), increased at 30-min intervals. When fish became fatigued, they fell back into the electrical field, which stimulated them to continue swimming until they were completely exhausted. The first velocity interval was the minimum possible in the tunnel and served as a 30-min acclimation period. The duration in minutes until the fish became exhausted and were forced past the electrical grid, wet weight, fork length, and tag information were recorded for each fish. Fish that exited the tunnel during the first 30 min were not included in the calculation of swimming performance. One-way analysis of variance (ANOVA) was used to determine differences in the fork lengths of tested fish attributable to feeding regime, temperature, and replicate treatment.

Critical swimming speed (CSS) was calculated using the formula

$$CSS = V_p + (V_f - V_p) \times m/30; \qquad (1)$$

V_f = water velocity at fatigue (cm/s);
V_p = penultimate velocity (cm/s);
m = time in minutes spent at fatigue velocity.

To standardize performance measurements for different sizes of fish, CSS was also converted to fish length per second by dividing by fork length. One-way ANOVA on performance indexes were then used to determine effects of temperature and feeding regime on CSS. It is recognized that the loading of 72 fish in the chamber at one time would have a significant hydrodynamic interference, thus making the results more relatively comparative within this study than comparative across different studies. The replication of each test six times gave a separation between groups of their relative swimming stamina.

Disease Challenge Tests

At the end of the 5-month cold-temperature rearing period, 96 fish from each of the six treatment groups were subdivided into six groups of 16 fish each. Five of these groups received a different inoculation dose of *Renibacterium salmoninarum* (Margee River strain), the causative organism for bacterial kidney disease; inoculation doses were 9.8×10^{-1}, $\times 10^{-2}$, $\times 10^{-4}$, $\times 10^{-6}$, and $\times 10^{-8}$ colony-forming units per 0.1 mL of broth. Each fish was injected intraperitoneally with 0.1 mL of inoculum after anesthesia with MS222. The sixth subgroup served as a control, in which each fish was sham inoculated with modified KDM2 (Evelyn 1977) sterile broth.

Each group of 16 fish was placed in a separate compartment with its own flow-through freshwater source and observed daily for mortalities for 77 d. All fish, including those previously unfed, were now fed daily. Ambient water temperatures ranged from 6.7°C at the start of the experiment to 7.5°C at termination. Dead fish and survivors were frozen at −20°C for later necropsy and immunofluorescent antibody tests to establish the presence and relative abundance of the *Renibacterium* organism in each group of fish.

TABLE 1.—Survival and mean sizes (±95% confidence intervals) of coho salmon *Oncorhynchus kisutch* reared at low temperatures for 150 d (N = 25 per group).

Group	Survival (%)	Length (mm)	Wet weight (g)	Condition factor
Pretreatment values				
		82 ± 2	5.65 ± 0.44	1.01 ± 0.02
Values after 150 d				
4.0°C fed	100	97 ± 3	10.00 ± 0.90	1.08 ± 0.03
2.0°C fed	100	97 ± 3	10.20 ± 1.10	1.10 ± 0.02
0.2°C fed	100	84 ± 4	6.47 ± 0.96	1.04 ± 0.03
4.0°C unfed	85	76 ± 4	3.92 ± 0.68	0.85 ± 0.02
2.0°C unfed	99	74 ± 4	3.93 ± 0.65	0.92 ± 0.05
0.2°C unfed	100	79 ± 3	4.62 ± 0.51	0.91 ± 0.02

Results and Discussion

Only the fish held at 4°C without food for 150 d experienced mortalities that could be attributed to rearing conditions. In the 5th month of exposure, 15% (52/350) of the starved fish held at 4°C died (Table 1). Survival in all other groups was at least 99%.

Temperature and starvation affected the final size and condition of the fish. Fed fish increased in total body weight by 15% (0.2°C), 82% (2.0°C), and 79% (4.0°C) from an initial weight of 5.7 g (Table 1). Among unfed fish, the fish held at 0.2°C decreased 18% in weight and those held at 2°C and 4°C decreased 31% in weight. Changes in gut and liver weights and in fork lengths followed the same trend as those given above for body weight.

Fish offered food at 0.2°C ate and grew only after the spring equinox. After this date (day 96), there was food present in the viscera and a measurable increase in fish size. The 15% increase in size during the last 54 d of rearing corresponds to a growth rate of 0.28%/d. Brett et al. (1969) calculated that sockeye salmon *Oncorhynchus nerka* should have nearly zero growth at 0°C, and Brett and Higgs (1970) extrapolated that the digestion rate in sockeye salmon approaches zero at 0°C. Growth at these low temperatures does occur but appears to be related to photoperiod.

Salinity Tests

Starvation for 150 d significantly decreased salinity tolerance, but temperature did not affect salinity tolerance (Table 2). The median tolerance (LC_{50}) of fed coho salmon to salt water remained near 30‰ from the beginning of the study until the end across all acclimation temperatures (Table 2). In contrast, salinity tolerance of starved fish dropped

TABLE 2.—Ninety-six-hour median lethal concentrations (LC_{50}) for salinity (‰) tolerance of coho salmon *Oncorhynchus kisutch* reared at low temperatures. Asterisks denote significant differences from controls (one-way analysis of variance, $P < 0.05$).

Temperature	Duration (d) 0	60	150
	Fed fish		
4°C	30.0	30.0	30.5
2°C			33.2
0.2°C		30.1	30.0
	Unfed fish		
4°C	30.0	29.0	24.8*
2°C			23.6*
0.2°C		29.4	24.8*

TABLE 3.—Critical swimming speeds (±95% confidence intervals) of juvenile coho salmon *Oncorhynchus kisutch* after 150 d of rearing at cold temperatures. All stamina tests were run after acclimation to 7°C. Asterisks denote significant differences from 4°C fed fish (one-way analysis of variance, $P < 0.01$* and $P < 0.001$**).

Temperature	Speed (cm/s)	Fork length/s
	Fed	
4°C	34.3 ± 2.0	3.70 ± 0.25
2°C	36.4 ± 4.1	3.89 ± 0.19
0.2°C	32.2 ± 1.4*	3.97 ± 0.15
	Starved	
4°C	25.1 ± 1.7**	3.23 ± 0.20**
2°C	23.7 ± 1.8**	3.09 ± 0.21**
0.2°C	25.0 ± 1.7**	3.25 ± 0.20**

significantly from 30 to about 24‰. This drop occurred during the last 90 d of rearing.

In the starved fish, salinity tolerance decreased in the last 90 d of the test, even though we expected increasing photoperiod to stimulate production of gill adenosine triphosphatase (ATPase) and therefore increased salinity tolerance. The lower condition factors and the inability to tolerate seawater imply that starved fish may be forced to spend another year in the stream before smoltification and migration can occur. Anadromous salmonids often reside 1 year longer in Alaskan streams than do salmonids in more southerly streams (Armstrong 1970; Crone and Bond 1976), probably because of a combination of reduced growing season and loss of condition in long and severe winters.

The synergistic effect of prolonged starvation and salinity tolerance is not well known; however, smaller salmonids are less able to osmoregulate in seawater than larger salmonids (Gould et al. 1985). Decreased quantities of gill ATPase are responsible. In addition, starvation is known to lower albumen (Love 1970) and plasma hormone levels (Donaldson et al. 1979), both of which are important for salinity adaptation (Hoar 1976; Folmar and Dickhoff 1979).

Stamina Tests

Fish that were fed during the controlled temperature exposures performed about 50% better ($P < 0.001$) than did starved fish in the stamina tunnel (Table 3). The critical swimming speed of fed fish ranged from 32 cm/s (fish reared at 0.2°C) to 36 cm/s (fish reared at 2°C). The differences in performance between the three temperature groups of fed fish were small, reflecting the small range of temperature treatments.

Fed fish reared at 2°C and 4°C performed significantly better ($P < 0.01$) than did fed fish reared at 0.2°C, but not when the effect of length was negated. When swimming speed was corrected for length of fish, the range of CSS was small for fed fish: 3.7 to 3.97 fork lengths/s, with the best performance from the fish reared at 0.2°C. This may reflect the more fusiform shape indicated by slightly lower condition factor. Body shape can affect drag and the more fusiform fish are known to have superior swimming performance (Taylor and McPhail 1985).

Unfed fish performed poorly in the stamina tests compared with fed fish (Table 3). The CSSs of starved fish ranged from 23.7 to 25.1 cm/s and were significantly different from those of the fed fish. There were no differences among the three temperature treatments (Table 3). Not all of the differences in CSSs between fed and starved fish can be attributed to size alone. The fed fish still outperformed the starved fish when CSSs were corrected for lengths ($P < 0.001$). The condition of the starved fish had deteriorated enough to have a disproportionate impact on CSSs.

The swimming stamina of fish after 150 d of rearing at low temperatures was affected primarily by two factors: size of fish and condition of fish. Both of these factors were affected more by feeding than by the different temperature treatments. Fork length has long been known to affect swimming speed of salmonids (Brett 1982). Glova and McInerney (1977) demonstrated that fish length is a critical factor in swimming speeds of juvenile coho salmon, especially at higher temperatures, and estimated critical swimming speeds (CSSs) of 3–4 lengths/s for coho salmon at 3°C and 75- to 100-mm fork length. Flagg et al. (1983) and Besner and

TABLE 4.—Cumulative mortality of fed and starved coho salmon *Oncorhynchus kisutch* 77 d after intraperitoneal inoculation of *Renibacterium salmoninarum* (N = 16 per group).

Inoculation dose[a]	Total number of fish dying per group of fish					
	Fed			Unfed		
	4°C	2°C	0.2°C	4°C	2°C	0.2°C
9.8×10^{-1}	0	0	2	12	9	12
9.8×10^{-2}	1	1	1	5	8	4
9.8×10^{-4}	1	0	1	3	8	5
9.8×10^{-6}	0	0	2	2	7	5
9.8×10^{-8}	2	0	0	4	7	5
Controls	0	0	0	7	9	3

[a]Number of colony-forming units, given in 0.1 mL.

Smith (1983) also found length-specific swimming speeds of 3.1–6.2 lengths/s for coho salmon of a similar size range. Our values (3.09–3.97 fork lengths/s) are similar.

When CSSs were corrected for fish length, stamina in starved fish was still significantly less than in fed fish. Stamina is reduced as starvation depletes major energy sources, such as glucose (Barton et al. 1988) and body lipids (Jezierska et al. 1982). Snyder (1980) found that coho salmon (7.1 cm) were unable to maintain their position in 45-cm/s velocity after only 3 weeks of starvation.

Swimming performance and size are important related factors that affect the survival of juvenile salmonids. Bams (1967) showed a direct relationship between sustained swimming performance and survival of predation in emergent sockeye salmon and proposed that, in the absence of severe environmental conditions, the most important component of juvenile salmon survival was stamina. Overwinter survival of coho salmon is directly related to length (Hartman et al. 1987). Larger body size enhances survival because more energy reserves can be stored.

Disease Challenge Tests

Unfed fish, whether injected with *Renibacterium salmoninarum* or not, showed higher mortalities than fed fish (Table 4). Few fed fish died and regressions of temperature or injection titers on mortality were not significant. All the injected, starved fish had high levels of bacterial kidney disease (+++ or greater antibody response), whereas the uninjected control fish had no measurable antibody response. Death in the disease-challenge fish, therefore, was caused by prolonged starvation rather than by disease.

Starvation had more of an effect on survival than did disease or temperature. Prolonged starvation can cause irreversible effects (Bilton and Robins 1973), even when feeding is resumed. In our study, many of the starved fish held in the disease challenge tanks died; those injected with bacteria and the controls died in equal numbers. Many fish that had been held for 5 months without food never resumed feeding. The starved fish were offered food after injection and ate voraciously. The first mortalities occurred 1 month after feeding resumed and continued at a steady pace for the next 2 months.

The most surprising result was the lack of clinical disease and mortality in the fed groups of fish when severely challenged with a high dose of the bacterium. Similar experiments with *R. salmoninarum* in fed rainbow trout *Oncorhynchus mykiss* reared at 11°C showed a mean death time of 30.5 d (range, 19–40 d) when fish were inoculated with 4.4×10^{-7} colony-forming units (McCarthy et al. 1984). Arctic and subarctic fishes may be resistant to the strain of *Renibacterium* used. There is no evidence that different *Renibacterium* isolates vary in virulence; however, recent studies with monoclonal antibodies have shown serological heterogeneity (Arakawa et al. 1987) among bacterial kidney disease isolates rather than homology, as was once thought (Bullock et al. 1974; McCarthy et al. 1984).

Implications for Overwinter Survival

Overwinter survival of salmonids appears to be related to fish size and condition (Reimers 1963; Hunt 1969) and availability of food (Elwood and Waters 1969). Starvation and long-term cold temperatures are two probable stressors for juvenile salmonids overwintering in streams in Alaska. Lack of food has a greater effect on survival because prolonged starvation not only affects the size of the fish but also their ability to cope with stress. Fish starved for 5 months lose their ability to tolerate full-strength seawater, and swimming stamina is decreased. These effects are not explained by size alone; starvation and temperature are interrelated in that fish at near-freezing temperatures may lose weight and experience a decline in condition more slowly than fish at slightly warmer temperatures. The impacts of food on overwinter survival therefore vary with temperature.

Acknowledgments

We thank Sally Short and Theodore Meyers of the Alaska Department of Fish and Game's Pathology Laboratory for performing the immunofluorescent antibody analyses and for assistance in preparing the manuscript.

References

Arakawa, C. K., J. E. Sanders, and J. L. Fryer. 1987. Production of monoclonal antibodies against *Renibacterium salmonarum*. Journal of Fish Diseases 10: 247–253.

Armstrong, R. H. 1970. Age, food, and migration of Dolly Varden smolts in southeastern Alaska. Journal of the Fisheries Research Board of Canada 27:991–1004.

Bams, R. M. 1967. Differences in performance of naturally and artificially propagated sockeye salmon migrant fry, as measured with swimming and predation tests. Journal of the Fisheries Research Board of Canada 24:1117–1153.

Barton, B. A., C. B. Schreck, and L. G. Fowler. 1988. Fasting and diet content affect stress-induced changes in plasma glucose and cortisol in juvenile chinook salmon. Progressive Fish-Culturist 50:16–22.

Besner, M., and L. S. Smith. 1983. Modification of swimming mode and stamina in two stocks of coho salmon (*Oncorhynchus kisutch*) by differing levels of long-term continuous exercise. Canadian Journal of Fisheries and Aquatic Sciences 40:933–939.

Bilton, H. T., and G. L. Robins. 1973. The effects of starvation and subsequent feeding on survival and growth of Fulton Channel sockeye salmon fry (*Oncorhynchus nerka*). Journal of the Fisheries Research Board of Canada 30:1–5.

Brett, J. R. 1982. The swimming speed of adult pink salmon, *Oncorhynchus gorbuscha*, at 20°C and a comparison with sockeye salmon, *O. nerka*. Canadian Technical Report of Fisheries and Aquatic Sciences 1143.

Brett, J. R., and D. A. Higgs. 1970. Effect of temperature on the rate of gastric digestion in fingerling sockeye salmon, *Oncorhynchus nerka*. Journal of the Fisheries Research Board of Canada 27:1767–1779.

Brett, J. R., J. E. Shelbourne, and C. T. Shoop. 1969. Growth rate and body composition of fingerling sockeye salmon, *Oncorhynchus nerka*, in relation to temperature and ration size. Journal of the Fisheries Research Board of Canada 26:2363–2394.

Bullock, G. L., H. M. Stuckey, and P. K. Chen. 1974. Corynebacterial kidney disease of salmonids: growth and serological studies on the causative bacterium. Applied Microbiology 23:811–814.

Crone, R. A., and C. E. Bond. 1976. Life history of coho salmon, *Oncorhynchus kisutch*, in Sashin Creek, southeastern Alaska. U.S. National Marine Fisheries Service Fishery Bulletin 74:897–923.

Donaldson, E. M., U. H. M. Fagerlund, D. A. Higgs, and J. R. McBride. 1979. Hormonal enhancement of growth. Pages 455–597 *in* W. S. Hoar, D. J. Randall, and J. R. Brett, editors. Fish physiology, volume 8. Bioenergetics and growth. Academic Press, New York.

Elwood, J. W., and T. F. Waters. 1969. Effects of floods on food consumption and production rates of a stream brook trout population. Transactions of the American Fisheries Society 98:253–262.

Evelyn, T. P. T. 1977. An improved growth medium for the kidney disease bacterium and some notes on using the medium. Bulletin de l'Office International des Epizooties 87:511–513.

Flagg, T. A., E. F. Prentice, and L. S. Smith. 1983. Swimming stamina and survival following direct seawater entry during parr–smolt transformation of coho salmon (*Oncorhynchus kisutch*). Aquaculture 32:383–396.

Folmar, L. C., and W. W. Dickhoff. 1979. Plasma thyroxine and gill NA^{+}–K^{+} ATPase changes during seawater acclimation of coho salmon, *Oncorhynchus kisutch*. Comparative Biochemistry and Physiology 63:329–332.

Glova, G. L., and J. E. McInerney. 1977. Critical swimming speeds of coho salmon (*Oncorhynchus kisutch*) fry to smolt stages in relation to salinity and temperature. Journal of the Fisheries Research Board of Canada 34:151–154.

Gould, R. W., and five coauthors. 1985. Seawater acclimation of premigratory (presmolt) fall chinook salmon: a possible new management strategy? Pages 15–19 *in* C. J. Sindermann, editor. Proceedings of the 11th U.S.–Japan meeting on aquaculture, salmon enhancement. NOAA (National Oceanic and Atmospheric Administration) Technical Report NMFS (National Marine Fisheries Service) 27.

Hartman G., J. C. Scrivener, L. B. Holtby, and L. Powell. 1987. Some effects of different streamside treatments on physical conditions and fish population processes in Carnation Creek, a coastal rain forest stream in British Columbia. Pages 330–372 *in* E. O. Salo and T. W. Cundy, editors. Streamside management: forestry and fisheries interactions. Institute of Forest Resources Contribution 57, University of Washington, Seattle.

Heifetz, J., M. L. Murphy, and K. V. Koski. 1986. Effects of logging on winter habitat of juvenile salmonids in Alaskan streams. North American Journal of Fisheries Management 6:52–58.

Hoar, W. S. 1976. General and comparative physiology, 2nd edition. Prentice Hall, Englewood Cliffs, New Jersey.

Hunt, R. L. 1969. Overwinter survival of wild fingerling brook trout in Lawrence Creek, Wisconsin. Journal of the Fisheries Research Board of Canada 26:1473–1483.

Jezierska, B., J. R. Hazel, and S. D. Gerking. 1982. Lipid mobilization during starvation in the rainbow trout, *Salmo gairdneri* Richardson, with attention to fatty acids. Journal of Fish Biology 21:681–692.

Koski, K. V., J. Heifetz, S. Johnson, M. Murphy, and J. Thedinga. 1984. Evaluation of buffer strips for protection of salmonid rearing habitat and implications for enhancement. Pages 138–155 *in* T. J. Hassler, editor. Proceedings of the Pacific Northwest stream habitat management workshop. Humboldt State University, Arcata, California.

Love, R. M. 1970. The chemical biology of fishes. Academic Press, New York.

McCarthy, D. H., T. R. Croy, and D. F. Amend. 1984. Immunization of rainbow trout, *Salmo gairdneri*

Richardson, against bacterial kidney disease: preliminary efficacy evaluation. Journal of Fish Disease 7:65–71.

Murphy, M. L., and five coauthors. 1984. Role of large organic debris as winter habitat for juvenile salmonids in Alaska streams. Pages 251–262 *in* Proceedings of the 64th annual conference of the Western Association of Fish and Wildlife Agencies. Western Association of Fish and Wildlife Agencies, Sacramento, California.

Reimers, N. 1963. Body condition, water temperature, and over-winter survival of hatchery-reared trout in Convict Creek, California. Transactions of the American Fisheries Society 92:39–45.

Silverstone, H. 1957. Estimating the logistic curve. Journal of the American Statistical Association 52:567–577.

Snyder, G. R. 1980. Effects of starvation on presmolt coho salmon. Doctoral dissertation. University of Idaho, Moscow.

Taylor, E. B., and J. D. McPhail. 1985. Variation in burst and prolonged swimming performance among British Columbia populations of coho salmon, *Oncorhynchus kisutch*. Canadian Journal of Fisheries and Aquatic Sciences 42:2029–2033.

Tschaplinski, P. J., and G. F. Hartman. 1983. Winter distribution of juvenile coho salmon (*Oncorhynchus kisutch*) before and after logging in Carnation Creek, British Columbia, and some implications for overwinter survival. Canadian Journal of Fisheries and Aquatic Sciences 40:452–461.

American Fisheries Society Symposium 19:155–164, 1997

Coastal Migrations of Arctic Ciscoes in the Eastern Beaufort Sea

W. A. BOND AND R. N. ERICKSON

Department of Fisheries and Oceans, Freshwater Institute
501 University Crescent, Winnipeg, Manitoba, R3T 2N6, Canada

Abstract.—Our understanding of the life history of Arctic ciscoes *Coregonus autumnalis* on the Beaufort Sea coast is incomplete because of a dearth of information on the extent and timing of migrations to the east of the Mackenzie River. In an attempt to narrow this information gap, a 3-year study of Arctic cisco summer coastal migrations was conducted in the Anderson River estuary (1989–1990) and on the west coast of Liverpool Bay (1991). Shore-based trap nets captured 31,158 Arctic ciscoes during the study, including age-0 fish, older juveniles, current-year spawners, and mature nonspawners (resting fish). Most of these fish probably originated from the Mackenzie River; however, indirect evidence of some level of spawning in the Anderson River suggests that the "single-source" theory in its strictest sense may not apply here. The results of the study are consistent with the theory that summer winds play an important role in the dispersal of age-0 Arctic ciscoes along the Beaufort Sea coast.

The Arctic cisco *Coregonus autumnalis* of the Mackenzie River system has a complex life history that includes extensive summer migrations along the Beaufort Sea coastline (Reist and Bond 1988; Bond and Erickson 1989). Our evolving understanding of these migrational patterns is most complete for the area west of the Mackenzie River, where many coastal studies have been conducted between 1970 and 1992. Perhaps the most significant contribution to this understanding has been the "single-source" concept presented by Gallaway et al. (1983). These authors rejected the conventional belief that Arctic ciscoes spawned in Alaska's Colville River, and hypothesized that the Colville River population is maintained not by local spawning but through recruitment of Mackenzie River fish by westward migration of young of the year (see Figures 1 and 2 for locations of places mentioned in the text). They suggested longshore coastal currents as a possible dispersal mechanism for the young ciscoes, thus implicating summer wind patterns in the process. Upon achieving first maturity, such fish return to the Mackenzie River to spawn. This theory has gained wide acceptance since its introduction and explains many of the observations reported by fishery workers west of the Mackenzie River.

A major gap in our understanding of Arctic cisco life history involves their utilization of coastal habitats east of the Mackenzie River, where few studies have been conducted. Young of the year and older Arctic ciscoes disperse eastward along the coast of Tuktoyaktuk Peninsula in early summer (Bond 1982; Lawrence et al. 1984). Some of these fish return to Tuktoyaktuk Harbour or the Mackenzie River delta in late summer for overwintering (Bond 1982). However, the eastward extent of these migrations and the periodicity of such movements is not clear. Large Arctic ciscoes have been captured along the entire coastline of the Tuktoyaktuk Peninsula (Lawrence et al. 1984), throughout Liverpool Bay (Bray 1975; Gillman and Kristofferson 1984), and from the vicinity of Baillie Islands (B. W. Worbets, Arctic Petroleum Operators Association, unpublished data). Young of the year, however, have not been documented east of McKinley Bay (Lawrence et al. 1984). Also, there is a question as to whether the single-source theory applies to Arctic ciscoes inhabiting the Liverpool Bay area or if other streams such as the Anderson River may serve as spawning areas for this species.

We address the issue of Arctic cisco utilization of coastal habitats located east of the Mackenzie River. We summarize the results of a 3-year study, the objective of which was to monitor the coastal movements of Arctic ciscoes at selected sites in the eastern Beaufort Sea and to describe their migrations as to seasonal timing and in terms of the length-frequency distribution, age composition, and sexual maturity. The study tests the hypothesis that Arctic ciscoes inhabiting Liverpool Bay and the Anderson River estuary during the summer months represent populations that spawn in areas other than the Mackenzie River. Based on wind patterns occurring over the southern Beaufort Sea in the past several years, the single-source theory and its wind-driven corollary would predict strong eastward movements of young-of-the-year Arctic ciscoes from the Mackenzie River in 1988 and 1991 and relatively poor eastward movements in 1989 and 1990. The results of this study are discussed in relation to these predictions.

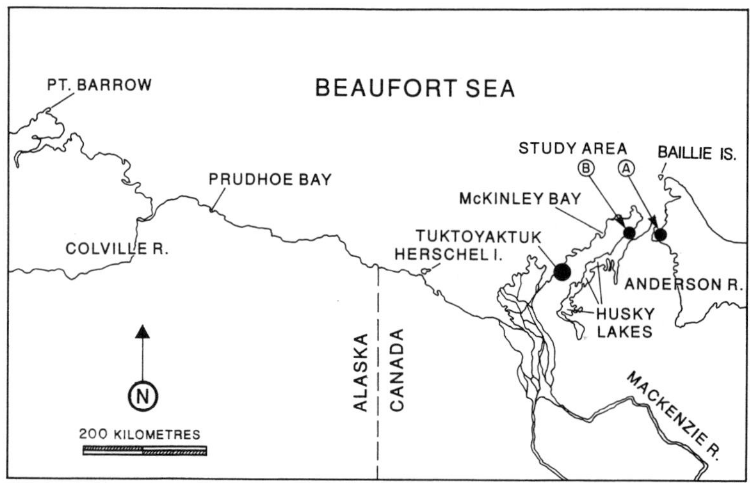

FIGURE 1.—The Beaufort Sea coastline, indicating location of sampling areas in Wood Bay (**A**) and Liverpool Bay (**B**).

Study Area

During 1989 and 1990, we worked in Wood Bay, located 175 km east of Tuktoyaktuk at the mouth of the Anderson River (Figure 1). The Anderson River drains an area of approximately 60,000 km^2, flows throughout the year, and is known to support populations of four other coregonines: broad whitefish *C. nasus*, lake whitefish *C. clupeaformis*, least cisco *C. sardinella*, and inconnu *Stenodus leucichthys* (Bond and Erickson 1991). Ice-out of the Anderson River occurs in May, but ice can persist in the northern half of Wood Bay until late June. Although the lower reaches of the Anderson River are deep enough to provide considerable overwintering habitat, Wood Bay itself is shallow and probably freezes to the bottom over much of its area. Summer water temperature and salinity in Wood Bay vary according to river discharge, wind patterns, and

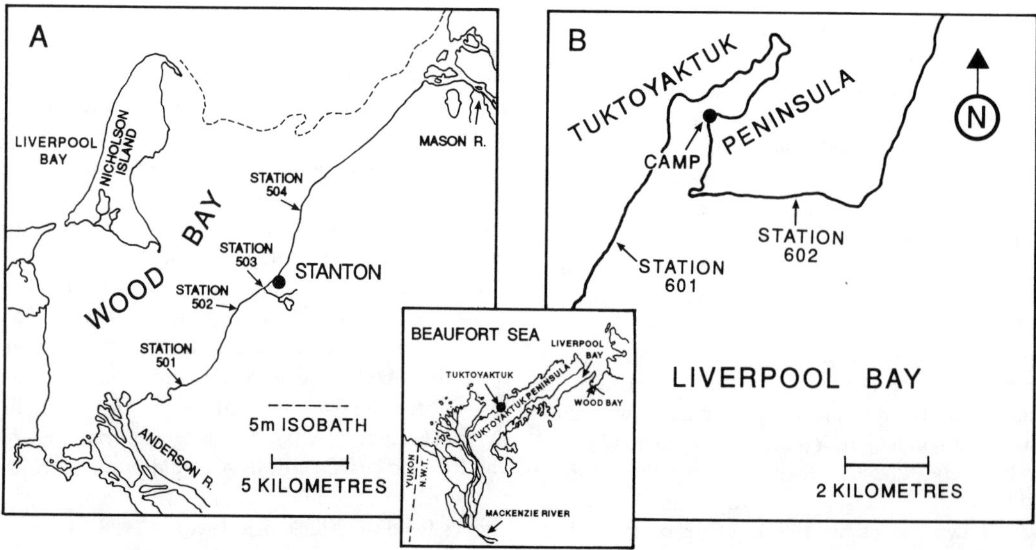

FIGURE 2.—Wood Bay (**A**) and Liverpool Bay (**B**) showing trap-net sites (at numbered stations) and location of the study area in relation to the Mackenzie River.

TABLE 1.—Details of trap-net sampling program for each year of the study (1989–1991) in the eastern Beaufort Sea.

Year	Station number	Sampling period	Lead length (m)	Total fishing effort (d)
1989	501	27 Jun–5 Sep	45	70
	502	14 Jul–5 Sep	30	52
1990	503	1 Jul–25 Aug	45	52
	504	30 Jun–25 Aug	45	55
1991	601	11–13 Jul	75[a]	43
		19–21 Jul		
		22 Jul–1 Sep		
	602	25 Jul–1 Sep	60[a]	38

[a]Longer lead necessitated by more gradually sloping shoreline.

distance from the river mouth. Maximum water temperature and salinity recorded at our study sites were 22.0°C and 13.9 g/kg in 1989 and 16.5°C and 20.9 g/kg in 1990.

The study shifted in 1991 to a site on Liverpool Bay, located approximately 50 km west of Anderson River on the south shore of the Tuktoyaktuk Peninsula (Figure 1). This location differed from Wood Bay in that no major river enters nearby. Liverpool Bay is an elongate water body about 12 km wide at our location, with a maximum depth exceeding 16 m. It connects with a large freshwater lake system (Husky Lakes) at its south end and with the open Beaufort Sea to the north. The bay receives Mackenzie River-influenced surface water from the Beaufort Sea and additional freshwater from Husky Lakes (Grainger 1975). Several small rivers (Miner, Kugaluk, Moose, and Smoke) also enter Liverpool Bay near its south end. Maximum water temperature and salinity recorded at our site in 1991 were 12.5°C and 13.1 g/kg. In the deeper parts of Liverpool Bay, salinity can remain below 30 g/kg throughout July and August (Grainger and Lovrity 1975).

Methods

The sampling gear consisted of a double cod end trap with a single lead extending perpendicularly to the shoreline and two 15.2-m wings set at 45° angles to the lead (Bond and Erickson 1989). In each year of the study, double cod end traps were established at two locations (Figure 2) to intercept fish migrating along the shoreline. Details relating to the sampling dates and net configurations in each year are described in Table 1.

Trap nets were checked at least daily. At each check, water temperature and salinity were recorded, trapped fish were transferred to a holding pen, and the trap was reset immediately. All fish were identified and counted, and fork length (FL ± 1 mm) was determined for some or all of the fish. In the case of Arctic ciscoes, separate counts and measured subsamples were obtained for fish less than 150 mm FL and greater than or equal to 150 mm FL. Ratios obtained from the measured subsamples were applied to actual counts of fish in these size ranges and the results combined to produce length-frequency descriptions for the total weekly catch by station and sampling area. Throughout the summer, the gonadosomatic index (GSI) was monitored for Arctic ciscoes greater than or equal to 330 mm long to confirm the presence or absence of current-year spawners in the catch. Gonads were weighed fresh (±1.0 g) and GSI determined from the formula (gonad weight × 100)/fish round weight.

Using the summed catches and combined fishing effort for the two sampling stations and the calculated length-frequency distributions, we determined catch per unit effort (CPUE) for each sampling area (i.e., year) without regard to lead length. Separate CPUE determinations were made for small (<150 mm), intermediate-sized (150–279 mm), and large (≥280 mm) Arctic ciscoes. In general, these size categories correspond to age-groups 0–1 (small) and 2–5 (intermediate-sized), and older fish, including all those of spawning size (large) (Bond and Erickson 1991). In this article, CPUE is expressed as the number of fish captured per day (24 h) within weekly time periods to eliminate short-term fluctuations and emphasize seasonal trends.

Because small Arctic ciscoes are not susceptible to capture in our trap nets until they have attained a length of about 50 mm, small-mesh beach seines (3.0-mm mesh) were used at various intervals to identify the presence of small ciscoes in early summer. Variable-mesh gill nets were used in 1990 (17–30 June) and 1991 (1–18 July) to sample large fish in the period before trap installation. Numbered Floy anchor tags (Type FD-68B) were applied in 1989 and 1990 to 3,285 Arctic ciscoes in Wood Bay to obtain information on the geographical extent of movement of individual fish.

Results

Length-Frequency Distribution

Small fish (<150 mm) dominated trap-net catches at Wood Bay in 1989, accounting for 76% of all Arctic ciscoes taken, whereas large individuals (≥280 mm) made up 21% (Figure 3). Large Arctic ciscoes were much more abundant in the 1990 Wood Bay catches, constituting 53% of the species total. Though less abundant than in 1989, small Arctic ciscoes still represented 34% of the catch in 1990.

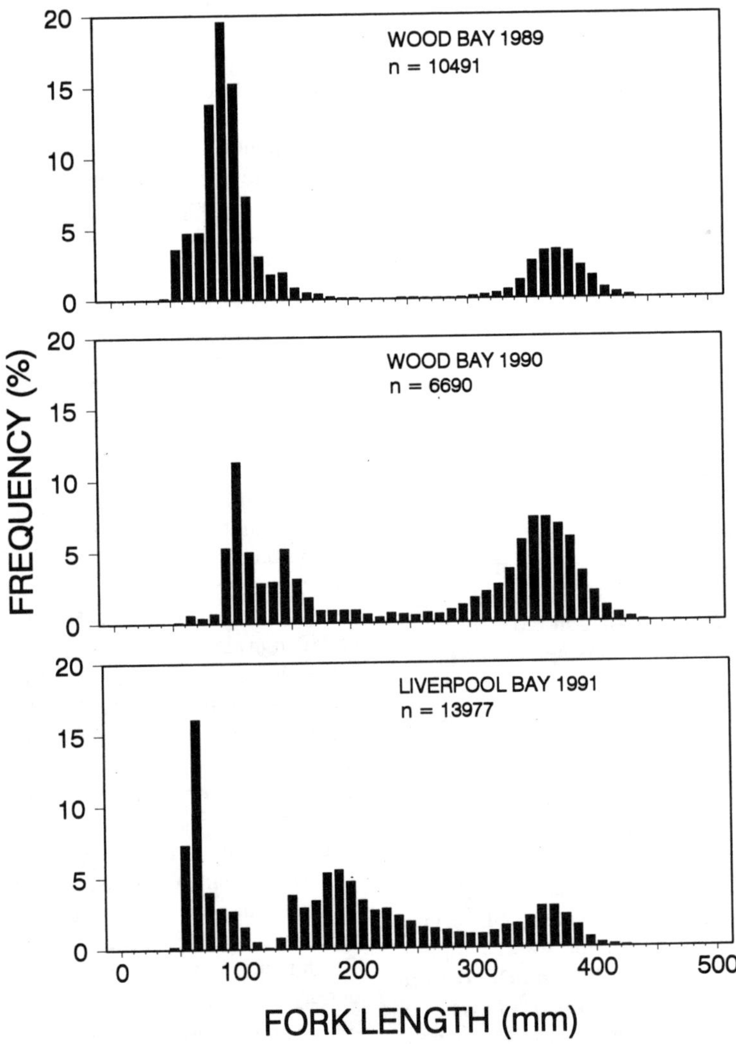

FIGURE 3.—Length-frequency distributions for Arctic ciscoes *Coregonus autumnalis* during each year of the study (1989–1991) in the eastern Beaufort Sea.

The length-frequency pattern for Arctic ciscoes in the 1991 Liverpool Bay catch differed markedly from that observed at Wood Bay (Figure 3). Although small fish were abundant, accounting for 40% of all Arctic ciscoes taken, the modal size range for this cohort was 50–69 mm (age 0) compared with 80–119 mm (age 1) in the previous two summers. Furthermore, a gap at 100–139 mm indicates an underrepresentation of age-1 fish in the Liverpool Bay samples. Another obvious difference when compared with the Wood Bay results was the prominence of intermediate-sized fish (150–279 mm), which constituted 39% of the Arctic ciscoes taken at Liverpool Bay. Within this group of intermediate-sized fish, the strongest representation was from individuals in the range of 160–209 mm (primarily age 3), which accounted for 57%. Large Arctic ciscoes made up 21% of the species total at Liverpool Bay, and, as in the previous years, this group was dominated by fish that were 340–389 mm long.

Seasonal Trends in Size Composition and Abundance

Small Arctic ciscoes (<150 mm).—An abundance of age-1 fish (80–119 mm) in the Anderson River

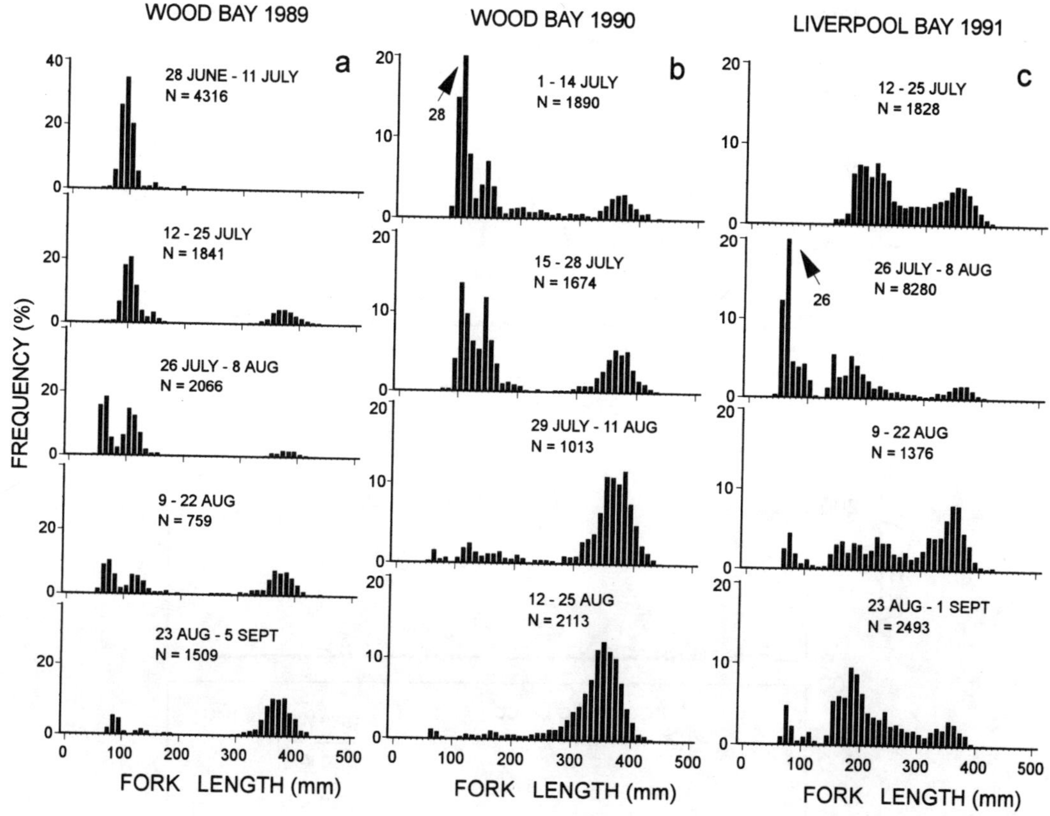

FIGURE 4.—Temporal changes in length-frequency distribution of Arctic ciscoes *Coregonus autumnalis* in Wood Bay during 1989 (**a**) and 1990 (**b**) and in Liverpool Bay during 1991 (**c**). (Note different scale for Wood Bay 1989.)

estuary produced high CPUE values for small Arctic ciscoes in early July 1989 (Figures 4a and 5a). Their abundance near the river mouth in early summer suggests that large numbers of age-0 fish of the 1988 year-class had overwintered within the Anderson River. By mid-July, catch rates for yearling ciscoes had dropped sharply, but the appearance of young of the year (50–69 mm) produced a minor abundance peak in late July to early August. Thereafter, CPUE for small Arctic ciscoes decreased substantially. A similar pattern was observed for small Arctic ciscoes in 1990 (Figures 4b and 5a), although lower CPUE values suggest that neither age-1 nor age-0 fish were as numerous as in 1989. Fewer than 100 age-0 Arctic ciscoes were captured in the trap nets at Wood Bay in 1990.

Although age-0 fish did not appear in Wood Bay trap nets until late July (Figure 4a), seining near Station 501 on 11 July 1989 produced a number of small Arctic ciscoes (N = 56; mean FL = 24.8 ± 4.1 mm; range = 18–37 mm). At this time, the north half of Wood Bay and much of Liverpool Bay were still ice-covered.

Small Arctic ciscoes were not present in Liverpool Bay until late July 1991 (Figure 4c). Although traps were not installed permanently until 22 July, daily seining at various locations failed to yield any small ciscoes until 28 July, the same day they first appeared in the traps. The small fish arriving at that time were young of the year (50–69 mm) whose appearance produced a major abundance peak in late July to early August (Figure 5a). After a major northwest storm on 3–4 August, CPUE was sharply reduced, although age-0 ciscoes continued to be captured in small numbers through the duration of the study.

Intermediate-sized Arctic ciscoes (150–279 mm).—Fish of this size are juveniles generally belonging to age-groups 2–5 inclusive. Intermediate-sized fish accounted for only 2.8% of Arctic ciscoes taken at Wood Bay in 1989 (Figure 3). Their somewhat greater abundance in 1990 (12.7%) appears to be

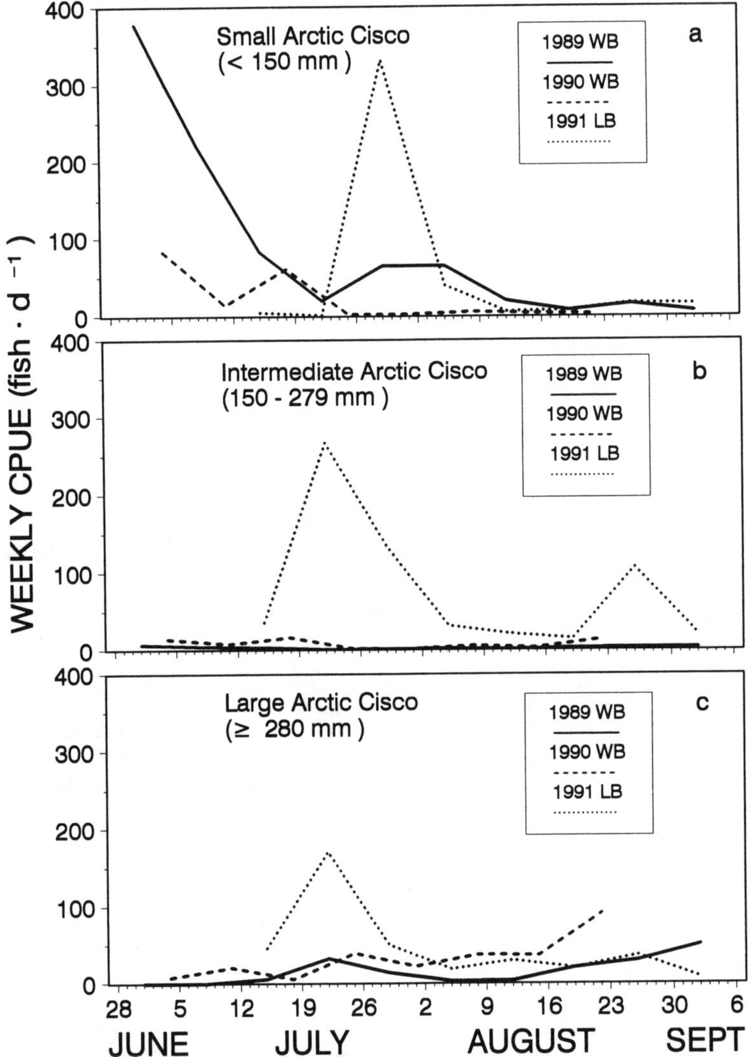

FIGURE 5.—Temporal changes in catch per unit effort (CPUE) for small (**a**), intermediate-sized (**b**), and large (**c**) Arctic ciscoes *Coregonus autumnalis* during each year of the study (1989–1991) in the eastern Beaufort Sea. Dates shown represent first day of weekly intervals.

attributable to the presence of members of the 1988 year-class (age 2; 130–159 mm) during July (Figure 4b).

Gill netting indicated that few intermediate-sized Arctic ciscoes were present in Liverpool Bay before mid-July 1991. However, large numbers began to appear in trap nets at that time, producing a major abundance peak during the latter half of the month (Figure 5b). These fish were mostly 170–239 mm in fork length (Figure 4c) and are thought to represent primarily age-groups 3 (1988 year-class) and 4 (1987). A second, smaller abundance peak occurred during the last week of August. Ciscoes contributing to the late August peak were somewhat smaller (150–219) than those captured in late July and probably were 2 and 3 years old.

Large Arctic ciscoes (≥280 mm).—Small individuals dominated early summer catches at Wood Bay, but the proportion of large ciscoes increased as summer progressed (Figure 4, a and b). Catch rates for large ciscoes were generally higher in 1990 (Figure 5c), when fishing occurred farther from the river

TABLE 2.—Prevalence of current-year spawners among female Arctic ciscoes *Coregonus autumnalis* with fork length greater than or equal to 330 mm by month for each year of the study (1989–1991) in the eastern Beaufort Sea.

	1989		1990		1991		Combined	
Month	*N*	Spawners	*N*	Spawners	*N*	Spawners	*N*	Spawners (% of sample)
Jun			35	12			35	12 (34.3)
Jul	26	10	51	17	136	59	213	86 (40.4)
Aug	58	2	50	3	36	5	144	10 (6.9)
Sep	14	0			8	0	22	0 (0)
Total	98	12	136	32	180	64	414	108 (26.1)

mouth. In both years, CPUE for large Arctic ciscoes increased in late August, suggesting a movement toward the Anderson River.

Gill-net results indicated that large Arctic ciscoes were already present in Liverpool Bay by 1 July 1991. Ciscoes that were gillnetted during 1–18 July (N = 222) ranged in fork length from 180 to 449 mm. The modal size interval was 370–379 mm and 65% of the catch was between 350 and 399 mm. Trap-net results identified a major movement of large Arctic ciscoes into and through the study area in late July (Figure 5c). Catch-per-unit-effort values dropped to relatively low levels in early August and decreased to near zero by early September.

Sexual Maturity

Most Arctic ciscoes in the Mackenzie River delta–Beaufort Sea area begin to spawn at about 350-mm FL, although some spawn before reaching that size. In the Mackenzie River, more than 98% of fish captured during spawning runs exceed 330 mm in length (Hatfield et al. 1972; J. N. Stein, C. S. Jessop, T. R. Porter, and K. T. J. Chang-Kue, Canada Department of the Environment, unpublished data). We estimate that 21% of all Arctic ciscoes captured during this study were mature fish. Based on an examination of 414 female ciscoes with fork lengths greater than or equal to 330 mm, mature fish included 29% current-year spawners and 71% mature nonspawners (fish that had spawned previously but would not have spawned in the year of capture). Within the group referred to in this article as large Arctic ciscoes (≥280 mm), 26% were classified as immature, including all those less than 330 mm long.

Current-year spawners tended to be more common in early summer (Table 2). In our sample of females at least 330 mm long, 40% of those examined during June and July, but only 6% of those in August and September, were considered current-year spawners. The means and ranges for GSI values attributed to female spawners by month during this study were June, 3.96 (2.15–7.22); July, 4.69 (2.08–8.85); and August, 9.15 (4.69–16.52). Female Arctic ciscoes captured at spawning time in the lower Mackenzie River display a mean GSI exceeding 23.0 (range: 19.69–31.27) (J. Reist, Canada Department of Fisheries and Oceans, personal communication).

Tag Returns

Floy tags were applied to 3,285 Arctic ciscoes in Wood Bay during 1989 and 1990. To date, only two recaptures have been reported, one in Liverpool Bay and one in Tuktoyaktuk Harbour. This study also recovered several fish tagged in other studies, including four that were tagged by staff of LGL Limited at McKinley Bay (Figure 1) in 1988; one of these was recaptured in Wood Bay in 1990, and the other three were recaptured in Liverpool Bay during 1991. Another fish, tagged 27 August 1985 (291 mm) at Prudhoe Bay, Alaska, was recaptured in Liverpool Bay on 21 August 1991 (417 mm). At least 1,000 km of coastal habitat separate the tagging and recapture sites for this fish.

Discussion

Mackenzie River Arctic ciscoes hatch in the spring, and most of the newly emerged young are thought to be swept quickly downstream by the spring flood, reaching the Mackenzie estuary in late May or June. Their subsequent dispersal into coastal habitats appears to be determined, at least partially, by early-summer wind patterns over the southern Beaufort Sea. Westward dispersal is apparently facilitated in years dominated by strong east winds (Fechhelm and Fissel 1988; Schmidt et al. 1989; Fechhelm and Griffiths 1990). In such years, young ciscoes can reach the Sagavanirktok and Colville rivers near Prudhoe Bay, Alaska, and thus avail themselves of the overwintering and rearing opportunities that those areas provide. Fechhelm and Griffiths (1990) estimate that such favorable conditions occur about 50% of the time. In summers lacking dominant wind patterns, or when west winds dominate, westward movement is more limited and recruitment of young into the Prudhoe Bay area is weak or nonexistent (Fechhelm and Griffiths 1990).

If east winds determine or enhance the westward dispersal of age-0 Arctic ciscoes from the Mackenzie River, then it follows that west wind conditions should promote their eastward movement. Our results suggest that in fact a correlation does exist between wind patterns and recruitment of age-0 Arctic ciscoes into the Wood Bay–Liverpool Bay area. For example, 1988 was a west-wind-dominated year featuring weak recruitment of young Arctic ciscoes into the Prudhoe Bay area (Fechhelm and Griffiths 1990). Although Wood Bay was not sampled in 1988, our 1989 results (high early-summer CPUE of age-1 fish) indicate a strong recruitment of age-0 ciscoes into that area in 1988, followed by overwintering in the Anderson River. Furthermore, the persistence of the 1988 recruits in the 1990 Wood Bay catch (age 2) and their strong representation at Liverpool Bay in 1991 (age 3) provide additional evidence of a strong eastward dispersal in 1988.

Similarly, 1991 was a year in which eastward dispersal would be expected to be favored. No recruitment of age-0 ciscoes was detected at Prudhoe Bay by early September 1991 (L. Moulton, MJM Research, personal communication), although a weak recruitment did occur in late September as evidenced by the collection of some age-1 ciscoes in June 1992 (W. B. Griffiths, LGL Limited, personal communication). Liverpool Bay results, on the other hand, suggest a substantial movement of young into that area in 1991. Although the high initial catch rates of late July to early August were not sustained, total catch of age-0 ciscoes still exceeded the numbers taken at Wood Bay during the previous two summers. Changes in local current patterns after the storm of 3–4 August 1991 may have altered the dispersal route of these small fish away from our sampling locations, thus accounting for their decreased abundance in August.

In contrast to the summers of 1988 and 1991, summer 1990 was dominated by strong easterly winds and featured heavy recruitment of young Arctic ciscoes into the Prudhoe Bay area (Griffiths, personal communication). At the same time, very weak recruitment was observed at Wood Bay, where the total catch of age-0 fish did not exceed 100 individuals. Underrepresentation of 1990 fish in the 1991 Liverpool Bay results (age 1) is further evidence that eastward dispersal of young ciscoes was weak in 1990.

The scarcity of intermediate-sized ciscoes in the 1989 and 1990 Wood Bay catches probably reflects the periodic nature of recruitment of young to the Anderson River. The years 1985–1987 were east-wind years, all featuring strong movements of age-0 fish into Prudhoe Bay (Fechhelm and Griffiths 1990). If recruitment to the Anderson River was weak in those years, one would expect to find weakness in age-groups 2–5 in 1989 and 1990. In fact, the gap in the Wood Bay length-frequency distribution at 150–279 mm represents primarily those age-groups. There also is a suggestion in our results that some age-1 ciscoes, having overwintered in the Anderson River, emigrate from that river during their second summer, perhaps to more favorable habitat in adjacent areas (e.g., Liverpool Bay). The disappearance of most age-1 Arctic ciscoes from our Wood Bay catches by mid-July, along with the absence of any indication of a late-summer return migration toward the Anderson River, tend to support the idea of such a movement.

The 1991 CPUE results indicate a strong movement of intermediate-sized fish past our Liverpool Bay location during the third week of July and a smaller peak in the last week of August. Because we can attribute no direction to the fish represented by these data, an interpretation of these results is difficult. It is not known whether these fish overwintered within the Mackenzie estuary or in the southern parts of Liverpool Bay. Both possibilities may have contributed to the results.

The movements of large Arctic ciscoes described in this study involve migrations of both current-year spawners and mature nonspawners. After spawning in the Mackenzie River, spent ciscoes move downstream to the Mackenzie River delta and estuary. The following summer, these fish disperse along the coastal margin to feed and rebuild their energy resources. Their westward migration along the Yukon coast begins in June (Bond and Erickson 1989) and extends beyond Herschel Island (Griffiths et al. 1975). Their return migration toward the Mackenzie River begins in late August (Bond and Erickson 1989) and continues after freeze-up (Griffiths et al. 1975).

Mature nonspawners represented a substantial portion of the catch of large Arctic ciscoes in all years of this study. At Wood Bay, they dominated the late August migrations of large ciscoes toward the Anderson River. There is no question but that the Anderson River provides overwintering habitat for mature nonspawning Arctic ciscoes.

At Liverpool Bay, mature nonspawners began to appear in the catches in late July, and members of this cohort were strong contributors to the peak of large Arctic ciscoes that passed our site at that time. As no late-summer abundance peak was detected for large Arctic ciscoes, we suspect that some mature nonspawners may overwinter near the southern end of Liverpool Bay.

Craig and Mann (1974) reported finding no direct evidence of previous spawning in Yukon coastal Arctic ciscoes classified as mature nonspawners. On the other hand, we have observed several cases on the Yukon coast (Bond and Erickson 1987), in Wood Bay, and in Liverpool Bay in which females had retained, within the coelom, varying quantities of eggs from the previous year's complement. Furthermore, at Wood Bay and Liverpool Bay, it was common to find mature nonspawners in extremely emaciated condition.

Current-year spawners leave their overwintering sites at the time of spring breakup and migrate quickly toward the Mackenzie River. Arctic cisco spawners begin to move upstream through the Mackenzie River delta as early as May, and the intensity of the migration increases through July (Stein, Jessop, Porter, and Chang-Kue, unpublished data). By early August, most current-year spawners have entered the Mackenzie River, and fish remaining in coastal habitats in late summer are primarily nonspawning individuals (Craig and Haldorson 1981; Bond 1982; Lawrence et al. 1984). Results of this study are consistent with previous observations. Current-year spawners were present in Wood Bay in late June and July and in Liverpool Bay during July. Their disappearance from the catches by early August indicates a movement out of the study area, presumably toward the Mackenzie River.

In summary, this study demonstrates that the summer coastal migrations of Arctic ciscoes in Wood Bay and Liverpool Bay involve young of the year, older juveniles, current-year spawners, and mature nonspawners. Although we feel that most of these fish originate from Mackenzie River spawning grounds, there is reason to believe that the single-source theory in its strictest sense does not apply in this area. For example, the capture of age-0 Arctic ciscoes near the mouth of the Anderson River on 11 July 1989 suggests spawning within this watershed. It seems unrealistic to expect that young-of-the-year ciscoes could have completed the migration from the Mackenzie River by such an early date. Furthermore, the identification of hybrids from Wood Bay involving crosses between Arctic ciscoes and other coregonid species (Reist et al. 1992) suggests some level of spawning in the Anderson River. More detailed work within the Anderson River watershed itself will be necessary to clarify the role of this river in the life history of Arctic ciscoes. Similarly, other rivers, such as the Kugaluk, should be examined in this context.

Results of our study are consistent with the theory that summer winds play an important role in the dispersal of young-of-the-year Arctic ciscoes away from the Mackenzie River as was suggested by Fechhelm and Fissel (1988) and Fechhelm and Griffiths (1990). We are unable at this time, however, to draw any conclusions regarding the relative significance of the Wood Bay–Liverpool Bay area and the Prudhoe Bay area as rearing areas for young Arctic ciscoes.

The southern end of Liverpool Bay, where the mixture of Beaufort Sea water and freshwater from the Husky Lakes creates extensive estuarine conditions under ice cover, represents an area that deserves a more intensive research effort. Such estuarine habitat is utilized for overwintering by Arctic ciscoes in other coastal areas, and on the basis of volume alone, the Liverpool Bay–Husky Lakes area appears capable of supporting large numbers of fish.

Acknowledgments

We thank the many people who assisted us in the field or through helpful discussions during the course of this study. The logistical support of the Polar Continental Shelf Project and the Science Institute of the Northwest Territories–Inuvik Research Centre is gratefully acknowledged. The study was funded in part by the Department of Fisheries and Oceans/ Inuvialuit Fisheries Joint Management Committee under the terms of the Western Arctic Land Claims settlement.

References

Bond, W. A. 1982. A study of the fishery resources of Tuktoyaktuk Harbour, southern Beaufort Sea coast, with special reference to life histories of anadromous coregonids. Canadian Technical Report of Fisheries and Aquatic Sciences 1119.

Bond, W. A., and R. N. Erickson. 1987. Fishery data from Phillips Bay, Yukon, 1985. Canadian Data Report of Fisheries and Aquatic Sciences 635.

Bond, W. A., and R. N. Erickson. 1989. Summer studies of the nearshore fish community at Phillips Bay, Beaufort Sea coast, Yukon. Canadian Technical Report of Fisheries and Aquatic Sciences 1676.

Bond, W. A., and R. N. Erickson. 1991. Fishery data from the Anderson River estuary, Northwest Territories, 1989. Canadian Data Report of Fisheries and Aquatic Sciences 849.

Bray, J. R. 1975. Marine fish surveys in the Mackenzie

Delta area. Fisheries Research Board of Canada, Manuscript Report Series 1326, Ottawa.

Craig, P. C., and L. H. Haldorson. 1981. Beaufort Sea barrier island–lagoon ecological process studies: Simpson Lagoon, part 4. Fish. Pages 384–678 *in* Environmental assessment Alaskan continental shelf, final report, volume 7. U.S. Bureau of Land Management and NOAA (National Oceanic and Atmospheric Administration) OCSEAP (Outer Continental Shelf Environmental Assessment Program), Boulder, Colorado.

Craig, P. C., and G. J. Mann. 1974. Life history and distribution of the Arctic cisco (*Coregonus autumnalis*) along the Beaufort Sea coastline in Alaska and the Yukon Territory. *In* P. McCart, editor. Life histories of anadromous and freshwater fishes in the western Arctic. Arctic Gas Biological Report Series 20(4), Calgary, Alberta.

Fechhelm, R. G., and D. B. Fissel. 1988. Wind-aided recruitment of Canadian Arctic cisco (*Coregonus autumnalis*) into Alaskan waters. Canadian Journal of Fisheries and Aquatic Sciences 45:906–910.

Fechhelm, R. G., and W. B. Griffiths. 1990. Effect of wind on the recruitment of Canadian Arctic cisco (*Coregonus autumnalis*) into the central Alaskan Beaufort Sea. Canadian Journal of Fisheries and Aquatic Sciences 47:2164–2171.

Gallaway, B., W. Griffiths, P. Craig, W. Gazey, and J. Helmericks. 1983. An assessment of the Alaskan stock of Arctic cisco (*Coregonus autumnalis*)—migrants from Canada? Biological Papers of the University of Alaska 21:4–23.

Gillman, D. V., and A. H. Kristofferson. 1984. Biological data on Pacific herring (*Clupea harengus pallasi*) from Tuktoyaktuk Harbour and the Liverpool Bay area, Northwest Territories, 1981 to 1983. Canadian Data Report of Fisheries and Aquatic Sciences 485.

Grainger, E. H. 1975. Biological productivity of the southern Beaufort Sea: the physical–chemical environment and the plankton. Beaufort Sea Project, Canada Department of the Environment, Technical Report 12, Victoria, British Columbia.

Grainger, E. H., and J. E. Lovrity. 1975. Physical and chemical oceanographic data from the Beaufort Sea, 1960 to 1975. Canada Department of the Environment, Fisheries and Marine Service, Technical Report 590, Ottawa.

Griffiths, W., P. Craig, G. Walder, and G. Mann. 1975. Fisheries investigations in a coastal region of the Beaufort Sea (Nunaluk Lagoon, Yukon Territory). *In* P. Craig, editor. Fisheries investigations in a coastal region of the Beaufort Sea. Arctic Gas Biological Report Series 34(2), Calgary, Alberta.

Hatfield, C. T., J. N. Stein, M. R. Falk, and C. S. Jessop. 1972. Fish resources of the Mackenzie River Valley, interim report 1, volume 1. Canada Department of the Environment, Fisheries Service, Winnipeg, Manitoba.

Lawrence, M. J., G. Lacho, and S. Davies. 1984. A survey of the coastal fishes of the southeastern Beaufort Sea. Canadian Technical Report of Fisheries and Aquatic Sciences 1220.

Reist, J. D., and W. A. Bond. 1988. Life history characteristics of migratory coregonids of the lower Mackenzie River, Northwest Territories, Canada. Finnish Fisheries Research 9:133–144.

Reist, J. D., J. Vuorinen, and R. A. Bodaly. 1992. Genetic and morphological identification of coregonid hybrid fishes from Arctic Canada. Polish Archives of Hydrobiology 39(3,4):551–561.

Schmidt, D. R., W. B. Griffiths, and L. R. Martin. 1989. Overwintering biology of anadromous fish in the Sagavanirktok River delta, Alaska. Biological Papers of the University of Alaska 24:55–74.

American Fisheries Society Symposium 19:165–174, 1997

Movements and Temperature Occupancy of Sonically Tracked Dolly Varden and Arctic Ciscoes in Camden Bay, Alaska

L. E. JARVELA
Minerals Management Service, Anchorage, Alaska 99508, USA

L. K. THORSTEINSON[1]
National Park Service, Anchorage, Alaska 99503, USA

Abstract.—Movement patterns of diadromous Dolly Varden *Salvelinus malma* and Arctic ciscoes *Coregonus autumnalis* in relation to thermal regime were determined in Camden Bay, Alaska. Six large Dolly Varden (total body length, L > 48 cm) and seven Arctic ciscoes (L > 33 cm) were tracked over distances (time) of 114 km (55 h 38 min) and 64 km (38 h 15 min), respectively. The directedness of movement varied between species and among individuals; most displayed an affinity for the shoreline. The distribution of directions of net movement by Dolly Varden was random, whereas that by Arctic ciscoes was nonrandom (true compass bearing, 253°T). The average gross ground speed of the Dolly Varden was 55.8 cm/s (range: 48.8–74.2 cm/s), or 1.06 L/s (range: 0.96–1.37 L/s). Net speeds ranged from 9.5 to 64.3 cm/s. The average gross ground speed of the Arctic cisco was 45.9 cm/s (range: 30.1–71.1 cm/s), or 1.04 L/s (range: 0.69–1.61 L/s). Their net speeds were 12.0–44.1 cm/s. The mean temperature of waters occupied by both species was 3.6°C (Dolly Varden, range: −0.5 to 6.5°C; Arctic cisco, range: 1.0 to 7.5°C). The habitat occupancy and movements by Dolly Varden and Arctic ciscoes are similar to those of related species using coastal habitats. Their use of the harsh physical environment of Camden Bay is interpreted as part of a behavioral pattern that optimizes the fishes' fitness and survival.

The availability, quantity, and quality of summer feeding habitat strongly influence the growth of fish populations in the Arctic (Gallaway 1990). Where food resources in freshwater are limited, many arctic salmonids make regular, usually annual, feeding migrations to marine waters (Craig 1989a). Exploitation of marine prey resources broadens their niche spaces, maximizes fitness, and promotes population stability (Gross 1987). Dispersal in marine waters also fosters colonization of new or ephemeral habitats (Quinn and Brodeur 1991). The arctic summer is brief, so the time available for feeding may be more important than prey abundance per se in terms of the quantity of energy reserves that a fish is able to assimilate (Dutil 1986). The primary foraging areas for most resident salmonid populations in northern Alaska and the Yukon are the relatively warm, brackish waters along the coast (Craig 1984). Essentially all of their annual energy supply comes from this habitat.

Large-scale oil and gas development along the Beaufort Sea coast has raised concerns about its effects on fish habitats and fish movements (Norton 1989; Gallaway et al. 1991). Assessment of the effects of such activities requires knowledge of species–habitat relationships (Wiens and Rotenberry 1981). Of particular concern are Dolly Varden *Salvelinus malma* and Arctic ciscoes *Coregonus autumnalis* because they are ecosystem dominants, are the most widely distributed salmonid species, and are important to the regional subsistence fisheries (Craig 1984, 1989b). Finally, international management concerns arise because of the transboundary migrations of these fishes (Everett and Wilmot 1990; Fechhelm and Griffiths 1990).

In 1988 we initiated a fish habitat-use investigation in the Beaufort Sea. One aspect of the research was to study the offshore component of the marine environment, to add to the existing information that is based mainly on data obtained at or near the shoreline. Our approach was to sample across the brackish-water zone into the marine-water zone using techniques that capture the dynamic temporal–spatial attributes of the habitat. Our hypothesis was that Dolly Varden and Arctic ciscoes prefer nearshore brackish, rather than offshore marine, waters. We report on temperature occupancy and short-term movements of these species in relation to these zones using ultrasonic telemetry. This is the first such application reported for North American arctic waters.

Study Area

Camden Bay is a 90-km-wide bight abutting the Arctic National Wildlife Refuge. Our study was conducted in and near Simpson Cove (Figure 1).

[1]Present address: U.S. Geological Survey, Biological Resources Division, 909 First Avenue, Suite 800, Seattle, Washington 98104, USA.

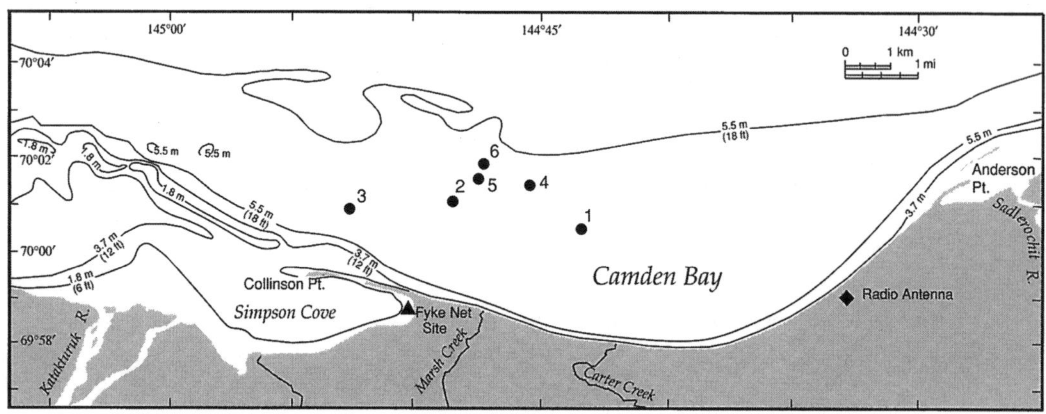

FIGURE 1.—Study area for ultrasonic tracking, Beaufort Sea, Alaska. Fish were captured at the fyke-net site and released at the locations indicated by the numbered, filled circles.

The bay lies in the center of the most marine-like section of the Beaufort Sea coast in Alaska (Gallaway et al. 1991). Runoff from local streams decreases markedly after spring breakup. The absence of barrier islands and the presence of comparatively deep water near shore facilitate vertical and lateral mixing by strong winds and currents, as well as intrusions of pack ice. Meteorological and oceanographic observations reveal a dynamic oceanographic regime during summer (Hale 1990, 1991). Currents are event dominated, with winds being the primary driving force. However, bathymetry, tides, and horizontal pressure gradients produced by runoff also influence local current patterns. Current speeds typically are less than 12 cm/s, but occasionally may exceed 40 cm/s.

The annual patterns of change of water properties and thermohaline structure in Camden Bay are similar to those reported elsewhere along the coast (Craig 1989a; Gallaway et al. 1991). In most areas, marine water (−1 to 3°C, 27–32 ppt) predominates during much of the year. However, in early summer a brackish water mass (5–10°C, 10–25 ppt) forms at the sea surface as a consequence of river runoff and insolation. Winds are influential in determining its geographic extent, thickness, and persistence. West winds hold the brackish water against the coast, preserving its integrity; east winds move it offshore, promoting its dissolution by mixing.

Methods

Details of our methods are presented in Thorsteinson et al. (1991). Most of our observations were made from an 11-m vessel. However, we also used a 5.5-m skiff when fish were followed in shallow water. Global Positioning System (GPS) receivers were the primary navigation instruments. Radar was used in 1990 during periods of intermittent satellite availability. Tests of the precision of GPS fixes while the vessel was stationary led us to record positions to the nearest 0.1 min of latitude and longitude in 1990 and to the nearest 0.01 min in 1991. A combination of GPS readings and radar ranges and bearings to the skiff were used to determine skiff positions. Position data were recorded at 5-min intervals during both years.

The V3T telemetry transmitters (VEMCO Ltd., Halifax, Nova Scotia) had temperature-dependent pulse rates and were individually calibrated at the factory. The transmitters were field checked by cross comparisons with an Applied Microsystems STD-12 salinity–temperature–depth (STD) recorder; agreement was within 0.5°C of the STD-12. In field trials at depths of 2 m or more, visual (VU meter) and aural transmitter detection thresholds were 0.57 and 0.45 km, respectively; a loud, clear signal was received at 0.21 km. Transmissions from waters about 1 m deep and close to shore were strongly degraded, and the detection threshold decreased to 0.1 km. Temperature data were recorded at 5-min intervals in 1990 and 1-min intervals in 1991. Miniature V2B pingers providing only position information were also used. The STD-12 was used to measure thermohaline structure at all release sites and along the tracks of fish when circumstances permitted. In 1991 we obtained surface current estimates during one tracking session using a 1-m^2 cruciform aluminum drogue suspended 1 m below the float.

TABLE 1.—Summary of fish tracking in 1990 and 1991 for Dolly Varden *Salvelinus malma* and Arctic ciscoes *Coregonus autumnalis* in Camden Bay. Fish 9 and 11 carried pinger-type transmitters; all others carried temperature-type transmitters.

Fish number	Total length (cm)	Start date	Track duration (min)
	Dolly Varden		
1	53.1	30 Jul 1990	166
2	54.0	31 Jul 1990	935
3	48.5	1 Aug 1990	292
4	56.9	3 Aug 1990	1,260
5	54.2	10 Aug 1990	140
6	53.5	11 Aug 1990	545
	Arctic Ciscoes		
7	42.9	12 Aug 1990	261
8	43.7	10 Aug 1991	577
9	43.7	12 Aug 1991	465
11	32.9	14 Aug 1991	45
12	44.1	16 Aug 1991	274
13	45.6	18 Aug 1991	330
14	45.6	20 Aug 1991	343

Dolly Varden and Arctic ciscoes were captured in a fyke net in eastern Simpson Cove. With the exception of one Arctic cisco equipped with a V2B pinger, large individuals (Table 1) were selected to minimize tag effects on swimming speed and behavior (Winter 1983; Mellas and Haynes 1985). Ultrasonic tags were attached externally to unanesthetized fish by pins passed through the tissue immediately behind the dorsal fin. Most fish were released 2–3 km offshore in Camden Bay proper; however, during 1991 encroaching pack ice caused us to release three fish in Simpson Cove (Figure 1). The offshore release area was selected on the basis of known occurrences of Dolly Varden and Arctic ciscoes, prevalence of water depths sufficient for the primary tracking vessel, and reducing the probability of tagged fish being recaptured by the fyke nets in Simpson Cove. All fish were acclimated to conditions at the release site before their release.

Ground speed was the primary variable used to evaluate fish movement rates; it is the vector sum of a fish's swimming speed and ambient currents and includes error terms attributable to the variable spatial relationship of the fish and vessel and navigation inaccuracies (Hawkins and Urquhart 1983). The vessel's and fish's positions were assumed to coincide in time and space with linear movements between successive position fixes. Raw ground-speeds data were first standardized to total body lengths per second (L/s) to detect excessive speeds. Studies of swimming performance of salmonids (Brett 1973; Beamish 1980; Williams and Brett 1987) indicate that the maximum swimming speed they can sustain for more than a few minutes is about 3 L/s. We used a screening criterion of 3.3 L/s, which included a 0.3 L/s allowance for favorable currents. When greater speeds occurred, examination of the fish's track plot usually indicated a suspect position fix, which was deleted. Speed was then recalculated over the longer time interval and the track replotted. The procedure was repeated until all individual speed estimates satisfied the criterion. Local averaging was then used to reduce remaining errors. The screened distance data were summed over either 1-h (1990) or 0.5-h (1991) segments of tracks, interpolating between position fixes when segment ends did not coincide with observed positions. Average ground speeds were calculated for each segment as well as for the entire track. The averaged time series data were screened for serial dependence with a runs test before statistical analyses. Only nonparametric procedures were used because of the small sizes and prevalent nonnormality and heteroscedasticity of the data sets (Berryman et al. 1988).

Directedness of fish movement was calculated by dividing the shortest water path between the starting and ending points of a track by the track's length. Circular statistics (Zar 1984) were used for tests of randomness of movements of each species, employing the compass bearings between the starting and ending positions of individual tracks. The level of significance of all tests was 0.05.

Transmitter pulses were digitally converted in real time to temperatures by the VEMCO receiver–decoder using individual transmitter calibration data programmed into the unit. Outliers produced by weak signals from the transmitters were removed from the raw temperature records and resulting gaps filled by interpolation before testing for autocorrelation with Statgraphics® time series analysis software.

The raw STD data were converted to engineering units, pressure-sorted into 0.1 m bins, and averaged. Empty bins were filled by interpolation. Vertical profiles of temperature and salinity were plotted. Drogue displacements were used to estimate current speed and direction.

Results

Habitat Attributes

Twenty-three STD casts were made in the study area in 1990 and 19 in 1991. Meteorological and oceanographic conditions at Camden Bay were

TABLE 2.—Summary of movements of Dolly Varden *Salvelinus malma* and Arctic ciscoes *Coregonus autumnalis* in Camden Bay. Net to gross ratio is an index of movement ranging from 0 (no directed movement) to 1 (unidirectional movement). Compass bearing is true degrees, based on starting and ending points of each track. Grand mean for ground speed is the total gross movement divided by the total time tracked; for compass bearing, it is the unweighted mean.

Fish number or statistic	Movement (km)		Net to gross ratio	Gross ground speed		Net ground speed (cm/s)	Compass bearing (°T)
	Gross	Net		(cm/s)	(L/s)		
Dolly Varden							
1	6.9	6.3	0.9	69.7	1.31	63.2	329
2	31.0	12.0	0.4	55.3	1.02	21.4	88
3	11.5	3.1	0.3	65.5	1.35	17.7	236
4	36.9	7.2	0.2	48.8	0.96	9.5	152
5	6.2	5.4	0.9	74.2	1.37	64.3	174
6	19.3	5.2	0.3	59.0	1.10	15.9	125
Grand mean				55.8	1.06	19.6	116
Arctic ciscoes							
7	7.7	7.0	0.9	48.9	1.14	44.1	231
8	10.4	4.2	0.4	30.1	0.69	12.0	256
9	11.5	5.4	0.5	41.1	0.94	19.2	268
11	0.9	0.4	0.5	34.4	1.04	15.9	120
12	11.7	4.2	0.4	71.1	1.61	25.7	272
13	8.3	3.3	0.4	41.9	0.92	16.7	248
14	12.6	7.2	0.6	61.5	1.35	34.7	278
Grand mean				45.9	1.04	23.0	253

markedly different between the two summers. In 1990 strong, persistent, easterly winds drove the pack ice far offshore. Relatively strong currents, weak vertical stratification, inshore upwelling, and marine waters typified the local oceanographic regime. In the uppermost 4 m, temperatures and salinities averaged 3.2°C and 29.9 ppt, respectively. The entire observed ranges in the upper 11 m of the water column were −1 to 6°C and 14–32 ppt, respectively. In 1991 comparatively weak and inconsistent east winds prevailed, so pack ice remained close to shore all summer. The water column usually was stratified, somewhat cooler, and much fresher than in 1990. Mean water temperature and salinity in the uppermost 4 m were 1.8°C and 19.0 ppt, respectively. Temperature and salinity ranges in the entire water column were 0–5°C and 15–23 ppt, respectively. In both years Simpson Cove was about 2°C warmer than Camden Bay proper, but salinities were similar. Drogue observations in Camden Bay on 13 August 1991 revealed a mean surface current speed estimate of 15.8 cm/s.

Movement Rates

Six Dolly Varden were tracked for 55 h 38 min over a total distance of 111.8 km (Table 2). The average gross ground speeds of the fish were quite similar; however, the 1-h average speeds of individuals varied by as much as 58 cm/s. Standard deviations of the gross ground speeds of the Dolly Varden ranged from 8.9 to 18.3 cm/s. The standardized ground speeds of fish 2, 4, and 6 were not significantly different (Kruskal–Wallis test: H_c = 3.159; df = 2, 0; $P > 0.10$). Alongshore (within 100 m) and offshore ground speeds of fish 2 and 4 were not significantly different (Mann–Whitney tests: $U' = 16$, $n_1 = 2$, $n_2 = 10$, $P > 0.2$; $U' = 46$, $n_1 = 9$, $n_2 = 10$, $P > 0.2$). We were able to estimate the swimming speed of fish 4 because it reversed direction of movement while swimming alongshore. We assumed that the alongshore current and fish's swimming speeds were constant and calculated the speeds from simultaneous equations. Fish 4's estimated swimming speed was 48 cm/s, and the current speed was 6 cm/s to the northeast. The swimming speed agrees closely with the fish's gross ground speed (48.8 cm/s), the current direction is consistent with an observation made during tracking that the vessel was drifting northeast parallel to the beach.

Seven Arctic ciscoes were tracked for 38 h 15 min over a total distance of 63.1 km (Table 2). Their average ground speeds varied more than those of the Dolly Varden. Gross ground speeds (0.5-h averages) of individual Arctic ciscoes varied by as much as 55 cm/s, a range quite similar to that observed for Dolly Varden, and their SDs ranged from 8.3 to 17.1 cm/s. Omitting fish 11, significant differences were measured in the standardized ground speeds of the Arctic cisco (Kruskal–Wallis test: $H_c = 40.35$, $\nu = 5$, $P < 0.001$). The speeds of fish 12 and 14 were higher than those of fish 8, 9, and 13 ($Q = 4.73$), whereas the intermediate speed

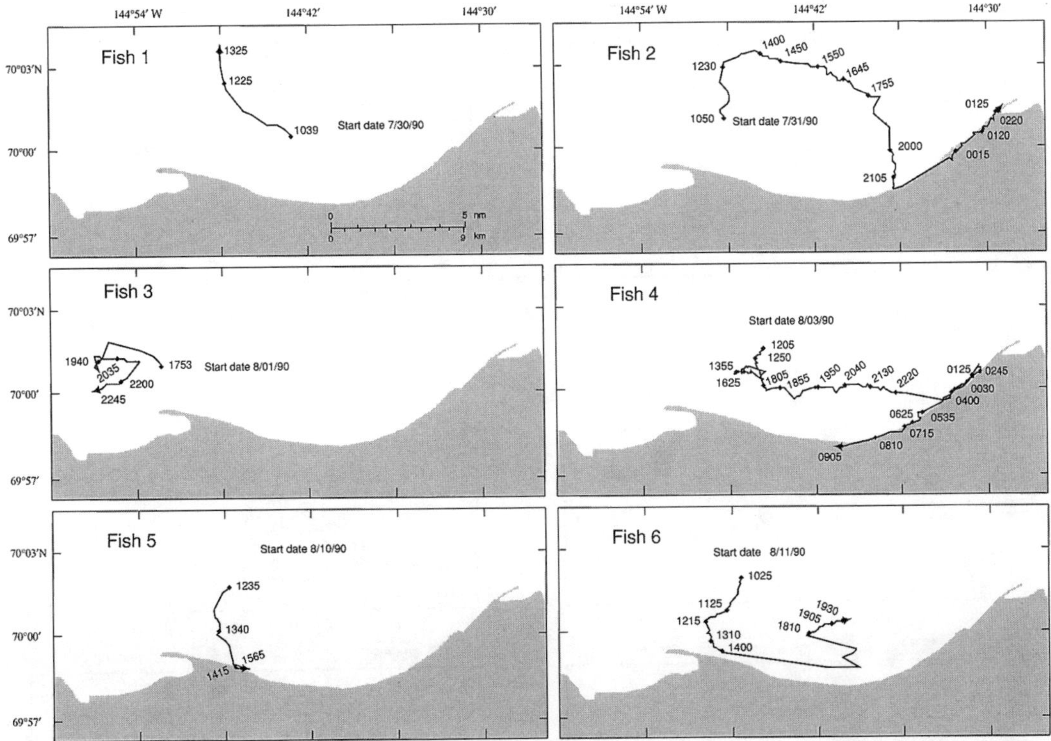

FIGURE 2.—Tracks of Dolly Varden *Salvelinus malma*, showing starting dates and representative times logged during tracking.

of fish 7 was not different from either of those two groupings (Q = 2.86) (multiple comparisons test: $Q_{0.05,6}$ = 2.936). The speed of fish 9 (small pinger) did not differ from those with larger transmitters (fish 7, 8, and 13), suggesting the absence of tag effects on swimming performance. However, there were significant differences in swimming speeds among several Arctic ciscoes with large transmitters (fish 12 and 14 versus fish 8 and 13, respectively).

Movement Patterns

The tracks of the tagged fish are shown in Figures 2 and 3. The compass bearings of the Dolly Varden (based on starting and ending positions) did not differ from random (Rayleigh test: R' = 0.49704, $N = 6, P > 0.2$). The compass bearings of the Arctic ciscoes (excluding the abbreviated track of fish 11) were between 231 and 278°T and nonrandom; their mean direction of movement was 253°T (Rayleigh test: N = 6, Z = 5.868, P < 0.001).

Temperature Occupancy

Because of the prevalence of autocorrelation in the temperature time-series data, our analysis was restricted to descriptive statistics.

The mean water temperature occupied by the Dolly Varden was 3.6°C (range: −0.5 to 6.5°C; N: 587). The frequency distribution of the pooled temperatures was bimodal (Figure 4). About 84% of the temperature occupancy by Dolly Varden was between 1 and 6°C; however, the fish used the entire observed range of temperatures. Comparisons of transmitted temperatures and STD data obtained during tracking suggest the Dolly Varden used only the uppermost 8 m of the water column and usually were in the uppermost 3–4 m. Some temperature records had "sawtooth" structures, indicating repeated vertical excursions by the fish.

The mean water temperature occupied by Arctic ciscoes was also 3.6°C (range: 1.2–7.4°C; N: 1,562). Approximately 95% of the temperature occupancy by Arctic cisco was between 2 and 6°C. The fre-

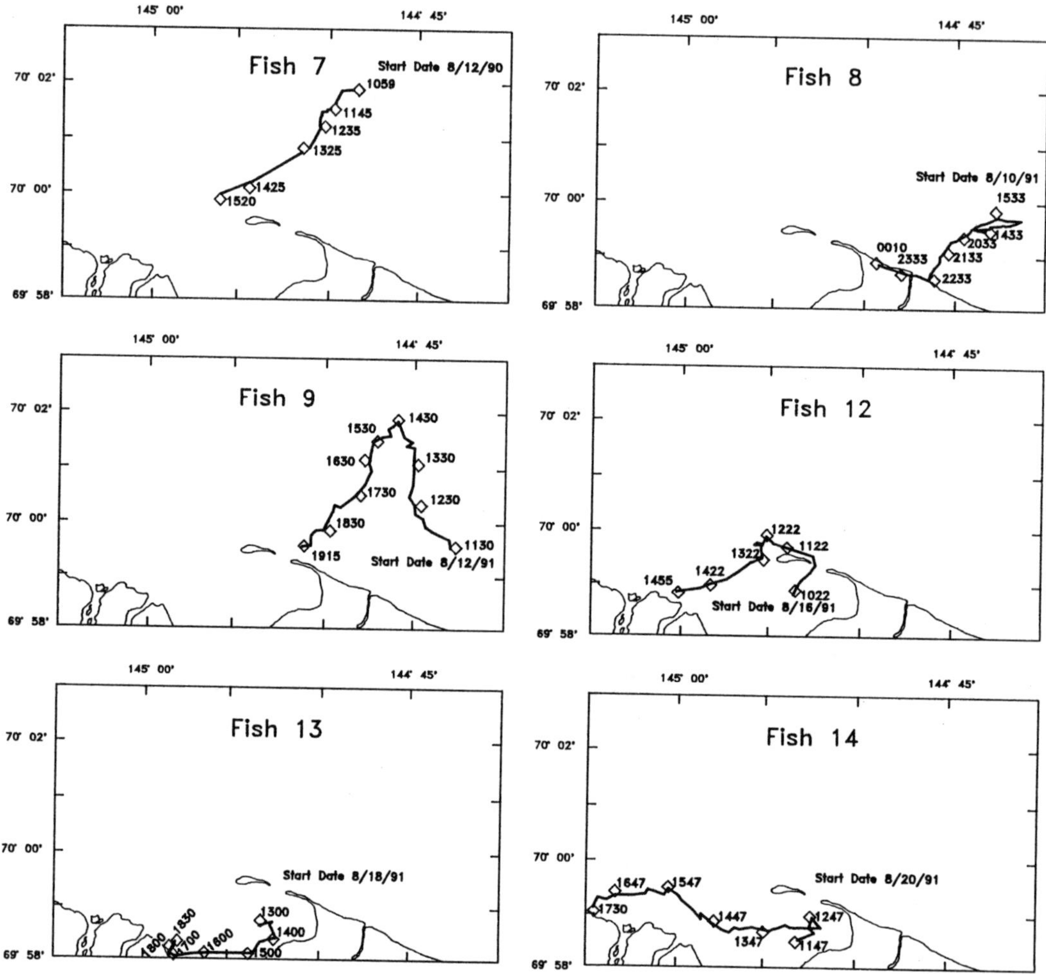

FIGURE 3.—Tracks of Arctic ciscoes *Coregonus autumnalis*, showing starting dates and representative times logged during tracking.

quency distribution of temperatures was weakly bimodal (Figure 4). A comparison of the pooled temperature data from the STD-12 and those from the fish indicate that the Arctic ciscoes used only the uppermost 6 m of the water column and usually were in the uppermost 3 m. The Arctic ciscoes did not occupy the coldest waters present.

Discussion

When sufficient data are available or ocean currents are weak relative to a fish's swimming speed, the correspondence between ground and swimming speeds can be good (Quinn 1988; Carey and Scharold 1990). The close agreement between the estimated swimming and gross ground speeds of fish 4 and 10 suggests that, at least for the longer tracks, ground speed corresponded quite closely to swimming speed.[2]

The grand and individual mean gross groundspeeds of the Dolly Varden were similar to those of steelhead *Oncorhynchus mykiss* and pink salmon *O. gorbuscha*, whereas the Arctic cisco speeds are lower (Table 3). However, given the size differences between species, standardized speeds provide a

[2]Fish 10 was a Dolly Varden equipped with a pressure transmitter. The fish was very lean and apparently in poor condition, so it was not included in the movement analysis. Fish 10's swimming speed during a 3-h period was estimated by vector subtraction of surface currents from its track over ground. The synoptic current data were obtained by successive drogue deployments near the fish while tracking was underway.

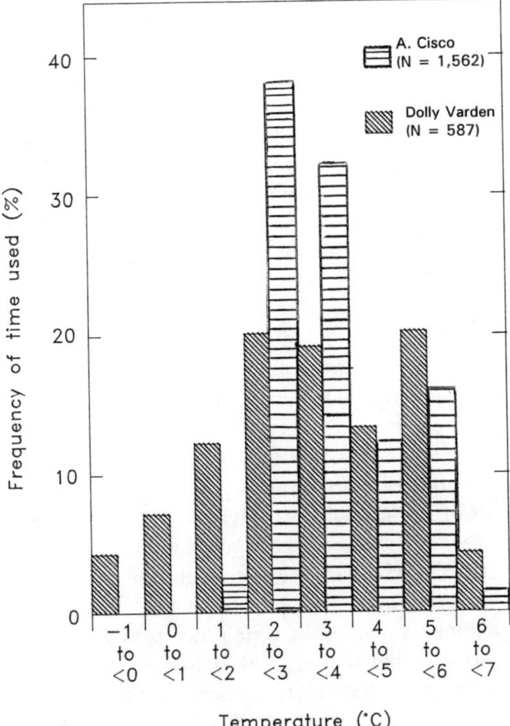

FIGURE 4.—Temperature occupancy of Dolly Varden *Salvelinus malma* and Arctic ciscoes *Coregonus autumnalis* in Camden Bay. The 1990 data were weighted by a factor of 5 to make them equivalent to the 1991 data. The temperatures used by Dolly Varden encompassed the entire range indicated by salinity–temperature–depth (STD) casts. In contrast, Arctic ciscoes did not use waters colder than 1°C. The upper limits of temperature occupancy of both species exceeded the maximum temperatures recorded from STD casts each year. These high values represent traverses of the plumes of streams occurring while fish followed shorelines.

more meaningful comparison. On that basis, the speeds of Dolly Varden and Arctic ciscoes in this study were greater than the published speeds of steelhead and cutthroat trout *O. clarki* and about the same as those of sockeye salmon *O. nerka*. Considering the accuracy of the estimates, our values and those of others approximate 1 L/s, the theoretical, optimal cruising speed of fish in the presence of currents (Trump and Leggett 1980). Presumably, swimming at that speed would be of selective advantage to fish attempting to assimilate and retain energy reserves while making long feeding dispersals or spawning migrations.

Data on Dolly Varden and Arctic cisco movement rates from arctic waters based on recaptures of tagged fish provide comparative information on net movement rates. Dempson (1984) estimated a net movement rate of 16.4 km/d during 5 d by an Arctic char *Salvelinus alpinus* in northern Labrador. Craig and Haldorson (1981) calculated an average net movement rate of 2.8 km/d for 30 Dolly Varden in the Beaufort Sea; they also noted that individual net movement rates, up to 78 km/d (90 cm/s), had been estimated from the region. Our estimates fell between these published values. The weighted group mean was 16.9 km/d (range: 8.2–54.6 km/d). The weighted mean net movement rate of the Arctic ciscoes we tracked was 19.9 km/d (range: 10.4–38.1 km/d). This value compares favorably with published estimates of 24 and 12.5 km/d (Nelson et al. 1987, cited by Fechhelm et al. 1989; Fruge et al. 1989).

The gross ground speeds of Dolly Varden and Arctic ciscoes we observed may typify movement rates of adult fish returning to freshwater. Age–length relations indicated that the Arctic ciscoes longer than 40.5 cm were 8–13 years old, and the

TABLE 3.—Gross ground speeds of salmonids. *N* is the number of fish tracked.

Species (*N*)	Total length (cm)	Group average speed (cm/s)	Group average speed (L/s)	Individual average speed (cm/s)	Individual average speed (L/s)	Reference
Dolly Varden (6)	48.5–56.9	55.8[a]	1.06	48.8–74.2	0.96–1.37	This study
Dolly Varden (1)				30.8		Armstrong and Reed (1971)
Cutthroat trout[b] (14)	31.0–40.0	33.6[a]	0.86	6.8–44.9	0.20–1.31	McCleave and LaBar (1972)
Steelhead (6)	72.0–89.0	57.4[a]	0.74	38.9–69.4	0.54–0.96	Ruggerone et al. (1990)
Sockeye salmon (13)		61.0	0.99			Blair and Quinn (1991)
Sockeye salmon (25)	60.6–73.1	64.7				Quinn (1988)
		66.8[c]	1.0[c]	31.9–108.0[c]	0.48–1.66[c]	
Pink salmon[d] (37)	41.0–60.0	48.3[e]				Martin et al. (1990)
Arctic cisco (7)	32.9–45.6	45.4	1.04	30.1–71.1	0.69–1.61	This study

[a]Weighted average; summed track lengths divided by summed times.
[b]Entire track. Tracks 1–4, 6–9, 14, 17, 19, 21, 23, 25.
[c]Swimming speed. Currents removed from ground speeds.
[d]Three control groups; sustained swimming toward stream.
[e]Grand mean of means weighted by numbers of fish in each control group.

Dolly Varden were 7–10 years old. Catch patterns from Camden Bay indicate that mature Dolly Varden and Arctic ciscoes vacate the area by mid-August (Palmer and Dugan 1990).

Directed movement by Dolly Varden was not evident from our data, whereas the Arctic cisco movements were nonrandom and westward. The dispersion seen in the Dolly Varden tracks could reflect variable current patterns (e.g., Hawkins and Urquhart 1983, p. 152). However, it seems more plausible that our sample included fish returning to rivers both east and west of the study area. The westward movement by Arctic ciscoes may represent a return migration of nonspawners to the Colville River. Both species tended to move toward shoreline edges (four of six Dolly Varden and all but one Arctic cisco). This behavior is consistent with the relatively heavy use of shoreline habitat observed in Simpson Lagoon (Craig et al. 1985) and Camden Bay (Fruge et al. 1989; Palmer and Dugan 1990).

Analysis of catches in the Prudhoe Bay area (Houghton et al. 1990) indicated that Dolly Varden and Arctic ciscoes prefer temperatures of 4–12°C during early July to mid-August. Much of the temperature occupancy by the Dolly Varden and Arctic ciscoes we tracked was below the lower limit of the apparently preferred temperature range at Prudhoe Bay, reflecting the paucity of warm water in 1990 and 1991.

Our data indicate that Dolly Varden and Arctic ciscoes selected the warmer halves of the available temperatures, which occurred at the sea surface. However, based on their horizontal movements, Dolly Varden did not avoid cold surface-waters. Fish 2, for example, swam through upwelled water having temperatures below 0°C; thus, the apparent preference for warmer waters may be less a reflection of temperature selection than of depth selection. The association of Arctic ciscoes and Dolly Varden with the surface layer was also observed in vertical gill-net catch patterns in Camden Bay (Fruge et al. 1989; Palmer and Dugan 1990). Elsewhere, Dolly Varden (Armstrong and Reed 1971), steelhead (Ruggerone et al. 1990), pink salmon (Martin et al. 1990), sockeye salmon (Stasko et al. 1976; Quinn and terHart 1987), and Arctic char (Johnson 1980) have displayed surface orientation.

Fechhelm and Gallaway (1984) hypothesized that the summer thermal structure of the Beaufort Sea induces shoreward movement by Arctic ciscoes. This occurs because the fishes' temperature preferenda lie at or above the highest water temperatures, which are attained inshore. The hypothesis can be broadened to include Dolly Varden and other amphidromous fishes, with the proviso that the expression of the bias may differ because of species-specific, age- and size-related physiological requirements and motivations (Quinn and Leggett 1987; Quinn and Brodeur 1991). Salinity probably has a similar influence because osmoregulatory abilities respond seasonally to photoperiod in Dolly Varden and Arctic char (Johnson and Heifetz 1988; Arneson et al. 1992). Shoreline affinity may also represent behavioral adaptation to declining osmoregulatory abilities with decreasing daylength. When brackish water is discontinuous or absent, close association with shorelines may promote survival of salmonids by ensuring that fish will encounter all available freshwater refugia.

In 1990 and 1991 little brackish water remained in Camden Bay after mid-July. Viewed solely on the basis of its thermohaline attributes, the habitat was of poor quality during both years, yet Dolly Varden and Arctic ciscoes were present in marine waters. We believe this represents a behavior that optimizes the joint conduct of their life processes (sensu Neill 1979) and that the net benefit to the animals was greater than that offered where preferred conditions occurred. Such feeding behavior has been reported for Arctic char (Johnson 1980) and Atlantic cod *Gadus morhua* (Rose and Leggett 1989).

Acknowledgments

This study was conducted while the authors were affiliated with the National Oceanic and Atmospheric Administration's Arctic Environmental Assessment Center, Anchorage, Alaska. Funding was provided by the Minerals Management Service (MMS), U.S. Department of the Interior (DOI), as part of the MMS Alaska Environmental Studies Program. The interpretations of data and opinions expressed by the authors do not necessarily reflect the views and policies of the DOI. The government does not approve, recommend, or endorse any proprietary material mentioned herein.

References

Armstrong, R. H., and R. D. Reed. 1971. A study of Dolly Varden in Alaska. Alaska Department of Fish and Game, Federal Aid in Fish Restoration, Project F-9-3-12, Annual Progress Report, 1970–71, Juneau.

Arneson, A. M., M. Halvorsen, and K. J. Nissen. 1992. Development of hypoosmoregulatory capacity in Arctic char (*Salvelinus alpinus*) reared under either continuous light or natural photoperiod. Canadian Journal of Fisheries and Aquatic Sciences 49:229–237.

Beamish, F. W. H. 1980. Swimming performance and oxygen consumption of the charrs. Pages 739–748 *in*

E. K. Balon, editor. Charrs: salmonid fishes of the genus *Salvelinus*. Dr. W. Junk, The Hague, Netherlands.

Berryman, D., B. Bobee, D. Cluis, and J. Haemmerli. 1988. Nonparametric tests for trend detection in water quality time series. Water Resources Bulletin 24: 545–556.

Blair, G. R., and T. P. Quinn. 1991. Homing and spawning site selection by sockeye salmon (*Oncorhynchus nerka*) in Iliamna Lake, Alaska. Canadian Journal of Zoology 69:176–181.

Brett, J. R. 1973. Energy expenditure of sockeye salmon, *Oncorhynchus nerka*, during sustained performance. Journal of the Fisheries Research Board of Canada 30:1799–1809.

Carey, F. G., and J. V. Scharold. 1990. Movements of blue sharks (*Prionace glauca*) in depth and course. Marine Biology 106:329–342.

Craig, P. C. 1984. Fish use of coastal waters of the Alaskan Beaufort Sea: a review. Transactions of the American Fisheries Society 113:265–282.

Craig, P. C. 1989a. An introduction to anadromous fish in the Alaskan arctic. Biological Papers of the University of Alaska 24:27–54.

Craig, P. C. 1989b. Subsistence fisheries at coastal villages in the Alaskan arctic, 1970–1986. Biological Papers of the University of Alaska 24:131–152.

Craig, P. C., and L. S. Haldorson. 1981. Fish. Pages 384–678 *in* S. R. Johnson and W. R. Richardson, editors. Beaufort Sea barrier island–lagoon ecological process studies. Final report, Simpson Lagoon. U.S. Department of the Interior, Bureau of Land Management and NOAA (National Oceanic and Atmospheric Administration) OCSEAP (Outer Continental Shelf Environmental Assessment Program), Boulder, Colorado.

Craig, P. C., W. B. Griffiths, L. Haldorson, and H. McElderry. 1985. Distributional patterns of fishes in an Alaskan arctic lagoon. Polar Biology 4:9–18.

Dempson, J. B. 1984. Identification of anadromous arctic char stocks in coastal areas of northern Labrador. Pages 143–162 *in* L. Johnson and B. Burns, editors. Biology of the Arctic char. University of Manitoba Press, Winnipeg.

Dutil, J. D. 1986. Energetic constraints and spawning interval in the anadromous Arctic charr (*Salvelinus alpinus*). Copeia 1986:945–955.

Everett, R. J., and R. L. Wilmot. 1990. Genetic stock structure of Arctic char (*Salvelinus alpinus*) from drainages to the Beaufort Sea in Alaska and Canada. Report of the U.S. Fish and Wildlife Service to Minerals Management Service and NOAA (National Oceanic and Atmospheric Administration), Anchorage, Alaska.

Fechhelm, R. G., and B. J. Gallaway. 1984. Temperature preference of juvenile Arctic cisco (*Coregonus autumnalis*) from the Alaskan Beaufort Sea, in relation to salinity and temperature acclimation. Pages 207–231 *in* U.S. Department of the Interior Bureau of Land Management and NOAA (National Oceanic and Atmospheric Administration) OCSEAP (Outer Continental Shelf Environmental Assessment Program) final report 23, Springfield, Virginia.

Fechhelm, R. G., and W. B. Griffiths. 1990. Effect of wind on the recruitment of Canadian Arctic cisco (*Coregonus autumnalis*) into the central Alaskan Beaufort Sea. Canadian Journal of Fisheries and Aquatic Sciences 47:2164–2171.

Fechhelm, R. G., J. S. Baker, W. B. Griffiths, and D. R. Schmidt. 1989. Localized movement patterns of least cisco (*Coregonus sardinella*) and Arctic cisco (*C. autumnalis*) in the vicinity of a solid-fill causeway. Biological Papers of the University of Alaska 24:75–106.

Fruge, D. J., D. W. Wiswar, L. J. Dugan, and D. E. Palmer. 1989. Fish population and hydrographic characteristics of Arctic National Wildlife Refuge coastal waters, summer 1988. U.S. Fish and Wildlife Service, Fairbanks, Alaska.

Gallaway, B. J. 1990. Factors limiting the growth of arctic anadromous fish populations. Pages 57–62 *in* R. M. Meyer and T. M. Johnson, editors. Fisheries oceanography—a comprehensive formulation of technical objectives for offshore application in the arctic. U.S. Department of the Interior, Minerals Management Service, OCS (Outer Continental Shelf) Study MMS 88-0042, Anchorage, Alaska.

Gallaway, B. J., W. J. Gazey, J. M. Colonell, A. W. Niedoroda, and C. J. Herlugson. 1991. The Endicott Development Project—preliminary assessment of impacts from the first major offshore oil development in the Alaskan arctic. Pages 42–80 *in* C. S. Benner and R. W. Middleton, editors. Fisheries and oil development on the continental shelf. American Fisheries Society Symposium 11, Bethesda, Maryland.

Gross, M. R. 1987. Evolution of diadromy in fishes. Pages 14–25 *in* M. J. Dadswell, editor. Common strategies of anadromous and catadromous fishes. American Fisheries Society Symposium 1, Bethesda, Maryland.

Hale, D. A. 1990. A description of the physical characteristics of nearshore and lagoonal waters in the eastern Beaufort Sea. Report of NOAA (National Oceanic and Atmospheric Administration) to U.S. Fish and Wildlife Service, Anchorage, Alaska.

Hale, D. A. 1991. A description of the physical characteristics of nearshore and lagoonal waters in the eastern Beaufort Sea, 1989. Report of NOAA (National Oceanic and Atmospheric Administration) to U.S. Fish and Wildlife Service, Anchorage, Alaska.

Hawkins, A. D., and G. G. Urquhart. 1983. Tracking fish at sea. Pages 150–166 *in* A. G. Macdonald and I. G. Priede, editors. Experimental biology at sea. Academic Press, London.

Houghton, J. P., C. J. Whitmus, and A. W. Maki. 1990. Habitat relationships of Beaufort Sea anadromous fish: integration of oceanographic information and fish-catch data. Pages 73–88 *in* R. M. Meyer and T. M. Johnson, editors. Fisheries oceanography—a comprehensive formulation of technical objectives for offshore application in the arctic. U.S. Department of the Interior, MMS (Minerals Management

Service), OCS (Outer Continental Shelf) Study MMS 88-0042, Anchorage, Alaska.

Johnson, L. 1980. The Arctic charr, *Salvelinus alpinus*. Pages 15–98 *in* E. G. Balon, editor. Charrs: salmonid fishes of the genus *Salvelinus*. Dr. W. Junk, The Hague Netherlands.

Johnson, S. W., and J. Heifetz. 1988. Osmoregulatory ability of wild coho salmon (*Oncorhynchus kisutch*) and Dolly Varden char (*Salvelinus malma*) smolts. Canadian Journal of Fisheries and Aquatic Sciences 45:1487–1490.

Martin, D. J., C. J. Whitmus, and L. A. Brocklehurst. 1990. Effects of petroleum contaminated waterways on migratory behavior of adult pink salmon. Pages 281–529 *in* U.S. Department of Commerce and U.S. Department of the Interior, OCSEAP (Outer Continental Shelf Environmental Assessment Program) Final Report 66, Anchorage, Alaska.

McCleave, J. D., and G. W. LaBar. 1972. Further ultrasonic tracking and tagging studies of homing cutthroat trout (*Salmo clarki*) in Yellowstone Lake. Transactions of the American Fisheries Society 101: 44–54.

Mellas, E. J., and J. M. Haynes. 1985. Swimming performance and behavior of rainbow trout (*Salmo gairdneri*) and white perch (*Morone americana*): effects of attaching telemetry transmitters. Canadian Journal of Fisheries and Aquatic Sciences 42:488–493.

Neill, W. H. 1979. Mechanisms of fish distribution in heterothermal environments. American Zoologist 19:305–317.

Nelson, T., T. C. Cannon, W. R. Olmstead, and K. C. Wiley. 1987. Surveys of domestic and commercial fisheries in the central and eastern Beaufort Sea. Pages 1–36 *in* Final report of the Endicott environmental monitoring survey, 1985, volume 7 to U.S. Army Corps of Engineers, Alaska District, Anchorage.

Norton, D. W., editor. 1989. Research advances on anadromous fish in arctic Alaska and Canada. Biological Papers of the University of Alaska 24:1–3.

Palmer, D. E., and L. J. Dugan. 1990. Fish population characteristics of Arctic National Wildlife Refuge coastal waters, summer 1989. U.S. Fish and Wildlife Service, Fairbanks, Alaska

Quinn, T. P. 1988. Estimated swimming speeds of migrating adult sockeye salmon. Canadian Journal of Zoology 66:2160–2163.

Quinn, T. P., and R. D. Brodeur. 1991. Intra-specific variations in the movement patterns of marine animals. American Zoologist 31:231–241.

Quinn, T. P., and W. C. Leggett. 1987. Perspectives on marine migrations of diadromous fishes. Pages 377–388 *in* M. J. Dadswell, editor. Common strategies of anadromous and catadromous fishes. American Fisheries Society Symposium 1, Bethesda, Maryland.

Quinn, T. P., and B. A. terHart. 1987. Movements of adult sockeye salmon (*Oncorhynchus nerka*) in British Columbia coastal waters in relation to temperature and salinity stratification: ultrasonic telemetry results. Pages 61–77 *in* H. D. Smith, L. Margolis, and C. C. Wood, editors. Sockeye salmon (*Oncorhynchus nerka*) population and future management. Canadian Special Publication of Fisheries and Aquatic Sciences 96.

Rose, G. A., and W. C. Leggett. 1989. Interactive effects of the geophysically-forced sea temperatures and prey abundance on mesoscale coastal distributions of a marine predator, Atlantic cod (*Gadus morhua*). Canadian Journal of Fisheries and Aquatic Sciences 46:1904–1913.

Ruggerone, G. T., T. P. Quinn, I. A. McGregor, and T. D. Wilkinson. 1990. Horizontal and vertical movements of adult steelhead trout, *Oncorhynchus mykiss*, in the Dean and Fisher channels, British Columbia. Canadian Journal of Fisheries and Aquatic Sciences 47: 1963–1969.

Stasko, A. B., R. M. Horrall, and A. D. Hasler. 1976. Coastal movements of adult Fraser River sockeye salmon (*Oncorhynchus nerka*) observed by ultrasonic tracking. Transactions of the American Fisheries Society 105:64–71.

Thorsteinson, L. K., L. E. Jarvela, and D. A. Hale. 1991. Arctic fish habitat use investigations: nearshore studies in the Alaskan Beaufort Sea, summer 1990. Report of NOAA (National Oceanic and Atmospheric Administration) to Minerals Management Service, Anchorage, Alaska.

Trump, C. L., and W. C. Leggett. 1980. Optimum swimming speed in fish: the problem of currents. Canadian Journal of Fisheries and Aquatic Sciences 37:1086–1092.

Wiens, J. A., and J. T. Rotenberry. 1981. Censusing and evaluation of avian habitat occupancy. Studies in Avian Biology 6:522–532.

Williams, I. V., and J. R. Brett. 1987. Critical swimming speeds of Fraser and Thompson River pink salmon (*Oncorhynchus gorbuscha*). Canadian Journal of Fisheries and Aquatic Sciences 44:348–356.

Winter, J. D. 1983. Underwater biotelemetry. Pages 371–395 *in* L. A. Nielson and D. L. Johnson, editors. Fisheries techniques. American Fisheries Society, Bethesda, Maryland.

Zar, J. H. 1984. Biostatistical analysis. Prentice-Hall, Englewood Cliffs, New Jersey.

American Fisheries Society Symposium 19:175–183, 1997

Movements of Postsmolt Anadromous Dolly Varden in Northwestern Alaska

ALFRED L. DECICCO
Alaska Department of Fish and Game, Sport Fish Division
1300 College Road, Fairbanks, Alaska 99701-1599, USA

Abstract.—Studies conducted from the early 1980s to the mid-1990s show that anadromous Dolly Varden *Salvelinus malma* in the Kotzebue Sound and Chukchi Sea drainages of Alaska exhibit complex movement patterns. Movements are framed within the context of annual seaward feeding migrations and fall migrations into freshwater for overwintering. Dolly Varden home for spawning but do not demonstrate fidelity to overwintering rivers. Movements associated with spawning differ from the annual migrations exhibited by nonspawners. The existence of summer and fall spawning groups further complicates the pattern. Summer spawners remain in freshwater during the year in which they spawn. If they overwintered in their home river, they move upstream to spawning areas during June or early July, remain on spawning grounds until early September, then descend to lower river areas where they join nonspawners to overwinter, remaining in freshwater for 20 months during the spawning cycle. If fish have overwintered in a river other than their natal river, they move to sea with nonspawners in June, travel directly to their home river, and move upstream in early July to join other summer spawners already on spawning grounds. Postspawning movements are similar to those of other summer spawners. Fall spawners feed at sea during summer, enter rivers in August, and travel directly to spring areas in headwater streams where they spawn and remain throughout the winter. Unlike summer spawners, fall spawners make annual migrations to sea. Stocks are mixed in major overwintering areas. A hypothesis based on ocean currents and life history patterns of Dolly Varden may explain the observed movements and mixing of stocks.

Anadromous Dolly Varden *Salvelinus malma* occur in coastal drainages throughout most of northern and western Alaska. They also occur throughout eastern Russia to Chaunsk Bay on the north coast of the Chukotsk Peninsula (Figure 1), where they are sympatric with anadromous *S. taranetzi* in some drainages (Behnke 1984; I. A. Chereshnev, Institute of Biological Problems of the North, personal communication).

Movement studies of Dolly Varden and Arctic char *S. alpinus* have been confined mainly to freshwater and nearshore areas (Armstrong 1974; Moore 1975; Craig and McCart 1976; Armstrong and Morrow 1980; Johnson 1980; Gyselman 1984; Dempson and Green 1985). Armstrong (1974) described complex movement patterns of southern-form Dolly Varden in southeastern Alaska. Sexually mature Dolly Varden in northwestern Alaska commonly overwinter in nonnatal rivers during years in which they have been to sea (DeCicco 1989). Most available literature suggests that while anadromous char are at sea, they do not travel far offshore. Exceptions have been reported from Kodiak Island, Alaska, where a Dolly Varden tagged in the Buskin River was recaptured across Shelikof Strait in Dakavak Bay, a distance of 160 km (S. Sonnichsen, Alaska Department of Fish and Game, personal communication), and from Kamchatka, where Dolly Varden were caught as far as 420 km offshore (Mishima 1975). Long-distance movements have been reported for both Dolly Varden and Arctic char. Jensen and Berg (1977) reported the longest distance traveled by an Arctic char tagged in the Vardnes River, Norway, as 940 km. The fish was recaptured in the Tuloma River, in the former USSR. Two other Arctic char were recaptured at distances of 500 and 400 km from the tagging location. Arctic char tagged at the outlet to Nauyuk Lake, Northwest Territories, Canada, have been recaptured at various locations in the Canadian archipelago, at distances of up to 500 km (Johnson 1989). An Arctic char tagged in the Ekalluk River, Northwest Territories, Canada, in 1979 was recaptured 3 years later in Shepherd Bay, 550 km to the east, and the longest distance traveled by an Arctic char in northern Labrador was 250 km (Dempson and Kristofferson 1987). In northwestern Alaska, a Dolly Varden tagged in a spawning area in the Noatak River system was recaptured at Point Hope 1 year later, at a distance of 485 km. Dolly Varden tagged in a Wulik River overwintering area in fall 1988 were recaptured in Norton Sound, St. Lawrence Island, and the Anadyr River, Russia. The longest distance traveled was 1,690 km (DeCicco 1992).

Like anadromous chars worldwide (Johnson

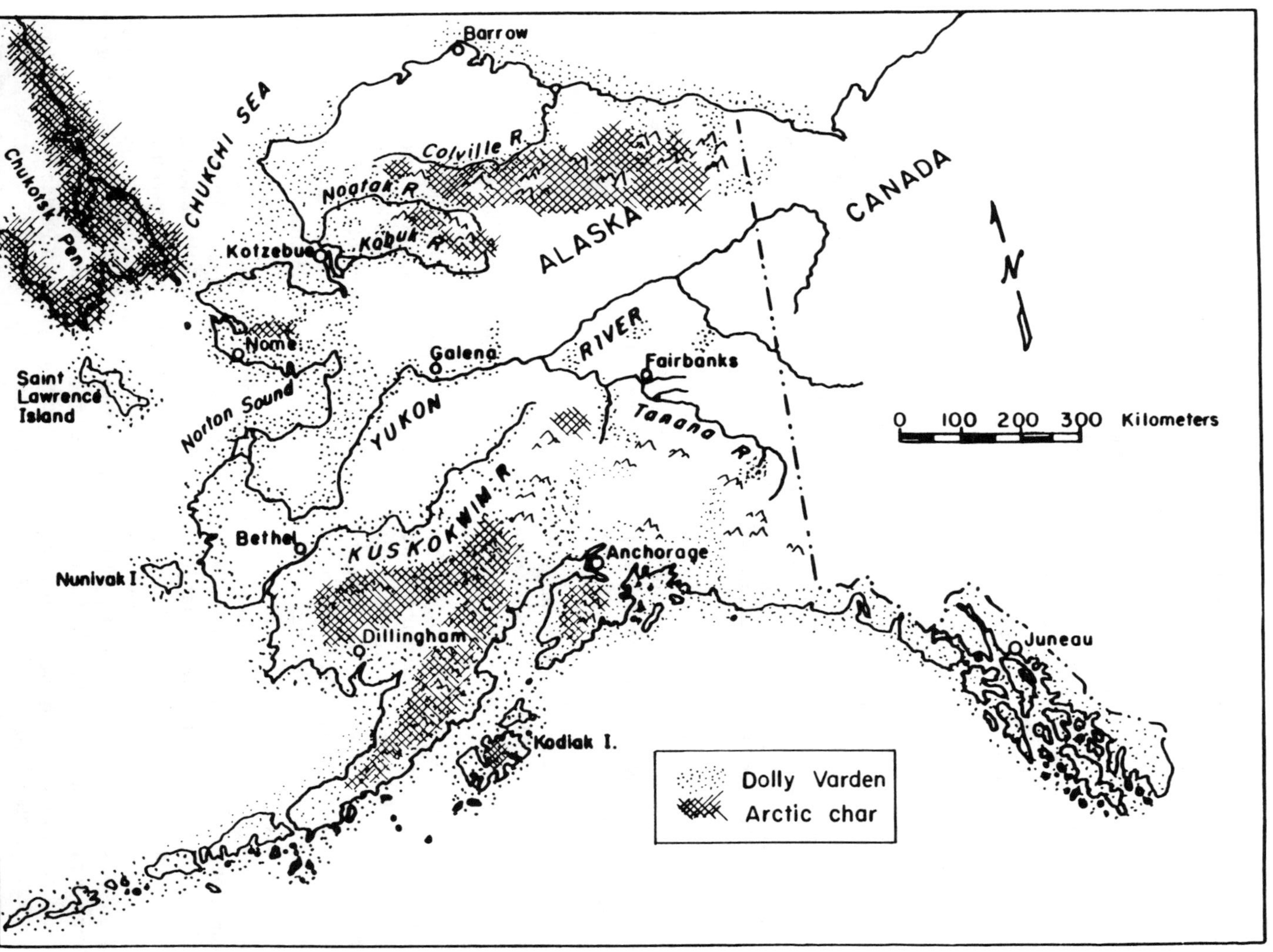

FIGURE 1.—The distribution of Arctic char *Salvelinus alpinus* and Dolly Varden *S. malma* in Alaska and on the Chukotsk Peninsula, Russia.

1980), Dolly Varden of northwestern Alaska spawn and overwinter in freshwater (Figure 2). After an initial freshwater residence of 2–5 years, they undertake annual summer feeding migrations into marine waters. Overwintering takes place in the lower Wulik, Kivalina and Noatak rivers, where sufficient groundwater provides suitable habitat throughout the arctic winter. Spawning areas are located in upstream areas and in tributary streams within the influence of springs.

This article is a review of existing data on movements of anadromous Dolly Varden in northwestern Alaska in relation to spawning and overwintering, describes long distance movements of Dolly Varden tagged in Alaska, and presents a framework whereby the observed movements may be explained.

Methods

Dolly Varden spawning and overwintering areas in the Noatak, Wulik, and Kivalina river drainages were located visually from a PA-18 aircraft. Within the study area (Figure 2), 4,620 anadromous northern-form Dolly Varden (Behnke 1980) were tagged between spring 1981 and fall 1983. Of these, 2,972 were tagged in spawning areas of the Noatak River drainage, 631 in spawning areas of the Wulik River, 46 in overwintering areas on the Wulik River, 60 in spawning areas on the Kivalina River, and 911 in overwintering areas on the Kivalina River. During October 1984, Telonics RB-5 radio transmitters were inserted in eight fall spawners in the Kugururok River. They were periodically tracked during the winter from a small aircraft using a telemetry receiver. In 1988 during September, 4,075 Dolly Varden were tagged in an overwintering area of the Wulik River. All fish were captured with beach seines or hook and line, measured for fork length, and tagged with individually numbered Floy FD-67 internal anchor tags. Tagged fish ranged in fork length from 230 to 870 mm. Subsistence fisheries near Kivalina, Noatak, and Kotzebue and the Kotzebue commercial fishery for chum salmon *Oncorhynchus keta* were sampled to obtain tag recoveries. Cooperation by local Native residents and sport anglers greatly aided tag recovery efforts. Movements were inferred from tag recovery, radio telemetry, and aerial observation of fish.

Results and Discussion

The recovery of 364 tags from 8,695 tagged fish (Table 1) in the study area and the observation of fish in spawning and overwintering areas indicate a complex movement pattern in relation to spawning and overwintering.

Seasonal Movements

The spring seaward migration of Dolly Varden from the Noatak River begins from late May to mid-June depending on the timing of ice breakup. The migration usually occurs with declining flow after spring high water. On the Wulik and Kivalina rivers, fish usually enter the sea during the third week of June but are sometimes delayed in the lagoon until early July if west winds move broken sea ice onshore. Kivalina residents report that larger Dolly Varden are the first to enter the sea, followed closely by smaller fish of mixed sizes. During 1989 Dolly Varden that overwintered in the Wulik River began moving seaward on 13 June and were being captured in Kivalina Lagoon through 3 July.

The fall migration from the sea into freshwater usually begins in mid-August and can extend into early October. Kivalina residents have observed that Dolly Varden migrate into the Wulik River by size in four overlapping runs, with the largest fish entering last. Most fish are in the river by the time it freezes, usually in early October; however, in 1986, few Dolly Varden were observed in overwintering areas during the last week of September, and schools of fish were reported actively moving upstream during mid-October, when much of the river was ice covered. Dolly Varden destined for Noatak River overwintering areas are taken annually in the Kotzebue Sound commercial salmon fishery during the last half of August, and I have taken them in Kotzebue Sound near the mouth of the Noatak River as late as 25 September.

Data suggest that Dolly Varden home for spawning, but stocks are mixed in rivers during the winter. Although direct evidence of return to natal streams is lacking (i.e., marked smolts recaptured while spawning in their stream of origin), homing is indicated by the recapture of 15 tagged, spawning fish. All had been tagged in spawning condition in earlier years and were recovered in the same spawning streams. The interval between tagging and recovery was 2 years in 13 cases and 4 years in 2 cases, supporting a general pattern of alternate-year spawning. Stocks are mixed in overwintering areas. Tag recoveries from Dolly Varden captured while overwintering in Wulik River included fish tagged while spawning in all major Noatak River spawning areas. Nonspawners tagged while overwintering in the Kivalina River have been recaptured overwin-

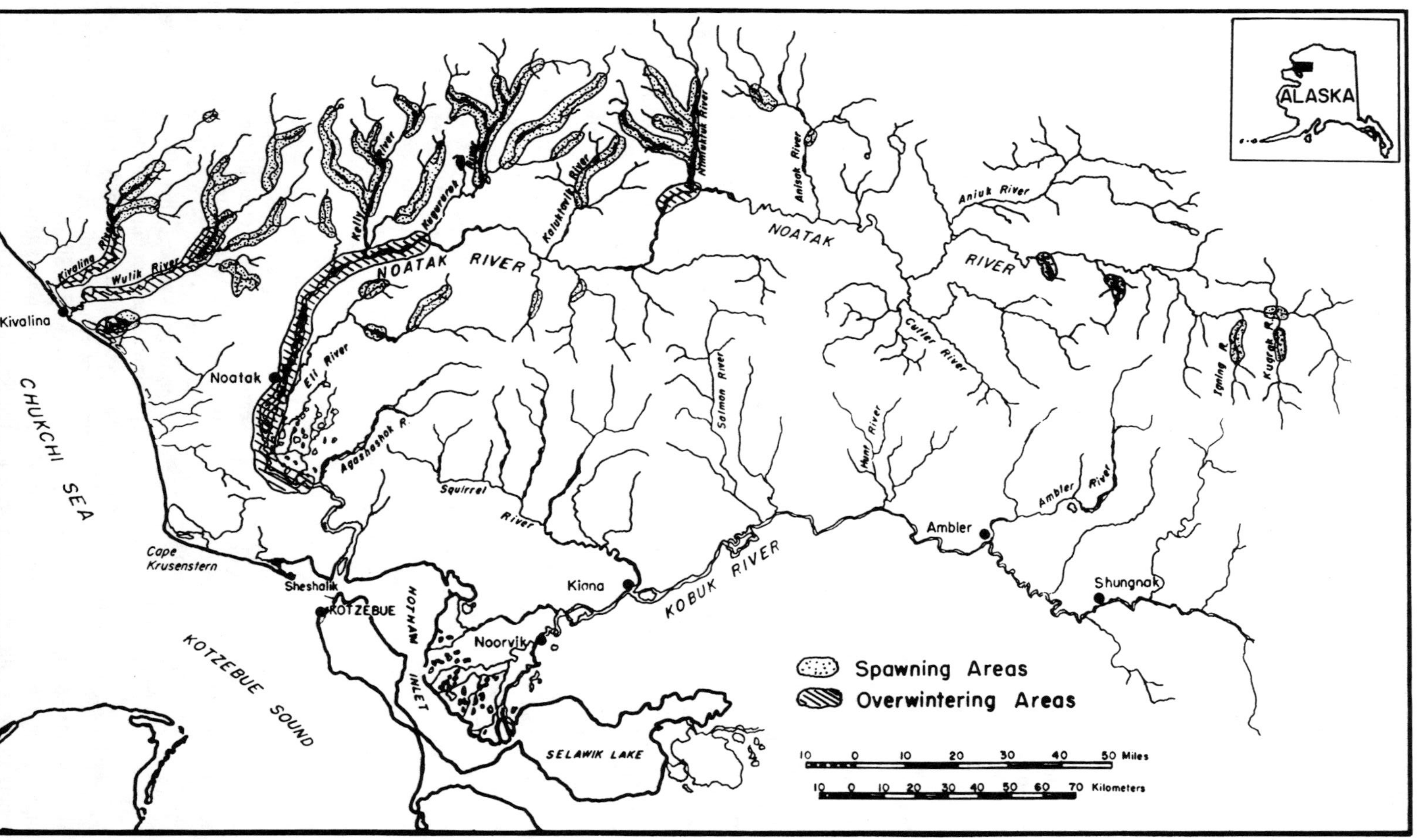

FIGURE 2.—The study area showing spawning and overwintering areas of Dolly Varden *Salvelinus malma* in the Wulik, Kivalina, and Noatak rivers.

TABLE 1.—Numbers of Dolly Varden *Salvelinus malma* tagged and recovered in the study area between 1980 and 1992. Abbreviations are NS = nonspawner, S = spawner, OW = overwintering area, and SP = spawning area.

Location	Wulik River		Noatak River	Kivalina River	
	NS	S	S	NS	S
Number tagged					
	4,121	631	2,972	911	60
Number recovered					
Wulik River (OW)	161	20	16	16	0
Wulik River (SP)	0	0	0	0	0
Noatak (OW)	3	1	42	1	0
Noatak (SP)	8	0	15	3	0
Kivalina (OW)	0	0	1	2	0
Kivalina (SP)	0	0	0	0	0
Kotzebue Sound	31	1	29	2	0
Norton Sound	1	0	0	0	0
Imuruk Basin	2	0	0	0	0
Pilgrim River	1	0	0	0	0
Saint Lawrence Island	1	0	0	0	0
Kobuk River	3	0	0	0	0
Point Hope	0	0	1	0	0
Wales	1	0	0	0	0
Russia	3	0	0	0	0
Total recovered	215	22	104	24	0

tering in the Wulik River and spawning in Noatak River tributaries. Nonspawners tagged in the Wulik River have been recaptured in Kotzebue and Norton sounds; in the Pilgrim, Kobuk, and Noatak rivers; near Teller and Wales on the Seward Peninsula; on Saint Lawrence Island; and in the Anadyr River, Russia, and near the Margje River, Russia. They have also been taken while spawning in tributaries of the Noatak River.

Summer and fall spawning populations exist in northwestern Alaska. They differ in seasonal movement patterns and in areas used for overwintering, but there is some overlap in timing of spawning and in spawning areas.

Summer spawners.—Most current-year summer spawners do not enter the sea, but move upstream from overwintering areas to enter tributary rivers in late June. By mid-July they are distributed throughout spawning areas. After spawning during late August or early September, spent fish move downstream where they join nonspawners and remain in lower river overwintering areas until the following spring, thus residing in freshwater for 20 months before returning to sea. A small number of summer spawners diverge from this general pattern by spawning earlier in the summer and then moving to sea. Two tagged fish were taken in Kotzebue Sound, one in late July and one in early August, after having been tagged in prespawning condition on the Nimiuktuk River only a few weeks earlier. It is assumed that these fish had spawned, but they were not available for examination. Early summer spawning has also been reported on the Kivalina River, where Dolly Varden have been observed actively spawning during early July but were absent from the area a few weeks later (P. Driver, Midnight Sun Lodge, personal communication). In addition, one Dolly Varden taken in Kotzebue Sound on 22 August 1983 contained retained eggs from a recent spawning act.

Because stocks are mixed for overwintering, some summer spawners are not in their home river system at the onset of their spawning year. These fish move to sea with nonspawners in spring and travel directly to their home river, ascending to spawning grounds by late July, where they join other summer spawners. One Dolly Varden recovered in July from a spawning area of the Kugururok River in the Noatak River system had been tagged 47 d earlier in the Wulik River, before the spring seaward migration. Less-direct evidence of this movement is suggested from June subsistence catches at Sheshalik Spit near Kotzebue, where local residents report "fat trout" with maturing gonads moving eastward toward the mouth of the Noatak River while "skinny trout" are moving west. Tag recoveries have shown the latter group to be fish that had spawned the previous year and were migrating from Noatak River overwintering areas.

Spawning takes place from the 3rd week of August through early September in main channels and smaller side channels within the influence of groundwater sources. Spent fish move downstream immediately after spawning to mix with nonspawners in main river overwintering areas. Of 42 tagged spawners later recovered from Noatak River overwintering areas, 23 were recaptured in spent condition during the same year as marked. Spent fish were also captured in the Noatak River near the mouth of the Kugururok River on 27 August 1983. On 13 September 1983, no fish were observed in summer spawning areas in the Noatak River drainage, while large numbers of spent Dolly Varden were present in the lower 24 km of the Kugururok River, well downstream from spawning areas.

Fall spawners.—After feeding at sea during the summer, fall spawners enter freshwater in August just before the earliest nonspawners move to overwintering areas. Only prespawning Dolly Varden were being taken in the Kotzebue Sound commercial salmon fishery through 15 August 1986, after which nonspawners dominated in catches. Fall spawners travel directly to upriver spawning areas,

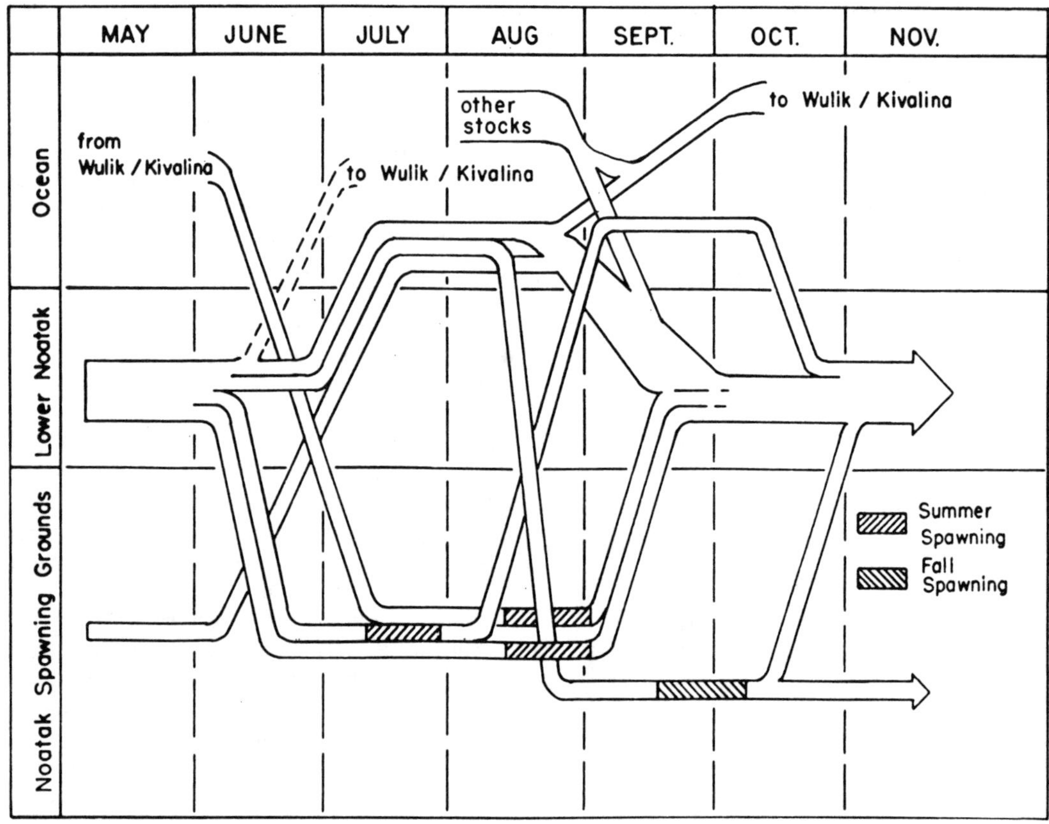

FIGURE 3.—Movements of Dolly Varden *Salvelinus malma* between Noatak River spawning and overwintering areas and the sea. Dashed lines indicate probable but undocumented movements.

which differ somewhat from areas used by summer spawners. Fall spawning occurs directly in spring-fed side channels and main channel springs from mid-September through early October. A prespawning female, not yet in spawning colors, taken 32 km upstream in the Kugururok River on 17 September 1979, contained eggs that were 4 mm in diameter. In contrast to summer spawners, most fall spawners remain near spawning areas or in other spring-influenced reaches of tributary streams to overwinter. Eight spent or partially spent fall spawners in the Kugururok River were implanted with radio transmitters on 1 October 1984. By 26 October, three had moved 5 km downstream, three had moved 24 km downstream, one remained within 1 km of the tagging location, and one had moved downstream into the Noatak River. Positions of the fish did not appreciably change by 29 November. On 23 March 1985, the fish were in the same general locations, although movements up to 2.4 km were noted. Radio-tagged Dolly Varden in the Anaktuvuk River on Alaska's north slope moved from only a few meters to 2.6 km between September 1980 and April 1981 (T. Bendock, Alaska Department of Fish and Game, personal communication). In this study, the four males remained nearest the spawning area, whereas the four females moved downstream. Three of the females were together just below a large spring 19 km from the nearest radio-tagged male. Movement patterns within a single river system are summarized in Figure 3.

Long Distance Movements

During August 1989, a Dolly Varden tagged in the Wulik River was recaptured 540 km upstream in the Anadyr River, near Markovo, Russia (Figure 4). The fish traveled 1,560 km during approximately 60 d, averaging 26 km/d. The fish was tagged on 17 September 1988 and would not have left the Wulik River until June 1989. A second Dolly Varden

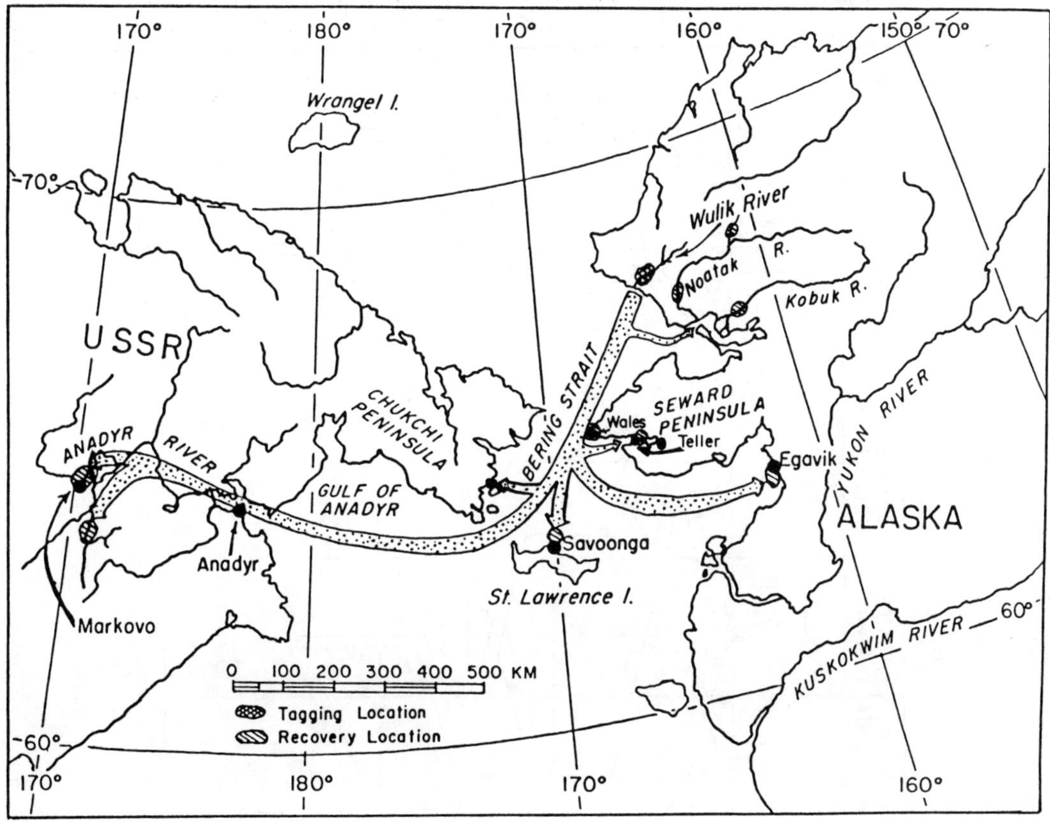

FIGURE 4.—Map showing long-distance movements of Dolly Varden *Salvelinus malma*, with tagging and recapture locations (updated from Figure 1 of DeCicco 1992).

tagged in the Wulik River in 1988 was recovered from the Anadyr River in 1990. The fish was recaptured 670 km upstream from the mouth, having traveled a total distance of 1,690 km in 14 months. A third tagged Dolly Varden was taken in Russian waters near the mouth of the Margje River in July 1992, approximately 425 km southwest from the Wulik River tagging location.

Other Dolly Varden tagged in 1988 have been recaptured in Alaskan waters long distances from the Wulik River. Two August 1989 recoveries were taken south of the Bering Strait, one near the mouth of the Egavik River, a distance of 570 km along the coast, and the other near Savoonga on Saint Lawrence Island, 180 km off the Alaskan mainland, 530 km from the Wulik River. They have also been taken in other Alaskan waters near the villages of Teller and Wales, in the Kobuk River, and in the Pilgrim River north of Nome.

The fish recovered in the Anadyr River in August 1989 was probably of Anadyr stock that had spent the previous winter in the Wulik River and had subsequently migrated to the Anadyr River to spawn. Anadromous Dolly Varden are reportedly uncommon in the Markovo area but some migrate past Markovo to spawn in Anadyr River tributaries 100–150 km upstream (Chereshnev, personal communication). There was no information provided on the sexual condition or size of this female Dolly Varden at capture; however, it was 522 mm in fork length when tagged, a size at which some Dolly Varden in northwestern Alaska first reach sexual maturity. Based on the location of capture, the other Russian tag recovery was also probably a fish of Anadyr stock that was to spawn in 1990. The stock identity of the fish recovered in Alaskan waters near and south of the Bering Strait is unknown; however, the fish recaptured in the Pilgrim River spawned there in 1992.

Surface current flow through the Bering Strait during the open-water period is in a northerly direction. Current speeds between 16 July and 28

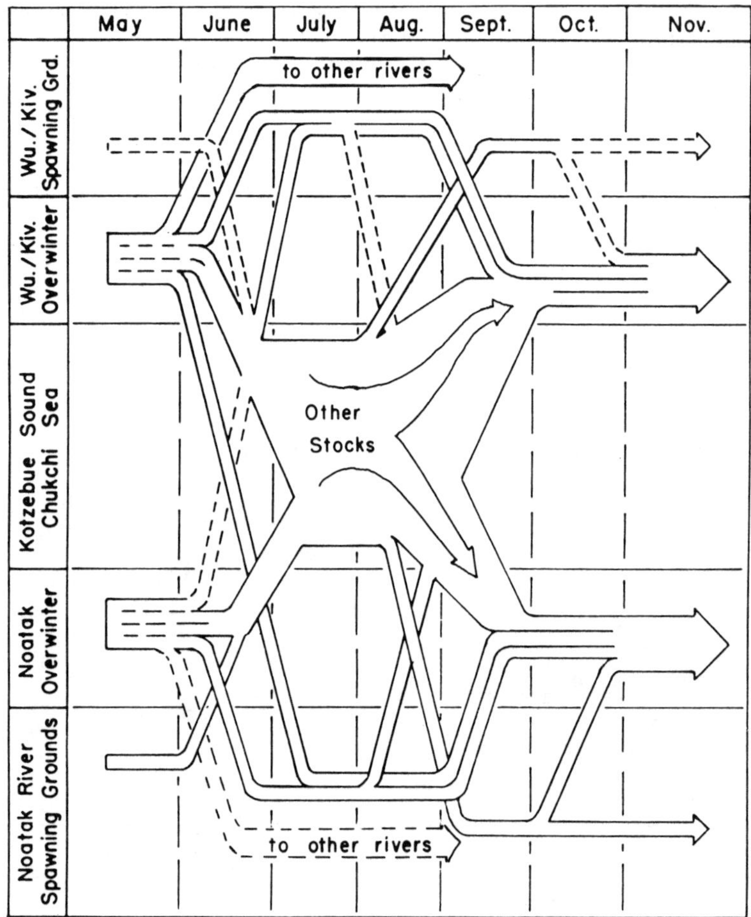

FIGURE 5.—Overall movement patterns of Dolly Varden *Salvelinus malma* between adjacent study-area river systems and the sea. Dashed lines indicate probable but undocumented movements. Abbreviations are Wu. = Wulik, Kiv. = Kivalina, and Grd. = Grounds.

August 1987 ranged from 25 cm/s to 85 cm/s, and although wind-driven current reversals have been documented, they typically occur after 1 September (R. Tripp, University of Washington, personal communication). In northwestern Alaska, Dolly Varden usually undertake three to five ocean migrations before reaching sexual maturity. Because fish do not demonstrate fidelity to overwintering areas, it is during this phase in their life history that fish, following food that may be carried by marine currents, may become distributed over a wide geographic area. The result could be a mixed northerly distribution of stocks after several ocean migrations. Upon reaching sexual maturity, a homeward migration would result in movements indicated by these tag recoveries. Straying has been considered as an alternative explanation for the observed movements; however, it seems unlikely that directed movements of the magnitude and speed observed would result from reproductive straying.

Anadromous Dolly Varden of northwestern Alaska exhibit movement patterns relating to overwintering, ocean residency, and spawning. Two spawning populations are thought to exist based on movement patterns, timing and location of spawning, and location of overwintering. The overall pattern of movement between and within adjacent river systems (Figure 5) is complex. Interdrainage movement is common, with some fish tagged in all major spawning areas of the Noatak River having been recaptured in the Wulik River overwintering area during years subsequent to spawning. Immature anadromous Dolly Varden also exhibit inter-drainage movement, which, in combination with marine currents and homing, is a probable explanation for observed long-distance movements.

Rounsefell (1958) ranked Dolly Varden slightly higher than Arctic char in relative degree of anadromy among salmonids, and Gyselman (1984) suggested that 250 km was the limit to which an Arctic char could migrate in one summer. Northern-form Dolly Varden are apparently well adapted to temporary residence in the marine environment. They are capable of, and may routinely undertake, long-distance ocean movements that are not necessarily coastal in nature. Although the amount of stock mixing is unknown, tag recoveries indicate that stocks from a wide geographic area along the Bering and Chukchi coasts are mixed both at sea and in overwintering areas.

Acknowledgments

The project from which these data were collected was partially funded by the U.S. Fish and Wildlife Service through the Federal Aid in Sport Fish Restoration Act (16 USC §§ 777 to 777k). I thank N. P. Novikov of the Pacific Research Institute of Fisheries and Oceanographics in Vladivostok, Russia, and I. A. Chereshnev and P. Gudkov of the Institute of Biological Problems of the North in Magadan, Russia, for information on tag recoveries. Ken Alt, W. Arvey, P. Houton, E. Adey, O. Knox, and J. Swan assisted with tagging. Phil Driver, Bob Uhl, and the people of Kivalina and Noatak kindly shared their considerable knowledge of Dolly Varden.

References

Armstrong, R. H. 1974. Migrations of anadromous Dolly Varden (*Salvelinus malma*) in southeastern Alaska. Journal of the Fisheries Research Board of Canada 31:435–444.

Armstrong, R. H., and J. E. Morrow. 1980. The Dolly Varden. Pages 99–140 *in* E. K. Balon, editor. Charrs: salmonid fishes of the genus *Salvelinus*. Dr. W. Junk Publishers, The Hague, Netherlands.

Behnke, R. J. 1980. A systematic review of the genus *Salvelinus*. Pages 441–480 *in* E. K. Balon, editor. Charrs: salmonid fishes of the genus *Salvelinus*. Dr. W. Junk Publishers, The Hague, Netherlands.

Behnke, R. J. 1984. Organizing the diversity of the Arctic charr complex. Pages 3–22 *in* L. Johnson and B. Burns, editors. Biology of the Arctic charr: proceedings of the international symposium on Arctic charr. University of Manitoba Press, Winnipeg.

Craig, P. C., and P. J. McCart. 1976. Fish use of near shore coastal waters in the western Arctic: emphasis on anadromous species. Pages 361–380 *in* D. W. Hood and D. C. Burrell, editors. Assessment of the Arctic marine environment, selected topics. University of Alaska, Institute of Marine Science, Occasional Publication 4, Fairbanks.

DeCicco, A. L. 1989. Movements and spawning of adult Dolly Varden charr (*S. malma*) in Chukchi Sea drainages of northwestern Alaska: evidence for summer and fall spawning populations. Physiology and Ecology Japan, Special Volume 1:229–238.

DeCicco, A. L. 1992. Long distance movements of anadromous Dolly Varden between Alaska and the U.S.S.R. Arctic 45:120–123.

Dempson J. B., and J. M. Green. 1985. Life history of anadromous arctic charr, *Salvelinus alpinus*, in the Fraser River, northern Labrador. Canadian Journal of Zoology 63:315–324.

Dempson, J. B., and A. H. Kristofferson. 1987. Spatial and temporal aspects of the ocean migration of anadromous Arctic char. Pages 340–357 *in* M. J. Dadswell, editor. Common strategies of anadromous and catadromous fishes. American Fisheries Society Symposium 1, Bethesda, Maryland.

Gyselman, E. C. 1984. The seasonal movement of anadromous Arctic charr at Nauyuk Lake, Northwest Territories, Canada. Pages 575–578 *in* L. Johnson and B. L. Burns, editors. Biology of the Arctic charr: proceedings of the international symposium on Arctic charr. University of Manitoba Press, Winnipeg.

Jensen, K. W., and M. Berg. 1977. Growth, mortality and migrations of the anadromous char, *Salvelinus alpinus*, L., in the Vardnes River, Troms, Northern Norway. Institute of Freshwater Research Drottningholm Report 56:70–80.

Johnson, L. 1980. The Arctic charr, *Salvelinus alpinus*. Pages 15–97 *in* E. K. Balon, editor. Charrs: salmonid fishes of the genus *Salvelinus*. Dr. W. Junk Publishers, The Hague, Netherlands.

Johnson, L. 1989. The anadromous Arctic charr, *Salvelinus alpinus*, of Nauyuk Lake, N.W.T. Canada. Physiology and Ecology Japan, Special Volume 1:201–227.

Mishima, S. 1975. A biological study of the anadromous Dolly Varden *Salvelinus malma* (Walbaum) distributed in the west coast off Kamchatka in summer season 1972–1974. Bulletin of the Faculty of Fisheries Hokkaido University 26:154–168.

Moore, J. W. 1975. Distribution movements and mortality of anadromous arctic char, *Salvelinus alpinus*, in the Cumberland Sound area of Baffin Island. Journal of Fisheries Biology 7:339–348.

Rounsefell, G. A. 1958. Anadromy in North American Salmonidae. U.S. Fish and Wildlife Service Fishery Bulletin 58:171–185.

American Fisheries Society Symposium 19:184–193, 1997

Interpopulation Variation in Growth Rates of Broad Whitefish

Ross F. Tallman

Department of Fisheries and Oceans
501 University Crescent, Winnipeg, Manitoba R3T 2N6, Canada

Abstract.—I analyzed data on length at age of broad whitefish *Coregonus nasus* from 23 locations in the arctic regions of Canada, Alaska, and Siberia to determine the effects on growth rate of age, climate, dispersion (over a geographical range), and distance from the sea. A multiple-regression analysis combining the effects of age, climate of resting area, latitude of spawning area, and dispersion and their interactions on fork length was significant ($P < 0.0001$, $R^2 = 0.90$). Age, climate, spawning latitude, and dispersion all had significant effects on length; distance from the sea did not. Climate and distance from the sea exhibited multicolinearity. Inclusion of a quadratic term to account for a curvilinear relationship between age and length while removing interactions resulted in only a slight increase in significance ($R^2 = 0.91$); a reduced model using climate, spawning latitude, and a square-root transformation of age gave the same result. Growth in Alaska appears to be more rapid than in Canada and Siberia. This effect is probably attributable to climatic moderation of Alaskan rivers entering the Pacific Ocean. The same effects were compared in an analysis using the von Bertalanffy growth function. All effects had a significant influence on growth rate ($P < 0.001$), with age and climate being the most important.

The broad whitefish *Coregonus nasus* is an arctic species that is distributed from west of Bathurst Inlet, along the arctic coast of the Canadian Northwest Territories and Alaska, to the western arctic coast of Russia. There is little published information on growth rates in the primary literature, particularly from populations in Canadian waters (Scott and Crossman 1973). Alt (1976) observed substantial variation in growth among stocks of Alaskan broad whitefish; he proposed that the observed variation was attributable to differences in the length of the growing season. Muth (1969) observed differences in growth rate between broad whitefish from the Mackenzie River and those of the Coppermine River. He also favored the hypothesis that growth rate differences were principally caused by climate. Berg (1948) suggested that there were two morphs of broad whitefish—fast growing and slow growing. In other anadromous species, such as Pacific salmon species, intraspecific differences in growth rates have been related to trade-offs between distance of the spawning grounds from the sea, size or gradient of the river system, and sexual selection favoring increased size for increased fecundity (Holtby and Healey 1986). The life history pattern of the broad whitefish could confound the analysis of growth patterns. Reist and Bond (1988) noted that broad whitefish typically migrate to nursery lakes distant from their spawning grounds. The migration probably evolved because the payoff from rapid growth in the nursery lakes offsets lost energy or mortality caused by the movement (Northcote 1978). The fish grow in this environment for several years before entering the spawning population. It is not known if the population of a rearing lake is composed of many stocks or only one. Thus, some samples would be expected to show a rapid-growth trajectory and greater variance of size at age than others because they are representative of a rearing population.

Since 1970 there have been several studies in the Canadian Arctic that have gathered growth information on broad whitefish, but most of these have not been published in the primary literature. This information could be useful for decisions regarding the viability of commercial fishing enterprises in the western Canadian Arctic (Stewart et al. 1994). The purpose of this article is to bring together the available information on growth of broad whitefish. The geographic variation in growth will be interpreted in terms of the growing season and migratory patterns (distance from the sea).

Methods

The data sources for this article (Table 1) pertain to 13 sites in Canada, 5 in Alaska, and 3 in Siberia (Figure 1). The emphasis on Canadian sources reflects their availability to the author and is not indicative of the amount of data collection that has occurred in the three countries. If possible, the raw data were used for analysis. Raw data were deduced or obtained directly from the reports or were supplied by authors. When this was not possible, mean values in a data source were used. A total of 3,824 records of age and length were available to be used in the analysis. The records from the Kara and

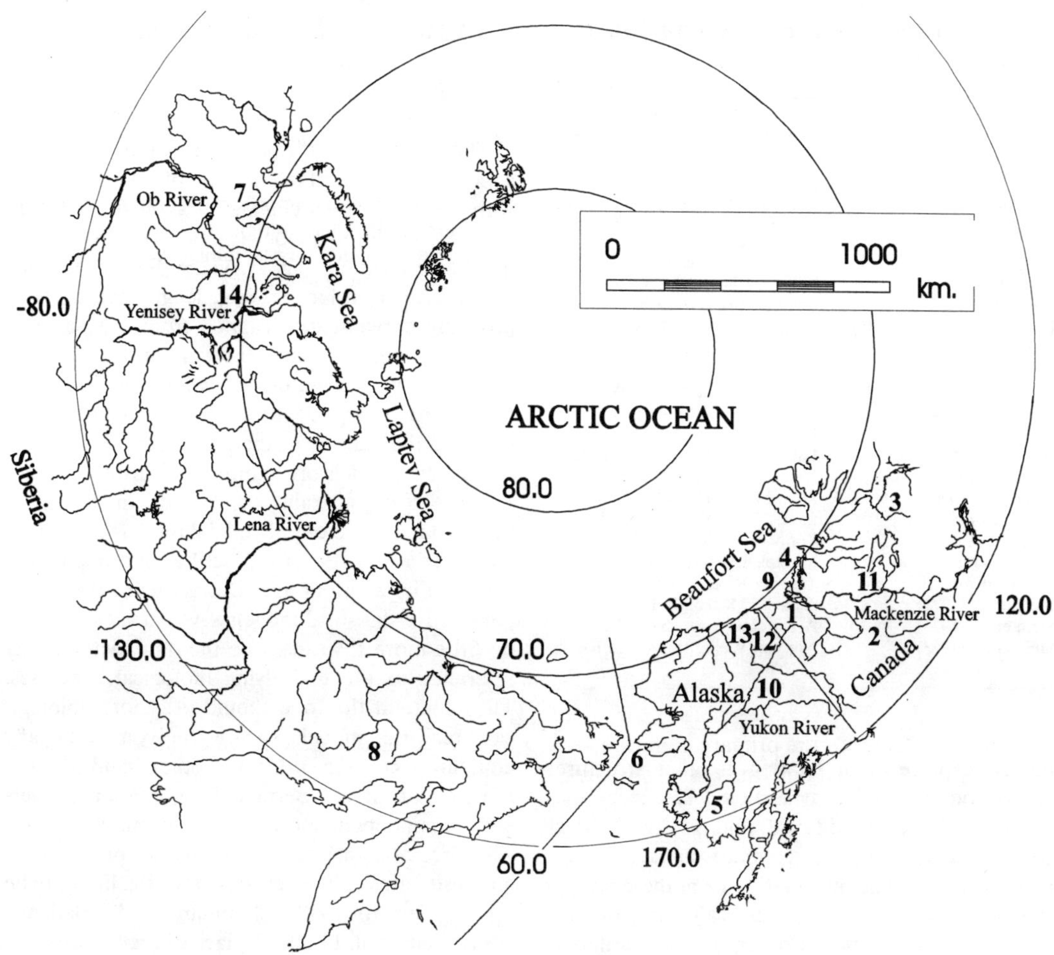

FIGURE 1.—Locations of data sources for length at age of broad whitefish *Coregonus nasus* in Canada, Alaska, and Siberia: (1) Aklavik, (2) Arctic Red River, (3) Coppermine River, (4) Freshwater Creek and Tuktoyaktuk Harbor, (5) Holitna River, (6) Imuruk Basin, (7) Kara River, (8) Kolyma River, (9) Ya Ya Esker lakes (6, 7, 17, 23, 24, 25), (10) Minto Flats, (11) Norman Wells, (12) Porcupine River, (13) Sagavanirktok River, and (14) Tanama River.

Kolyma rivers gave total length rather than fork length. Fork length was estimated from the length–weight relationship given in Muth (1969):

$$\text{Weight} = 0.003\ (\text{Length})^{3.41}. \qquad (1)$$

The effect of climate on length at age was approximated at two locations. One location was the latitude of the site where the sample was taken, which was considered the probable spawning site. The other location was the latitude of the river entry into the sea, which was considered indicative of the resting and growing area. After preliminary analysis (see "Results" below), the latitude of the resting area was replaced with a more direct climate variable. Climate effects were approximated by dividing the number of degree-days below 18°C by the mean July temperature isotherm at the latitude of a river mouth. The climate effect and spawning-location effect were used in subsequent analyses. Distance from the sea was taken as the length of the river course from the point of field capture to the mouth of the river. A measure of dispersion from the longitudinal center of the species distribution was included in the model. This was the number of degrees of longitude from the center of the distribution, arbitrarily set at 130°E.

Age was taken from scales. In most studies there was no attempt to validate the aging technique by

TABLE 1.—Data sources for length at age of broad whitefish *Coregonus nasus* from Canada, Alaska, and Siberia. The Ya Ya Esker is a feature in the Mackenzie River delta (N.W.T. = Northwest Territories).

Source	Location
Alt (1976)	Minto Flats, Alaska
Alt (1976)	Porcupine River, Alaska
Alt (1976)	Imuruk Basin, Alaska
Alt (1976)	Sagavanirktok River, Alaska
Alt (1976)	Holitna River, Alaska
Berg (1948)	Kara River, Siberia
Berg (1948)	Kolyma River, Siberia
Bond (1982)	Tuktoyaktuk Harbor, N.W.T.
Bond and Erickson (1985)	Freshwater Creek, N.W.T.
Hatfield et al. (1972a, 1972b)	Arctic Red River, N.W.T.
Hopky and Ratynski (1983)	Tuktoyaktuk Harbor, N.W.T.
Machniak (1976)	Lake 6, Ya Ya Esker, N.W.T.
Machniak (1976)	Lake 7, Ya Ya Esker, N.W.T.
Machniak (1976)	Lake 17, Ya Ya Esker, N.W.T.
Machniak (1976)	Lake 23, Ya Ya Esker, N.W.T.
Machniak (1976)	Lake 24, Ya Ya Esker, N.W.T.
Machniak (1976)	Lake 25, Ya Ya Esker, N.W.T.
Muth (1969)	Coppermine River, N.W.T.
Muth (1969)	Mackenzie River, N.W.T.
Popov (1976)	Tanama River, Siberia
Stein et al. (1973)	Aklavik, Mackenzie River, N.W.T.
Stein et al. (1973)	Arctic Red River, N.W.T.
Stein et al. (1973)	Norman Wells, Mackenzie River, N.W.T.

mark–recapture or by looking at other structures such as otoliths or fin ray sections. The exceptions are the studies of Bond (1982) and Bond and Erickson (1985), which provide comparisons of scale age and otolith age. The information from these studies indicates that at younger ages scales and otoliths give similar values, but at older ages scales underestimate age relative to otoliths. I made no adjustment in my calculations for possible underestimates of age.

A multiple regression model was used. Neter et al. (1983) provide detailed methodology of the computations involved in this procedure. The preliminary model (model 1) was

$$Y_i = \beta_0 + \beta_1 X_{1i} + \beta_2 X_{2i} + \beta_3 X_{3i} + \beta_4 X_{4i} + \beta_5 X_{5i} + \varepsilon_i; \tag{2}$$

Y_i = length of fish i;
β_0 = intercept of the multiple regression (each $X_{ji} = 0$);
β_1 = effect of age on length i;
β_2 = effect of distance from the sea on length i (DFS);
β_3 = effect of latitude of the resting location on length i (RESTLOC);
β_4 = effect of latitude of the spawning location on length i (SPAWNLOC);
β_5 = effect of dispersion on length i (DISPERSE);
ε_i = error in length i, assuming errors are random and normally distributed with a mean 0 and a variance σ.

The second model (model 2) included all interactions and replaced the latitude of resting location with the more direct climate measure (CLIMATE, $\beta_{3'}$). The reduced model (model 3) introduced a quadratic term for age (AGESQR), recognizing that the relationship between length and age was probably curvilinear. Distance from the sea was removed from this model. Model 4 did not include distance from the sea or dispersion; age was transformed to its square root. All calculations were based on the Statistical Analysis System (SAS) General Linear Models procedure (SAS Institute 1989).

There is considerable controversy in the literature about the use of polynomial models employing parametric statistics to describe interpopulation variation in length at age. Detractors of the use of parametric statistical models say that these models do little more than describe the data, without any reference to the underlying biological processes. They contend that one should use more "biologically realistic" models, such as the von Bertalanffy equation (Rao 1958). On the other hand, the parameters of the von Bertalanffy equation are inherently nonindependent: a change in one parameter will force a change in the others. No precise variance estimates on the parameters or the lines can be made (Cerrato 1990; Chouinard and Mladenov 1991; Chen et al. 1992). The lack of precise variance estimates means that to make comparisons between populations researchers must resort to approximations—such as simultaneous comparisons of two or all three of the von Bertalanffy parameters (Kingsley 1979; Bernard 1981) or the use of likelihood ratio tests (Kimura 1980). As Cerrato (1990) points out, both approaches are approximate ones; they are taken from linear statistical theory, and their validity when applied to the von Bertalanffy equation depends on the degree of bias and nonnormality in the parameter estimates caused by the nonlinearity of the model. In addition, both approaches are characterized in terms of asymptotic properties for which no small-sample theory exists. Finally, both approaches handle unequal and unknown error variances in an approximate way.

These concerns led me to focus on polynomial models, which I believe, are adaptable to any situation and have reliable and proven methods for estimating the effects of different factors and making comparisons between lines. Nevertheless, for comparison, I reanalyzed the data to generate von Bertalanffy growth function (VBGF) parameters for each population. The SAS program NLIN (SAS Institute 1989) was used to do successive approximations until a convergence criterion was met. The maximum number of iterations was set at 500.

The VBGF is expressed as

$$L_t = L_\infty(1 - \exp(-K(t - t_o))); \qquad (3)$$

t = age;
L_t = length at age;
L_∞ = the asymptotic length;
K = the Brody growth coefficient;
t_0 = the theoretical age at which L_t is zero (Ricker 1975).

Standard nonlinear optimization techniques of curve fitting were used to estimate the coefficients and their associated standard error (Cerrato 1990).

The nonlinear formulation of the VBGF means that a general linear model could not be used for analysis of covariance. Instead, an analysis of the residual sum of squares (ARSS) was employed to compare VBGF among populations grouped according to their distance from the sea, spawning location, resting location, dispersal from 130°E, and climate (Chen et al. 1992). For each treatment the samples were divided into two groups, representing warmer and colder climate, closer or more distant from the sea, spawning location above or below 68°N, resting location above or below 69°N, and dispersal greater than or less than 95° from the center of the range. Because there was no way to model the response to an effect as a smooth curve (as one could do using linear models), the two-group mode was used to approximate modal values of the effect.

Procedures for the ARSS were as follows: (1) residual sum of squares (RSS) and associated degrees of freedom (DF) of the VBGF were calculated for each sample; (2) the resultant RSS and DF of each sample were added to yield summed RSS and DF; (3) data of all samples were pooled to calculate the RSS and DF of a total VBGF; and (4) the F-statistic was calculated as

$$F = \frac{\dfrac{\text{RSS}_p - \text{RSS}_s}{\text{DF}_{\text{RSS}_p} - \text{DF}_{\text{RSS}_s}}}{\dfrac{\text{RSS}_s}{\text{DF}_{\text{RSS}_s}}} = \frac{\dfrac{\text{RSS}_p - \text{RSS}_s}{3(K - 1)}}{\dfrac{\text{RSS}_s}{N - 3K}}; \qquad (4)$$

RSS_p = RSS of each VBGF fitted by pooled growth data;

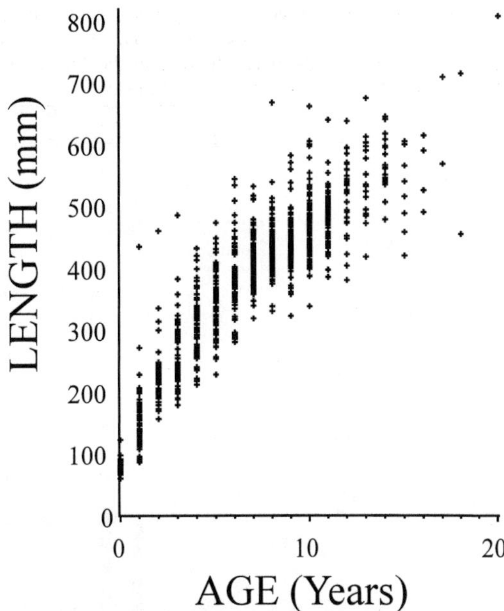

FIGURE 2.—Variation in length at age for 23 populations of broad whitefish *Coregonus nasus* from Alaska, Canada, and Siberia. Length is in millimeters and age is in years.

RSS_s = sum of RSS of each VBGF fitted to growth data for each sample;
N = total sample size;
K = number of samples in the comparison.

To test for a difference between the samples, the calculated F value was compared with an F with DFs of $3(K - 1)$ (numerator) and $N - 3K$ (denominator).

Results

The variation in length at age for all populations in the study is shown in Figure 2. After age 0, the range in length at age was quite consistent across ages at approximately 200–550 mm. The pattern of growth suggests a simple linear relationship between length and age. Simple linear regression lines fitted to the data suggest that Alaskan rivers have higher growth rates than do Canadian or Siberian rivers (Figures 3–5). The Coppermine River in Canada has one of the slowest growth rates of all populations. The results were similar whether mean lengths at age for all populations or raw data from a reduced number of locations were used in the regression analysis.

The results from model 1 were highly significant

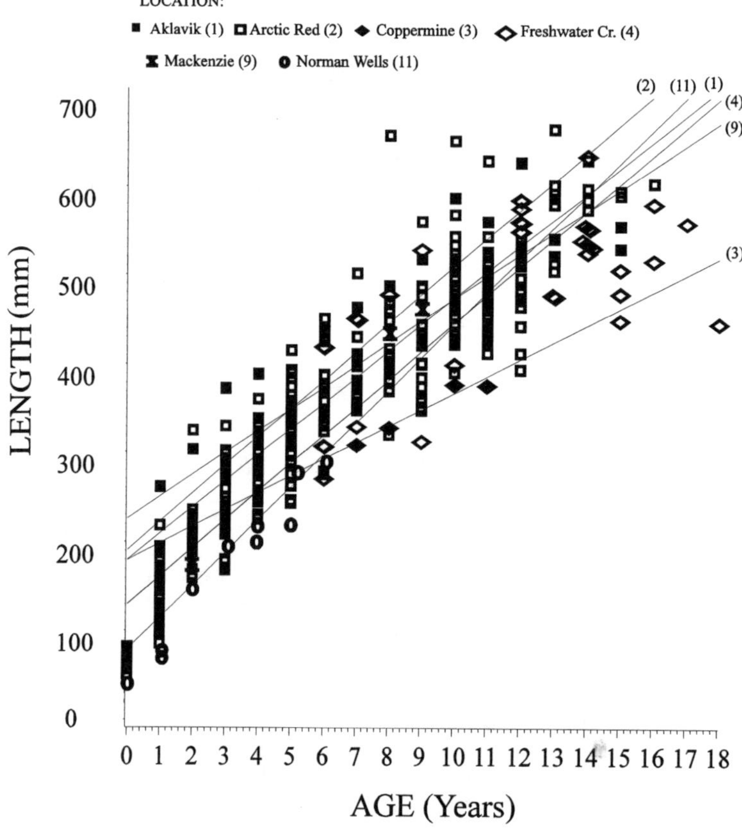

FIGURE 3.—Simple linear regressions of length (mm) versus age (years) for Canadian river populations of broad whitefish *Coregonus nasus*.

($P < 0.0001$), accounting for 86% of the variation in length at age (Table 2). Age, distance from the sea, and dispersion had significant effects on fish length. The latitude of the resting location was nonsignificant. With all parameters held constant except dispersion, length declined at a rate of 0.78 mm per degree of longitude away from 130°E.

The results from model 2 were also highly significant ($P < 0.0001$). The inclusion of interactions and use of more direct measures of climate improved the accountability in variation of length at age to 90% (Table 2). Age, climate, spawning location, and dispersion, as well as the interactions between them, had significant effects on fish length. Distance from the sea, and interactions between distance from the sea and other variables, were nonsignificant. Age accounted for a large portion of the variability in the data, followed by spawning location and dispersion. Latitude of spawning location was negatively correlated with size. With all other parameters held constant, length declined 96 mm per degree increase in latitude. There was considerable multicolinearity between distance from the sea and the climate measure. I decided not to use distance from the sea in the following models. The residuals of the regression against age suggested that a linear model might not be appropriate (Figure 6). To account for curvilinearity, I examined two alternative models: one that adds a quadratic term for age and the other employing a square-root transformation of age.

Interactions in the quadratic model (model 3) were minimal (Table 2). This model gave highly significant results ($P < 0.0001$). Despite the exclusion of interactions, the R^2-value increased only from 0.90 (model 2) to 0.91 (model 3). Age, age-

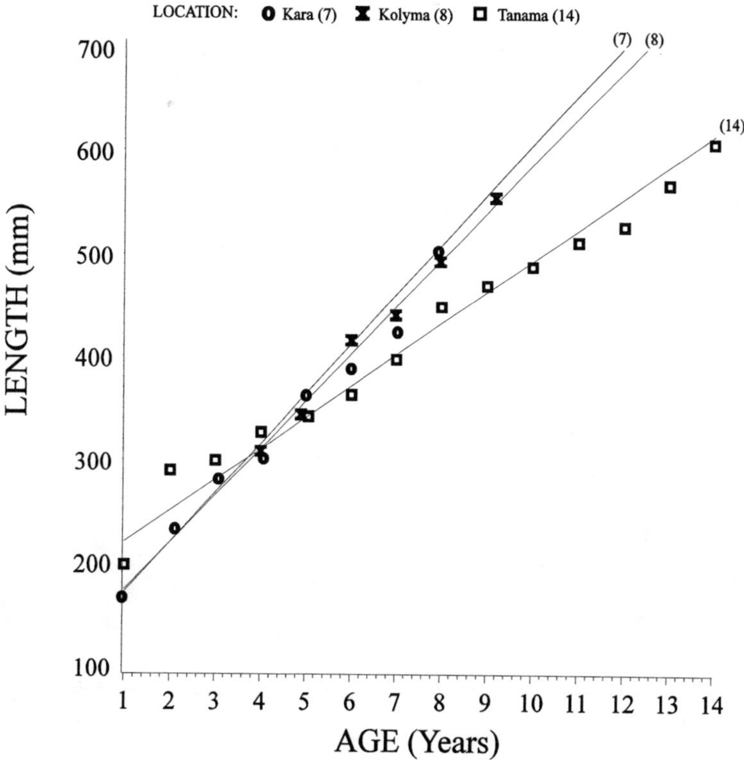

FIGURE 4.—Simple linear regressions of length (mm) versus age (years) for Alaskan river populations of broad whitefish *Coregonus nasus*.

squared, climate, spawning location, and dispersion had significant effects on fish length in model 3.

The effect of dispersion was nonsignificant in the square-root transformation model. Thus, I recalculated the regression using only climate, spawning location, transformed age, and the interaction between climate and spawning location (model 4, Table 2). The regression was highly significant ($P < 0.0001$, $R^2 = 0.91$). After transformed age, climate and the climate-by-spawning location interaction accounted for the greatest amount of the variation in length. Age accounted for a large portion of the variability in all analyses.

In a comparison of the four models (Table 2), the intercept is shown to be highly variable; this is not a concern, because in multiple regression the intercept β_0 represents the mean response when the independent variable equals 0 (Neter et al. 1983). Otherwise, β_0 does not have any particular meaning as a separate term in the regression model. Age was by far the most reliable indicator of size. Models 2 and 4 had significant interactions; model 3 did not. The curvilinear nature of the regression (caused by the quadratic term for age) may have reduced interactions to a minimum in model 3. Model 1, with the lowest R^2, is the least reliable of the group; interactions were not assessed in this model.

The von Bertalanffy growth function (VBGF) for each population is summarized in Table 3. In five populations—Coppermine, Kara, Kolyma, and Tanama rivers and Ya Ya L24—convergence could not be made after 500 iterations. In the converged functions, L_∞ ranged from 352 to 781 mm. All K values were between 0 and 1. All but four of the t_0 values were less than zero.

Table 4 shows the analysis of the residual sum of squares (ARSS) comparisons (see "Methods" for explanation) among populations differing in distance from the sea, spawning location, resting location, dispersal, and climate. All of these effects were

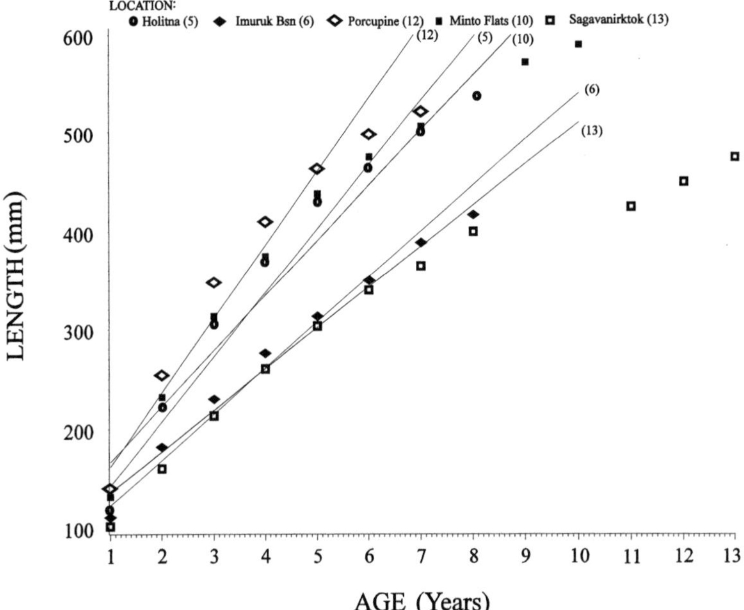

FIGURE 5.—Simple linear regressions of length (mm) versus age (years) for Siberian river populations of broad whitefish *Coregonus nasus*.

significant ($P < 0.001$), the greatest effects being climate and resting location. The parameter estimation program did not converge for the effects of spawning location and resting location. Therefore, I concluded that climate was the most important environmental effect in my analysis.

TABLE 2.—Estimated β-values, interactions, and coefficients of determination (R^2) for growth models of broad whitefish *Coregonus nasus* from Canada, Alaska, and Siberia. Bold values were not significant; only interactions shown were significant. Blank spaces indicate no test.

Model characteristic	Model 1	Model 2	Model 3	Model 4
Intercept (β_0)	151.80	5,148	10.06	711.08
AGE (β_1)	32.65	30.97	54.04	
DFS (β_2)	0.02	**96.96**	**−0.00**	
RESTLOC (β_3)	**−0.14**			
CLIMATE ($\beta_{3'}$)		−40.36	−0.06	−1.40
SPAWNLOC (β_4)	**1.09**	−95.92	2.99	−9.02
DISPERSE (β_5)	−0.77	16.12	−0.65	
AGESQR			−1.72	
XPRIME				136.30
Interactions		$\beta_{3'} \times \beta_4$ $\beta_{3'} \times \beta_5$	Minimal	$\beta_{3'} \times \beta_4$
R^2	0.86	0.90	0.91	0.91

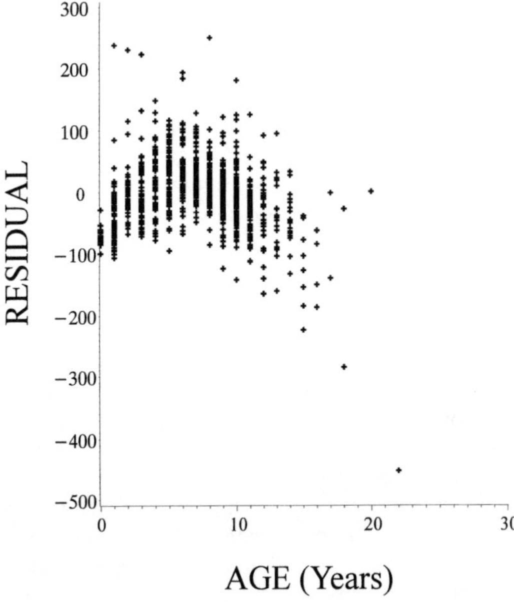

FIGURE 6.—Residuals of analysis of regression for age effect on length of broad whitefish *Coregonus nasus*. Age is in years.

TABLE 3.—Summary of von Bertalanffy results for 23 populations of broad whitefish *Coregonus nasus* from the arctic regions of Canada, Alaska, and Siberia. Standard errors are in parentheses. Asterisks indicate pooled data from two sources at one location (see Table 1). Dashes indicate no data.

Location	L_∞ (mm)	K	t_0	N	Convergence (iterations)
Aklavik	652 (68)	0.117 (0.034)	−1.780 (0.049)	55	70
Arctic Red River*	742 (88)	0.105 (0.023)	−1.350 (0.040)	72	76
Coppermine River	593 (−)	0.073 (−)	−3.280 (−)	197	No
Freshwater Creek	642 (21)	0.111 (0.008)	−1.110 (0.008)	358	74
Holitna River	631 (10)	0.228 (0.009)	0.108 (0.010)	100	37
Imuruk Basin	632 (49)	0.124 (0.018)	−0.612 (0.010)	100	79
Kara River	1,598 (−)	0.035 (−)	−2.165 (−)	100	No
Kolyma River	1,230 (−)	0.059 (−)	−0.789 (−)	60	No
Mackenzie River	532 (19)	0.213 (0.026)	>0.001 (0.052)	100	38
Minto Flats	685 (11)	0.196 (0.009)	−0.100 (0.011)	10	49
Norman Wells	459 (183)	0.138 (0.089)	−0.816 (0.039)	20	62
Porcupine River	601 (8)	0.298 (0.011)	0.120 (0.012)	90	32
Sagavanirktok River	509 (1,206)	0.124 (0.817)	−4.139 (1.194)	130	186
Tanama River	1,748 (−)	0.023 (−)	−4.527 (−)	160	No
Tuktoyaktuk Harbor*	496 (7)	0.205 (0.009)	−0.771 (0.012)	587	37
Ya Ya L17/L23*[a]	781 (66)	0.146 (0.068)	−0.936 (0.228)	11	80
Ya Ya L24	435 (−)	0.147 (−)	−1.525 (−)	8	No
Ya Ya L25	352 (0)	0.251 (0.017)	−0.504 (0.036)	11	34
Ya Ya L6/L7*	486 (12)	0.434 (0.189)	3.540 (1.281)	8	21

[a]Ya Ya = Ya Ya Esker, which is a feature in the Mackenzie River delta; L = lake.

Discussion

Model 3, which incorporates age, latitude of spawning location, dispersion from 130°E, a climatic effect, and a quadratic term to allow the regression to curve with changing age, appears to be the most realistic and useful model to describe variation in length among broad whitefish populations. In this model, the interactions among effects are minimal; thus, interpretation and prediction using the model are much simplified. The parameter estimate for the intercept and the rates of change attributable to the main effects in the model are intuitively reasonable. For example, the change in the mean length resulting from an increase of 1 year in age (with all other effects held constant) is 54 mm. Among all the models, moderation of climate appears to be associated with higher growth rate.

TABLE 4.—Results of an analysis of residual sum of squares for comparisons of the effects of climate, distance from the sea (DFS), longitudinal dispersion (DISPER), latitude of spawning area (SPL), and latitude of resting area (REL) on von Bertalanffy growth functions (VBGFs) for broad whitefish *Coregonus nasus* from Canada, Alaska, and Siberia. In this analysis, RRS_p = residual sum of squares of the VBGF fitted to pooled growth data = 3,609,922; RRS_s = sum of the residual sum of squares of each VBGF fitted to growth data for each sample; DF_p = degrees of freedom (pooled) = 1,385; and DF_s = degrees of freedom (sum) = 1,382. Asterisks indicate nonlinear parameter estimations that did not converge.

Comparison	RSS_s	F	P
CLIMATE	3,159,964	65	<0.001
DFS	3,270,243	48	<0.001
DISPER	3,274,173	47	<0.001
SPL*	3,368,435	33	<0.001
REL*	3,095,977	77	<0.001

The analysis of the residual sum of squares also indicated that moderation of climate is the most important effect on growth; however, this does not discount the importance of distance from sea and dispersal. Resting location and spawning location probably are important, but because the parameter estimation program did not converge, the actual importance of these effects cannot be assessed.

The VBGFs for individual populations are interesting if one considers them under the criteria of Chen et al. (1992) that L_∞ should be reasonably close to the maximum fish length (Moreau 1987); that t_0 should be smaller than 0 so that fish of age 0 could have a positive length (Moreau 1987); and that K should vary between 0 and 1 for fishes with long life spans (Pauly 1978). Among the convergent data sets, the L_∞ fall into two groups: those near 500 mm or less and those 600 mm and more. Those L_∞ 600 mm and more are reasonable asymptotic values for growth. Of the lower L_∞ group (500 mm or less), several are from lake populations in the outer Mackenzie River delta and one is from Tuktoyaktuk Harbor; these areas may be used by juvenile fish for rearing (Hopky and Ratynski 1983; Reist and Bond 1988). Two of the lower-L_∞ group are from the Mackenzie River in the mainstem and at Norman Wells, Northwest Territories; these were collected during the summer (Hatfield et al. 1972a, 1972b), and thus, could also be dominated by juveniles. The remaining population in this group is from the Saganivirtok River and has a very low growth rate (Figure 4). Otherwise, the K and t_0 values are generally reasonable for describing the growth data.

The Mackenzie River represents an unusual feature in that it transports a large amount of heat from the temperate zone to the Arctic. If climate were a major factor, one would expect that the growth would be greater in Mackenzie River stocks compared to others that have river mouths at similar latitudes.

The results from polynomial model analysis suggest that the climate of the resting area has a significant impact on growth. Distance from the sea and climate of the resting area are probably linked because longer rivers where longer spawning migrations might take place, such as the Mackenzie, also originate farther south and thus transport more heat to their river deltas. This in turn moderates the climate. The latitude of capture appears to be related to growth rate, as does the dispersion from 130°E. The interpretation of these results are more difficult, but latitude of spawning location could represent a climate effect for stocks that migrate to an upstream location well in advance of spawning. Growth could be enhanced if a long period is spent in the river system. The effect of dispersion may reflect the marginal conditions for survival at the edge of the species range. Populations at the edge of the range are under stress and hence growth rates are lower. The range of growth among Canadian, Alaskan, and Siberian stocks of broad whitefish appears similar; however, Canadian stocks are on the edge of the species range.

The utility of my analysis depends upon the data that are used. Unfortunately, indications are that scales tend to underestimate the age of fish, particularly older members of the population (W. Bond, Canadian Department of Fisheries and Oceans, personal communication). If scales produce a consistent bias for all studies, my conclusions regarding climate at the river mouth and spawning location are, nevertheless, valid. However, if the bias is location dependent (e.g., Canadian versus Alaskan studies), then a problem exists. There is a need for age validation studies for broad whitefish.

Acknowledgments

J. Reist and W. Bond of the Canadian Department of Fisheries and Oceans (DFO) provided helpful advice and encouragement. G. Chouinard of the Canadian DFO and J. Allard of the Université de Moncton provided statistical advice. The manuscript was reviewed by D. Bodaly and W. Bond of the Canadian DFO.

References

Alt, K. 1976. Age and growth of Alaskan broad whitefish, *Coregonus nasus*. Transactions of the American Fisheries Society 105:526–528.

Berg, L. S. 1948. Freshwater fishes of the U.S.S.R. and adjacent countries, volume 1, 4th edition. Oldbourne Press, London.

Bernard, D. R. 1981. Multivariate analysis as a means of comparing growth in fish. Canadian Journal of Fisheries and Aquatic Sciences 38:233–236.

Bond, W. A. 1982. A study of the fishery resources of Tuktoyaktuk Harbour, southern Beaufort Sea coast, with special reference to life histories of anadromous coregonids. Canadian Technical Report of Fisheries and Aquatic Sciences 1119.

Bond, W. A., and R. N. Erickson. 1985. Life history studies of anadromous coregonid fishes in two freshwater lake systems on the Tuktoyaktuk Peninsula, Northwest Territories. Canadian Technical Report of Fisheries and Aquatic Sciences 1336.

Cerrato, R. M. 1990. Interpretable statistical tests for growth comparisons using parameters in the von Bertalanffy equation. Canadian Journal of Fisheries and Aquatic Sciences 47:1416–1426.

Chen, Y., D. A. Jackson, and H. H. Harvey. 1992. A comparison of von Bertalanffy and polynomial functions in modelling fish growth rate. Canadian Journal of Fisheries and Aquatic Sciences 49:1228–1235.

Chouinard, G. A., and P. V. Mladenov. 1991. Comparative growth of sea scallop (*Placopecten magellanicus*) in the southern Gulf of St. Lawrence. Pages 261–267 *in* J. C. Therriault, editor. The Gulf of St. Lawrence: small ocean or big estuary. Canadian Special Publication of Fisheries and Aquatic Sciences 113.

Hatfield, C. T., J. N. Stein, M. R. Falk, and C. S. Jessop. 1972a. Fish resources of the Mackenzie River valley. Department of the Environment, Fisheries Service, Interim Report 1, Volume 1, Winnipeg, Manitoba.

Hatfield, C. T., J. N. Stein, M. R. Falk, C. S. Jessop, and D. N. Shepherd. 1972b. Fish resources of the Mackenzie River valley. Department of the Environment, Fisheries Service, Interim Report 1, Volume 2, Winnipeg, Manitoba.

Holtby, L. B., and M. C. Healey. 1986. Selection for adult size in female coho salmon (*Oncorhynchus kisutch*). Canadian Journal of Fisheries and Aquatic Sciences 43:1946–1959.

Hopky, G. E., and R. A. Ratynski. 1983. Relative abundance, spatial and temporal distribution, age and growth of fishes in Tuktoyaktuk Harbour, N.W.T., 28

June to 5 September, 1981. Canadian Manuscript Report of Fisheries and Aquatic Sciences 1713.

Kimura, D. K. 1980. Likelihood methods for the von Bertalanffy growth curve. U.S. National Marine Fisheries Service Fishery Bulletin 77:765–776.

Kingsley, M. C. S. 1979. Fitting the von Bertalanffy growth equation to polar bear age–weight data. Canadian Journal of Zoology 57:1020–1025.

Machniak, K. 1976. The aquatic resources of the Ya Ya Esker in the Mackenzie River delta. Aquatic Environments Limited (Consultants Report 3) to Department of Fisheries and Oceans, Winnipeg, Manitoba.

Moreau, J. 1987. Mathematical and biological expressions of growth in fishes: recent trends and further developments. Pages 81–113 *in* R. C. Summerfelt and G. E. Hall, editors. The age and growth of fish. Iowa State University Press, Ames.

Muth, K. M. 1969. Age and growth of the broad whitefish, *Coregonus nasus*, in the Mackenzie and Coppermine rivers, Northwest Territories. Journal of the Fisheries Research Board of Canada 26:2252–2256

Neter, J., W. Wasserman, and M. H. Kutner. 1983. Applied linear regression models. Irwin, Homewood, Illinois.

Northcote, T. G. 1978. Migratory strategies and production in freshwater fishes. Pages 326–329 *in* S. D. Gerking, editor. Ecology of freshwater fish production. Blackwell Scientific Publications, Oxford, UK.

Pauly, D. 1978. A preliminary compilation of fish length growth parameters. Berichte aus dem Institut fuer Meereskunde an der Christian-Albrechts Universitaet, Kiel 55.

Popov, P. A. 1976. The growth and onset of sexual maturation of the broad whitefish, *Coregonus nasus*, and the Ob whitefish, *Coregonus lavaretus pidschian*, of the Tanama River. Journal of Ichthyology 16:414–419.

Rao, C. R. 1958. Some statistical methods for comparison of growth curves. Biometrics 14:1–17.

Reist, J. D., and W. A. Bond. 1988. Life history characteristics of migratory coregonids of the lower Mackenzie River, Northwest Territories, Canada. Finnish Fisheries Research 9:133–144.

Ricker, W. E. 1975. Computation and interpretation of biological statistics of fish populations. Fisheries Research Board of Canada Bulletin 191.

SAS Institute. 1989. SAS/STAT user's guide, version 6, 4th edition. SAS Institute, Cary, North Carolina.

Scott, W. B., and E. J. Crossman. 1973. Freshwater fishes of Canada. Fisheries Research Board of Canada Bulletin 184.

Stein, J. N., C. S. Jessop, T. R. Porter, and K. T. J. Chang-Kue. 1973. Fish resources of the Mackenzie River valley. Department of the Environment, Fisheries Service, Interim Report 2, Winnipeg, Manitoba.

Stewart, D. B., R. A. Ratynski, L. M. J. Bernier, and D. J. Ramsey. 1994. A fishery development strategy for the Canadian Beaufort Sea–Amundsen Gulf Area. Canadian Technical Report of Fisheries and Aquatic Sciences 1910.

American Fisheries Society Symposium 19:194–207, 1997

Population Dynamics of Broad Whitefish in the Prudhoe Bay Region, Alaska

BENNY J. GALLAWAY AND ROBERT G. FECHHELM
LGL Alaska Research Associates, Inc.
4175 Tudor Centre Drive, Suite 101, Anchorage, Alaska 99508, USA

WILLIAM B. GRIFFITHS
LGL Limited
9768 Second Street, Sidney, British Columbia V8L 3Y8, Canada

JOHN G. COLE
LGL Alaska Research Associates, Inc.

Abstract.—A decade of monitoring size and age structure and catch per unit effort of broad whitefish *Coregonus nasus* in the Prudhoe Bay region of Alaska has begun to yield an understanding of the life history of this species and the factors, including the effects of offshore developments, that govern population levels. The Endicott Causeway was constructed in spring 1985, before breakup of sea ice. Before 1985, abundance levels were high and the population was dominated by juveniles, especially the 1979 year-class. The local population reached its lowest levels for the period of record in 1985–1987, and the size structure became markedly bimodal with the 1979 year-class constituting most of the fish in the large-size mode. When these fish reached maturity in 1988, they disappeared from the population. Population levels then increased rapidly to pre-1985 levels, and the size structure that emerged at the end of 1992 greatly resembled the size structure observed in 1982. Evidence is presented showing that changes in abundance and size structure do not appear to be attributable to a causeway effect and that a density-dependent life history strategy may be implicated. We question the appropriateness of steps that have been taken to mitigate the perceived causeway effects on broad whitefish.

The broad whitefish *Coregonus nasus* is widely distributed in the fresh and brackish waters of the coastal zone of northern Alaska and the Yukon and Northwest Territories of Canada (Scott and Crossman 1973). Within this region, the species exhibits a discontinuous distribution. One center is associated with the Mackenzie River and delta area, the other is in western Alaska (Figure 1).

Because of its size and depth, the Mackenzie River flows throughout the year, providing extensive spawning and overwintering habitat. The delta and the Tuktoyaktuk Peninsula extending east of the delta are characterized by extensive networks of deep freshwater lakes interconnecting with each other and the Beaufort Sea through stream and river channels.

These lakes are used by broad whitefish both as summer feeding areas and, if deep enough, as overwintering habitat (Bond 1982; Bond and Erickson 1985). Some of the fish may also move in and out of these lakes over the summer and into nearshore coastal areas to feed. These emigrations take place both under the ice and after breakup (Bond and Erickson 1985). Lake habitats are thought to be critical for the survival and rearing of a significant number of the broad whitefish found in the Mackenzie River system (Bond and Erickson 1985).

This life history pattern suggests that the broad whitefish is primarily a freshwater species that uses the coastal zone as a migration corridor and an alternate feeding habitat under suitable (low-salinity) conditions.

The second center of the broad whitefish distribution in arctic North America occurs in western Alaska in the area bounded by the Meade and Sagavanirktok rivers (Figure 1). This region is separated from the Mackenzie River region by about 400 km. Even the largest rivers (e.g., the Colville River) of the western Alaskan Arctic are not big or deep enough to sustain year-round flow (Craig 1989), and there are few networks of deep, freshwater lakes as compared with the Mackenzie River region. Those that do occur are restricted to the western part of the overall range in northern Alaska, the subregion lying between the Colville and Meade rivers.

The Sagavanirktok River, located about 80 km east of the Colville River (Figure 1), harbors a disjunct spawning population of broad whitefish (McCart et al. 1972; Bendock 1979). There are no deep, freshwater lake systems around its delta, and

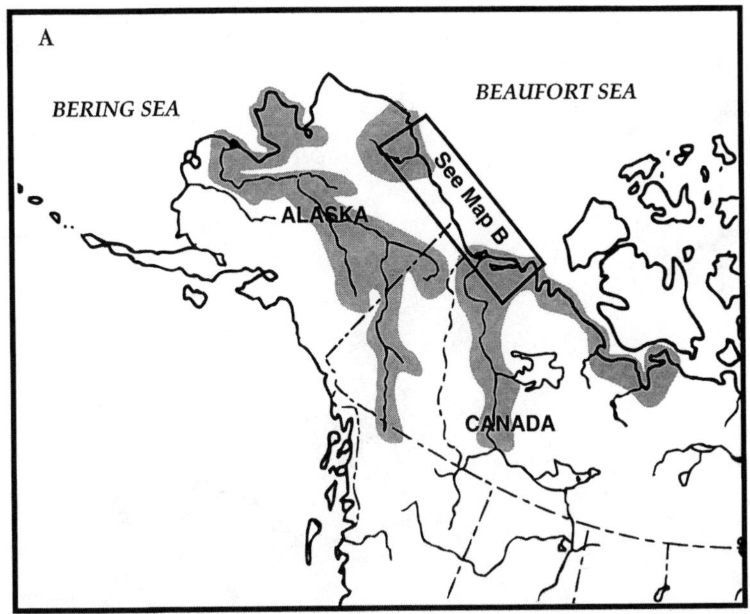

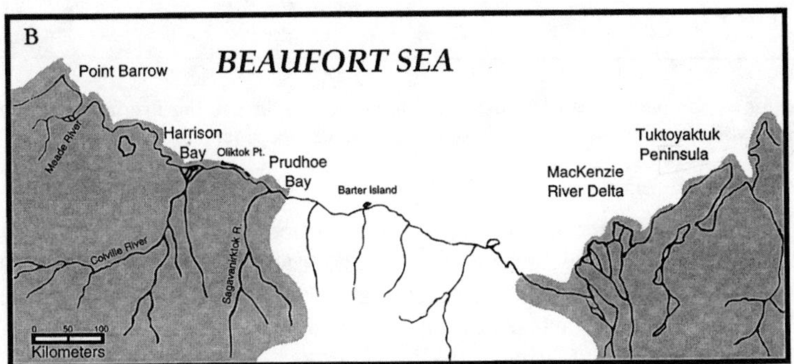

FIGURE 1.—Distribution (shaded areas) of broad whitefish *Coregonus nasus* in North America (**A**) and in the Alaskan Arctic, bounded by the Meade and Sagavanirktok rivers (**B**) in the 1980s.

freshwater overwintering habitat is restricted to the relatively few deep pools in the delta and upstream river channels. Craig (1989) has estimated that the amount of freshwater habitat available to fish in the winter in the Sagavanirktok and similar rivers may be as little as 1–5% of the amount available during the summer. Undoubtedly, overwintering habitat provides a severe constraint on the fish populations.

Additionally, the absence of freshwater lakes suitable for providing summer feeding habitat could be another constraint on the broad whitefish population in the Sagavanirktok River. In this area, the fish must use the low-salinity zone around the mouth of the river for rearing. Juvenile broad whitefish cannot tolerate salinity levels above 15–20 ppt for long periods of time (de March 1989). Although the estuarine area characterized by low salinity levels is larger earlier in the summer as compared with later, suitable rearing habitat is not extensive on either a temporal or spatial basis.

Concerns and Issues

For more than a decade, the Sagavanirktok River broad whitefish population, which is not exploited by any fishery, has been under scrutiny because of its proximity to the Endicott Causeway. This causeway was constructed at the mouth of the river in early 1985 to support an offshore oil and gas development (Figure 2). The concern has been that this

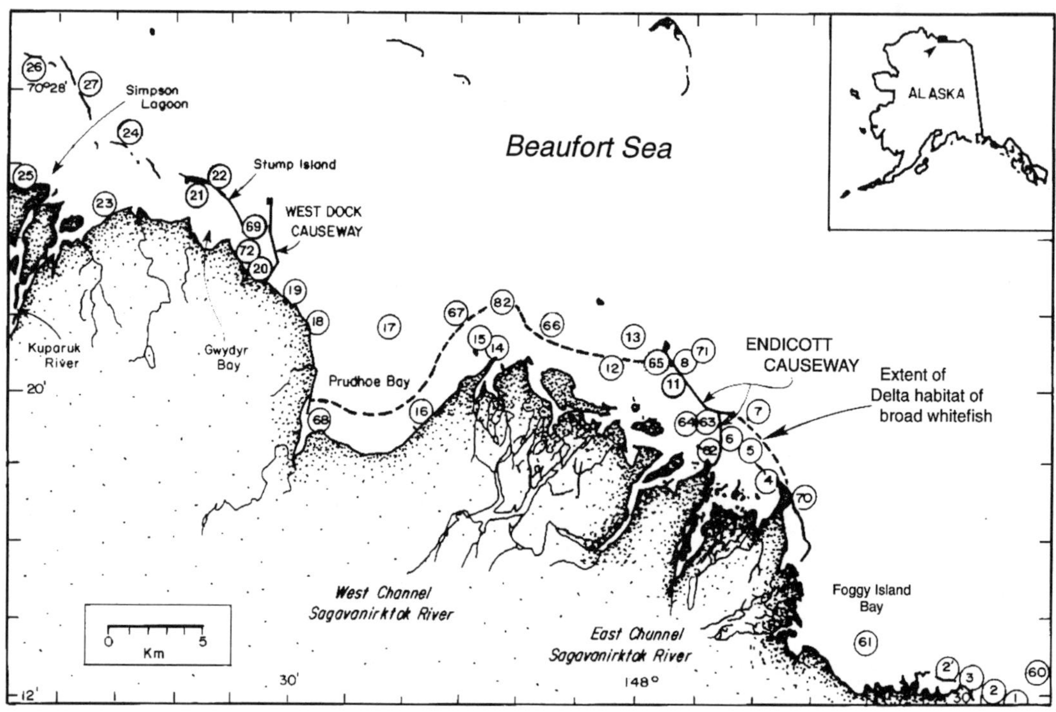

FIGURE 2.—Location of sampling sites (designated by numbered circles) in the Prudhoe Bay region. Delta rearing habitat used by broad whitefish *Coregonus nasus* is shown by the dashed line.

causeway might adversely affect the species' ecology. Causeways can, at times, alter coastal hydrography causing nearshore waters to become colder and more saline because of a wake effect at their tips (Niedoroda and Colonell 1990; Gallaway et al. 1991). Depending upon the extent and severity of the hydrographic effects, causeways could reduce the already limited amount of summer rearing habitat, leading to reductions in the local population of broad whitefish.

Population declines attributable to causeway effects on habitat could directly affect future oil and gas development in the Arctic (USACE 1980, 1984). The perception that the Endicott Causeway has had adverse effects on broad whitefish habitat and populations based upon data for the period 1982–1987 has already contributed to the requirement for additional breaching of this causeway (Hachmeister et al. 1991).

In this article, we expand the observations on the Sagavanirktok River broad whitefish population with 5 years of additional data. The database now covers more than a decade, 1982–1992. The data show that the population subsequently rebounded from a period of low abundance and a change in size structure that was coincident with construction of the causeway. The population returned to a precauseway state in 1990, and it appears that another population decline and change in size structure is in progress.

We propose that the observed changes in population level and size structure are occurring independent of any causeway effect and are probably reflective of natural fluctuations. Our underlying message is that long-term studies are necessary for understanding the population dynamics of Arctic fishes and that such understanding is required before changes can be attributed to development effects.

Study Area

The Sagavanirktok River empties into the Alaska Beaufort Sea at approximately 70°15′N, 147°50′W (Figure 2). Seaward of the delta is a shallow shelf of sediment (≤1.5 m deep) extending about 16 km along the coast and 3–4 km seaward. The Endicott Causeway, constructed in 1984–1985, is a solid-fill jetty located between the two main channels of the Sagavanirktok River (Figure 2). A curved, 6-km-long offshore segment connecting two oil production islands is attached to the mainland by a 2.5-km-long segment with two breaches. Prudhoe Bay, which lies immediately west of the Sagavanirktok River delta, is approximately 7 km across, with a maximum depth of about 2.5 m. At the western margin of Prudhoe Bay is the West Dock Causeway (Figure 2). It is a 4-km-long, solid-fill jetty that terminates in a seawater intake facility. Much of the coastline west of West Dock Causeway is protected by an offshore chain of barrier islands that extends 60 km to the west. The coastline east of the Sagavanirktok River delta is exposed and shallow (≤1.5 m deep within 2 km of the shore).

Methods

Field Sampling

A composite of the sampling sites that have been occupied over the last decade (1982–1992) is shown in Figure 2. The sampling sites occupied and the collecting methods used are summarized by Colonell and Gallaway (1990) for the period 1982–1984; by Gallaway et al. (1991) for 1985–1987; by Griffiths et al. (1992) for 1985–1989; and by Gallaway et al. (1997, this volume) for 1990–1991.

Fish were collected during all years using fyke nets consisting of double cod-end traps (1.27-cm stretched nylon mesh), two 15.2-m wings (2.5-cm stretched nylon mesh), and a common 60-m lead (2.5-cm stretched nylon mesh) attached to the shore. Fyke nets are passive sampling devices; their use can lead to a strong bias in abundance estimates because high catches may reflect strong movement as opposed to high abundance. Except during periods of bad weather, sampling continued 24 h/d throughout each summer (usually late June through mid to late September), with nets being serviced at approximately the same time each day. The nets were emptied, and captured fish were placed into floating holding pens. Fish were anesthetized in a dilute solution of tricaine methane sulfonate (MS-222), identified, counted, measured for fork length, and, during 1982–1984 and 1988–1992, small fish (120–250 mm) were marked and large fish (>250 mm) were individually anchor tagged. Fish less than 120 mm long were not marked because handling caused excessive mortality. Most individuals were held in the floating pens until the effects of the anesthetic had worn off and then released.

Catch per Unit Effort by Region and Size

Abundance levels were indexed using catch per unit effort (CPUE) expressed as number of fish caught per net per 24 h. Catch per unit effort was calculated separately for two subregions within the study area: the area west of West Dock (eastern Simpson Lagoon) and the Sagavanirktok River delta, which serves as the primary rearing habitat for broad whitefish; the latter is delineated by the dashed line on Figure 2. Although some sampling was conducted in Foggy Island Bay (Figure 2), none of these stations were sampled for more than 3 of the 11 years.

Not all stations within each subregion were sampled during all or even most of the years. We selected those that had the longest history as index stations. The index stations selected west of West Dock were 20, 22, and 23 (Figure 2). The sampling record for these stations begins in 1984 and extends through 1992. Stations 20 and 23 were sampled each year except for 1987. Although nets were operated at station 22 in all years, no effort was conducted in 1987 during the reference period defined below.

Stations 4, 6, 12, and 14 were used to index CPUE within the Sagavanirktok River delta subregion. These four stations were the only ones sampled in the 1982 baseline study; station 14 was sampled in each year of 1982–1992 except in 1987; stations 4 and 6 were sampled in all years except 1983 and 1984; and station 12 was sampled in all years except 1983, 1984, and 1987.

The net at station 14 consistently collected larger numbers of broad whitefish than any of the other nets. This posed a particular problem for 1983 and 1984 when this was the only index station sampled. Using the CPUE from this net alone would overestimate the annual CPUE index for these years, so the least-squares linear regression between $\log_e$ CPUE for station 14 and $\log_e$ CPUE for all four index stations during 1982, 1985, 1986, and 1988–1992 was used to estimate the total CPUE for 1983 and 1984.

We then established a reference time period over which CPUE would be calculated. Fishing at some stations during some years began as early as mid-June, coincident with open water in the river delta. In other years, sampling was not initiated at these and other stations until the first week in July. Likewise, sampling was terminated at different times among years—as early as 1 September or as late as 28 September, coincident with freeze-up. We selected the period 1 July to 15 August as the reference period. Initiation of sampling at one or more of the index stations was never later than the first week in July. Most broad whitefish moved out of the coastal zone and back into the rivers by mid-August.

In each subregion, total catch of broad whitefish,

within each 20-mm length interval across the entire length range, was accumulated for the reference period and divided by the total effort associated with the catch to yield catch per net per 24 h for each size interval; these were summed to obtain the total CPUE.

Population Estimates

Mark–recapture studies of small broad whitefish (120–250 mm) were conducted in 1982–1984 and during 1988–1992. The total catch from all stations was recorded daily, along with the number of fish marked or tagged and released, and the number of previously marked or tagged fish that were recaptured.

Because of small recapture-sample sizes, the Gazey-Staley (1986) method was used to estimate population size for the length range of marked fish in 1982 and for the period 1988–1992. These were years in which we had conducted the requisite field studies. Using this approach, the population was estimated using the sequential Bayes computational algorithm provided in Gazey and Staley (1986). For more detail on this method, see Gallaway et al. (1997). For large fish, there were too few recaptures of tagged fish to enable a population estimate from mark–recapture analysis.

In 1983 and 1984 the population size was estimated by Moulton et al. (1986) using the more traditional Schnabel (1938) method. Estimates derived from the two methods for the common size range of fish (120–250 mm) were compared using correlation analysis based upon the mark–recapture data for 1982 and 1988–1992. A strong correlation would provide justification for using the 1983–1984 population estimates calculated by a different method.

Other Analyses

The results of the analyses described above raised questions that could be evaluated using data compiled in the Arctic Anadromous Fish Databank (Slaybaugh et al. 1989). We have maintained this databank over the course of the Endicott Monitoring Program. These supplemental analyses included a compilation of large broad whitefish tagged east and west of West Dock and the number of these fish that were recaptured in these subregions and in the Colville River; a review of the daily CPUE patterns at stations in the Prudhoe Bay region during 1983–1987; and the size composition of the catch in early and late season of 1984.

TABLE 1.—Population estimates (N) of broad whitefish *Coregonus nasus* and 95% confidence intervals obtained using the Bayes computational algorithm (Gazey and Staley 1986) compared to those obtained using the Schnabel (1938) traditional method. These statistics are for broad whitefish in the Prudhoe Bay region, 1982–1984 and 1988–1992.

Year and method	N	Lower limit	Upper limit
1982			
Bayesian	188,620	157,820	227,780
Schnabel	172,298	143,282	213,432
1983			
Schnabel	66,300		
1984			
Schnabel	25,800	22,500	32,300
1988			
Bayesian	88,885	59,333	118,444
Schnabel	84,555	63,428	118,496
1989			
Bayesian	168,555	122,139	210,806
Schnabel	155,649	122,988	201,394
1990			
Bayesian	431,046	345,940	524,320
Schnabel	432,341	360,397	533,842
1991			
Bayesian	206,738	166,316	235,789
Schnabel	200,812	171,292	240,548
1992			
Bayesian	138,275	117,300	160,800
Schnabel	133,227	116,062	115,379

Results

Population estimates for fish 120–250 mm long in the Sagavanirktok River subregion for 1982–1984 and 1988–1992 ranged from a low of 25,800 in 1984 to a high of 432,341 in 1990 (Table 1). The estimates yielded by the Bayes computational algorithm were about the same as those obtained using the Schnabel method applied to the same data. Results of correlation analysis showed the two estimates were significantly ($P < 0.001$) correlated ($r^2 = 0.998$). We concluded from these results that the population estimates obtained by Moulton et al. (1986) using the Schnabel method could be used to estimate population size in 1983 and 1984 and that these estimates could be directly compared to those we obtained in 1982 and 1988–1992 using the Bayes algorithm.

In 1983 and 1984 only one of the index stations (net 14) was sampled. In years when all four index stations were sampled, more broad whitefish were characteristically caught at net 14 than at any of the other nets. The observed relationship between $\log_e$ CPUE for net 14 versus total $\log_e$ CPUE was

$$\log_e(\text{Total CPUE}) = -0.928 + 1.194(\log_e \text{net 14 CPUE}). \quad (1)$$

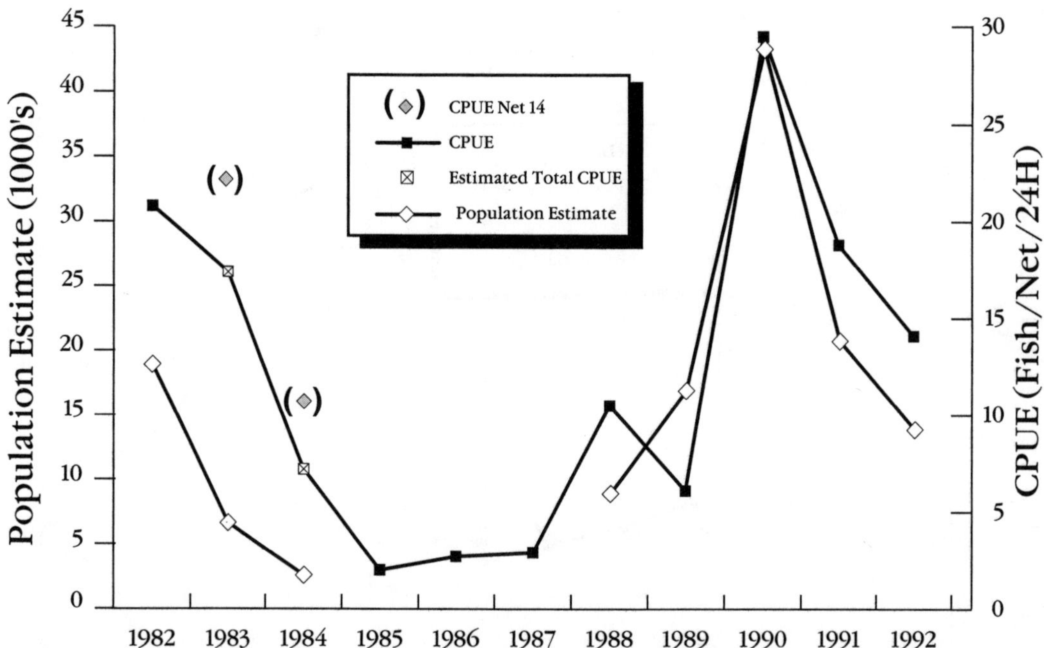

FIGURE 3.—Comparison of abundance trends reflected by population estimates and catch per unit effort values (CPUE, number of fish per net per 24 h) for fish 120–250 mm long from the Sag Delta region, 1982–1992. Total CPUE values for 1983 and 1984 were estimated from the observed relationship between CPUE at net 14 and total CPUE.

The observed relationship was significant (P = 0.012) and, with an r of 0.83, accounted for 69% of the variance. This relationship was used to estimate total CPUE for 1983 and 1984 for comparison to other years.

The abundance trends reflected by the population estimates for fish 120–250 mm long were similar to those trends reflected by CPUE for the same-size fish (Figure 3). The correlation (r = 0.79) between population estimates and CPUE was significant (P = 0.02), with the observed relationship accounting for 62% of the sample variance. We concluded from these results that CPUE could reasonably be used to index broad whitefish abundance.

Abundance

Total CPUE of broad whitefish in the Sagavanirktok River delta subregion declined over the period 1982–1987, especially between 1983 and 1985 (Figure 4). After 1987, abundance began to increase and reached peak levels in 1990 and 1991. This period of high abundance was followed by another marked decline in abundance in 1992 (Figure 4).

The initial decline in abundance was well underway before the construction of the Endicott Causeway in 1985. Although the increase in abundance coincided with the decision to require additional breaching in the causeway to mitigate perceived habitat effects, the breaches were not installed until 1994 (Figure 4).

Total CPUE of broad whitefish in the subregion west of West Dock did not exhibit the marked fluctuations observed in the Sagavanirktok River delta subregion (Figure 4). Lowest abundance levels were on the order of 6 fish per net per 24 h (1984 and 1992), and the highest levels were on the order of about 15 fish per net per 24 h (Figure 4). During the 7-year period from 1985 to 1991, CPUE was remarkably constant (12–15 fish per net per 24 h). Except for the period 1985–1986, CPUE in the West Dock subregion was consistently lower than CPUE observed in the Sagavanirktok River delta subregion.

Population Size Structure

The length-frequency distribution for broad whitefish collected in the Sagavanirktok Delta region (Figure 5) was bimodal in 1982, with both modes falling in the juvenile size range (<250 mm). Fish in subadult and adult size ranges (≥250 mm)

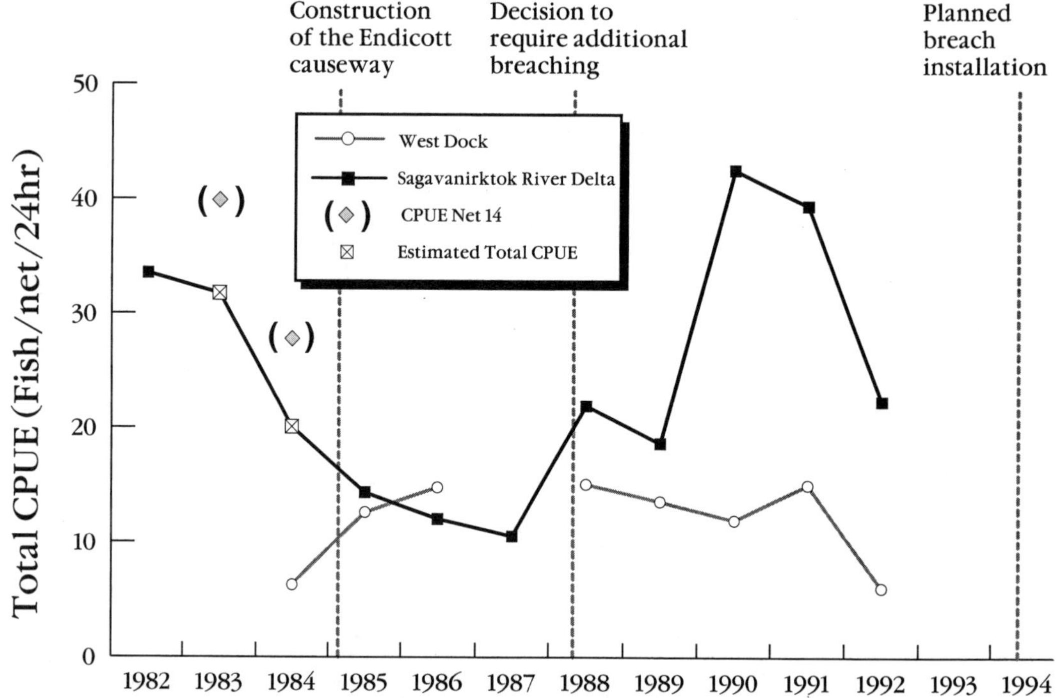

FIGURE 4.—Abundance trends (catch per unit effort, CPUE) of broad whitefish *Coregonus nasus* in the Sagavanirktok River delta and west of West Dock, 1982–1992, in relationship to the construction at the Endicott Causeway. The mitigation action believed necessary in early 1988 did not occur until 1994.

were scarce, constituting less than 7.6% of the total catch. Based upon the sharp length-delineation of cohorts that characterize the younger year-classes of broad whitefish, coupled with otolith age determinations of subsequent studies (Fechhelm et al. 1992; Griffiths et al. 1992), we determined that the larger mode in 1982 represented 3-year-old fish of the 1979 year-class. From 1982 to 1984, fish in all length-classes below the 1979 year-class were generally well represented.

During the period of low abundance that occurred during 1985–1987, a different bimodal distribution was observed (Figure 5). Representatives of the 1979 year constituted an upper mode consisting of large fish and the lower mode was dominated by fish in the length range of age-0 and age-1 fish. The scarcity of fish between the two modes suggests that there was little survival beyond age 2 during these years (Figure 5).

The 1979 year-class was well represented as subadults (ages 3 to 5) from 1982 to 1987. These fish appear as the upper mode in Figure 5, Sag Delta; growing from approximately 220 mm in 1982 to 380 mm in 1987. However, this mode virtually disappeared in 1988 when the 1979 year-class reached nine years of age (Figure 6). Their disappearance corresponds with the onset of maturity for broad whitefish in this region (LGL Alaska 1990, 1991, 1992a, 1992b).

The disappearance of the 1979 year-class in 1988 was also coincident with the beginning of a sustained increase in population size, and virtually all the fish collected after 1988 were in the juvenile size range (Figure 6). In 1990, a particularly strong year-class in the age-1 length range dominated the catches, greatly elevating the size of the population. These fish, representative of the 1989 year-class, remained abundant in 1991 when a second good recruitment of age-1 fish appeared (Figure 6). The recruitment of the 1989 and 1990 year-classes resulted in the high abundance observed in 1990 and 1991.

The 1989 year-class exhibited good survival, and as age-3 fish in 1992 they constituted the upper mode of a bimodal size distribution, where both modes fell in the juvenile size range (Figure 6). Large fish (≥250 mm) were not abundant. This distribution is much like that observed 10 years earlier in 1982 (Figures 5 and 6).

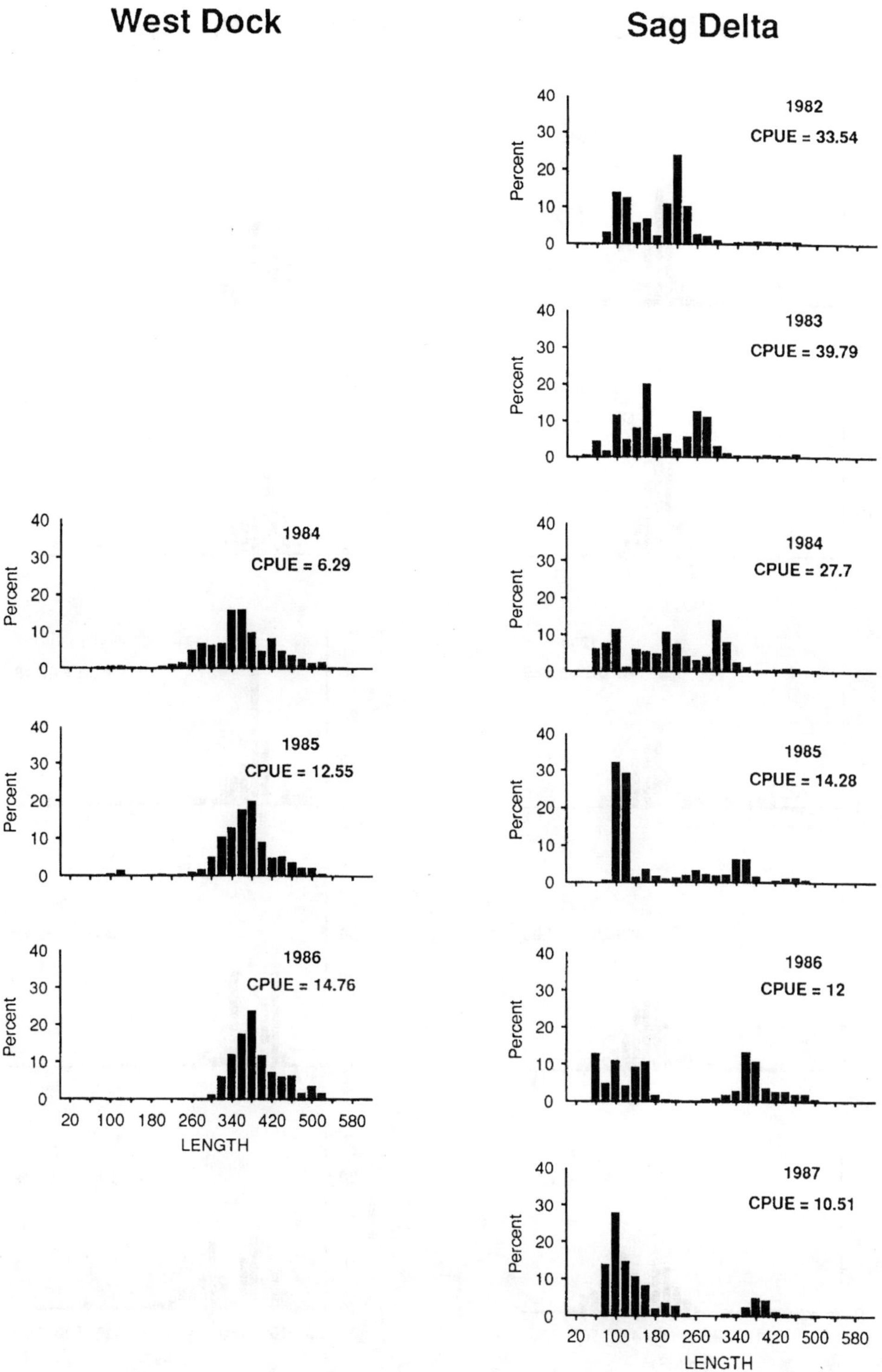

FIGURE 5.—Length-frequency patterns (length in mm) of broad whitefish *Coregonus nasus* at West Dock and Sagavanirktok River delta subregions of the study area, 1982–1987. Annual catch per unit effort (CPUE) values (number of fish per net per 24 h) are shown for each subregion and year.

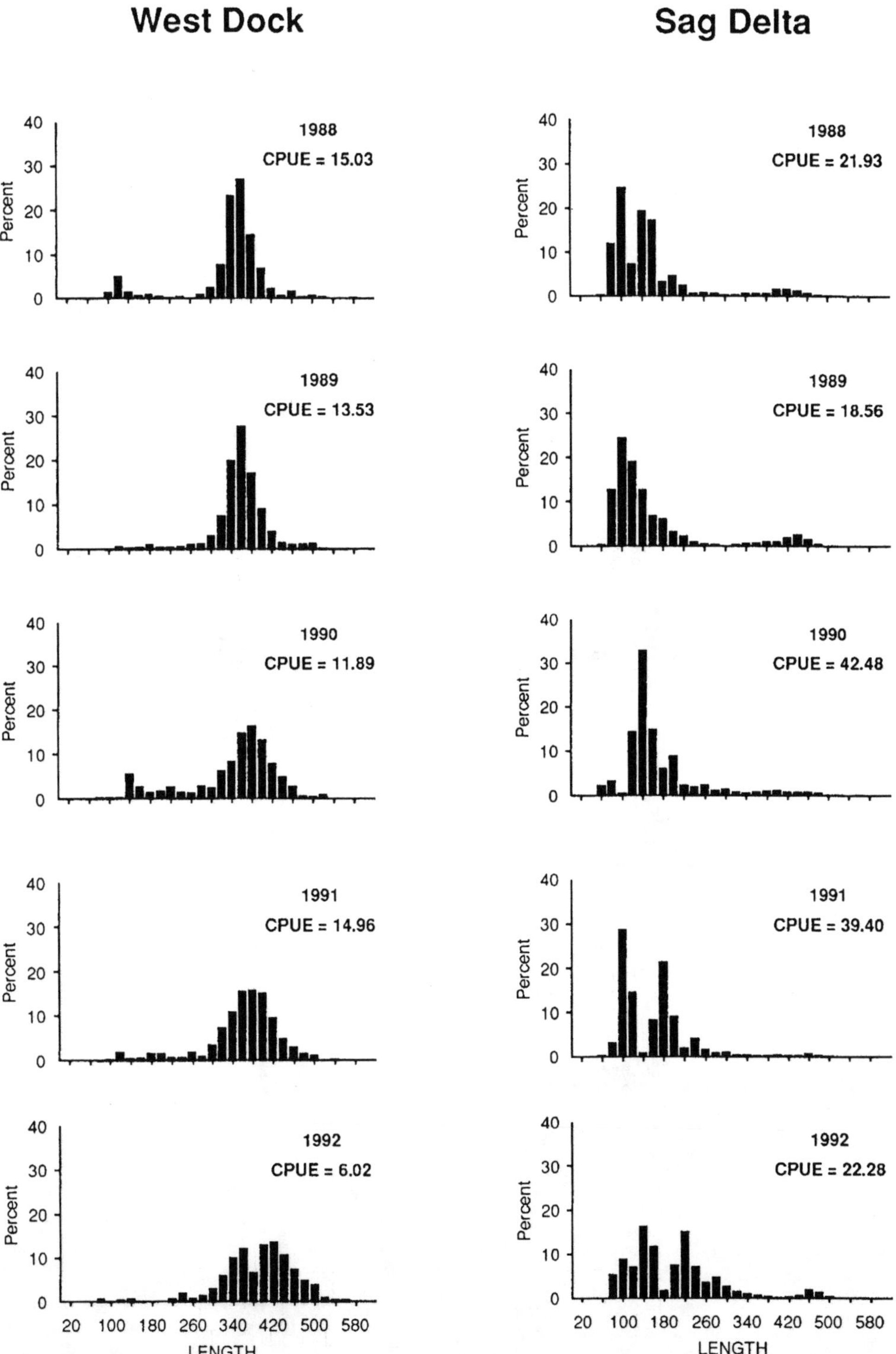

FIGURE 6.—Length-frequency patterns (length in mm) of broad whitefish *Coregonus nasus* at West Dock and Sagavanirktok River delta subregions of the study area, 1988–1992. Annual catch per unit effort (CPUE) values (number of fish per net per 24 h) are shown for each subregion and year.

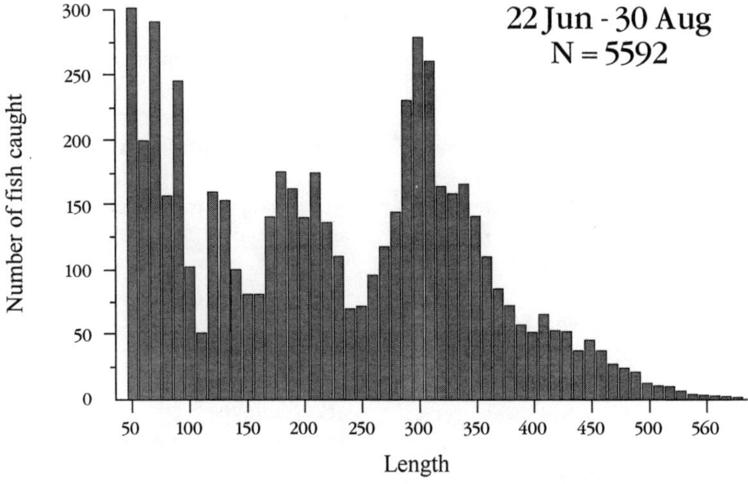

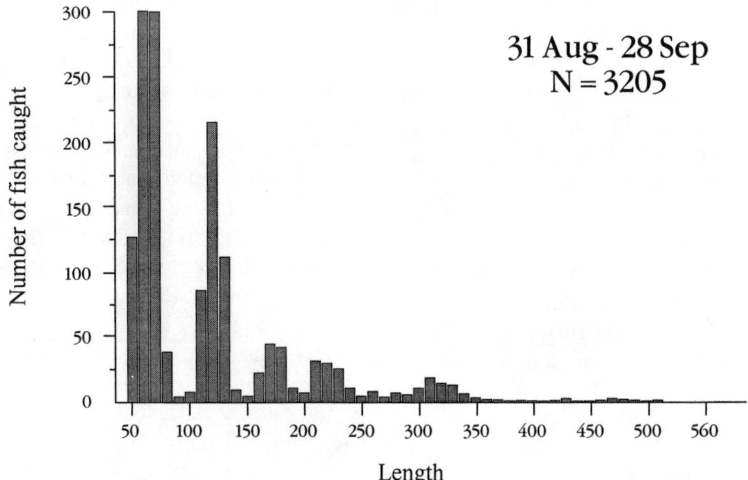

FIGURE 7.—Length-frequency (length in mm) and total catch (N) of broad whitefish *Coregonus nasus* in early (22 June to 30 August) and late (31 August to 28 September) summer 1984. Data are for fish collected in the Prudhoe Bay and Sag Delta area bounded by the dashed line in Figure 2.

Review of the daily CPUE patterns within years showed that small (especially age-0), but not large, broad whitefish often entered the coastal zone late in the summer during the 1982–1987 period of population decline. The most pronounced of these observed events occurred in September 1984. After the fish had moved into the river in mid-August, large numbers of small, but not large, broad whitefish moved back into the coastal zone in early September (Figure 7). They traveled around the shoreline of Prudhoe Bay to the west, ultimately reaching as far as the east base of West Dock (Moulton et al. 1986). When these fish entered the coastal zone, salinity was about 10‰, which is low for this time of year. However, a marine intrusion occurred beginning around 14 September, inundating the bay with salinity above the lethal limits (15–20‰, de March 1989) for small broad whitefish. These fish probably perished.

In contrast to the Sagavanirktok River delta subregion, size structure of the population west of West Dock exhibited little variation (Figures 5 and 6). In this area, the broad whitefish samples were dominated by large fish in about the same size range each year (Figures 5 and 6). These data are consistent with the idea that these are subadult and nonspawning adult fish from the Colville River system that disperse as far as the eastern end of Simpson Lagoon during the summer feeding period.

Recapture patterns of tagged broad whitefish in our study area and in the Colville River suggest that there is little exchange of large fish east and west of West Dock and that the fish west of West Dock may originate in the Colville River. During the period 1985–1991, 9,153 large broad whitefish were tagged and released in the Prudhoe Bay region. Of these, 44% (4,014) were tagged west of West Dock and 56% (5,139) were tagged east of West Dock, including in Foggy Island Bay. Of the fish tagged and released west of West Dock, 3% (122 fish) were recaptured; of those, 85% (104 fish) were recaptured in the same subregion and 15% (18 fish) were recaptured east of West Dock. Similarly, about 3% of the 5,139 fish marked and released east of West Dock were recaptured; of those, 89% (134 fish) were recaptured in the same subregion and 11% (17 fish) were recaptured west of West Dock.

Twenty-nine broad whitefish tagged in our study area were recaptured in the fall–winter Colville River cisco fisheries; 23 (79%) were tagged and released west of West Dock, and 6 (21%) were tagged and released east of West Dock. Following Sokal and Rohlf (1969), we tested for equality of the percentages represented in the release and recapture samples. The recapture sample had a significantly higher ($P < 0.0001$) proportion of fish tagged west of West Dock than did the release sample.

Discussion

The abundance and size structure patterns of the broad whitefish collected in the Sagavanirktok River delta region during 1982–1992 appeared to be largely separate and independent from those patterns for fish west of West Dock. The catch per unit effort of broad whitefish west of West Dock was low and relatively constant, and the collections consisted almost entirely of large fish in subadult and adult size ranges. Based upon the tag return data, there was only moderate exchange of large broad whitefish between the two subregions.

We think that most of the fish from the subregion west of West Dock are of Colville River origin. The tag return data show that at least some fish from our study area, especially those from the area west of West Dock, overwinter in the Colville River. However, the need to know the origin of these fish is diminished because they were not abundant enough to contribute substantially to the observed dynamics of the population in the Sagavanirktok River delta.

The question of main interest is, What caused the period of low abundance and change in size structure that occurred in the Sagavanirktok River delta in 1985–1987 after construction of the Endicott Causeway? Two hypotheses have been suggested: (1) the changes are related to effects of the Endicott Causeway and (2) the fluctuations reflect natural cycles.

Hachmeister et al. (1991) present the case for a habitat effect, based on data from the period 1982–1987. They believe that, although the causeways have little effect on habitat in early summer, marked effects occur in middle and late summer because of interruption of alongshore flow of water and deflection of river plumes offshore. Hachmeister et al. (1991) suggested that this effect would be greatest in years having a high frequency of easterly winds.

Hachmeister et al. (1991) observed that growth of broad whitefish was positively associated with the frequency of west winds and river discharge, with growth varying on an annual basis. They (Hachmeister et al. 1991:93) stated that, "because the difference in environmental conditions found in 'east wind' and 'west wind' years is amplified by the presence of the causeways, it is likely that some of the differences in observed growth rates can be attributed to causeway effects on habitat."

Hachmeister et al. (1991) also described the change in broad whitefish age structure. They reported that by 1987, four year-classes, all of which had been well represented at age 0, were virtually absent from the population. They (Hachmeister et al. 1991:93) did not attribute a cause to this trend but stated that "the trend is suggestive of an unstable population that could be vulnerable to habitat changes."

Based upon the results of the 1982–1987 studies described in Hachmeister et al. (1991), in early 1988 the U.S. Army Corps of Engineers required that larger breaches would have to be constructed in the Endicott Causeway as a mitigative measure. For economic reasons, industry accepted the mitigation requirement in early 1991, after an extended period of negotiations. The breaches were installed in 1994.

Since this time, Griffiths et al. (1992) quantified the effects of the Endicott Causeway on temperature and salinity in the delta, the distributional response of broad whitefish to these habitat changes, and the effects of the changes on their growth and condition. They found that the causeway-induced hydrographic changes have corresponded with significant changes in the local distribution of broad whitefish within the delta but that the changes have not been sufficient to significantly affect their growth. The finding of a lack of a growth effect on broad whitefish is consistent with the data shown in Figure 12 of Hachmeister et al. (1991).

A remaining question is whether growth is a sensitive enough measure to detect sublethal effects. Griffiths et al. (1992) and Fechhelm et al. (1992) show that broad whitefish growth is strongly governed by temperature and that if the fish are exposed to a net reduction of only 1°C over the summer, this small change results in an estimated 20% reduction in growth. Growth is a sensitive indicator of habitat quality and is well suited for impact assessment. Had the fish not been able to avoid exposure to the observed causeway effects on temperature, an effect would have been obvious.

We have shown that (1) the size of the population (which was low in 1985–1987) has once more grown to 1982 levels and is probably in the midst of another decline; and (2) the size structure of the population returned in 1992 to the initial state observed in 1982. These responses are not related to any mitigative measure, for none had been implemented until 1994. Furthermore, the observed "recoveries" cannot be attributed to an extended series of "west wind" years as described by Hachmeister et al. (1991), because 4 of the 5 years since 1987 were dominated by winds having an easterly component. The strength and persistence of the easterly winds observed during the peak in broad whitefish abundance in 1990 were as strong as any observed during the 11-year period from 1982 to 1992.

The requirement for additional breaching in the Endicott Causeway to mitigate perceived adverse effects on broad whitefish habitat and populations appears questionable. Based upon analysis of population parameters (e.g., population level and structure, and growth), we see little basis for an effects determination. We believe that population parameters are sensitive to habitat change and are well suited to address effects of habitat reduction. We concur with Robertson (1991) that impact predictions may be appropriately based on expected changes in habitat, but impact assessments are best made by study of the populations involved.

We think that observed fluctuations in abundance and size structure will ultimately be shown to be a density-dependent response of the population to a limiting amount of overwintering habitat. The data to-date are consistent with a cyclical, bimodal population structure that might result from density-driven phenomena as theorized for other Arctic fish populations by Johnson (1976, 1981, 1983), Power (1978), and Hammer (1989). The size distribution pattern seen in 1982 when the population was high and only small fish were present was observed once more in 1992. A low population and size distribution pattern observed in 1985–1987 (when large and small but no intermediate-sized fish were present) was also seen in the same region in the mid-1970s (Bendock 1979). Disappearance of large fish from our samples at age of maturity (age 9) corresponded to increased survival and population growth. The documentation of two different size-frequency patterns, each of which was again seen after a 10-year interval, and the dominance of the 1979 and 1989 year-classes suggest a 10-year cycle.

Acknowledgments

We thank B. Wilson for critically reviewing the manuscript and V. J. Chhipa for computer assistance. Additional thanks to Jean Erwin, G. Fain Hubbard, Peggy Kircher, Mike Smolen, and Glen Klinkhart for helping to develop the manuscript and graphics. Environmental monitoring programs were funded by owners of the Duck Island Unit, which is operated by BP Exploration (Alaska) Inc. Environmental monitoring programs from 1982 to 1987 were conducted under the joint supervision of the Alaska Department of Fish and Game, the U.S. Environmental Protection Agency, the National Marine Fisheries Service, the North Slope Borough, the U.S. Army Corps of Engineers, and the U.S. Fish and Wildlife Service. Studies conducted since 1988 were under the auspices of the North Slope Borough. The conclusions and opinions expressed in this article do not necessarily represent those of the above-mentioned individuals or organizations.

References

Bendock, T. 1979. Beaufort Sea estuarine fishery study. Pages 670–729 *in* Environmental assessment of the Alaskan continental shelf. Final Report of the principal investigators, volume 4. U.S. Bureau of Land Management and NOAA (National Oceanic and Atmospheric Administration), OCSEAP (Outer Continental Shelf Environmental Assessment Program), Boulder, Colorado.

Bond, W. A. 1982. A study of the fishery resources of Tuktoyaktuk Harbour, southern Beaufort Sea coast, with special reference to life histories of anadromous coregonids. Canadian Technical Report of Fisheries and Aquatic Sciences 1119.

Bond, W. A., and R. N. Erickson. 1985. Life history studies of anadromous coregonid fishes in two freshwater lake systems on the Tuktoyaktuk Peninsula, Northwest Territories. Canadian Technical Report of Fisheries and Aquatic Sciences 1336.

Colonell, J. M., and B. J. Gallaway, editors. 1990. An assessment of marine environmental impacts of West Dock causeway. Report by LGL Alaska Research Associates, Inc. and Environmental Science and Engineering, Inc. for Prudhoe Bay Unit Owners represented by Arco Alaska, Inc., Anchorage, Alaska.

Craig, P. C. 1989. An introduction to anadromous fishes in the Alaska Arctic. Biological Papers of the University of Alaska 24:27–54.

de March, B. G. E. 1989. Salinity tolerance of larval and juvenile broad whitefish (*Coregonus nasus*). Canadian Journal of Zoology 67:2392–2397.

Fechhelm, R. G., R. E. Dillinger, Jr., B. J. Gallaway, and W. B. Griffiths. 1992. Modelling of in situ growth and temperature relationships of yearling broad whitefish in Prudhoe Bay, Alaska. Transactions of the American Fisheries Society 121:1–12.

Gallaway, B. J., R. G. Fechhelm, and W. J. Gazey. 1997. Estimating abundance of arctic anadromous fish in the coastal Beaufort Sea during summer: is an open population model necessary? Pages 274–286 *in* J. Reynolds, editor. Fish ecology in Arctic North America. American Fisheries Society Symposium 19, Bethesda, Maryland.

Gallaway, B. J., W. J. Gazey, J. M. Colonell, A. W. Niedoroda, and C. J. Herlugson. 1991. The Endicott Development Project—preliminary assessment of impacts from the first major offshore oil development in the Alaskan Arctic. Pages 42–80 *in* C. S. Benner and R. W. Middleton, editors. Fisheries and oil development on the continental shelf. American Fisheries Society Symposium 11, Bethesda, Maryland.

Gazey, W. J., and M. J. Staley. 1986. Population estimation from mark–recapture experiments using a sequential Bayes algorithm. Ecology 67:941–951.

Griffiths, W. B., B. J. Gallaway, W. J. Gazey, and R. E. Dillinger. 1992. Growth and condition of Arctic cisco and broad whitefish as indicators of causeway-induced effects in the Prudhoe Bay region, Alaska. Transactions of the American Fisheries Society 121: 557–577.

Hachmeister, L. E., D. R. Glass, and T. C. Cannon. 1991. Effects of solid-fill gravel causeways on the coastal central Beaufort Sea environment. Pages 81–96 *in* C. S. Benner and R. W. Middleton, editors. Fisheries and oil development on the continental shelf. American Fisheries Society Symposium 11, Bethesda, Maryland.

Hammer, J. 1989. Freshwater ecosystems of polar regions: vulnerable resources. Ambio 198:7–22.

Johnson, L. 1976. Ecology of arctic populations of lake trout, *Salvelinus namaycush*, lake whitefish, *Coregonus clupeaformis*, Arctic char, *S. alpinus*, and associated species in unexploited lakes of the Canadian Northwest Territories. Journal of the Fisheries Research Board of Canada 33:459–488.

Johnson, L. 1981. The thermodynamic origin of ecosystems. Canadian Journal of Fisheries and Aquatic Sciences 38:571–590.

Johnson, L. 1983. Homeostatic characteristics of single species of fish stocks in arctic lakes. Canadian Journal of Fisheries and Aquatic Sciences 40:987–1024.

LGL Alaska. 1990. The 1988 Endicott Development Fish Monitoring Program, volume II: recruitment and population studies, analysis of 1988 fyke net data. Report by LGL Alaska Research Associates, Incorporated for BP Exploration (Alaska), Incorporated and the North Slope Borough, Anchorage.

LGL Alaska. 1991. The 1989 Endicott Development Fish Monitoring Program, volume II: analysis of 1989 fyke net data. Report by LGL Alaska Research Associates, Incorporated for BP Exploration (Alaska), Incorporated and the North Slope Borough, Anchorage.

LGL Alaska. 1992a. The 1990 Endicott Development Fish Monitoring Program, volume II: analysis of fyke net data. Report by LGL Alaska Research Associates, Incorporated for BP Exploration (Alaska), Incorporated and the North Slope Borough, Anchorage.

LGL Alaska. 1992b. The 1991 Endicott Development Fish Monitoring Program, volume II: analysis of 1991 fyke net data. Report by LGL Alaska Research Associates, Incorporated for BP Exploration (Alaska), Incorporated and the North Slope Borough, Anchorage.

McCart, P., P. Craig, and H. Bain. 1972. Report on fisheries investigations in the Sagavanirktok River and neighboring drainages. Report by Aquatics Environments Limited for Alyeska Pipeline Service Co., Bellevue, Washington.

Moulton, L. L., and seven coauthors. 1986. 1984 Central Beaufort Sea fish study. Waterflood monitoring program fish study. Report by Entrix, Inc., LGL Ecological Research Associates, Inc., and Woodward-Clyde Consultants for Envirosphere, Co., Anchorage, Alaska.

Niedoroda, A. W., and J. M. Colonell. 1990. Beaufort sea causeways and coastal ocean dynamics. Pages 509–516 *in* S. K. Chakrabartl, H. Maeda, C. Aage, and F. G. Nielsen, editors. Proceedings of the 9th international conference of offshore mechanics and Arctic engineering. American Society of Mechanical Engineers, New York.

Power, G. 1978. Fish population structure in arctic lakes. Journal of the Fisheries Research Board of Canada 35:53–59.

Robertson, S. B. 1991. Habitats versus populations: approaches for assessing impacts on fish. Pages 97–108 *in* C. S. Benner and R. W. Middleton, editors. Fisheries and oil development on the continental shelf. American Fisheries Society Symposium 11, Bethesda, Maryland.

Schnabel, Z. E. 1938. The estimation of the total fish population of a lake. American Mathematical Monthly 45:348–352.

Scott, W. B., and E. J. Crossman. 1973. Freshwater fishes of Canada. Fisheries Research Board of Canada Bulletin 184.

Slaybaugh, D. K., B. J. Gallaway, and J. S. Baker. 1989.

The databank for Arctic anadromous fish: description and overview. Biological Papers of the University of Alaska 24:4–26.

Sokal, R. L., and F. J. Rohlf. 1969. Biometry. The principles and practice of statistics in biological research. Freeman, San Francisco.

USACE (U.S. Army Corps of Engineers). 1980. Final environmental impact statement, Prudhoe Bay oil field, waterflood project. U.S. Army Corps of Engineers, Alaska District, Anchorage.

USACE (U.S. Army Corps of Engineers). 1984. Final environmental impact statement, Prudhoe Bay oil field, Endicott development project. U.S. Army Corps of Engineers, Alaska District, Anchorage.

American Fisheries Society Symposium 19:208–213, 1997

Experimental Introduction of Arctic Grayling to a Rehabilitated Gravel Extraction Site, North Slope, Alaska

CARL R. HEMMING

Alaska Department of Fish and Game, Habitat and Restoration Division
1300 College Road, Fairbanks, Alaska 99701, USA

Abstract.—I evaluated survival, growth, and reproductive success of 210 Arctic grayling *Thymallus arcticus* introduced to a rehabilitated gravel extraction site in the Kuparuk River oilfield, Alaska. The purpose of this study was to determine if a reproducing population could be established in a deepwater gravel extraction site connected to a small tundra stream system. I recaptured 68 Arctic grayling in two seasons (1990 and 1991) of sampling, demonstrating that at least 32% of the introduced population survived at least one winter. A 1991 mark–recapture survey generated an adjusted Peterson estimate of 56 grayling, which would represent a 2-year survival rate of 27%. Tagged fish recaptured in 1990 and 1991 grew rapidly after introduction. The results demonstrate that small tundra streams lacking suitable overwintering habitat for most fish species can support introduced Arctic grayling when connected to a deepwater gravel extraction area, but reproduction may be limited at such sites.

Overwintering habitat is a limiting factor for North Slope fish populations (Schmidt et al. 1989). Winter stream habitat available to fish may be reduced by 95% from that found during the ice-free season (Craig 1989). Freshwater overwintering habitats have been described as deep lake basins, spring areas, and riverine pools that retain under-ice water. In the North Slope oilfield (Colville to Sagavanirktok rivers), known fish-overwintering habitat is limited to several deep, isolated pools in the lower Sagavanirktok and Kuparuk rivers and to more extensive under-ice areas in the Colville River delta (Figure 1).

In an article classifying Beaufort Sea drainage stream types, Craig and McCart (1975) identify tundra streams (smaller runoff streams as compared with larger groundwater-fed rivers) as areas suitable for spawning and summer feeding, but not for overwintering, by Arctic grayling *Thymallus arcticus*. Tundra streams freeze to the bottom in winter, eliminating overwintering habitat. If an Arctic grayling is to use a tundra stream during summer, it must have access to a large river or lake with under-ice water during winter. Several tundra streams exist in the oilfield but have no access to overwintering areas (i.e., they are isolated).

Flooded, gravel extraction sites, excavated to provide material for development of oilfield infrastructure, are similar to lake basins used by overwintering fish. These artificial basins are deep enough to retain large quantities of under-ice water with dissolved oxygen concentrations at or near saturation during the ice-covered season (Hemming 1988).

A deep, gravel extraction basin in an isolated tundra stream could provide suitable overwintering habitat. To test this idea, a habitat enhancement project was proposed. The project area consisted of channel connections between Kuparuk Mine site B, an abandoned gravel extraction area, and East Creek, a small tundra stream that discharges to the Beaufort Sea (Figure 2). The project was completed in May 1989 and provided permanent, open-water channels between stream habitat and the flooded basin.

This project provided an opportunity for determining if Arctic grayling could survive and reproduce in a system consisting of a gravel extraction site and small tundra stream. I evaluated the survival, growth, and reproductive success of Arctic grayling transplanted into this experimental site in 1989, during the first 2 years (1990–1991) after the introduction.

Study Area

The Kuparuk River oilfield is within the Arctic Coastal Plain physiographic province (Wahrhaftig 1965). The area is characterized by low topographic relief and numerous shallow lakes and lake basins. Climatic conditions are severe, with mean daily temperatures of 0–9°C during summer (June through mid-September) and well below freezing during the remainder of the year. The mean annual wind speed is 22 km/h. Arctic Coastal Plain annual precipitation is less than 25 cm.

Kuparuk Mine site B is a former gravel extraction site that was flooded in 1978 when water from East Creek, an adjacent small tundra stream, filled the excavated area. The 3.7-ha pond, consisting of adjoining 1.3- and 2.4-ha basins, has a maximum depth of 11.3 m and a mean depth of 7.1 m (Hemming

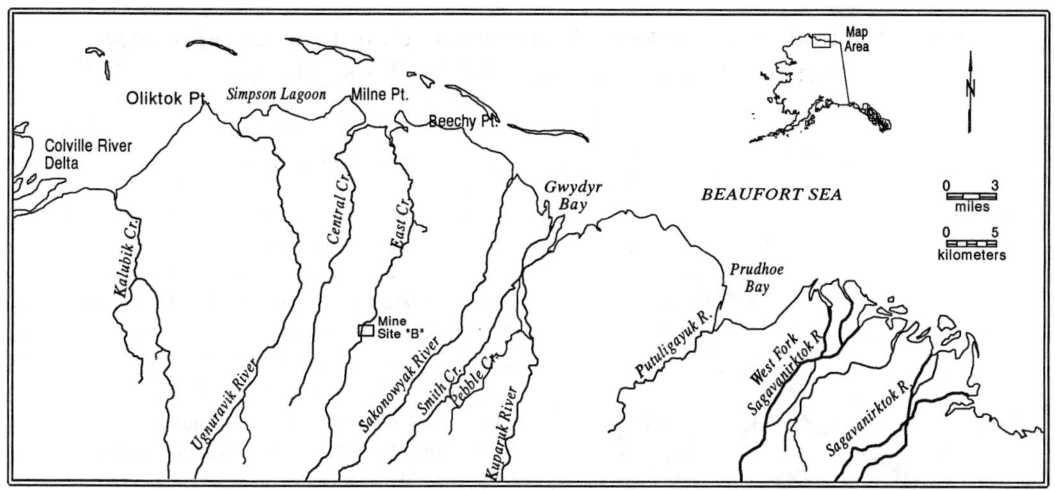

FIGURE 1.—Prudhoe Bay–Kuparuk oilfield area and the location of Kuparuk Mine site B.

1988). In 1986 and 1987 I sampled Kuparuk Mine site B with gill nets and minnow traps; ninespine sticklebacks *Pungitius pungitius* and broad whitefish *Coregonus nasus* were captured.

Before site rehabilitation, a narrow band of wetland separated Kuparuk Mine site B and East Creek; surface water flowed between them only during high water. In May 1989 three connection channels were excavated at Kuparuk Mine site B. An 18 × 24-m inlet channel was excavated to a depth of 1.8 m between East Creek and the site. Two similar-sized channels were excavated between the two adjacent lake basins, forming an island (Figure 2). The excavated channels provide a permanent connection between East Creek and Kuparuk Mine site B during the open-water season.

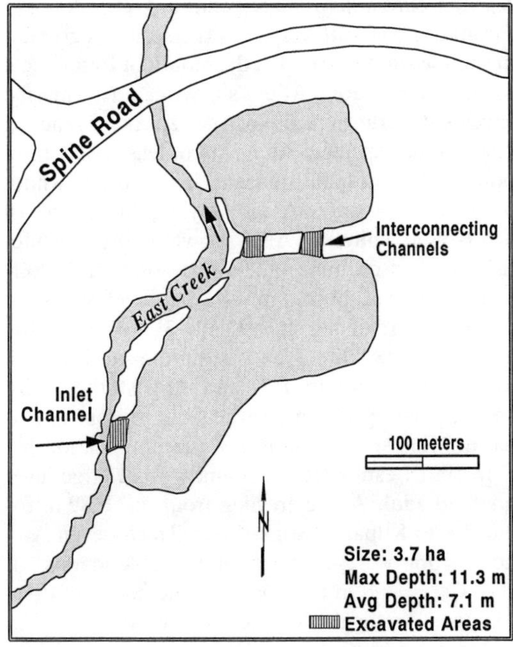

FIGURE 2.—Kuparuk Mine site B rehabilitation plan.

East Creek is a small, single-channel tundra stream with a beaded configuration. The beads are a series of wide ponds or pools separated by very narrow, often poorly defined channels. East Creek empties into Simpson Lagoon between the Colville and Kuparuk rivers. Peak discharge in the watershed (about 28.3 m^3/s) generally occurs during the first week in June. By mid-June discharge decreases to less than 0.3 m^3/s, and by late summer, channel sections between pools may become intermittent. The system is ice covered by mid-September and discharge ceases during the winter. East Creek is 26 km long; Kuparuk Mine site B is located 16 km upstream from the mouth. A drainage area of 132 km^2 provides runoff above Kuparuk Mine site B. East Creek is crossed by the main east–west road system in the Kuparuk River oilfield, 40 m downstream from mine site B.

Methods

Arctic grayling transplanted to Kuparuk Mine site B were captured on 21–25 June 1989 at seven locations accessible by road within the Sagavanirktok River drainage near Happy Valley Creek. Most

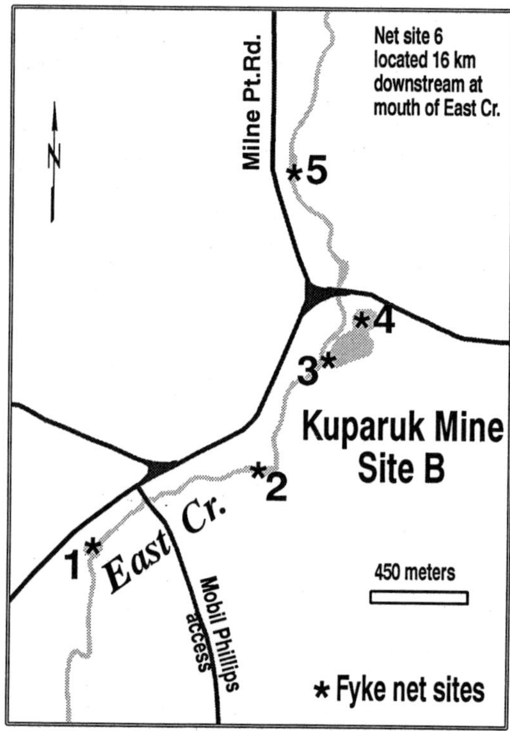

FIGURE 3.—Fyke-net sample sites in East Creek and Kuparuk Mine site B, 1990 and 1991.

fish were captured in fyke nets, either at the mouths of tundra streams tributary to the Sagavanirktok River or in backwater areas or sloughs associated with the river floodplain. Each fish was marked with individually numbered anchor tags, measured to the nearest 1 mm for fork length, and scale-sampled for age determination. Two hundred ten Arctic grayling, 176–399 mm long (mean = 283 mm, SD = 52), were transported to Kuparuk Mine site B and released on 26–27 June 1989.

Fyke nets and angling gear were used to capture fish in Kuparuk Mine site B and East Creek at various times in 1989, 1990, and 1991. Each fyke net was 3.7 m long, with two 1.2-m-square entrance frames, five hoops, and a 1.8-m cod end. I placed the nets at six road-accessible locations, including sites upstream and downstream from Kuparuk Mine site B, at the inlet channel, and in the pond (Figure 3). With few exceptions, the nets were checked daily. Netting duration was recorded to determine catch per unit effort for each species at each net site.

Each captured fish was identified and released at the capture site; the fork length of each fish was measured, except for ninespine sticklebacks. Arctic grayling were examined for tags or tag wounds and scars; those without tags were re-marked. Ninespine sticklebacks were too numerous to count at most net sites; their abundance was estimated using a 1.5-cm-diameter fine-mesh scoop. The number of scoops required to remove all ninespine sticklebacks from the net was multiplied by the number of individuals in a single scoop.

In 1991, a single-census, mark–recapture procedure (adjusted Peterson method, Ricker 1975) was used to estimate Arctic grayling abundance in Kuparuk Mine site B. Fish were double-marked with adipose fin clips and anchor tags in June and July. The recapture event was in August. I obtained information on movements of tagged fish by evaluation of recapture data.

Results

In 1989–1991, 82.7 net-days of trapping effort occurred at selected locations in East Creek and Kuparuk Mine site B. Ten fish species were captured; in numerical order of abundance, they were as follows: 221,248 (estimated) ninespine sticklebacks; 96 Arctic grayling; 63 broad whitefish; 32 least ciscoes *Coregonus sardinella*; 9 Arctic ciscoes *C. autumnalis*; 7 fourhorn sculpin *Myoxocephalus quadricornis*; 3 rainbow smelt *Osmerus mordax*; 1 round whitefish *Prosopium cylindraceum*; 1 Dolly Varden *Salvelinus malma*; and 1 Arctic flounder *Pleuronectes glacialis*. Species richness was greatest at sample site 6, located at the mouth of East Creek, in an estuarine area. At this site we captured marine species (fourhorn sculpin and Arctic flounder); anadromous species (Arctic cisco, least cisco, rainbow smelt, ninespine stickleback, and broad whitefish); and freshwater Arctic grayling. Six species of fish (round whitefish, Arctic grayling, broad whitefish, least cisco, ninespine stickleback, and Dolly Varden) were captured in the East Creek system at the five upstream net sites (Figure 3). In June Arctic grayling were only found upstream of Kuparuk Mine site B, but in July and August they were absent from upstream locations but present in Kuparuk Mine site B and at downstream locations.

In 1991 I estimated the abundance of large juvenile and adult Arctic grayling from the 1989 introduction to Kuparuk Mine site B. Thirty-seven Arctic grayling were captured and double-marked in East Creek and in the pond in June and July 1991. Eighteen Arctic grayling were captured at the inlet to the pond in August 1991; 12 were marked fish. The estimated population of Arctic grayling in Ku-

TABLE 1.—Growth rate of Arctic grayling *Thymallus arcticus* transplanted to Kuparuk Mine site B during 23–25 June 1989 and recaptured in 1990 and 1991.

Length category (mm)	Number of fish	Average fork length (mm)		Average growth (mm/year)
		At tagging	At recapture	
<200	6	192	336	67
200–250	11	221	316	58
250–300	5	285	347	35
>300	8	331	366	26

paruk Mine site B was 56 (95% confidence interval, 33–100). Based on this population estimate, the 2-year survival rate for the 210 grayling introduced to Kuparuk Mine site B was 27% (95% confidence interval, 16–48).

Growth information was obtained from 30 recaptured Arctic grayling retaining anchor tags. I found rapid growth rates among all recaptured Arctic grayling, with those smaller at the time of transplant showing the greatest increase in length (Table 1). With one exception (a fish that was 238 mm when tagged and grew 105 mm/year), growth rates were fairly uniform among similar-sized fish at introduction. The length-frequency distribution of Arctic grayling when transplanted compared with those of the catches in 1990 and 1991 also indicates rapid growth (Figure 4).

There was little evidence of reproductive success by Arctic grayling introduced into Kuparuk Mine site B. No age-0 fish were captured in 1990. In 1991, two age-1 fish, produced in 1990, were captured: one in Kuparuk Mine site B in June and another in the inlet channel to the site in August. Also, in August 1991 three 38-mm age-0 fish were captured at the mouth of East Creek.

Recaptures of tagged Arctic grayling indicate upstream movement from Kuparuk Mine site B shortly after breakup, followed by downstream movement toward the pond. On 25–29 June 1990, seven fish were recaptured at net sites downstream of the initial capture location. In 1991, a week earlier than in 1990 (19–23 June), three fish were recaptured at sites other than their initial capture location; all had moved from Kuparuk Mine site B to a net site 1.6 km upstream on East Creek. Six Arctic grayling recaptured in July–August 1991 had moved downstream from a location (the previous month) 1.6 km above Kuparuk Mine site B and to the pond. Two Arctic grayling, recaptured multiple times during 1991, first moved upstream from Kuparuk Mine site B (19–21 June) and later (20 August) returned to the site. Both were adult fish (>300 mm); one was a ripe female at the time of initial capture.

Discussion

With the exception of the age-0 and age-1 fish found in 1991, all Arctic grayling captured in Kuparuk Mine site B and East Creek are assumed to be individuals transplanted to the site in 1989. There are three reasons for this assumption: (1) sampling before the transplant with gill nets and minnow traps failed to capture Arctic grayling; (2)

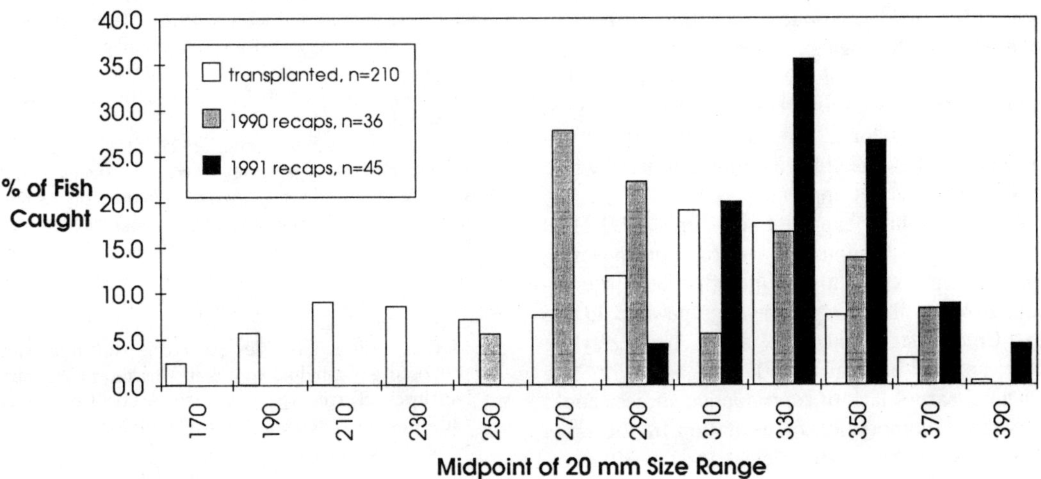

FIGURE 4.—Length-frequency distribution of Arctic grayling *Thymallus arcticus* transplanted to Kuparuk Mine site B in 1989 and of those recaptured in 1990 and 1991.

all fish captured in 1990 and 1991 were in the length range to be expected of fish transplanted in 1989; and (3) the majority of untagged fish (84%) had identifiable tag scars or wounds. Based on this assumption, 32% of the transplanted Arctic grayling were recaptured after at least one winter in Kuparuk Mine site B. Kuparuk Mine site B provides the only known overwintering habitat in the East Creek system; therefore, the documented survival of introduced Arctic grayling can be directly attributed to the deepwater habitat in the former gravel extraction site.

Arctic grayling grew rapidly in the first 2 years after introduction. Of the 30 recaptured fish retaining anchor tags after at least 1 year, average annual growth was 47 mm. These fish were 4–12 years old at the time of recapture, based on scale age determinations. Craig and Poulin (1975) reported growth of approximately 40 mm/year for Arctic grayling in a small tundra stream draining into the Kavik River near Prudhoe Bay; they described this growth rate as among the fastest reported for arctic populations.

In the Kavik River study (Craig and Poulin 1975), all mature fish exceeded 295 mm. In the East Creek system, 45 large fish captured in 1991 were 293–390 mm long (average = 336 mm, SD = 23 mm). Considering the measured growth rate among East Creek fish and the size distribution of the 1991 catch, all fish would have reached mature size (300 mm) by the 1992 spawning season.

The 1991 population estimate indicated that 27% of the Arctic grayling introduced to Kuparuk Mine site B survived from June 1989 until August 1991, the equivalent of 52% annual survival. Based on direct observations of fishing at the site and discussion with oilfield personnel, mortality attributable to fishing was negligible. No piscivorous fish and few avian or terrestrial predators were observed in the study area; predation, therefore, was unlikely as an important factor in the mortality of the introduced fish. The reason for an annual mortality rate of about 50% is not apparent.

Arctic grayling captured in June 1991 had spawned (flaccid abdomen) or were in prespawning (running ripe) condition. Although these observations indicate that Arctic grayling spawned in the East Creek system, only three age-0 Arctic grayling were captured in August 1991.

The apparent lack of reproductive success could result from competition or predation by the large population of ninespine sticklebacks in Kuparuk Mine site B and East Creek. Skaugstad (1989) found little or no survival of Arctic grayling sac fry in interior-Alaska ponds containing threespine sticklebacks *Gasterosteus aculeatus*. Adult Arctic grayling may, however, use ninespine sticklebacks as food (deBruyn and McCart 1974). The large population of ninespine stickleback in the East Creek system may provide an abundant food source for adult Arctic grayling.

Management Implications

I identified 324 ha of deep gravel excavations within the North Slope oilfields. Eight sites, consisting of more than 200 ha total, are located adjacent to small tundra streams similar to East Creek. It is possible to connect these sites to the adjacent streams, increasing the amount of overwintering habitat. The results of this study indicate that Arctic grayling introduced to such systems will overwinter and grow rapidly, but persistence of an introduced population may be limited by low-to-moderate survival and poor reproductive success.

Acknowledgments

Alvin G. Ott provided overall project direction and assisted with a significant portion of the field data collection. Roger Post, Jack Winters, and Richard Shideler also assisted with various phases of this project. Financial assistance for this project was provided by the Prudhoe Bay and Kuparuk River Oil Production Units through a grant to the Alaska Department of Fish and Game, Habitat Division.

References

Craig, P. C. 1989. An introduction to anadromous fishes in the Alaskan Arctic. Biological Papers of the University of Alaska 24:27–54.

Craig, P. C., and P. J. McCart. 1975. Classification of stream types in Beaufort Sea drainages between Prudhoe Bay, Alaska and the Mackenzie Delta, Northwest Territories, Canada. Arctic and Alpine Research 7:183–198.

Craig, P. C., and V. A. Poulin. 1975. Movements and growth of Arctic grayling (*Thymallus arcticus*) and juvenile Arctic charr (*Salvelinus alpinus*) in a small arctic stream, Alaska. Journal of the Fisheries Research Board of Canada 32:689–697.

deBruyn, M., and P. McCart. 1974. Life history of the grayling (*Thymallus arcticus*) in Beaufort Sea drainages in the Yukon Territory. Pages 1–39 *in* P. J. McCart, editor. Fisheries research associated with proposed gas pipeline routes in Alaska, Yukon and Northwest Territories. Canadian Arctic Gas Study Ltd./Alaskan Arctic Gas Study Co. Biological Report Series 15, Calgary.

Hemming, C. R. 1988. Aquatic habitat evaluation of flooded North Slope gravel mine sites (1986–1987). Alaska Department of Fish and Game, Habitat Division, Technical Report 88–1, Juneau.

Ricker, W. E. 1975. Computation and interpretation of the biological statistics of fish populations. Fisheries Research Board of Canada Bulletin 191.

Schmidt, D., W. Griffiths, and L. R. Martin. 1989. Overwintering biology of anadromous fish in the Sagavanirktok River delta, Alaska. Biological Papers of the University of Alaska 24:55–74.

Skaugstad, C. 1989. Evaluation of arctic grayling enhancement: a cost per survivor estimate. Alaska Department of Fish and Game, Division of Sport Fish, Fisheries Data Series 96, Juneau.

Wahrhaftig, C. 1965. Physiographic divisions of Alaska. U.S. Geological Survey, Professional Paper 482, Washington, DC.

American Fisheries Society Symposium 19:214–223, 1997

Growth or Reproduction: The Life History Gamble of Landlocked Arctic Char

H. H. Parker

Markhuset, Gambles Lane, Woodmancote, Cheltenham, GL52 4PU, England

Abstract.—Populations of landlocked Arctic char *Salvelinus alpinus* often exhibit multimodal population structures with mature individuals occurring in all modes. Examples from the literature and a comparative survey of populations on Ellesmere Island, Northwest Territories, Canada, are used to show that these modes are usually associated with different habitats. A qualitative mathematical model is used to demonstrate how overall lifetime fitness can be optimized through differing allocation of resources between growth, reproduction, and survival. In the case where mature individuals exist within each habitat, the opportunity exists for the phenotype to adopt alternative life history strategies, and the model predicts that these will be optimized in fitness terms at different growth rates. Although not constituting rigorous proof, evidence supporting this argument is provided by the natural systems, where significant variation in size and growth rate at a given age is a measurable characteristic.

Populations of Arctic char *Salvelinus alpinus* show a great variation in average size of individuals. The range extends from stunted dwarf populations with an average size of a few grams to anadromous stocks averaging several kilograms per individual (Johnson 1980). Many populations also exhibit a multimodal structure in which more than one morph or size mode occurs sympatrically (e.g., Balon 1980; Jonsson and Hindar 1982; Johnson and Burns 1984; Sparholt 1985; Riget et al. 1986; Walker et al. 1988; Kawanabe et al. 1989; Parker and Johnson 1991). In most cases considerable variation in external color, meristic counts, age at maturity, habitat preferences, and diet are associated with each morph.

In the majority of the above cases it has not been possible to distinguish genetically between the morphs, and the authors have concluded that the morphs correspond to different elements of the same population. Balon (1984a, 1984b) argued that such a situation is an example whereby epigenesis (the interaction among gene products, cells, tissues, organs, and the environment in the formation of the organism) has caused the formation of an "altricial ⇌ precocial dynamic state."

With the altricial ⇌ precocial dynamic state, it is argued that a constant epigenetic tendency to specialize and form a precocial individual is being counteracted by a juvenilization process that recreates the altricial individual. The precocial state is more specialized and hence vulnerable to environmental change, whereas the altricial state allows a retreat to the more generalized and flexible juvenile lifestyle. Precocial (normal, specialized) and altricial (dwarf, generalist) char thus correspond to different life history states, which, with sufficient environmental stability, could eventually achieve reproductive isolation and sympatric speciation.

Balon proposes that the mechanism for the formation of these altricial ⇌ precocial pairs is saltatory ontogeny. Saltatory ontogeny requires that structures align their rates of development to become complete simultaneously and that a new function is initiated by a rapid transition from one stabilized state to another. In this article a simple model is presented to link this concept to different life history options for utilization of scarce resources. The example of a bimodal, landlocked char population is used as the simplest system for which the argument can be demonstrated, with the dwarf and normal char corresponding to stabilized states at different trophic levels.

Methods

Borup Fjord Arctic Char

The model is designed to illustrate the situation in which an individual Arctic char can occupy a succession of different habitats during the course of its ontogeny. This idea is not new but perhaps has been best expressed for Arctic char from Takvatn, Norway, where good evidence is presented for a relict anadromous life cycle (Klemetsen et al. 1989). Arctic char at different stages and sizes are associated with different habitats (Figure 1).

The same general pattern was clearly visible in the unexploited Arctic char populations studied at Borup Fjord, Ellesmere Island, 81°N, in 1988 and 1991 (Parker and Johnson 1991). The smaller, dwarf fish, occupying the more marginal habitats, retained their parr markings; whereas the larger, normal fish had the general characteristics of smolts, being silvery and without parr marks. In

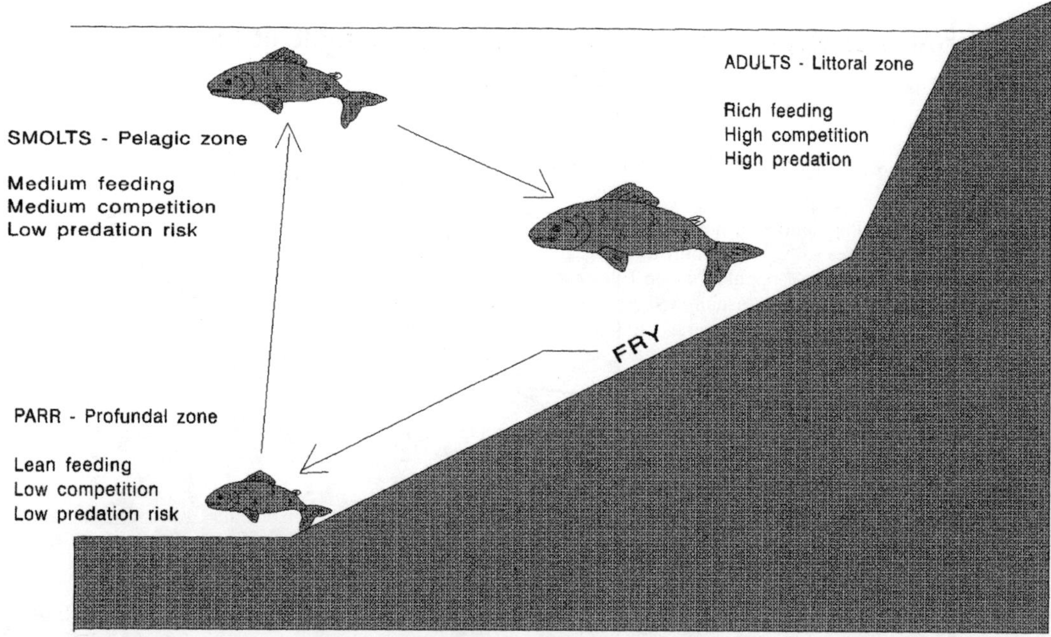

FIGURE 1.—Schematic model of the life cycle related habitat shifts by Arctic char *Salvelinus alpinus* in Takvatn under high population density (adapted from Klemetsen et al. 1989).

their juvenile stages, those fish destined to become normals were externally indistinguishable from mature dwarfs, and both types occupied the same habitats.

Additional important features of these unexploited populations at the northernmost extreme of land were that reproductive individuals were caught within all size modes sampled and that lifetime energetic investment in reproduction was extremely low. This low level of surplus energy is entirely consistent with the findings of Venne and Magnan (1989), who examined the variation of the main life history parameters of Arctic char along a north–south gradient. Growth rate and relative fecundity decreased with increasing latitude; but longevity, age at maturity, and relative egg diameter increased. The Borup Fjord data (Parker and Johnson 1991) fit these trends well, with the exceptions that the population age-classes never reached the stage where 50% spawned in any given year and relative egg diameter was lower than that found farther south. Although it is possible that the readings for relative egg size are affected by the fact that the measurements were taken early in the growing season, both observations are nevertheless consistent with a picture of decreasing surplus energy with latitude.

The relevance of the existence of mature fish within each size mode is that an opportunity for coexistent, alternative life history strategies results. If, as is probable, the minimal available surplus energy is a sufficiently severe energetic constraint to cause an individual to be able to spawn only once in a lifetime, then an individual spawning in one mode is precluded from spawning in another. Alternative life history strategies are therefore effectively defined by the mode in which an individual spawns.

The Model

As explained above, the model makes the important simplifying assumption that an individual only spawns once in a lifetime, but it should be realized that even if this were not the case, it would not preclude the existence of alternative life history strategies based on similar mechanisms. The strategies proposed here should therefore be regarded as extremes of a continuum of possible alternative life history strategies.

The second main assumption is that an individual is capable of altering its growth rate and time of reproduction in order to optimize its overall lifetime fitness while subject to physiological and external constraints such as population density, avail-

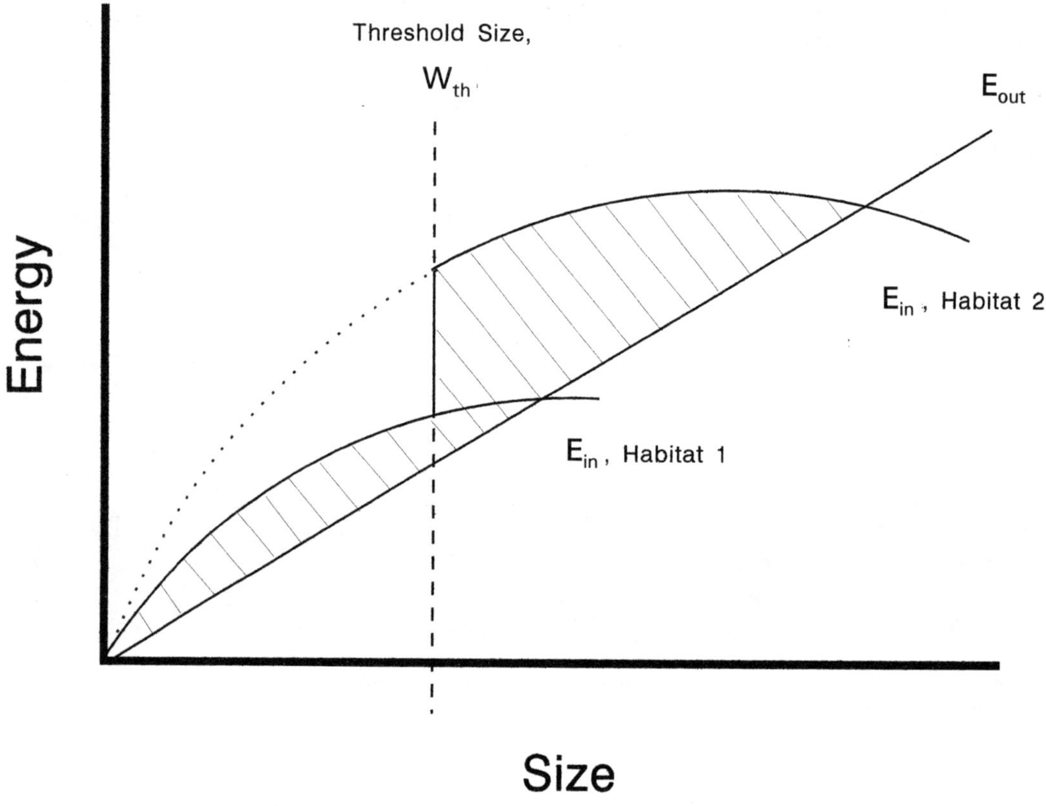

FIGURE 2.—Energy available (E_{in}) and energy costs (E_{out}) versus size for dwarf (1) and normal (2) habitats. The richer feeding of habitat 2 is only available once an individual reaches threshold size (W_{th}).

ability of food, and predation pressure. It is important to realize that no attempt is made to predict the mechanisms for this ability; the model merely analyzes the effect that this might have on observable characteristics of the population.

A third assumption of the model is that habitats are occupied by individuals of given sizes. Thus, a small fish is precluded from the richer habitats by a dominance hierarchy or increased risk of predation or both (Werner et al. 1983). Thus, for a population consisting of dwarf and normal subpopulations that simultaneously occupy two separate habitats, an individual must obtain a threshold size, W_{th}, (Figure 2) in order to exploit the richer, normal habitat.

The fourth major feature of the model is that it defines a habitat in epigenetic terms via the constraints that it imposes on the main life history parameters of the animal living within it. In the natural systems, each morph is usually defined in terms of location or diet. Examples include the profundal, pelagic, and littoral habitats defined for Arctic char from Takvatn by Klemetsen et al. (1989); or the piscivorous, benthic, and planktonic habitats defined for Thingvallavatn fish by Sandlund et al. (1989). For a mathematical model, however, a verbal description of habitat is inadequate and a numerical description is required.

Inevitably, a mathematical definition of habitat lacks the descriptiveness of a verbal one, and in this case it is defined somewhat abstractly in terms of tradeoffs between growth, reproduction, and survival. A poor or lean habitat will consist of a severe set of tradeoffs, whereas a good or rich habitat has a relatively benign set of tradeoffs. Figure 2 illustrates the relative amounts of energy available to an individual within two different habitats. Habitat 1, corresponding to the dwarfs, has little energy available, whereas habitat 2, corresponding to the normals, has more. However, an individual is not able to benefit from the increased energy available in habitat 2 until it reaches the threshold size, W_{th}.

The first fundamental tradeoff is modeled by von Bertalanffy growth curves, where large size and access to a greater share of the resources is being traded against higher maintenance costs. Von Bertalanffy growth curves can be regarded as the result of a balance between resources obtained, E_{in}, and resources used to meet the "cost of living," E_{out}. The balance ($E_{in} - E_{out}$) is the amount of resources available for growth and reproduction. Von Bertalanffy (1960) argued on physiological grounds that catabolism, the breakdown of tissues, proceeded at a rate proportional to body weight, W, whereas the rate of anabolism, synthesis of body tissues, was assumed proportional to body surface area. Assuming that all available energy is then applied to somatic growth, this leads to the equation

$$\frac{dW}{dt} = k_2W^b - k_1W^a \quad (1)$$

$b = 2/3$, $a = 1$; t = time; k_1 and k_2 = constants.

If integrated, the sigmoid growth curve results. A more accurate model might use other values for a and b, measured from experimental data (Reiss 1989), but because the aim of this model is to be conceptual rather than quantitative, arithmetic convenience is a greater virtue than accuracy.

Other relationships between growth, reproduction, and survival are also apparent and form an essential part of the model. Details follow, and the reader is referred to Parker and Johnson (1991) for the data on which they are based.

Growth versus reproduction.—The number of viable offspring is related to body size at spawning. For an all-female population, the number of viable offspring at spawning is approximately proportional to $(W)^{1.2}$.

Mortality versus reproduction.—As discussed, it is assumed that the char spawn only once. This is a fundamental constraint on an individual's life history, because once an individual has spawned, it ceases to contribute to the population in fitness terms because it has no further offspring.

Growth rate versus mortality.—The major difficulty in most fish studies is the lack of information relating to mortality, and other than an ability to discern basic trends, the Borup Fjord study is unfortunately no different. Over and above the normal problems of sampling bias within any given habitat, determining mortality rates is almost intractably complicated for multimodal populations. Assuming that the model of ontogenetic succession as presented here is accepted, there are two predominant reasons for this intractability.

First, individuals within different habitats behave differently and thus have different susceptibilities to fishing methods. Pelagic fish, for example, generally exist at low density and are thus difficult to catch, whereas piscivorous fish live in relatively localized areas and swim at relatively high speeds, thus making them extremely vulnerable to most fishing methods. Second, when an age-class recruits from one habitat to the next, and given the first reason above, it is impossible to determine what proportion of the age-class has recruited to the next habitat and what proportion has suffered genuine mortality.

Two age-versus-size plots from Parker and Johnson (1991) are illustrated in Figure 3. Although these plots are typical of many fish populations, they are of completely unexploited populations. The largest fish are not generally the oldest, indicating that some fish grow faster than others.

For the reasons above, accurate quantitative data on the real mortality rates are not available, so the model has to make do with an educated guess followed by an analysis of the robustness of the general result to variations in the expressions used. The example chosen is an adaptation from Sibly and Calow (1986) and uses

$$Z = 0.1 + \frac{u^2}{5}; \quad (2)$$

Z = instantaneous mortality and $u = (1/W)\ dW/dt$ as defined in equation 1.

The relationship between growth rate and mortality is modeled by the term u, with the $1/W$ term indicating reduced mortality at larger size.

The survival at time t is defined as

$$S(t) = \exp[-\textstyle\int Z(t) \cdot dt]. \quad (3)$$

The model examines the two life history strategies—reproductive and growth—available to the small, or dwarf, individual (Figure 4). The individual adopting the reproductive strategy spawns as a dwarf. In this case, the individual's overall lifetime fitness is optimized by altering the age and size at maturity, subject to the constraints of its habitat, until the maximum number of viable offspring are obtained. The individual adopting the growth strategy as a dwarf spawns subsequently as a normal fish. In this case, the first part of its life history strategy is to reach the threshold size, W_{th}, of the normal habitat with minimum mortality. The second phase of its life history strategy is to optimize age and size at maturity within the normal habitat.

Standard numerical methods were used to optimize the value of k_2 and hence growth rate, which maximized the individual's overall lifetime fitness in

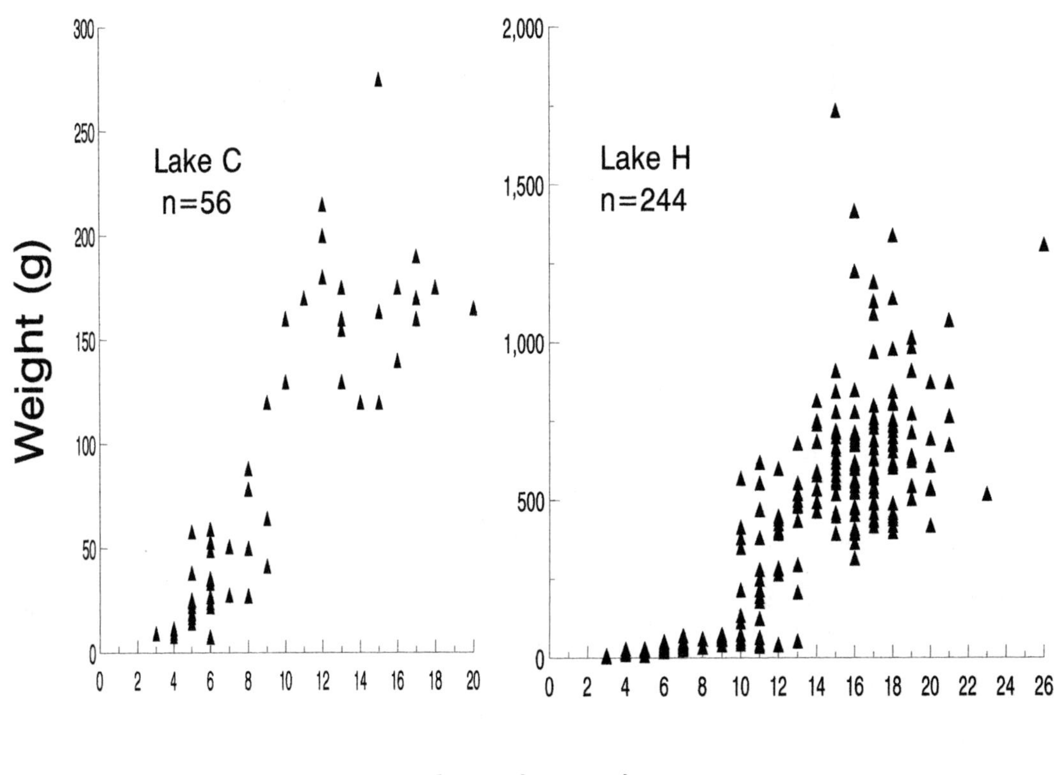

FIGURE 3.—Age versus weight for populations of Borup Fjord Arctic char *Salvelinus alpinus* in 1988 (from Parker and Johnson 1991).

accordance with the two strategies identified. The computed examples used in this article assume that the von Bertalanffy parameter, $k_1 = 1$, which defines the cost of living of a phenotype, is constant for the hypothetical phenotype examined. The parameter k_2 is allowed to vary, but only subject to the mortality penalties defined above.

Results

Reproductive Strategy

The combined effect of increasing size and decreasing survival leads to a local fitness maximum for any value of k_2 (Figure 5). In addition, there is an optimum k_2 that pertains to the maximum possible fitness that can be obtained if any value of k_2 is physiologically possible. In this case, the optimal value of k_2 is 3.28 (Figure 6).

Growth Strategy

The optimum value of k_2 for the growth strategy obviously depends on the value of the threshold size, W_{th}, under consideration. Consider size–survival curves for two different values of k_2 (Figure 7). The crossover point corresponds to where one k_2 becomes less risky than the other and shows that a low k_2 will prevent a slow-growing fish from ever reaching threshold size. The values of the optimal growth-rate parameter, k_2^*, required to achieve various threshold sizes, W_{th}, with minimum mortality, S_{max}^*, are in Table 1. Any threshold size greater than the optimal size at maturity required by the reproductive strategy requires a higher k_2^* and hence a higher growth rate.

Comparison with the Natural System

It is predicted from the above that overall lifetime fitness is maximized for the growth strategy at a higher growth rate than for the reproductive strategy during the phase when the individual occupies the dwarf habitat, which agrees with a number of observations. For example, Snorrason et al. (1989) concluded that piscivorous-char from Thingvallavatn, Iceland, were derived from juveniles that

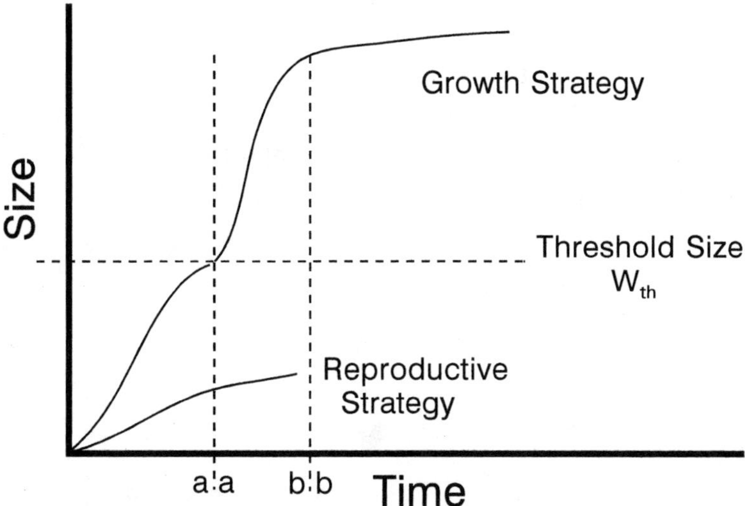

FIGURE 4.—Two life history options available to the dwarf individual. The dwarf adopting the reproductive strategy optimizes growth and the timing of reproduction within habitat 1 in order to maximize its number of offspring. The growth strategy is associated with achieving recruitment to habitat 2 with minimum mortality and, subsequently, optimizing growth and timing of reproduction within habitat 2. The growth interval aa–bb corresponds to a period of saltatory ontogeny (Balon 1984a, 1984b) as the individual shifts from one habitat to the next.

were slightly, but significantly, larger on average than planktonic-char of the same age. Similarly, visual examination of the growth trajectories in Figure 3 may indicate that in Lake C only the larger fish at ages 7–8 are likely to recruit to the normal subpopulation. Lake H data are somewhat confused by the probable existence of a missing morph, probably planktonic, which was not successfully sampled. (The evidence and reasons for this are not relevant to this article.) Finally, Parker and Johnson (1991) noted that for the four lakes studied at Borup Fjord, a high proportion of normal char correlated well with high average growth rate during the first few years of development.

Model Sensitivity

The model is sensitive to the exact nature of the growth–survival tradeoffs. Nevertheless, as long as the individual is restricted to one habitat and growth cannot be enhanced without detriment to survival, the result that there *is* a growth rate that optimizes fitness under the reproductive strategy remains. Furthermore, as long as the threshold recruitment size, W_{th}, is larger than the optimum size at maturity predicted for the reproductive strategy, a higher growth rate will be required for an individual to adopt the growth strategy.

Discussion

The main prediction of the model, allowing for its assumptions, is that an individual capable of adjusting its growth rate and timing of reproduction is capable of successfully adopting alternative life history strategies. In the case of dwarf char in a bimodal population, optimizing overall lifetime fitnesses for two separate strategies results in two optimal growth rates.

Unfortunately, the model presented here does not allow quantitative predictions to be made because of the lack of sufficiently accurate data. Instead, the model is intended as a vehicle to express a concept in qualitative terms. Without accurate data for input into the model, any predictions it makes are likely to be meaningless. Because quantitative prediction is not possible, simplicity and clarity of expression are greater virtues than a doomed attempt at accuracy.

The qualitative nature of the model and also its sensitivity to the exact nature of tradeoffs between life history parameters is mirrored in nature by examples of sexual dimorphism and sex ratio differences frequently observed in Arctic char and other salmonids (Johnson 1980; Jonsson and Hindar 1982; Parker and Johnson 1991). In the Borup Fjord study (Parker and Johnson 1991), males

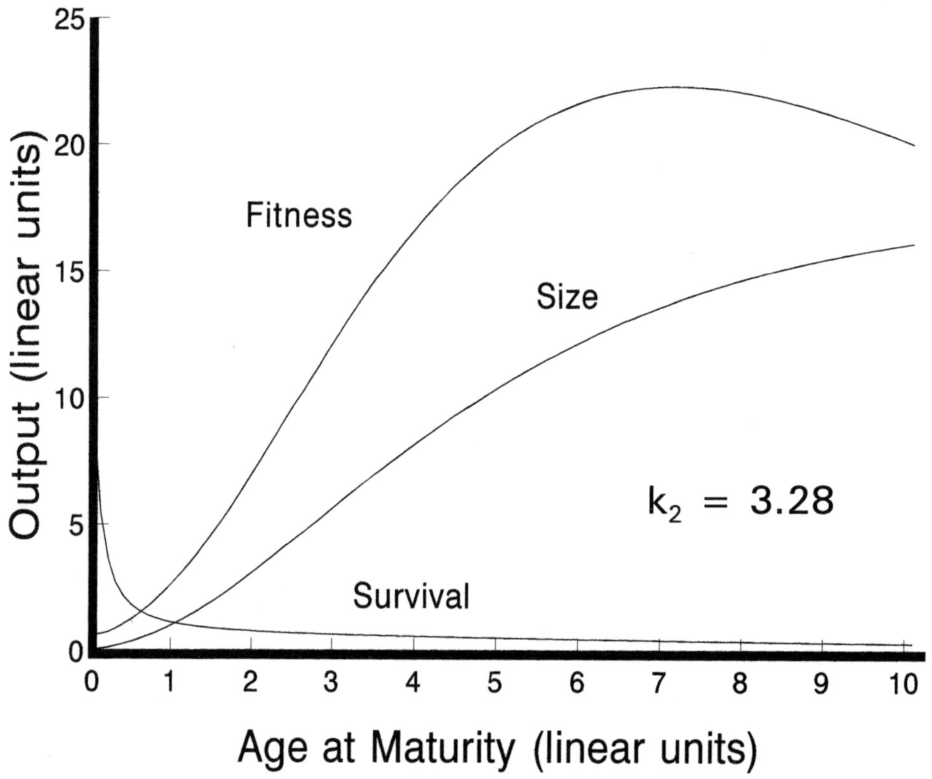

FIGURE 5.—Typical behavior of survival, fitness, and size with age at maturity for a dwarf individual adopting a reproductive strategy. The model calculates the fitness achieved for all possible ages at maturity. The optimal age at maturity is associated with the maximum calculated fitness value. To achieve all graphs on the same axes, fitness and survival have been multiplied by a factor of 10 and size has been divided by 2. Within the constraints of the model, the absolute values have no significance. Relative values are sufficient for determining the optimizations sought in the model.

spawned more frequently than females and gonadosomatic index decreased with size for males, but increased for females. However, the modal size of dwarf males was less than that of females, with the situation reversed for normal char. It is probable that these observations are related and that the slightly different reproductive strategies adopted by opposite sexes sharing the same environment might be leading to different optimal solutions to life history tradeoffs.

The model life history tradeoffs are based on an amalgam of ideas presented elsewhere. Sibly and Calow (1986) argue that knowledge of phenotypic physiology has much to offer in understanding the evolution of adaptations. In particular, they are concerned with physiological adaptations of resource acquisition and use because, they argue, this is not only the basis of the way that organisms function physiologically but also the basis of their form (allocation of resources between different structures) and behavior (allocation of resources between different activities). They also state that any one physiological process has to operate within the context of others because the resources available must be shared. This leads to a requirement for the organism to optimize the use of its limited resources between conflicting demands, and tradeoffs between physiological processes must occur. At the simplest level, resources can be used to invest in the major life history parameters of growth, reproduction, or survival (reducing mortality risk); but resources used for one cannot be used for the others.

Another important contribution to the conceptual argument is the term "ontogenetic niche" (Werner and Gilliam 1984), which refers to the patterns of resource use that develop as an organism increases in size from hatching to its maximum.

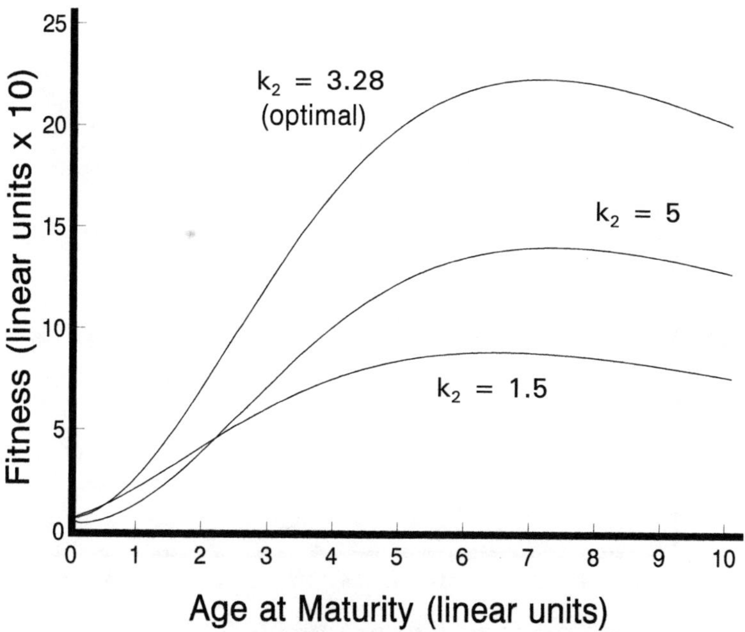

FIGURE 6.—Comparison of relative fitness for the reproductive strategist given different levels of investment in growth. One value of k_2 (= 3.28) achieves a greater optimum value of fitness than all other values of k_2.

Fundamentally, each semidiscrete habitat associated with any given Arctic char morph (Figure 1) can be viewed as a separate ontogenetic niche.

The conceptual framework of the model combines the above two ideas and proposes that the interaction between an organism and its habitat (or niche) can be defined in terms of tradeoffs in the three life history parameters of growth, reproduction, and mortality. If it is then accepted that an individual is capable of adopting a succession of separate niches during its ontogeny, each new niche can be defined by a new set of tradeoffs among the life history parameters. The multimodal Arctic char population can be viewed as a special case of this ontogenetic succession, where mature individuals can exist within each niche, allowing alternative life history strategies to develop.

Although this article looks at the effects of an organism's ability to control its growth rate and age and size at maturity, it makes no predictions as to the mechanism that is allowing this to occur. Field observation does, however, offer some clues. The reproductively mature dwarf, for example, looks like a juvenile salmonid parr and has an entirely different coloration and jaw shape from the normal. A probable process has previously been recognized as the phenomenon of heterochrony.

Heterochrony is usually defined as a process of phyletic change in ontogeny where the timing of onset or rate of development of a feature is accelerated or retarded relative to an ancestor's ontogeny (de Beer 1958; Gould 1977). The term has, however, also been applied in other contexts. Bruton (1989) uses it on the level of whole-organism ontogeny to describe changes in the timing of thresholds in life history. These changes allow the organism to match the appropriate life history style with the appropriate environment for the right length of time to take advantage of favorable conditions for that particular life interval or to await favorable conditions in the next life interval. Similarly, Noakes et al. (1989) discuss heterochronic

TABLE 1.—Optimal growth-rate parameter, k_2^*, and optimal survival parameter, S_{max}^*, for an individual Arctic char *Salvelinus alpinus* growing to a given threshold size, W_{th}; k_1 is the von Bertalanffy parameter. Optimal values of k_2^* and S_{max}^* depend on the W_{th} that is to be reached before recruitment to habitat 2 occurs.

k_1	W_{th}	k_2^*	S_{max}^*
1	75	4.45	0.0103
1	50	3.95	0.0195
1	28	3.28	0.0425

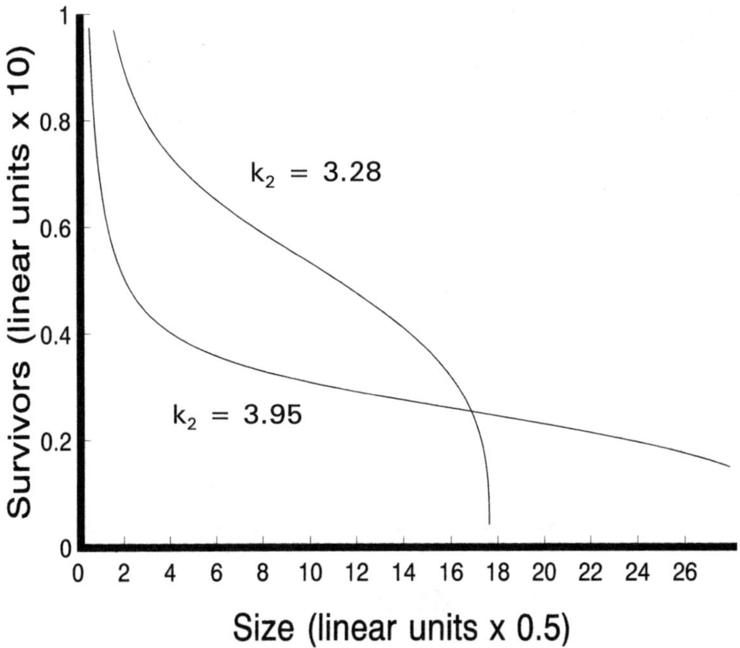

FIGURE 7.—Relative chances of survival to a given size assuming two different levels of investment in growth, k_2. In order to reach sizes above that of the crossover point, an individual will experience less risk of mortality if the higher growth rate, $k_2 = 3.95$, is adopted. For clarity, only two values of k_2 are shown. For any given size, however, there is a value of k_2 which optimizes the chance of surviving to that size.

shifts in ontogeny in relation to niche utilization as a proximate mechanism for the widespread phenomenon of phenotypic variation within salmonine species. It is these interpretations at a whole-organism level that are perhaps the most relevant to this article. In the model, heterochrony is manifest at the most fundamental level in that an individual's overall lifetime fitness is optimized by variation in timing of reproduction and somatic growth.

Balon (1984a, 1984b) also cites heterochronic processes as a mechanism. Balon's concept of saltatory ontogeny is also demonstrated in this model by the period of rapid growth that results while the dwarf char is recruiting to the normal subpopulation. (This period of fast growth is shown as the interval aa–bb in Figure 4.)

The ecological implications of heterochrony as described by such terms as saltatory ontogeny and altricial $\rightleftharpoons$ precocial dynamic states and its manifestation in multimodal Arctic char populations were originally explored by Parker and Johnson (1991). Verbally, the importance of multimodal populations is that Arctic char are able to exploit a wide range of potential habitats without being dependent on any one of them. In the event of a population crash within one habitat, recruits and offspring from the other habitats will be able to re-exploit the failed habitat when it recovers. A significantly more-stable population results, which must be important in the notoriously variable arctic climate.

Acknowledgments

I thank K. W. Hankinson, Royal Air Force, of the Joint Services Expeditions to Ellesmere Island, 1988 and 1991, for his support as expedition leader; J. C. Ellis-Evans of British Antarctic Survey, Director of Studies, for his unstinting support; and R. M. Nisbet, University of California at Santa Barbara, for his encouragement of a somewhat individual line of research and for his comments on a first draft of the manuscript. Financial assistance was provided by the British Ecological Society in the form of a Small Ecological Project Grant and by the Institution of Electrical Engineers in the form of a traveling scholarship.

References

Balon, E. K., editor. 1980. Charrs: salmonid fishes of the genus *Salvelinus*. Dr. W. Junk, The Hague, Netherlands.

Balon, E. K. 1984a. Life histories of Arctic charrs: an epigenetic explanation of their invading ability and evolution. Pages 109–141 *in* L. Johnson and B. Burns, editors. Biology of the Arctic charr. Proceedings of the international symposium on Arctic charr. University of Manitoba Press, Winnipeg.

Balon, E. K. 1984b. Patterns in the evolution of reproductive styles in fishes. Pages 35–53 *in* G. W. Potts and R. J. Wootton, editors. Fish reproduction: strategies and tactics. Academic Press, London.

Bruton, M. N. 1989. The ecological significance of alternative life history styles. Pages 503–553 *in* M. N. Bruton, editor. Alternative life-history styles of animals. Kluwer Academic Publishers, Dordrecht, Netherlands.

de Beer, G. R. 1958. Embryos and ancestors, 3rd edition. Clarendon Press, Oxford.

Gould, S. J. 1977. Ontogeny and phylogeny. Belknap Press, Cambridge, Massachusetts.

Johnson, L. 1980. The Arctic charr, *Salvelinus alpinus*. Pages 15–98 *in* E. K. Balon, editor. Charrs: salmonid fishes of the genus *Salvelinus*. Dr. W. Junk, The Hague, Netherlands.

Johnson, L., and B. Burns, editors. 1984. Biology of the Arctic charr. Proceedings of the international symposium on Arctic charr. University of Manitoba Press, Winnipeg.

Jonsson, B., and K. Hindar. 1982. Reproductive strategy of dwarf and normal Arctic charr (*Salvelinus alpinus*) from Vangsvatnet Lake, western Norway. Canadian Journal of Fisheries and Aquatic Sciences 39:1404–1413.

Kawanabe, H., F. Yamazaki, and D. L. G. Noakes, editors. 1989. Biology of charrs and masu salmon. Physiology and ecology Japan, Special Volume 1.

Klemetsen, A., P.-A. Amundsen, H. Muledal, S. Rubach, and J. I. Solbakken. 1989. Habitat shifts in a dense resident Arctic charr, *Salvelinus alpinus*, population. Pages 187–200 *in* Kawanabe et al. (1989).

Klemetsen, A. and P. Grotnes. 1980. Coexistence and immigration of two sympatric Arctic charr. Pages 757–763 *in* E. K. Balon, editor. Charrs: salmonid fishes of the genus *Salvelinus*. Dr. W. Junk, The Hague, Netherlands.

Noakes, D. L. G., S. Skulason, and S. S. Snorrason. 1989. Alternative life-history styles in Salmonine fishes with emphasis on Arctic charr, *Salvelinus alpinus*. Pages 329–346 *in* M. N. Bruton, editor. Alternative life-history styles of animals. Kluwer Academic Publishers, Dordrecht, Netherlands.

Parker, H. H., and L. Johnson. 1991. Population structure, ecological segregation and reproduction in non-anadromous Arctic charr, *Salvelinus alpinus* (L), in four unexploited lakes in the Canadian high Arctic. Journal of Fish Biology 48:123–147.

Reiss, M. J. 1989. The allometry of growth and reproduction. Cambridge University Press, Cambridge, UK.

Riget, P. F., K. H. Nygaard, and B. Christensen. 1986. Population structure, ecological segregation and reproduction in a population of Arctic char (*Salvelinus alpinus*) from Lake Tasersuaq, Greenland. Canadian Journal of Fisheries and Aquatic Sciences 43:985–992.

Sandlund, O. T., and five coauthors. 1989. Reproductive investment in polymorphic Arctic charr of Thingvallavatn. Pages 383–392 *in* Kawanabe et al. (1989).

Sibly, R. M., and P. Calow. 1986. Physiological ecology of animals. Blackwell Scientific Publications, Oxford, UK.

Snorrason, S. S., and five coauthors. 1989. Shape polymorphism in Arctic charr, *Salvelinus alpinus*, in Thingvallavatn, Iceland. Pages 393–404 *in* Kawanabe et al. (1989).

Sparholt, H. 1985. The population, survival, growth, reproduction and food of Arctic charr, *Salvelinus alpinus* (L.), in four unexploited lakes in Greenland. Journal of Fish Biology 26:311–330.

Venne, H., and P. Magnan. 1989. Life history tactics in landlocked Arctic charr (*Salvelinus alpinus*): a working hypothesis. Pages 239–248 *in* Kawanabe et al. (1989).

von Bertalanffy, L. 1960. Principles and theory of growth. Pages 137–259 *in* W. W. Nowinski, editor. Fundamental aspects of normal and malignant growth. Elsevier, Amsterdam.

Walker, A. F., R. B. Greer, and A. S. Gardner. 1988. Two ecologically distinct forms of Arctic char *Salvelinus alpinus* (L.) in Loch Rannoch, Scotland. Biological Conservation 43:43–61.

Werner, E. E., and J. F. Gilliam. 1984. The ontogenetic niche and species interactions in size-structured populations. Annual Review of Ecological Systems 15: 393–425.

Werner, E. E., J. F. Gilliam, D. J. Hall, and G. G. Mittelbach. 1983. An experimental test of the effects of predation risk on habitat use in fish. Ecology 64:1540–1548.

American Fisheries Society Symposium 19:224–228, 1997

Identification of Arctic and Bering Ciscoes in the Colville River Delta, Beaufort Sea Coast, Alaska

JOHN W. BICKHAM

LGL Ecological Genetics, Inc.
1410 Cavitt Street, Bryan, Texas 77901, USA and
Texas A&M University, Department of Wildlife and Fisheries Sciences, College Station, Texas 77843, USA

JOHN C. PATTON

LGL Ecological Genetics, Inc.

SHAMONE MINZENMAYER

Texas A&M University

L. L. MOULTON

MJM Research
5460 N.E. Tolo Road, Bainbridge Island, Washington 98110, USA

BENNY J. GALLAWAY

LGL Ecological Genetics, Inc.

Abstract.—Arctic ciscoes *Coregonus autumnalis* and Bering ciscoes *C. laurettae* were obtained from the Colville River delta during fall 1990 to assess the degree of morphological and genetic differentiation of these two forms. Specimens initially were identified to species based on external features. Subsequently, gill raker counts were found to be distinctly bimodal, with Bering ciscoes having 21–24 and Arctic ciscoes having 24–30 gill rakers. Mitochondrial DNA (mtDNA) haplotypes were also different between the two species. All Bering ciscoes had the BB haplotype as defined by *Dde* I and *Hinf* I restriction patterns of the 16S and NADH dehydrogenase-1 (NDH-1) regions of the mtDNA. Arctic ciscoes had the AA, BA, or CA haplotype, except for two specimens that had the BB haplotype characteristic of Bering ciscoes. The two BB Arctic ciscoes had gill raker counts of 26 and 28, which are characteristic of Arctic ciscoes. Arctic and Bering ciscoes were also assayed electrophoretically for the glycerol-3-phosphate dehydrogenase (G3PDH-1) locus. Four alleles were identified, of which two showed strong frequency differences between the two species. One of the two Arctic ciscoes with the BB mtDNA haplotype had a G3PDH-1 genotype common to Arctic ciscoes and the other had a genotype found at low frequency in both species. It is concluded that Arctic and Bering ciscoes possess diagnostic morphological characters that allow for their accurate separation in the fishery. Moreover, only a low level of apparent introgression of mtDNA, from Bering ciscoes into Arctic ciscoes, was detected. Our results confirm that these taxa represent distinct species as currently recognized.

The Arctic cisco *Coregonus autumnalis* is an anadromous salmonid that spawns in the tributaries of the Mackenzie River in Canada (Gallaway et al. 1983). Coastal habitat includes the nearshore, low-saline, neritic waters of the Beaufort Sea from Victoria Island (Canadian Northwest Territories) to Point Barrow (Alaska). The two most important coastal feeding and overwintering areas are the deltas of the Colville River in Alaska and the Mackenzie River in the Northwest Territories. This species and the broad whitefish *C. nasus* are the two most economically important fishes along the Alaskan north slope. Native subsistence fishing and commercial fishing for Arctic ciscoes take place in the Colville River delta (Gallaway et al. 1983).

The most closely related taxon to the Arctic cisco ostensibly is the Bering cisco *C. laurettae*, which spawns primarily in the Yukon River system and whose coastal distribution overlaps that of the Arctic cisco from Point Barrow to the Colville River (Craig 1989). These taxa are superficially very similar and their specific status has been questioned (Dillinger 1989). Interpretations of the population trends in the Arctic cisco fishery at the Colville River is complicated by the presence of Bering ciscoes. The mechanisms of transport of Arctic and Bering ciscoes from their spawning areas to the Colville River delta are distinctly different. Arctic ciscoes are transported by longshore currents from the Mackenzie River as young of the year (Gallaway et al. 1983; Fechhelm and Fissel 1988), whereas Bering ciscoes migrate from the Yukon River to the

Colville River as juveniles or adults. Thus, the correct interpretation of the causes of population fluctuations in the fishery requires the ability to accurately distinguish between these two forms. The purpose of this study is to assess the degree of differentiation between Arctic and Bering ciscoes using morphological, biochemical, and molecular characteristics taken from field-identified specimens and to assess the level of genetic introgression and the taxonomic status of these species.

Methods

Specimens were collected from the Colville River delta using gill nets during fall 1990. Individuals were identified as Bering ciscoes (N = 20) or Arctic ciscoes (N = 49) based on the presence of small black dots with faint halos on the body, or white spots on the fins, or both; these are visible in Bering ciscoes and absent in Arctic ciscoes. Fish were frozen whole and shipped to the laboratory, where they were dissected to remove tissues (heart, liver, and muscle) for allozyme and DNA analyses. The first gill arch was removed from each fish and the gill rakers counted as described by Dillinger (1989). A small piece of the liver of each fish was homogenized for protein electrophoresis. The enzyme glycerol-3-phosphate dehydrogenase (G3PDH-1; Enzyme Commission number [EC] 1.1.1.8) was analyzed (for all fish except three Arctic ciscoes) as described by Bickham et al. (1989) and Morales et al. (1993). Mitochondrial DNA (mtDNA) was examined for all specimens using the polymerase chain reaction restriction-enzyme assay as described in Cronin et al. (1993). A region of approximately 2,000 base pairs (2 kb), including part of the 16S ribosomal gene and all of the NADH dehydrogenase-1 (NDH-1; EC 1.6.99.3) gene (and four transfer RNAs), was amplified using the primers LGL 381 and LGL 563 (primer sequences reported in Cronin et al. [1993]). The 2-kb amplification products were cut with two restriction enzymes, *Dde* I and *Hin*f I, whose recognition sequences are CTNAG and GANTC, respectively. Digestion products were separated by electrophoresis in 2% agarose gels, stained with ethidium bromide, and photographed with polaroid film on an ultraviolet light box.

Results

Gill raker counts ranged from 24–30 (mean = 27) for Arctic ciscoes and 21–24 (mean = 22) for Bering ciscoes (Figure 1). There was one Bering cisco and two Arctic ciscoes with gill raker counts of 24, the area of overlap in ranges of the two species.

Mitochondrial DNA haplotypes were distinctly different for Arctic and Bering ciscoes. Two restriction enzymes (*Dde* I and *Hin*f I) gave diagnostic fragment patterns for the two species. Three patterns (A, B, and C) were observed when Arctic cisco DNA was cut with *Dde* I, but only the B pattern was observed in Bering cisco DNA. Although *Dde* I produces approximately seven bands, only two bands were variable. In the B fragment pattern, the largest band was approximately 600 base pairs (bp) and the third largest was approximately 300 bp. In the A pattern, the largest band was 550 and the third largest band was 300 bp. In the C pattern, the largest band was 600 and the third largest was about 280 bp. The second largest band was not variable and any variation in the smaller bands was not resolved. Two patterns (A and B) were observed with *Hin*f I in Arctic cisco DNA but, only the B pattern was observed in Bering cisco DNA. Four bands were observed in the A pattern, with the largest being approximately 1,000 bp. In pattern B, the 1,000-bp band was missing and was replaced with bands of approximately 600 and 400 bp. Thus, all Bering ciscoes had BB haplotypes for these two enzymes. Arctic ciscoes had haplotypes AA (N = 35), BA (N = 6), CA (N = 6), or BB (N = 2).

Genotype and allele frequencies were computed for Arctic and Bering ciscoes for the G3PDH-1 locus and analyzed using BIOSYS-1. Allele frequencies for Arctic and Bering ciscoes, respectively, were A: 0.272, 0.275; B: 0.511, 0.225; C: 0.054, 0.400; and D: 0.163, 0.100. Mean heterozygosity (H) was 0.636 for Arctic ciscoes and 0.704 for Bering ciscoes. Chi-square tests for Hardy–Weinberg equilibrium (using Levene correction for small sample sizes) gave P less than 0.001 for Arctic ciscoes and P equals 0.257 for Bering ciscoes.

Discussion

The difference in gill raker counts in our samples is consistent with previous observations of lower but overlapping counts for Bering ciscoes compared with Arctic ciscoes (Dymond 1943; McPhail 1966; Dillinger 1989). This finding and the subtle, but consistent, difference in color pattern between the two indicate that these taxa are morphologically distinct, at least in their geographic area of overlap.

Previous genetic studies have also shown Arctic and Bering ciscoes to be distinct. Bickham et al. (1989) and Morales et al. (1993) found no fixed differences in allele frequencies between these taxa,

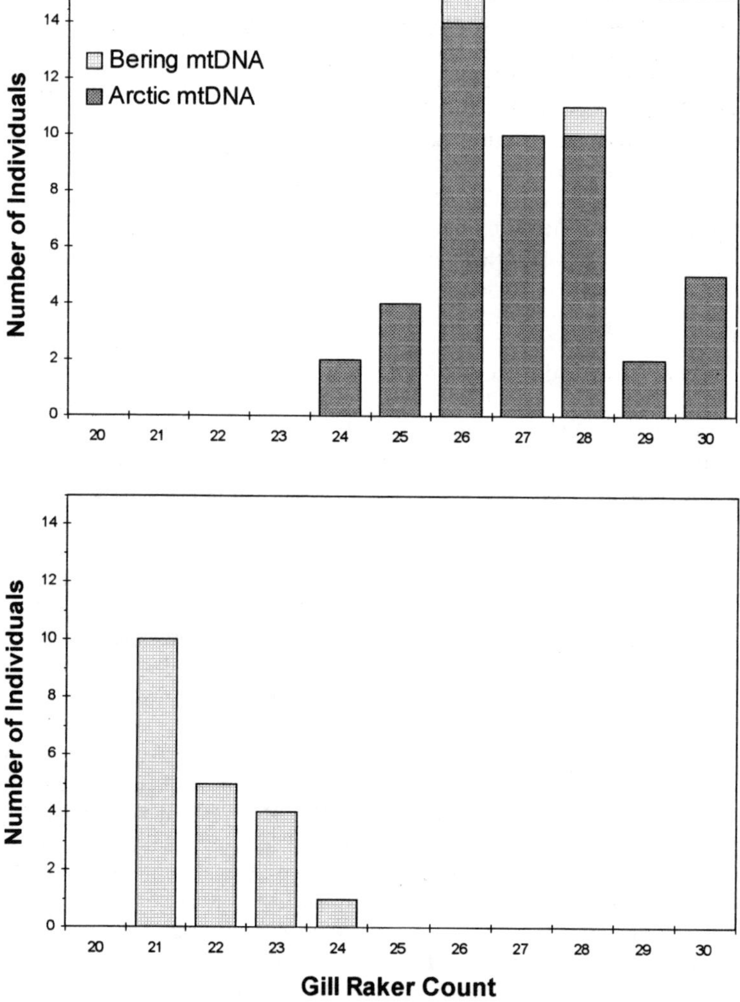

FIGURE 1.—Gill raker counts and mitochondrial DNA (mtDNA) haplotypes (see "Results" for haplotype details) for 69 fish taken from the Colville River fishery in fall 1990. Individuals were identified in the field as Arctic cisco *Coregonus autumnalis* (top panel) or Bering cisco *C. laurettae* (bottom panel) based on external coloration (see "Methods" for explanation).

but they noted strong frequency differences at two loci (mannose-6-phosphate isomerase [MPI; EC 5.3.1.8] and G3PDH-1). Our samples showed allele frequencies for G3PDH-1 similar to those found in the previous studies, which is another indication that fish were generally assigned correctly to their respective taxa. Lockwood and Bickham (1992) examined genome size variation in Arctic ciscoes taken from Beaufort Sea coastal aggregations and concluded that the observed pattern of variation (lower average genome sizes in the western region and higher genome sizes in the eastern) resulted from inclusion of higher numbers of Bering ciscoes in the western samples. Thus, Bering ciscoes were thought to have a lower genome size than Arctic ciscoes.

In our samples, all fish identified as Bering ciscoes had BB mtDNA haplotypes, whereas fish identified as Arctic ciscoes mostly had mtDNA haplotypes (AA, BA, or CA) not found in Bering ciscoes. However, two specimens identified as Arctic ciscoes had BB haplotypes. Thus, these two specimens might be animals of hybrid origin or were misidentified.

The two Arctic ciscoes with the Bering mtDNA haplotype (BB) had gill raker counts of 26 and 28 (Figure 1), well within the range of Arctic ciscoes and outside that of Bering ciscoes. Additionally, these fish had G3PDH-1 genotypes of BB and AB; the frequencies of BB in Arctic and Bering ciscoes was 0.4 and 0, respectively; the frequencies of AB genotypes was 0.045 and 0.1, respectively. Thus, one of the two fish of questionable identity had an Arctic cisco G3PDH-1 genotype and the other had a genotype found in low frequency in both species. Our results indicate that these fish were not Bering ciscoes misidentified as Arctic ciscoes. Rather, these fish were probably of hybrid origin possessing a Bering cisco female ancestor. These female ancestors probably were not the mothers of either of these fish; it is likely that they were more distant ancestors.

Mitochondrial DNA is a particularly sensitive genetic marker for revealing past hybridization events because it is maternally inherited and clonal (Brown 1983; Wilson et al. 1985; Avise 1986). After a few generations of backcrossing to one of the parental species, the nuclear genotypes of the descendants of a hybrid cross quickly come to resemble one of the parental species. However, when mtDNA is introduced by hybridization it is retained in any unbroken female lineage. Because of these features, mtDNA has been used successfully to track introgressive hybridization in a number of vertebrate species, such as mule deer *Odocoileus hemionus* and white-tailed deer *O. virginianus* (Carr et al. 1986; Cronin et al. 1988, 1991; Ballinger et al. 1992; Carr and Hughes 1993).

The frequency of fish with hybrid ancestry is apparently low among Arctic ciscoes. In our sample, only 2 of 49 Arctic ciscoes (4%) had Bering cisco haplotypes. It would be difficult for such hybridization events to occur because fish of these two species at the Colville River delta must go in opposite directions to reach the mouths of the spawning rivers (the Mackenzie and Yukon rivers). Nevertheless, such straying could occur and the absence of fixed differences in allozymes (Bickham et al. 1989; Morales et al. 1993) could result from a past history of low-level hybridization.

We therefore conclude that hybridization is rare between Arctic and Bering ciscoes and that these taxa are best considered to be different species as presently recognized. Furthermore, it is concluded that consistent morphological differences between these two species are present and that the fisheries biologists at the Colville River are able to identify these species with virtually 100% accuracy.

Acknowledgments

We thank the North Slope Borough for granting a collecting permit. BP Exploration (Alaska) Inc. provided most of the funding for this project. Partial funding was also provided by the Minerals Management Service and the U.S. Fish and Wildlife Service.

References

Avise, J. C. 1986. Mitochondrial DNA and the evolutionary genetics of higher animals. Philosophical Transactions of the Royal Society of London Series B Biological Sciences 312:325–342.

Ballinger, S. W., L. H. Blankenship, J. W. Bickham, and S. M. Carr. 1992. Allozyme and mitochondrial DNA analysis off a hybrid zone between white-tailed and mule deer (*Odocoileus*) in west Texas. Biochemical Genetics 30:1–11.

Bickham, J. W., S. M. Carr, B. G. Hanks, D. W. Burton, and B. J. Gallaway. 1989. Genetic analysis of population variation in the Arctic cisco using electrophoretic, flow cytometric, and mitochondrial DNA restriction analyses. Biological Papers of the University of Alaska 24:112–122.

Brown, W. M. 1983. Evolution of animal mitochondrial DNA. Pages 62–88 *in* M. Nei and R. K. Koehn, editors. Evolution of genes and proteins. Sinauer, Sunderland, Massachusetts.

Carr, S. M., S. W. Ballinger, J. N. Derr, L. H. Blankenship, and J. W. Bickham. 1986. Mitochondrial DNA analysis of hybridization between sympatric white-tailed and mule deer in west Texas. Proceedings of the National Academy of Sciences of the United States of America 83:9576–9580.

Carr, S. M., and G. A. Hughes. 1993. The direction of hybridization between species of North American deer (*Odocoileus*) as inferred from mitochondrial cytochrome B sequences. Journal of Mammalogy 74: 331–342.

Craig, P. C. 1989. An introduction to anadromous fishes in the Alaskan Arctic. Biological Papers of the University of Alaska 24:27–54.

Cronin, M. A., M. E. Nelson, and D. F. Pac. 1991. Spatial heterogeneity of mitochondrial DNA and allozymes among populations of white-tailed deer and mule deer. Journal of Heredity 82:118–127.

Cronin, M. A., W. J. Spearman, R. L. Wilmot, J. C. Patton, and J. W. Bickham. 1993. Mitochondrial DNA variation in chinook salmon (*Oncorhynchus tshawytscha*) and chum salmon (*O. keta*) detected by restriction enzyme analysis of polymerase chain reaction (PCR) products. Canadian Journal of Fisheries and Aquatic Sciences 50:708–715.

Cronin, M. A., E. R. Vyse, and D. G. Cameron. 1988. Genetic relationships between mule deer and white-tailed deer in Montana. Journal of Wildlife Management 52:320–328.

Dillinger, R. E. 1989. An analysis of the taxonomic status of the *Coregonus autumnalis* species complex in North America, and an investigation of the life histories of whitefishes and ciscoes (Pisces; Coregoninae) in North America and Eurasia. Doctoral dissertation. Memorial University of Newfoundland, St. John's, Newfoundland.

Dymond, J. R. 1943. The coregonine fishes of northwestern Canada. Transactions of the Royal Canadian Institute 24:171–232, Toronto.

Fechhelm, R. G., and D. B. Fissel. 1988. Wind-aided recruitment of Canadian Arctic cisco (*Coregonus autumnalis*) into Alaskan waters. Canadian Journal of Fisheries and Aquatic Sciences 45:906–910.

Gallaway, B. J., W. B. Griffiths, P. C. Craig, W. J. Gazey, and J. W. Helmericks. 1983. An assessment of the Colville River delta stock of Arctic cisco—migrants from Canada? Biological Papers of the University of Alaska 21:4–23.

Lockwood, S. F., and J. W. Bickham. 1992. Geographic trends in DNA content in Beaufort Sea coastal populations of Arctic cisco (*Coregonus autumnalis*) identified by flow cytometry. Transactions of the American Fisheries Society 121:13–20.

McPhail, J. D. 1966. The *Coregonus autumnalis* complex in Alaska and northwestern Canada. Journal of the Fisheries Research Board of Canada 23:141–148.

Morales, J. C., J. W. Bickham, J. N. Derr, and B. J. Gallaway. 1993. Genetic analysis of population structure in Arctic cisco (*Coregonus autumnalis*) from the Beaufort Sea. Copeia 1993:863–867.

Wilson, A. C., and 10 coauthors. 1985. Mitochondrial DNA and two perspectives on evolutionary genetics. Biological Journal of the Linnean Society 26:385–400.

American Fisheries Society Symposium 19:229–239, 1997

Characteristics of a Lightly Exploited Population of Arctic Grayling in the Sinuk River, Seward Peninsula, Alaska

ALFRED L. DECICCO, MARGARET F. MERRITT, AND ALLEN E. BINGHAM
Alaska Department of Fish and Game, Sport Fish Division
1300 College Road, Fairbanks, Alaska 99701, USA

Abstract.—A lightly exploited population of Arctic grayling *Thymallus arcticus* in the Sinuk River on Alaska's Seward Peninsula was studied annually during 1989–1991. Growth and population structure were compared to those of other populations of Arctic grayling residing in Seward Peninsula and interior Alaska streams. The population in the Sinuk River consists of high proportions of large, old fish. These Arctic grayling grow rapidly up to age 7 (about 440 mm in fork length), whereupon growth slows appreciably and mortality is negligible. Sinuk River Arctic grayling attained a theoretical maximum length of 469 mm (SE = 4). Faster growth was exhibited by Arctic grayling in the Sinuk River (Brody growth coefficient = 0.46, SE = 0.04) than in most other populations studied in rivers on the Seward Peninsula and in interior Alaska. Scale ages ranged to 15 years and otolith ages to 18 years. Abundance of Arctic grayling (>324 mm) was estimated at 1,464 fish (SE = 274) in 1989 and at 1,453 fish (SE = 296) in 1990 from mark–recapture experiments in a 40-km section of the 75-km river. Estimated survival from 1989 to 1990 was 1.0. Evidence suggests that the Arctic grayling population in the Sinuk River is static and experiences low recruitment. Characterizations of this lightly exploited population will help managers better understand the potential impacts of sport fishing on Arctic grayling population structures.

The Seward Peninsula–Norton Sound area of western Alaska has many streams that support Arctic grayling *Thymallus arcticus* sport fisheries. In 1982–1991, Arctic grayling constituted an average of 21% of the total reported freshwater sport fish harvest for the Seward Peninsula–Norton Sound area. Concerns regarding the stock status of Arctic grayling led to regulation changes in 1988 that reduced the daily bag limit of Arctic grayling on the Seward Peninsula to five per day, five in possession, with only one greater than 380 mm in total length. These concerns also prompted the Alaska Department of Fish and Game (ADFG) to initiate research studies aimed at defining sustainable yield for several Arctic grayling stocks in Seward Peninsula streams. Although fishing pressure can be substantial at road-accessible streams, there are also less-accessible Arctic grayling populations that sustain little or no exploitation. The Seward Peninsula is the only area in Alaska outside of Bristol Bay that regularly produces trophy-sized (≥1.36 kg) Arctic grayling. Of 119 Arctic grayling registered with the ADFG Trophy Fish Program between 1967 and 1991, 30 (25%) were from the Seward Peninsula, and of those, 15 came from the Sinuk River.

Management of exploited populations of Arctic grayling in Seward Peninsula streams is hampered by lack of information on the intrinsic characteristics of a typical unexploited population. Such information would help managers better understand observed population structures, possible stock–recruitment relationships, and ultimately the potential productivity of Arctic grayling populations. Based on the population composition and dynamics of unexploited populations, fishery managers may want to enact regulations that have the best chance of retaining these characteristics in populations not yet heavily exploited, or of restoring exploited populations.

Accordingly, a study was initiated by the ADFG on the uppermost 40 km of the Sinuk River, located approximately 40 km west of Nome (Figure 1). The angler tag-return rate from Arctic grayling tagged in this section of the Sinuk River from 1989 through 1991 (0.6%) was less than that for neighboring streams (1.0–2.1%). Characteristics of this lightly exploited population are assumed to more closely resemble those of an unexploited population than of one subject to more intense harvest.

Stock assessment studies initiated on this river in 1989 continued annually, with the objectives of estimating abundance, survival, age and length compositions, mean length at age, and growth. This article relates estimates of abundance, survival, composition, and growth of Arctic grayling in the Sinuk River to characteristics of an unexploited fishery and compares these estimates to exploited Arctic grayling fisheries in Seward Peninsula and interior Alaska streams. Several hypotheses regarding the mechanisms influencing the population structure and dynamics are examined.

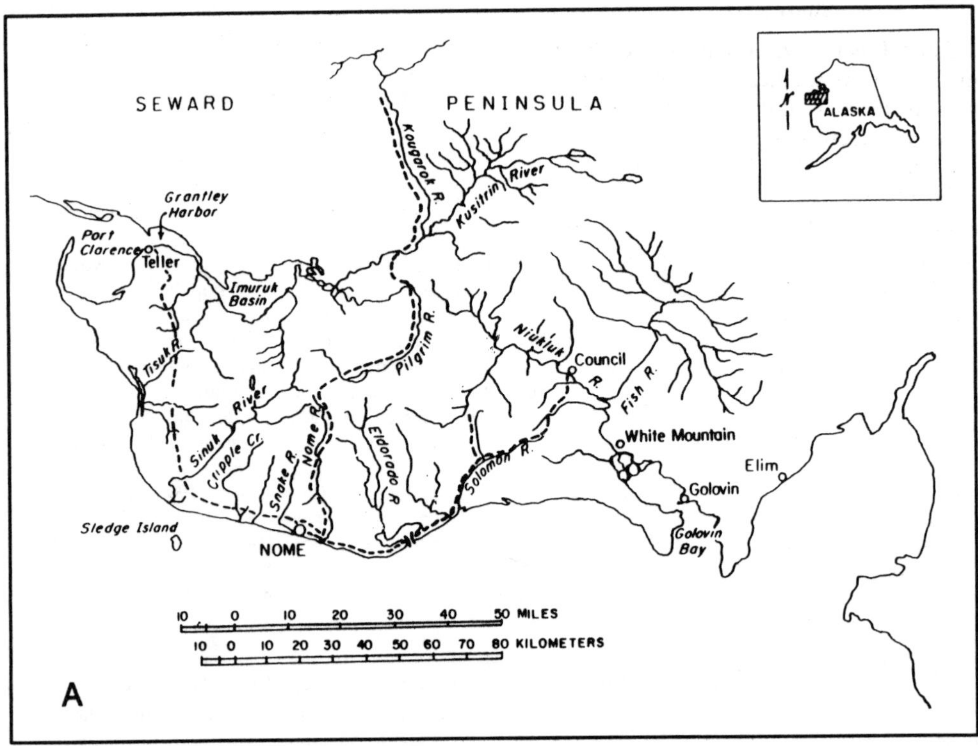

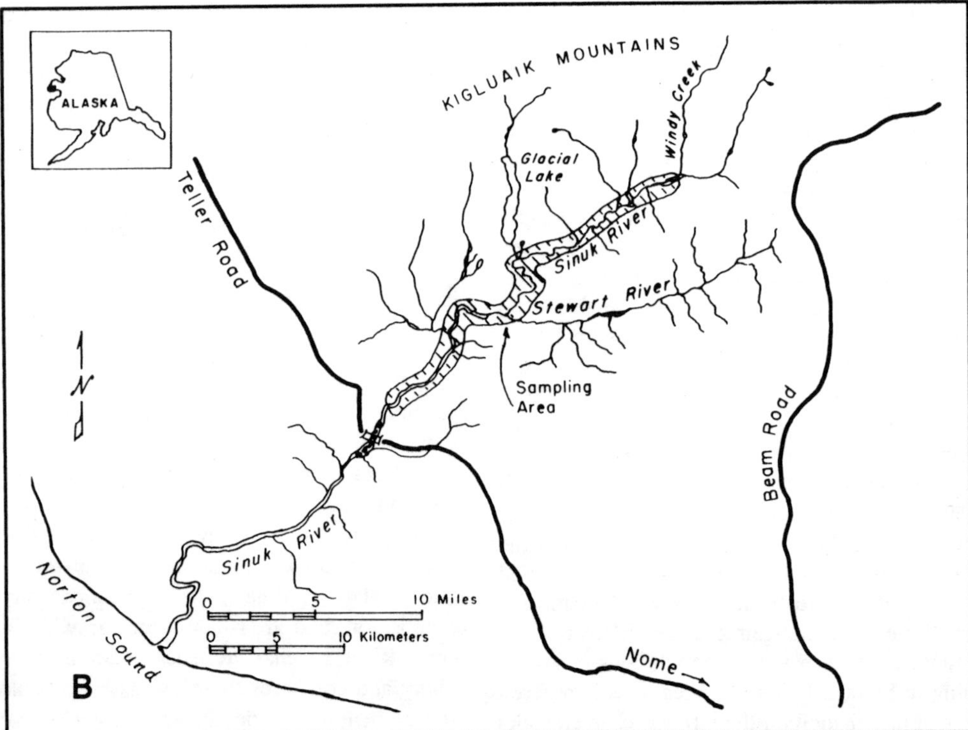

FIGURE 1.—The southern Seward Peninsula, Alaska, showing the Sinuk River and other river systems (**A**) and the 40-km section of the Sinuk River sampled during 1989–1991 (**B**).

Study Area

The Sinuk River is 75 km long and drains a 794-km^2 area of the southeast side of the Kigluaik Mountains on the Seward Peninsula (Figure 1). The river flows in a southwesterly direction and enters the Bering Sea at 64°35′N, 166°15′W. Major tributaries of the Sinuk River include the catchment of Glacial Lake and the Stewart River. Access, hence angling pressure, is limited to the vicinity of the bridge crossing at the Nome–Teller Road. The lower river and 2 km upstream from the bridge are navigable by jet boat during times of moderate to high water levels. Difficult access has resulted in virtually no angling pressure in the upper reaches.

The section of the Sinuk River from a point 2 km downstream from its confluence with Windy Creek to 2 km upstream from the Teller Road bridge (40 km) was studied in 1989–1991; the section between the Teller Road bridge and Norton Sound was sampled during 1988–1990. Flow is rapid in the upper reaches of the river, with large boulder fields where the river flows through glacial moraines.

Mean monthly air temperatures in Nome are below freezing in October through April and range up to a high of 15.5°C during July and August. Mean monthly precipitation is about 2.5 cm in October through April and ranges up to 10 cm in July and August. Daylight is continuous during late May through late July.

Methods

Sampling Gear and Techniques

On the Sinuk River, Arctic grayling were sampled from an inflatable raft using hook and line. Both spin and fly-fishing techniques were employed using a variety of artificial lures. Sampling during each year occurred over a 5–6 d period in early August.

Each Arctic grayling captured was measured to the nearest 1-mm fork length (FL). Fish longer than 150 mm were tagged with individually numbered Floy FD-67 internal anchor tags, inserted such that the "T" anchor locked between the base of adjacent dorsal fin rays. Each fish was also marked with a partial fin clip. Approximately 25% of a ventral fin or the entire adipose fin was removed with scissors. A different fin was partially clipped each year to provide a permanent mark and to assess tag loss.

To determine age, scales were taken from the left side of the fish, approximately midway between the dorsal fin and the lateral line down from the posterior insertion of the dorsal fin. Scales were cleaned and pressed in acetate. Ages were determined from scales by counting annuli from the acetate impressions using a microfiche reader. All scale impressions were read twice by the same reader. If readings were not the same, the scale was read a third time. When two readings agreed, this was taken to be the age of the fish. If the reader could not replicate the age of the fish in three readings, the age sample was discarded. Regenerated scales were not aged. Otoliths from fish that died during sampling were collected and stored dry in envelopes. They were immersed in a clearing agent (loess solution) and surface read with reflected light using a binocular microscope. Ages from otoliths were determined by counting the translucent rings, following procedures of Nordeng (1961). Ages for individual fish for which both scale and otolith readings existed were compared; however, the small number of otolith samples precluded direct statistical comparisons.

Parameter Estimation

Mark–recapture experiments were conducted to estimate the abundance and survival rate of Arctic grayling in 1989 and 1990. The parameter estimates apply to all Arctic grayling in the population that were greater than 324 mm. This minimum size limit represents the smallest fish from the 1989 mark that was recaptured during the 1990 sampling event and alleviates biases from unequal capture probabilities relating to size of fish and growth recruitment among years.

The 1989 mark event and the 1990 recapture event were used to calculate the abundance of Arctic grayling during 1989 using the modified Petersen estimator of Chapman (1951). The hypergeometric model variance estimator presented in Seber (1982, p. 60) was used to estimate the variance of the abundance estimate. The assumptions necessary for an unbiased estimate of population abundance (see Seber 1982) were addressed either by the methods used to conduct the experiments or by direct evaluation using a variety of hypothesis tests. In particular, the nonparametric test of Robson and Flick (1965) was used to test for significant growth recruitment between years. Additionally, two-sample Kolmogorov–Smirnov tests (Conover 1980) were used to test hypotheses regarding unequal capture or recapture probabilities relating to fish size. Losses from mortality were assumed to affect the marked and unmarked portions of the population similarly.

The 1989 mark event along with the mark–recapture events of 1990 and 1991 were used to estimate the 1990 abundance and the survival rate from 1989 to 1990. The modified Jolly–Seber (Jolly 1965; Seber 1965) model estimators of Buckland (1980) were used to calculate these estimates. The assumptions necessary for unbiased parameter estimates for this model (see Seber 1982) were addressed by study design or evaluation of the capture-history database. A three-sample Anderson–Darling test (Scholz and Stephens 1987) was used to evaluate capture probabilities related to size of fish. Additionally, the nonparametric test described by Pollock et al. (1990) and provided in the program JOLLYAGE was used to further evaluate the hypothesis of equal capture probability across size-classes.

Unfortunately, the effects of recruitment and mortality (which are accounted for in the Jolly–Seber model) upon the capture and recapture length distributions during each year cannot be separated from possible effects of unequal capture probabilities. As such, the tests described above were only used as general guidelines in evaluating the possible need to stratify the experiments according to fish length.

Composition of the Sinuk River Arctic grayling population was described by use of frequency histograms of the captured fish. The full range of fish captured was used in constructing the frequency histograms. No adjustment was made for possible differences in capture probabilities among length or age groups outside the range of fish used in the mark–recapture experiments.

Growth characteristics of Arctic grayling populations on the Seward Peninsula were estimated by fitting a von Bertalanffy-type growth model (Ricker 1975) to length-at-scale-age data using nonlinear least-squares regression analysis (Marquardt 1963). Fit of the data to von Bertalanffy models was examined using the coefficient of determination (R^2) following equation 1 in Kvalseth (1985). A separate model was developed for length-at-age data using otoliths taken from Arctic grayling in the Sinuk River. Ages estimated from scale and otolith samples from the Sinuk River were compared to evaluate potential aging error. The parameters of the von Bertalanffy model were compared between these two data sources, as well as with other Arctic grayling populations. All comparisons among the various models were made using Hotelling's T^2 multivariate statistical procedure by Bernard (1981) as modified by Cerrato (1990).

Results

Population Abundance and Survival Estimates

In 1989 during the mark event, 138 tagged Arctic grayling were released in the Sinuk River. During 1990, we examined 236 Arctic grayling, of which 22 were marked in 1989. In 1991, we captured 325 Arctic grayling; 40 carried 1990 marks and 31 carried 1989 marks.

The abundance of Arctic grayling greater than 324 mm in a 40-km section of the Sinuk River in 1989, estimated from data collected in 1989 and 1990, was 1,464 fish (SE = 274). This estimate is similar to the 1990 abundance estimate of 1,453 fish (SE = 296), estimated from data collected in 1989, 1990, and 1991. Survival between August 1989 and August 1990 was estimated at 1.0 for Arctic grayling greater than 324 mm.

Composition Estimates

An examination of the length distribution of Arctic grayling populations most recently sampled on the Seward Peninsula shows limited length ranges (Figure 2). Ninety percent of the Arctic grayling in the Sinuk River were greater than 350-mm FL. The largest fish sampled from the Sinuk River was 528-mm FL. From scale readings, higher proportions of Arctic grayling aged 10 years and older were found in the Sinuk River (20%) than in other rivers (which ranged from 0.5% to 4.5%) (Figure 3). The oldest Arctic grayling sampled from the Sinuk River was aged at 15 years using scales.

Growth

For Sinuk River Arctic grayling the theoretical mean maximum length (L_∞) was estimated at 469 mm (SE = 4); the estimated Brody growth coefficient (K) was 0.46 (SE = 0.04); and the theoretical age at length equal to zero (t_o) was estimated at 1.20 years (SE = 0.21). The asymptotic length was estimated very precisely, indicative of the appreciable numbers of old fish in the sample. Precision of the estimated theoretical age at length equal to zero was low, resulting from a lack of young fish in the sample. Although the model follows length-at-scale-age data (Figure 4), an appreciable degree of variation in length is not explained by the model (R^2 = 0.44, Table 1).

Based on ages determined from scales, growth was more rapid for Arctic grayling sampled from the Sinuk River than for Arctic grayling sampled from other Seward Peninsula streams, except for the Fish and Niukluk rivers (Table 1). Insufficient numbers of old Arctic grayling present in samples collected from the Kuzitrin, Nome, Pilgrim, and Snake rivers contributed to the lack of precision for the asymptotic length estimates for these populations. The two-by-two Hotelling's T^2 tests of each model to the model for the Sinuk River samples

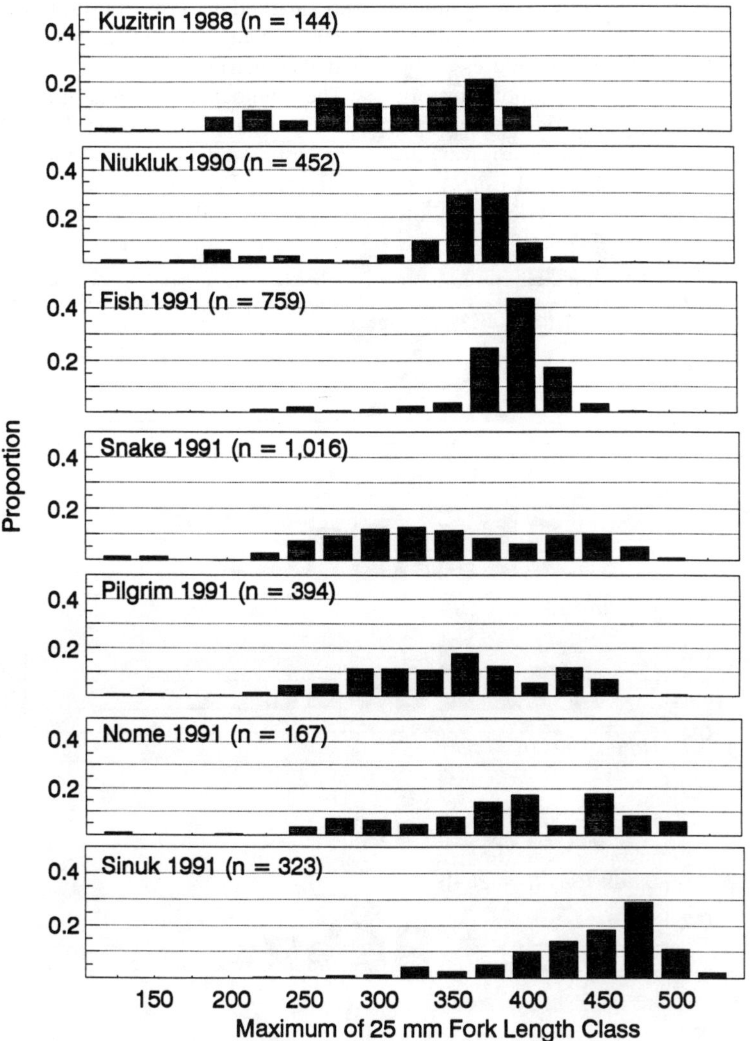

FIGURE 2.—Incremental length compositions of Arctic grayling *Thymallus arcticus* from seven Seward Peninsula rivers in the year most recently sampled.

indicated that all models were significantly different from the Sinuk River model (Table 2, Figure 5).

Length at age was examined for tagged Arctic grayling 1 year and 2 years after initial scale samples were obtained (Figure 6). Rate of growth slowed with increasing length; little increase in length occurred after Arctic grayling reach about 440 mm.

Aging Error

Age estimates from scales were lower than those from otoliths in Arctic grayling from the Sinuk River (Figure 7). Parameter estimates for the von Bertalanffy growth model were compared for ages derived from otoliths and scales. Parameter estimates from the model using scale ages were similar to those of the model from otolith ages (otolith model: $N = 21$, $t_o = -0.93$ [SE = 1.1], $L_\infty = 492$ mm [SE = 31], and $K = 0.18$ [SE = 0.051]; scale model: $N = 21$, $t_o = -0.16$ [SE = 1.0], $L_\infty = 504$ mm [SE = 52], and $K = 0.23$ [SE = 0.092]). The Hotelling's T^2 statistic comparing these two models indicated no significant difference ($T^2 = 5.55$; df = 3, 38; $P = 0.175$).

Discussion

The Sinuk River is the least-exploited area sampled by the ADFG on the Seward Peninsula. Arctic

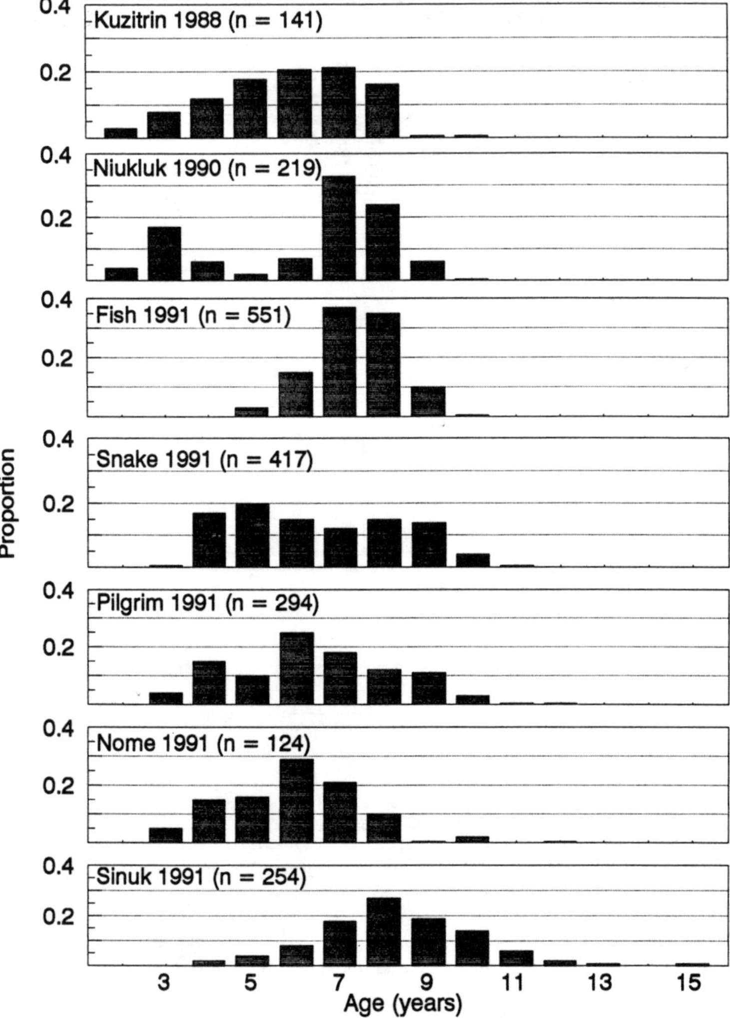

FIGURE 3.—Age compositions of Arctic grayling *Thymallus arcticus* from seven Seward Peninsula rivers in the year most recently sampled.

grayling attained large maximum size, lived long, and had low annual mortality. These characteristics may be typical of unexploited Arctic grayling populations on the Seward Peninsula. All other Arctic grayling populations in rivers sampled on the Seward Peninsula are exploited to a greater extent.

Abundance estimates apply only to the size range indicated and are thought to be unbiased, but scale ages are probably biased low. Otolith analysis yields considerably higher ages than scale analysis in Arctic grayling from northern latitudes (Tripp and McCart 1974; Craig and Poulin 1975; Sikstrom 1983; Skopets and Prokop'yev 1990; Merritt and Fleming 1991; Skopets 1991). Our data support this observation and call into question the use of scales as a descriptor of age structure in Arctic grayling populations with large proportions of old fish. Arctic grayling from the Sinuk River have been aged to 15 years using scales, suggesting that much older fish may be present. Our otolith sample is small because it consists only of fish incidentally killed during sampling.

We think our samples generally represented length ranges of fish present within the section of river sampled. Small Arctic grayling (150–300 mm) are commonly found in the main stem of interior Alaska rivers and are commonly caught in the Tanana River drainage (Clark et al. 1991) with hook-

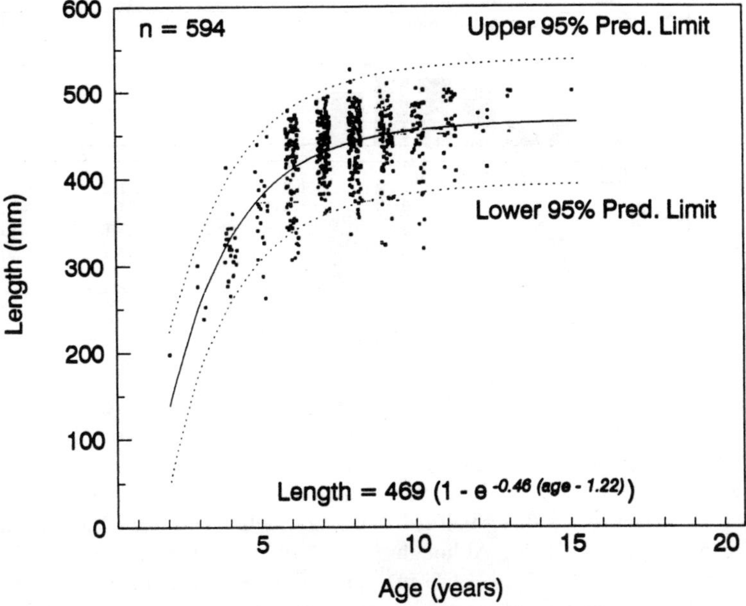

FIGURE 4.—Scatter plot of length-at-age data collected from Arctic grayling *Thymallus arcticus* sampled from the Sinuk River on the Seward Peninsula during 1989–1991, along with the fitted von Bertalanffy growth parameters (Pred. = predicted).

and-line gear, as used in this study. Small Arctic grayling were not common in rivers sampled on the Seward Peninsula; however, they were present in higher proportions in the Kuzitrin, Snake, Pilgrim, and Nome rivers than in the Sinuk River. Higher proportions of small fish contributed to the better fit of length-at-age data to growth models for these rivers. This is because small fish have not yet expe-

TABLE 1.—Estimated von Bertalanffy growth parameters for age-at-length data collected from various Seward Peninsula and interior Alaska populations of Arctic grayling *Thymallus arcticus*. Values of R^2 (coefficient of determination) were not calculated for von Bertalanffy models used in the interior region. The parameters are defined as follows: t_o = theoretical age at length equal to zero; L_∞ = theoretical mean maximum length; and K = the Brody growth coefficient.

River	Sampling years	Sample size	t_o	SE	L_∞	SE	K	SE	R^2
				Seward Peninsula					
Fish	1988, 1990, 1991	1,287	0.85	0.06	401	2	0.47	0.02	0.71
Kuzitrin	1988	174	−0.87	0.10	574	45	0.12	0.02	0.95
Niukluk	1988–1990	1,511	1.10	0.08	403	4	0.41	0.02	0.62
Nome	1988, 1991	180	0.31	0.24	541	23	0.23	0.03	0.85
Pilgrim	1988, 1990, 1991	720	−1.30	0.37	547	23	0.14	0.02	0.74
Snake	1991	865	−0.72	0.11	566	15	0.15	0.01	0.87
Sinuk	1989–1991	594	1.20	0.21	469	4	0.46	0.04	0.44
				Interior					
Chena[a]	1986–1988	4,301	−1.72	0.11	538	21	0.10	0.01	
Salcha[b]	1986–1988	1,198	−0.42	0.16	489	19	0.16	0.02	
Chatanika[b]	1986–1988	1,469	−1.01	0.20	375	11	0.19	0.02	
Fielding Lk.[c]	1986–1992	5,506	0.04	0.14	469	22	0.20	0.02	
Piledriver Sl.[d]	1991	2,156	−1.01	0.28	367	22	0.19	0.03	

[a]From Clark (1990).
[b]From Clark et al. (1991).
[c]From Clark (1993).
[d]From Fleming (1991).

TABLE 2.—Hotelling's T^2 test summaries comparing the fitted von Bertalanffy growth parameters from various Seward Peninsula populations of Arctic grayling *Thymallus arcticus* with the fitted parameters for the Sinuk River population, Seward Peninsula, Alaska.

River	Sampling years	Sample size	T^2	df	*P*-value
Fish	1988, 1990, 1991	1,287	1,487	3 and 1,877	<0.001
Kuzitrin	1988	174	36.00	3 and 764	<0.001
Niukluk	1988–1990	1,511	1,408	3 and 2,101	<0.001
Nome	1988, 1991	180	24.84	3 and 770	<0.001
Pilgrim	1988, 1990, 1991	720	68.95	3 and 1,310	<0.001
Snake	1991	865	90.09	3 and 1,455	<0.001
Sinuk	1989–1991	594			

rienced the variety of environmental factors that influence size, so their lengths are not as varied at age as are those of larger fish.

We have several hypotheses regarding the relative lack of small-sized Arctic grayling in samples from the Sinuk River. It is possible that small fish occupy tributaries of the Sinuk River. Although extensive effort was not made to find small fish in the Sinuk River drainage, sampling of its major tributary, the Stewart River, occurred in 1989 using hook and line and few fish were found (two pink salmon *Oncorhynchus gorbuscha* and two Arctic grayling were observed). Arctic grayling were not observed near the outlet of Glacial Lake nor in the upper 0.5 km of the stream draining the lake. Small Arctic grayling were not caught or observed in the Sinuk River downstream from the Teller Road bridge during sampling trips in 1988 and 1990.

Recruitment could be limited through density-independent events such as high river discharge during critical times of fry emergence and rearing. Holmes et al. (1986) and Clark (1992) found that flood events during June in the Chena River negatively impacted recruitment of Arctic grayling. It is likely that the Sinuk River experiences early-summer high-water events in some years that might result in similar impacts. Strong and weak cohorts would be expected to be present in the population if river discharge was limiting recruitment. This was not evident in the age structure of the population; however, errors in age determination from scales may be obscuring our ability to detect cohort strength. If river discharge was the main factor limiting recruitment, similar effects would be expected in nearby drainages such as the Snake and Nome rivers. These drainages contained larger proportions of young Arctic grayling than did the Sinuk River, suggesting that river discharge was not a significant factor in limiting recruitment during the years of our study.

Overwintering habitat is probably limited in Seward Peninsula streams, although probably not to the degree observed on Alaska's north slope (Craig and McCart 1974; Bendock 1981, 1982; Bendock and Burr 1984). Reduced winter habitat may force small Arctic grayling into areas inhabited by larger Arctic grayling and Dolly Varden *Salvelinus malma*, where they may be subject to increased risk of predation. In such an environment, it would be advantageous for young fish to grow rapidly, particularly in their first year of life, to avoid being eaten by larger fish. This hypothesis embraces the concept of genetic selection for rapidly growing prerecruits. Given some bias in ages determined from scales, length-at-age data and the growth model of Arctic grayling from the Sinuk River tend to support this hypothesis.

Johnson (1976) found that unexploited populations of lake trout *S. namaycush*, lake whitefish *Coregonus clupeaformis*, and Arctic char *S. alpinus* in lakes of the Northwest Territories, Canada, typically exhibited a bimodal length structure, with many age-classes being represented in each length mode. He suggested that this population structure indicated the long-term existence of a state close to equilibrium and theorized that the dominant portion of the population (the large-length mode) was exerting a suppressive force on the small-length mode. When larger fish were removed from the population, smaller recruits rapidly joined the dominant group. A similar mechanism may be operating in the Sinuk River Arctic grayling population and in other lightly exploited or unexploited Arctic grayling populations on the Seward Peninsula.

The Sinuk River Arctic grayling population is dominated by large fish, making up a unimodel length distribution with a mean fork length of about 450 mm. The small-length mode is probably represented only by the annual production of age-0 fish and hence not readily available to the sampling gear used in this study. Large fish may be limiting the survival of small fish, either by predation or by

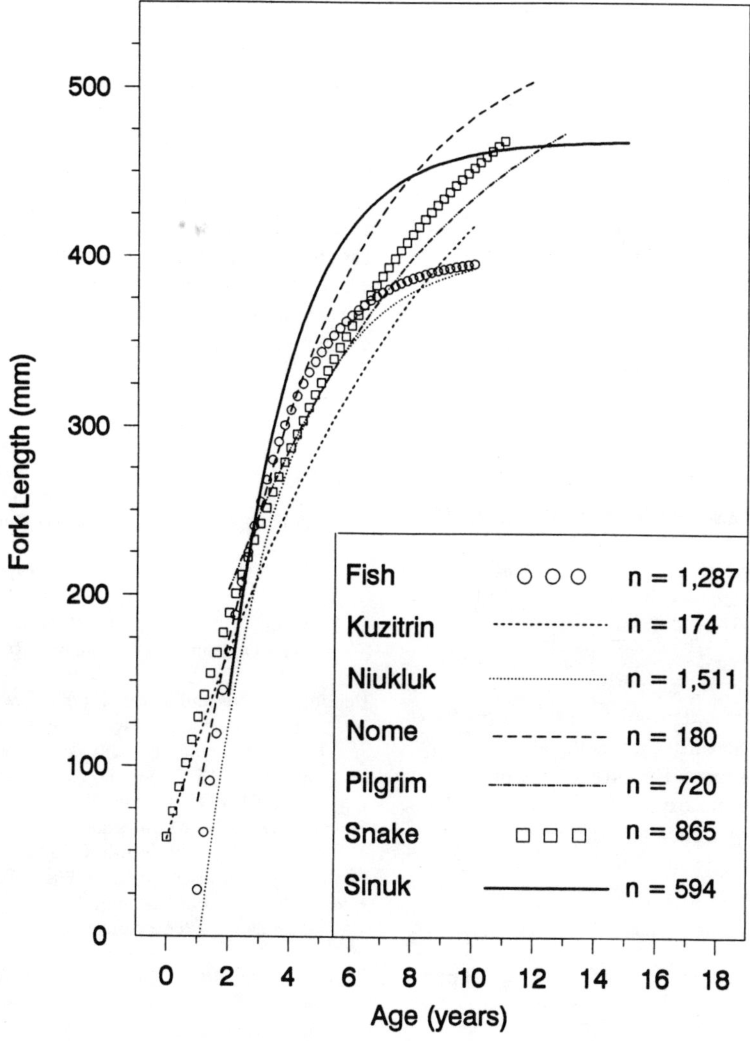

FIGURE 5.—Plots of the fitted von Bertalanffy growth parameters for various populations of Arctic grayling *Thymallus arcticus* sampled from seven river systems on the Seward Peninsula, 1988–1991.

forcing small fish into less suitable habitats. The predominance of large (old) fish in the population suggests an accumulation of energy within the large-length mode. The removal of large fish from the population would allow some young recruits to the large-length mode. Rapid growth of fish under age 6 suggests that this process is quite fast. Energy requirements for reproduction by large fish are probably such that little energy is available for somatic growth in most years. These factors result in the exhibited growth pattern and in the difficulty of using scales to determine ages of old fish.

Fisheries managers must be concerned with the consequences of harvest, including the modifications of population structure induced by exploitation. The size structure of an unexploited population may be rapidly modified when fishing is allowed under liberal regulations. Clady et al. (1975) found alterations in small-length modes when large smallmouth bass *Micropterus dolomieu* were removed from lakes previously unfished. In the case of the upper 40-km section of the Sinuk River, light harvest (a few large Arctic grayling taken within a relatively long period of time) has not altered the population's characteristics from those of an unexploited population. Estimated

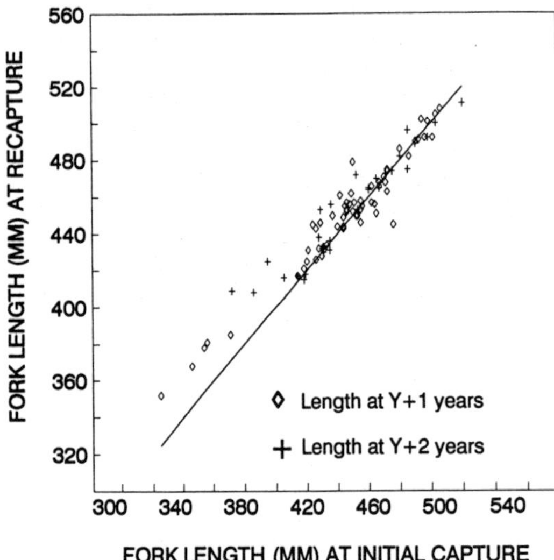

FIGURE 6.—Length of Arctic grayling *Thymallus arcticus* recaptured 1 and 2 years after initial sampling (Y). The dark line represents zero growth, 1989–1991.

growth parameters and population structure of Arctic grayling in Seward Peninsula streams might mimic those observed for Arctic grayling in the Sinuk River at lighter exploitation levels than those they currently experience.

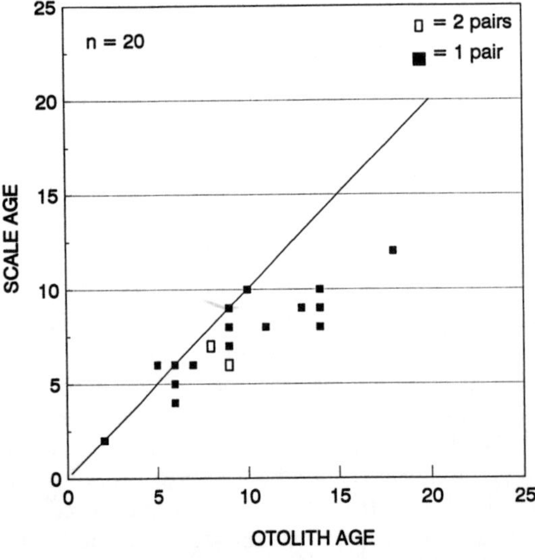

FIGURE 7.—Paired scale and otolith ages (in years) from Arctic grayling *Thymallus arcticus* incidentally killed during sampling on the Sinuk River, Seward Peninsula, during 1989–1991.

Acknowledgments

This research was partially supported by the Federal Aid in Sport Fish Restoration Act (16 U.S.C. §§ 777 to 777k) contracts F-10-5, 6-8-2; F-10-6, R-3-2(d); and F-10-7, R-3-2(d). We thank John H. Clark for supervision and administration of the research program. M. Amberg, B. Arvey, B. Clark, K. Reeves, and K. Rose assisted in fieldwork. S. Case assisted in typing and formatting the manuscript. The Bureau of Land Management provided helicopter support. S. Meyer and J. C. George offered valuable editorial comments.

References

Bendock, T. N. 1981. Inventory and cataloging of Arctic area waters. Alaska Department of Fish and Game, Federal Aid in Fish Restoration Project F-9-13, 22(G-I-I), Annual Performance Report 1980–1981, Juneau.

Bendock, T. N. 1982. Inventory and cataloging of Arctic area waters. Alaska Department of Fish and Game, Federal Aid in Fish Restoration Project F-9-14, 23(G-I-I), Annual Performance Report 1981–1982, Juneau.

Bendock, T. N., and J. Burr. 1984. Inventory and cataloging of Arctic area waters. Alaska Department of Fish and Game, Federal Aid in Fish Restoration Project F-9-15, 24(G-I-I), Annual Performance Report 1983–1984, Juneau.

Bernard, D. 1981. Multivariate analysis as a means of comparing growth in fish. Canadian Journal of Fisheries and Aquatic Sciences 38:233-236.

Buckland, S. T. 1980. A modified analysis of the Jolly Seber capture-recapture model. Biometrics 36:419–435.

Cerrato, R. M. 1990. An interpretable statistical test for growth comparisons using parameters in the von Bertalanffy equation. Canadian Journal of Fisheries and Aquatic Sciences 47:1416–1426.

Chapman, D. G. 1951. Some properties of the hypergeometric distribution with applications to zoological censuses. University of California Publications in Statistics 1:131–160, Berkeley.

Clady, M. D., D. E. Campbell, and G. P. Cooper. 1975. Effects of trophy angling on unexploited populations of smallmouth bass. Pages 425–429 *in* H. Clepper, editor. Black bass biology and management. Sport Fishing Institute, Washington, DC.

Clark, R. A. 1990. Stock status of Chena River Arctic grayling during 1990. Alaska Department of Fish and Game Fishery Data Series 91-35, Anchorage.

Clark, R. A. 1992. Influence of stream flows and stock size on recruitment of Arctic grayling (*Thymallus arcticus*) in the Chena River, Alaska. Canadian Journal of Fisheries and Aquatic Sciences 49:1027–1034.

Clark, R. A. 1993. Stock assessment of Arctic grayling in Fielding Lake during 1991 and 1992. Alaska Department of Fish and Game Fishery Data Series 93-2, Anchorage.

Clark, B., D. M. Fleming, and W. P. Ridder. 1991. Stock assessment of Arctic grayling in the Salcha, Chatanika, and Goodpaster rivers. Alaska Department of Fish and Game Fishery Data Series 91-15, Anchorage.

Conover, W. J. 1980. Practical nonparametric statistics, second edition. Wiley, New York.

Craig, P. C., and P. McCart. 1974. Fall spawning and overwintering areas of fish populations along proposed pipeline routes between Prudhoe Bay and the Mackenzie Delta (1972-1973). Pages 1–36 *in* P. McCart, editor. Fisheries research associated with proposed gas pipeline routes in Alaska, Yukon and Northwest Territories. Canadian Arctic Gas Study, Ltd. Biological Report Series 15(3), Calgary, Alberta.

Craig, P. C., and V. Poulin. 1975. Movements and growth of Arctic grayling and juvenile Arctic char in a small arctic stream, Alaska. Journal of the Fisheries Research Board of Canada 32:689–697.

Fleming, D. F. 1991. Stock assessment of Arctic grayling in Piledriver Slough, 1991. Alaska Department of Fish and Game Fishery Data Series 91-71, Anchorage.

Holmes, R. A., W. P. Ridder, and R. A. Clark. 1986. Distribution, abundance, and natural history of the Arctic grayling in the Tanana drainage. Alaska Department of Fish and Game, Federal Aid in Fish Restoration Project F-10-1, 27(G-8-1), Annual Report of Progress, 1985–1986, Juneau.

Johnson, L. 1976. Ecology of Arctic populations of lake trout, *Salvelinus namaycush*, lake whitefish, *Coregonus clupeaformis*, Arctic char, *S. alpinus*, and associated species in unexploited lakes of the Canadian Northwest Territories. Journal of the Fisheries Research Board of Canada 33:2459–2488.

Jolly, G. M. 1965. Explicit estimates from capture–recapture data with both death and immigration-stochastic model. Biometrika 52:225–247.

Kvalseth, T. O. 1985. Cautionary note about R^2. American Statistician 39:279–285.

Marquardt, D. W. 1963. An algorithm for least-squares estimation of nonlinear parameters. Journal of the Society of Industrial and Applied Mathematics 11: 431–441.

Merritt, M. F., and D. F. Fleming. 1991. Evaluations of various structures for use in age determination of Arctic grayling. Alaska Department of Fish and Game Fishery Manuscript 91-6, Anchorage.

Nordeng, H. 1961. On the biology of char (*Salmo alpinus* L.) in Salangen, North Norway. 1. Age and spawning frequency determined from scales and otoliths. Nytt Magasin for Zoologi 10:67–123.

Pollock, K. H., J. D. Nichols, C. Brownie, and J. E. Hines. 1990. Statistical inference for capture–recapture experiments. Wildlife Monographs 107.

Ricker, W. E. 1975. Computation and interpretation of biological statistics. Fisheries Research Board of Canada Bulletin 191.

Robson, D. S., and W. A. Flick. 1965. A non-parametric statistical method for culling recruits from a mark–recapture experiment. Biometrics 21:936–947.

Scholz, F. W., and M. A. Stephens. 1987. K-Sample Andersen–Darling tests. Journal of the American Statistical Association 82:918–924.

Seber, G. A. F. 1965. A note on the multiple recapture census. Biometrika 52:249–259.

Seber, G. A. F. 1982. The estimation of animal abundance and related parameters, second edition. Charles Griffin and Co., Ltd., London.

Sikstrom, C. B. 1983. Otolith, pectoral fin ray and scale age determinations for Arctic grayling (*Thymallus arcticus*). Progressive Fish-Culturist 45:220–223.

Skopets, M. B. 1991. Biological features of a subspecies of Arctic grayling in northeast Asia. II. The Alaskan grayling *Thymallus arcticus signifer*. Journal of Ichthyology 31(1):87–102.

Skopets, M. B., and N. M. Prokop'yev. 1990. Biological characteristics of a subspecies of the Arctic grayling in northeast Asia. I. The Kamchatkan grayling *Thymallus arcticus mertensi*. Journal of Ichthyology 30(4):43–58.

Tripp, D. B., and P. McCart. 1974. Life histories of grayling (*Thymallus arcticus*) and long nose suckers (*Catastomus catastomus*) in the Donnelly River System, Northern Territories. Pages 1–91 *in* P. McCart, editor. Life histories of anadromous and freshwater fish in the western Arctic. Canadian Arctic Gas Study, Ltd., Biological Report Series 20(1), Calgary, Alberta.

American Fisheries Society Symposium 19:240–249, 1997

Population Genetic Structure of Dolly Varden from Beaufort Sea Drainages of Northern Alaska and Canada

REBECCA J. EVERETT[1]
U.S. Fish and Wildlife Service, Fish Genetics Laboratory
1011 East Tudor Road, Anchorage, Alaska 99503, USA

RICHARD L. WILMOT[2]
U.S. Geological Survey, Biological Resources Division, Alaska Science Center
1011 East Tudor Road, Anchorage, Alaska 99503, USA

CHARLES C. KRUEGER
Cornell University, Department of Natural Resources
College of Agriculture and Life Sciences, Fernow Hall, Ithaca, New York 14853, USA

Abstract.—Dolly Varden *Salvelinus malma* spawn, rear, and overwinter in freshwater tributaries to the Beaufort Sea and migrate to coastal waters to feed. Oil and gas development activities that alter the freshwater or marine environments could affect char populations and the local fisheries they support. The purpose of this study was to describe the genetic relationships among Dolly Varden populations from Beaufort Sea tributaries. Allozyme electrophoresis was used to analyze variation at 49 loci (21 were polymorphic) in 27 collections made from 11 river drainages. Average heterozygosity observed was 0.038 (range: 0.016–0.052). Average percent of loci that were polymorphic was 19%. Overall heterogeneity G-tests indicated highly significant differences among collections ($P < 0.001$). Cluster analysis of Nei's genetic distance sometimes formed groups of geographically proximate collections; however, correspondence between genetic and geographic distances was weak in hierarchical groupings. Differences among collections within and between river systems were observed. On average, 91% of the genetic variation occurred among individuals within collections, whereas 8% was attributable to differences among river systems and 1% to differences within systems. Resident char from Sadlerochit Spring were genetically distinct from char in other collections due to low variability rather than to the presence of unique alleles. Genetic data indicated that multiple populations of char occur along the arctic coast of Alaska and Canada, often organized by major river system, and that more than one population may occur within river systems. Human activities that affect critical habitats, such as areas used for spawning or overwintering, must be considered with respect to the effects on individual populations rather than on a generalized char "population" that uses the Beaufort Sea.

Dolly Varden *Salvelinus malma* (herein also referred to as char) of the Beaufort Sea of Alaska and Canada migrate between freshwater tributaries and coastal waters, and are important in subsistence fisheries in both the United States and Canada (Craig 1989a). Oil and gas development activities that change either the freshwater or marine environments could affect char populations and the fisheries they support. The char's life history and migration patterns in this region reflect an adaptive response to the severity of the habitat and to seasonal and annual variation in habitat availability.

Char in this region spend most of their lives in tributaries of the Beaufort Sea, where they spawn, rear, and overwinter. Spawning occurs in autumn in areas often associated with springs in or near the Brooks Range (McCart 1980), and the eggs hatch in spring. Juvenile char live in tributaries until 3–5 years of age, when they begin to migrate to coastal areas to feed (Craig 1977a; McCart 1980). Mature and immature char return to freshwater in late summer and early fall and overwinter in tributaries as marine waters become supercooled (<0°C) and highly saline. These char typically do not spawn until age 5 or 6 (Bain 1974; Craig 1977a). Although an individual can spawn repeatedly over its lifetime, sufficient energy reserves are rarely acquired in one season to permit spawning every year.

Overwintering habitat is critical to the survival of Dolly Varden in this arctic environment. Only about 2% of the freshwater habitat—deep river pools and springs—remains available to fish by the end of winter (Craig 1989b), and these areas are shared by Dolly Varden of all life history stages. If an overwintering area fails (e.g., anoxia or freezing)

[1]Deceased.

[2]Present address: National Marine Fisheries Service, Alaska Fisheries Science Center, Auke Bay Biological Laboratory, 11305 Glacier Highway, Juneau, Alaska 99801, USA.

fish are unable to move to a different refuge because connecting stream segments freeze to the bottom. Therefore, the availability of deep-pool habitat in tributaries is critical to the viability of Dolly Varden populations.

Char populations are potentially vulnerable to habitat perturbations such as those caused by oil and gas development activities. Though Dolly Varden populations are adapted to the physical extremes of the Arctic by long life and repeated spawning, they also rely on the availability of overwintering areas and on their ability to migrate. In freshwater habitats, water removal from overwintering sites for well drilling, road construction, or other human use could deplete the limited overwintering resource used by all life history stages of char. Construction of river crossings, channelization, or removal of material from the rivers could affect migratory corridors and spawning substrate quality. If separate populations occur in coastline tributaries, the effects of development activities could cause local extinctions and the loss of genetic diversity.

In this study, allozyme electrophoresis was used to determine if genetic data supported the contention that different populations of Dolly Varden occurred within the major river drainages of the Beaufort Sea area and to refute the null hypothesis that char from this region exist as a single panmictic population. The amount and pattern of genetic variation is described and compared within and among char collections from 11 river systems.

Methods

Collections

Char were collected at 27 sites on 11 tributaries to the Beaufort Sea in 1985, 1986, and 1987 by U.S. Fish and Wildlife Service personnel (Table 1; Figure 1). Taxonomic designation of char from this region as Dolly Varden follows the recommendations of Reist et al. (1997, this volume). Electrofishing and minnow traps were used to collect juveniles from anadromous populations and juveniles and adults from stream-resident populations. Anadromous spawning adults were not used in this study because of the extreme logistical difficulties of sampling fish during the spawning season in this region. Sample sizes at sites ranged from 15 to 97 fish. Char were also collected from upstream of the waterfall on the Babbage River (site 26), upstream Firth River (site 23), Shublik Spring (site 9, Canning River drainage), and Sadlerochit Spring (site 12); these fish were presumed to have a stream-resident life history and were included in the analysis for comparison to anadromous populations.

Electrophoretic Procedures and Locus Designations

Horizontal starch-gel electrophoresis with histochemical staining was used to identify protein products of allozyme loci (Aebersold et al. 1987). Enzymes were examined in white muscle, liver, and eye tissues (Table 2). Allozyme nomenclature follows that suggested by Shaklee et al. (1990). Allelic product mobilities were designated as gel migration distances relative to the product of the most common allele. The most common allele expressed at a locus was always designated as **100*. The *PGM-3,4** locus was variable, but not scored reliably. Putative variation was observed at *βHA** and *XDH-1**, but phenotypes were scoreable only in collections made in 1985 and 1986, when live fish were transported to the laboratory. Consistently good resolution was not obtained in 1987 collections because samples were frozen at −20°C rather than in liquid nitrogen or dry ice before shipping. Thus, out of a possible 53 loci for 24 enzymes, 49 loci that encode 22 enzymes were scored reliably.

Statistical Procedures

Conformance to Hardy–Weinberg expectations was assessed using a chi-square statistic. For each population, chi-square values from all variable loci were summed and the total was compared to a chi-square distribution. Variation at isoloci that share alleles (e.g., *sAAT-1,2**) cannot be assigned to a specific locus. This variation was assigned arbitrarily to one locus for purposes of data analysis and not examined for conformance to Hardy–Weinberg expectations.

Genetic differences among samples were assessed with percent of polymorphic loci, heterozygosity calculations, *G*-tests, and genetic distance coefficients (*D*). Data from different collections within a river system, made either at different sites or in different years, were sometimes pooled. Data were pooled when *P* was greater than or equal to 0.01, indicating that collections were not significantly different based on a *G*-test (Sokal and Rohlf 1981; see criteria of Shaklee and Phelps 1990). Because of pooling, the data set was reduced to 16 collections from the original 27 sites; thus, some collections included data from fish collected at more than one site within a river system (e.g., data from collections at three sites in the Firth River system were com-

TABLE 1.—Collections of Beaufort Sea Dolly Varden *Salvelinus malma* sampled from Alaska and Canada in 1985–1987 with site location (Universal Transverse Mercator, UTM), number of samples (*N*), and date collected. Site numbers correspond to those used in Figure 1. Numbers (#) under the collections from the Hulahula and Babbage rivers refer to the pooled sets of data used for analysis in Tables 3–6.

Collections	Site number	UTM coordinates			*N*	Date
		Zone	Latitude	Longitude		
Anaktuvuk River	1	5W	7624250	574000	40	May 1986
Sagavanirktok River						
Ivishak River	2	6W	7663300	463800	50	Sep 1986
Echooka River	3	6W	7683800	486800	24	Apr 1986
Ribdon River	4	6W	7615250	453500	40	May 1986
Lupine River	5	6W	7652130	443000	48	Aug 1987
Kavik River	6	6W	7691000	517100	40	Sep 1986
Canning River						
	7	6W	7652250	553250	27	May 1986
	8	6W	7719500	527300	70	Aug 1987
Shublik Spring	9	6W	7699300	533800	59	Aug 1987
	10	6W	7687800	536000	62	Aug 1987
Marsh Fork	11	6W	7663000	540000	29	May 1986
Sadlerochit Spring	12	6W	7731300	600300	62	Aug 1987
Hulahula River						
#1	13	6W	7740000	609500	15	Oct 1985
#2	14	6W	7711000	602000	37	Oct 1985
#2	15	6W	7690500	595300	59	Oct 1985
#1	16	6W	7734300	608500	97	Aug 1987
Aichilik River						
	17	7W	7694000	413500	40	Sep 1986
	18	7W	7683500	417300	70	Aug 1987
Egaksrak River	19	7W	7702500	435800	41	May 1986
Kongakut River						
	20	7W	7712000	471700	40	Sep 1986
	21	7W	7666000	463000	90	Aug 1987
Firth River						
	22	7W	7625300	507500	64	Aug 1987
	23	7W	7610250	494000	47	Aug 1987
Joe Creek	24	7W	7646500	502300	50	Sep 1986
Babbage River						
#1	25	7W	7625300	579300	53	Aug 1987
#2	26	7W	7619500	575500	21	Aug 1987
Canoe River	27	7W	7612500	592500	35	Sep 1986

bined). The amount of genetic variation within the 16 collections was described by percent of polymorphic loci and the average percent of heterozygous loci per individual (*H*). Expected heterozygosity for each locus was calculated with the observed allele frequencies in each collection and was based on expected random mating proportions (Hardy–Weinberg). Isoloci were treated as a single locus for these calculations (total of 42 loci or locus combinations).

Allele counts by locus were compared statistically by contingency table analysis with *G*-tests to test heterogeneity between pairs of collections (Sokal and Rohlf 1981). Because of the robustness of the test, only cells with expected values less than 1.0 were combined. The critical value used to reject the null hypothesis (no differences) for the *G*-tests was increased to account for the increase in type I error when multiple tests of the same hypothesis were made (Cooper 1968).

Pairwise genetic distance (*D*) data (Nei 1972) were calculated based on polymorphic loci only. When missing data occurred for loci, distances were calculated by assuming the allele frequencies of other collections within the same drainage (*sAAT-4** of Ribdon was assumed the same as that for Lupine, *sAAT-4** and *sAAT-1,2** of Hulahula #1 was assumed to be the same as Hulahula #2, and *GAPDH-3** of Canoe was assumed to be the same as the average between Babbage #1 and #2). Unweighted pair-group method using arithmetic averages (UPGMA) cluster analysis of the distance coefficients was used to construct a dendrogram (Sneath and Sokal 1973).

Total gene diversity (H_T) was partitioned to estimate within-collection (H_S), between-collection (D_{ST}), and relative gene diversity (G_{ST}; Nei 1973; Chakraborty 1980). Sample data were analyzed hierarchically by sites within drainages and by collections from different drainages.

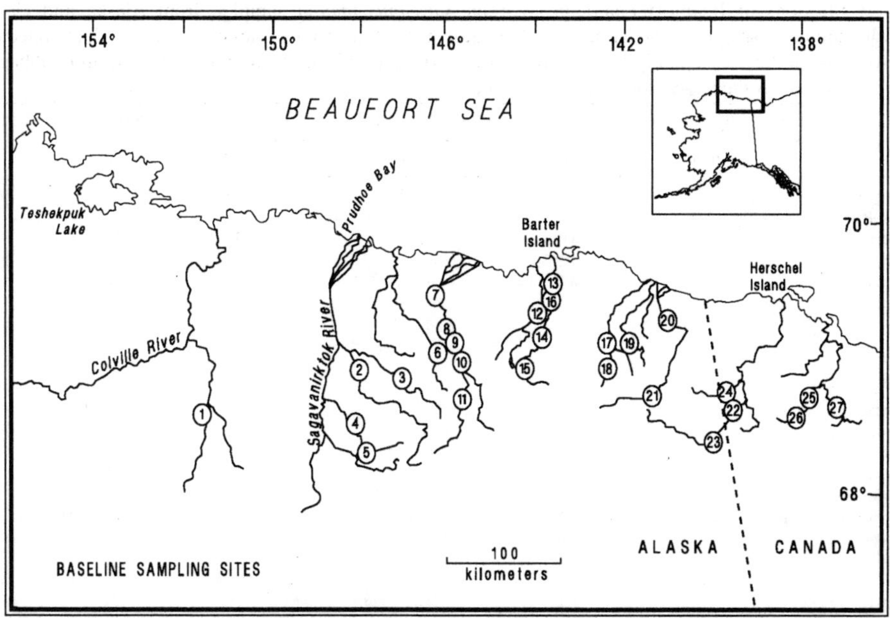

FIGURE 1.—Sampling sites for Dolly Varden *Salvelinus malma* collected from Beaufort Sea tributaries of Alaska and Canada in 1985–1987. Site numbers correspond to locations in Table 1.

Results

Genetic Variability within Collections

Forty-nine loci coding for 22 enzymes in three tissues were scored in the collections (Table 3). Of the 49 loci examined, 21 were variable and 28 were monomorphic. The 21 polymorphic loci included six isoloci (*sAAT-1,2**; *GPI-B1,2**; *sIDHP-1,2**; *sMDH-A1,2**; *sMEP-1,2**; and *sIDDH-1,2**). No evidence of departure from expected genotypic distributions (Hardy–Weinberg proportions) was observed in any collection when tests for all loci were summed in the analyses. Individual loci were out of equilibrium, but no more than expected as a result of type I error. Percent of polymorphic loci ranged from 7.1 to 28.6% (average 19%, SE = 6.7%; Table 4). Average heterozygosity ranged from 1.6 to 5.2% and over all collections was 3.8% (SE = 1.02%). The lowest heterozygosity (1.6%) was observed in resident char collected from Sadlerochit Spring. No allele substitutions were observed among stocks at any locus.

Genetic Variability among Samples

Data sets from collections of char made at different sites or in different years were not different within the Aichilik, Canning, Firth, and Kongakut drainages, nor between the Ivishak and Echooka rivers within the Sagavanirktok system, and thus these data sets were pooled ($P > 0.01$; Table 5). Similarly, char from Hulahula River sites 13 and 16 and from sites 14 and 15 were not different; however, when data from these char were pooled (Hulahula #1 and Hulahula #2, respectively) and compared, differences were detected (Table 5). Within the Sagavanirktok system, tests of the pooled data from the Ivishak and Echooka rivers against the data from char collected from the Ribdon and Lupine rivers revealed significant differences. Differences also were observed among the three collections from within the Babbage River system. No differences were observed in those *G*-tests that compared resident char from the Canning River (site 9, Shublik Springs) and Firth River (site 23) to anadromous char within each system. However, upstream resident fish from Babbage River (site 26) were different from the downstream collection (site 25) and char from a tributary (Canoe River site 27). After combining similar data sets (see "Methods"), 16 collections were used in subsequent analyses. The 16 collections were significantly different from each other, based on a *G*-test of all stocks and summed over all variable loci ($G = 1{,}237$; df = 143; $P \ll 0.001$).

TABLE 2.—Enzymes, International Union of Biochemistry-Nomenclature Committee (IUBNC) numbers, and loci examined in samples of Beaufort Sea Dolly Varden *Salvelinus malma* collected from northern Alaska and Canada in 1985–1987. Buffers include AC (Clayton and Tretiak 1972), pH 6.1 and pH 6.8; AC+ was AC with NAD; RW (Ridgway et al. 1970), pH 8.2; and EBT (was similar to that of Boyer et al. 1963), pH 8.5. Tissues include muscle (M), liver (L), and eye (E). Pairs of loci numerically separated by commas (e.g., *sAAT-1,2**) were electrophoretically indistinguishable isoloci (Allendorf and Thorgaard 1984).

Enzyme or other protein	IUBNC number	Loci	Buffer	Tissue
Adenylate kinase	2.7.4.3	*AK-1**	AC 6.8	M
Alcohol dehydrogenase	1.1.1.1	*ADH-1**	RW	L
Aconitate hydratase	4.2.1.3	*sAH-1**	AC 6.8	L
Aspartate aminotransferase	2.6.1.1	*sAAT-3*, sAAT-4**	RW	E, L
		*sAAT-1,2**	RW	M
Creatine kinase	2.7.3.2	*CK-1*, CK-2**	RW	M
		*CK-5**	RW	E
Fumarate hydratase	4.2.1.2	*FH**	AC 6.8	M
β-*N*-Acetylhexosaminidase	3.2.1.52	*βHA**	AC 6.8	L
Glucose-6-phosphate isomerase	5.3.1.9	*GPI-B1,2**	RW	M
		*GPI-A1**	RW	M
Glutathione reductase	1.6.4.2	*GR-1**	RW	L
Glyceraldehyde-3-phosphate dehydrogenase	1.2.1.12	*GAPDH-3*, GAPDH-4**	AC 6.1	E
Glycerol-3-phosphate dehydrogenase	1.1.1.8	*G3PDH-1*, G3PDH-2**	AC 6.1 RW	M, L
Cytosol nonspecific dipeptidase	3.4.13.18	*PEPA-1**	EBT	M
Isocitrate dehydrogenase	1.1.1.42	*mIDHP-1*, mIDHP-2**	AC 6.8	M
		*sIDHP-1,2**	AC 6.1, 6.8	L, E
Lactate dehydrogenase	1.1.1.27	*LDH-A1*, LDH-A2**	RW	M
		LDH-B1, LDH-B2**	RW	E
		*LDH-C1**		
		*LDH-B2**	RW	L
Tripeptide aminopeptidase	3.4.11.4	*PEPB-1**	EBT	M
Malate dehydrogenase	1.1.1.37	*sMDH-A1,2**	AC 6.1	M
		*sMDH-B1,2**	AC 6.1	M
Malic enzyme (NADP[†])	1.1.1.40	*mMEP-1*, mMEP-2*, sMEP-1,2**	AC 6.1	M
Phosphoglucomutase	5.4.2.2	*PGM-1*, PGM-2**	RW	M
		*PGM-3,4**	AC 6.1	L, M
Phosphogluconate dehydrogenase	1.1.1.44	*PGDH**	AC 6.8	M, L, E
L-Iditol 2-dehydrogenase	1.1.1.14	*sIDDH-1,2**	RW	L
Superoxide dismutase	1.15.1.1	*sSOD-1**	RW	L
Triose-phosphate isomerase	5.3.1.1	*TPI-1*, TPI-2*, TPI-3*, TPI-4**	TG	E
Xanthine dehydrogenase	1.1.1.204	*XDH-1**	RW	L

Cluster analysis of Nei's genetic distances between the char collections identified some groups of collections with close genetic affinity that were also geographic neighbors (Figure 2). For example, char from the Aichilik River, from the middle of the coastal area sampled, were grouped with char from the Kongakut River, located to the east approximately 10 km (Figure 1). Similarly, char from the Ivishak–Echooka, Kavik, and Canning rivers formed a group that represented collections from the western region of the coastal area sampled (Figures 1 and 2). However, the correspondence between genetic and geographic distances was weak in the hierarchical groupings. For example, the western group identified above was next linked to char from Babbage #1, Babbage #2, and Canoe, the easternmost sites in Canada. Also, char from the Lupine and Ribdon rivers of the Sagavanirktok system in the west were grouped together but were joined next to char from the Firth River in the east. The greatest pairwise genetic distance was between the Sadlerochit Spring char (a resident form) and all other collections.

Variation among individuals within collections accounted for 91% of the total gene diversity observed (Table 6). Most of the remaining observed variation (8%) was attributable to differences among the 11 drainages, with a minor component attributable to differences within drainages (1%).

Discussion

Amount and Pattern of Genetic Variation

The average heterozygosity observed in Dolly Varden (H = 0.038) from Beaufort Sea drainages was in the range typical of fish species in general (H = 0.051; Nevo et al. 1984). More variation as measured by average heterozygosity was observed in Beaufort Sea Dolly Varden than in Arctic char *Salvelinus alpinus* of North America, Ireland, Swe-

TABLE 3.—Gene frequencies of variable loci in 16 collections or pooled collections of Dolly Varden *Salvelinus malma* sampled in 1985, 1986, and 1987 from the Beaufort Sea area of Alaska and Canada. Variants of duplicated loci were arbitrarily assigned to one locus of the pair. Only frequencies of alternate alleles other than *100** are given. Names of loci (abbreviated here) are in Table 2. No data (ND) were available from some collections at some loci. Codes for collections are AI = Aichilik, AN = Anaktuvuk, B1 = Babbage #1, B2 = Babbage #2, CN = Canning, CA = Canoe, EG = Egaksrak, FI = Firth, H1 = Hulahula #1, H2 = Hulahula #2, IV = Ivishak and Echooka, KA = Kavik, KO = Kongakut, LU = Lupine, RI = Ribdon, SA = Sadlerochit Spring. See Table 1 for more details of collections.

Locus	Allele name and *N*	Collections															
		AI	AN	B1	B2	CN	CA	EG	FI	H1	H2	I V	KA	KO	LU	RI	SA
*sAAT-4**	*33**	0.012	0.025	0.000	0.000	0.031	0.000	0.014	0.079	0.019	ND	0.000	0.000	0.000	0.000	ND	0.000
	N	80	40	53	21	179	33	35	76	80		22	37	85	45		44
*sAAT-1,2**	*75**	0.082	0.092	0.000	0.000	0.092	0.000	0.049	0.050	0.050	ND	0.034	0.000	0.088	0.044	0.050	0.133
	*129**	0.000	0.000	0.000	0.000	0.000	0.000	0.000	0.008	0.000	ND	0.000	0.000	0.000	0.000	0.000	0.000
	N	85	38	53	21	206	35	41	130	80		72	40	85	45	40	45
*sAH-1**	*115**	0.194	0.292	0.019	0.000	0.216	0.029	0.243	0.211	0.220	0.239	0.174	0.137	0.177	0.211	0.112	0.011
	*130**	0.265	0.305	0.490	0.559	0.270	0.271	0.200	0.328	0.220	0.294	0.291	0.400	0.275	0.322	0.300	0.889
	N	85	36	52	17	183	35	35	128	91	46	72	40	85	45	40	45
*GAPDH-3**	*Null**	0.345	0.066	0.067	0.150	0.246	ND	0.500	0.177	0.440	0.675	0.225	0.294	0.286	0.233	0.270	0.000
	N	82	38	52	20	183		32	107	91	40	71	34	84	45	37	45
*GPI-B1,2**	*55**	0.000	0.050	0.010	0.000	0.002	0.000	0.000	0.019	0.000	0.000	0.007	0.000	0.000	0.000	0.077	0.000
	N	85	40	52	21	211	35	41	129	95	51	74	40	85	45	39	45
*GPI-A1**	*96**	0.241	0.100	0.154	0.000	0.123	0.371	0.171	0.306	0.158	0.140	0.088	0.012	0.282	0.411	0.333	0.156
	N	85	39	52	21	211	35	41	129	95	50	74	40	85	45	39	45
*mIDHP-1**	*220**	0.018	0.000	0.000	0.000	0.003	0.000	0.000	0.019	0.000	0.010	0.014	0.025	0.012	0.022	0.050	0.000
	N	85	37	53	21	149	35	35	129	95	51	74	40	85	45	40	45
*sIDHP-1,2**	*80**	0.122	0.000	0.000	0.000	0.035	0.000	0.014	0.053	0.027	0.073	0.007	0.100	0.059	0.000	0.000	0.000
	N	85	39	53	21	184	35	35	131	91	48	73	40	85	45	40	45
*LDH-C1**	*97**	0.054	0.000	0.000	0.000	0.005	0.014	0.071	0.027	0.011	0.059	0.027	0.012	0.035	0.000	0.077	0.000
	N	83	35	53	21	188	35	35	132	95	51	73	40	85	45	39	45
*sMDH-A1,2**	*128**	0.000	0.044	0.000	0.000	0.000	0.000	0.000	0.000	0.000	0.000	0.000	0.000	0.000	0.000	0.000	0.000
	N	82	34	53	21	182	35	35	128	93	52	72	40	85	40	40	45
*sMEP-1,2**	*69**	0.000	0.000	0.000	0.000	0.000	0.000	0.000	0.000	0.000	0.019	0.000	0.000	0.000	0.000	0.000	0.000
	N	85	40	53	21	197	35	35	132	95	54	74	40	85	45	40	45
*PGDH-1**	*95**	0.000	0.000	0.000	0.000	0.000	0.000	0.000	0.000	0.000	0.000	0.007	0.000	0.000	0.011	0.012	0.000
	N	85	40	53	21	197	35	35	132	95	51	74	40	85	45	40	45
*PGM-2**	*88**	0.000	0.000	0.000	0.000	0.000	0.000	0.000	0.004	0.021	0.029	0.000	0.000	0.000	0.000	0.000	0.000
	N	85	35	51	21	161	28	39	128	95	51	50	40	85	45	40	45
*sIDDH-1,2**	*43**	0.035	0.125	0.000	0.000	0.088	0.000	0.014	0.015	0.086	0.000	0.087	0.000	0.041	0.011	0.000	0.000
	N	85	36	53	21	188	34	35	131	87	17	23	40	85	44	40	44
*sSOD-1**	*115**	0.053	0.000	0.000	0.000	0.026	0.015	0.057	0.004	0.086	0.093	0.028	0.000	0.024	0.089	0.112	0.000
	*87**	0.000	0.000	0.056	0.043	0.000	0.028	0.000	0.000	0.000	0.000	0.000	0.000	0.000	0.000	0.000	0.000
	N	85	35	53	21	212	35	35	132	93	54	72	40	85	45	40	45

den, and Norway (H = 0.00–0.024; Ferguson 1981; Kornfield et al. 1981; Andersson et al. 1983; Hindar et al. 1986). Most Arctic char sampled for the studies listed above were from landlocked lake populations, but anadromous and small, resident (dwarf) stocks were also included in some cases. The amount of variation in Beaufort Sea Dolly Varden was closer to the average reported for anadromous salmonids (H = 0.041; Gyllensten 1985; and tabulated in Altukhov and Salmenkova 1991) than for Arctic char.

Evidence for separate populations within a drainage occurred in three (Babbage, Hulahula, and Sagavanirktok) of the seven river systems where more than one collection had been taken. Within the Babbage River system, the Canoe River collection was significantly different from the two Babbage River collections, similar to observations by Reist (1989). The isolated population above the waterfall on the Babbage River (Babbage #2, Site 26) was different from the other two collections within that system (primarily because of fixation of the common allele at *GPI-A1**) but was most similar to the collection below the falls (Babbage #1, site 25; Figure 2). The differences observed between collections from upstream and downstream in the Hulahula River were not expected, because of the lack of hydrographic complexity in the river system. Perhaps the series of overwintering and spawning pools (20–30 km apart) that characterize this system serves to isolate local stocks. Evidence for separate char populations also occurred within the Sagavanirktok River system (Ivishak–Echooka, Lupine, and Ribdon rivers), even though several stocks may rely on the same overwintering area in the Ivishak River (see Yoshihara 1973; Furniss 1974, 1975).

TABLE 4.—Expected average percent of heterozygosity per locus (*H*) and percent of loci examined that were polymorphic in 16 populations of Dolly Varden *Salvelinus malma* sampled from tributaries of the Beaufort Sea in Alaska and Canada in 1985–1987. The average value of *H* over all populations was weighted by sample size (SEs for average *H* and average percent of polymorphic loci are in parentheses).

Drainage and site	Year	*N*	*H* (%)	Polymorphic loci (%)
Aichilik River	1986, 1987	85	5.04	23.8
Colville River				
Anaktuvuk	1986	40	3.81	19.0
Babbage River				
#1	1987	53	2.48	11.9
#2	1987	21	2.43	7.1
Canoe River	1986	35	2.51	9.8
Canning River				
5 Sites	1986, 1987	212	4.13	26.2
Egaksrak River	1986	41	4.32	21.4
Firth River				
3 Sites	1986, 1987	132	4.29	28.6
Hulahula River				
#1	1985, 1987	95	4.57	23.8
#2	1985	54	4.66	22.5
Kavik River	1986	40	3.14	14.3
Kongakut River	1986, 1987	85	4.54	21.4
Sadlerochit Spring	1987	45	1.63	7.1
Sagavanirktok River				
Ivishak, Echooka	1986	74	3.62	26.2
Lupine River	1987	45	4.36	19.0
Ribdon River	1986	40	5.16	22.0
Total		1,097		
Average (SE)			3.79 (1.02)	19.0 (6.7)

Based on these data, more than one population of Dolly Varden within each Beaufort Sea river drainage should be anticipated when assessing ecological and genetic risks from development or exploitation.

River systems that exhibited no genetic differences among Dolly Varden collections included those with multiple collections from the same site in different years and from multiple sites. Allele frequencies were similar in collections from geographically close sites sampled in two different years (Aichilik, Kongakut, and Hulahula rivers). No significant differences over all loci were observed among three collections from sites #22–24 (Figure 1) on the Firth River system (this study; Table 5). However, in another study a significant difference at a single locus was reported between two char collections from the Firth River (Reist 1989). No divergence was detected among collections from multiple sites in the Canning River drainage, although one of the collections contained resident Dolly Varden (#9, Shublik Spring).

The genetic divergence observed among Dolly Varden populations of different river systems probably has been maintained by homing behavior (e.g., Furniss 1974; Craig and McCart 1975). Although Dolly Varden are known to overwinter in nonnatal drainages (e.g., Craig 1977a; Armstrong 1984), the pattern of relationships among stocks indicated that sufficient isolation exists to permit genetic differentiation of populations. The differentiation observed among Dolly Varden in this study was in accord with that of a previous study that documented genetic differences between char from the Babbage River, Firth River, and two Mackenzie River tributaries (Reist 1989).

The gene diversity observed among anadromous Dolly Varden from the western Beaufort Sea in this study (G_{ST} = 0.09) was similar to that reported for other anadromous salmonids (Gyllensten 1985) and corresponded to what has been considered as local differences (Ryman 1983). The level of between-stock diversity among Beaufort Sea Dolly Varden (9%) was less than that of nonmigratory Arctic char (14–53%; Kornfield et al. 1981; Andersson et al. 1983; Hindar et al. 1986), typically from isolated

TABLE 5.—Heterogeneity tests (*G*) of allozyme data among collections of Beaufort Sea Dolly Varden *Salvelinus malma* sampled at different sites and different years (1985–1987) within a river system. A value of *P* greater than 0.01 was used as the criterion for pooling data for further analyses. Degrees of freedom reflect the number of variable loci in the comparisons. Site numbers refer to those used in Table 1 and Figure 1.

Collection	Site numbers	Year	*G*	df	*P*
Aichilik	17, 18	1986, 1987	12.45	11	0.330
Babbage	25, 26, 27	1986, 1987	46.95	7	<0.001
Canning	7, 8, 9, 10, 11	1986, 1987	19.31	8	0.013
Firth	22, 23, 24	1986, 1987	11.64	13	0.458
Hulahula	13 and 16 pooled versus 14 and 15 pooled	1985, 1987	29.41	9	<0.001
Kongakut	20, 21	1986, 1987	5.71	10	0.839
Sagavanirktok					
Ivishak, Echooka	2, 3	1986	11.99	8	0.152
Ivishak and Echooka pooled versus Ribdon, Lupine	2 and 3 pooled versus 4, 5	1986, 1987	53.60	12	<0.001

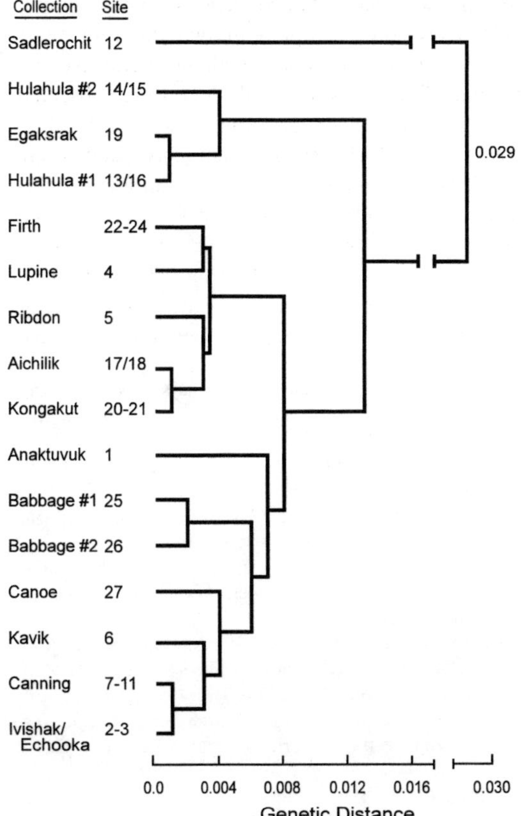

FIGURE 2.—Dendrogram of cluster analysis of Nei's genetic distances based only on polymorphic loci calculated between collections of Dolly Varden *Salvelinus malma* from Beaufort Sea tributaries in northern Alaska and Canada in 1985–1987. Site numbers refer to those used in Table 1 and Figure 1.

lake habitats. The 9% between-stock diversity we observed corresponded to the level of differentiation (5–15%) described as moderate in various taxa (Wright 1978).

Life History Differences

The general pattern of genetic variation among collections was not strongly associated with anadromous versus stream-resident life history types. Typically, more similarity was observed between collections of resident and anadromous char in the same drainage than to resident populations elsewhere. Although the Babbage River #2 collection (above the waterfall) and Sadlerochit Spring char (from a spring area not known to include anadromous individuals; Craig 1977b) were genetically distinct and probably reproductively isolated from other populations. The apparent divergence of these populations was related to the low level of variability ($0.016 < H < 0.024$) observed rather than to the presence of unique alleles. In the Babbage River system, small, resident Dolly Varden presumed to be from above the waterfall have been observed spawning with large, anadromous Dolly Varden below the falls (Bain 1974). This unidirectional downstream gene flow probably has prevented the two populations from diverging further. The resident Dolly Varden of the Shublik Spring collection were not detectably different from other collections in the Canning River drainage even though these char are isolated by a falls (McCart and Craig 1973).

In most other salmonids that have been studied, only a small percentage of the divergence among populations was attributable to the life history distinction between resident and migratory forms, for example, rainbow trout *Oncorhynchus mykiss* (Allendorf and Utter 1979), brown trout *Salmo trutta* (Ryman and Ståhl 1981), and Arctic char (Hindar et al. 1986). Resident Dolly Varden of the Beaufort Sea area probably arose independently in various drainages where conditions were unfavorable or impossible for migration. Nordeng (1983) found that some individuals in a brood of Arctic char reared in a hatchery and then released matured early and remained as small residents, while others

TABLE 6.—Gene diversity analysis among populations of Dolly Varden *Salvelinus malma* from Beaufort Sea tributaries of Alaska and Canada. The average values represent data from all 16 collections from the 11 river systems sampled during 1985–1987.

		Absolute gene diversity				Relative gene diversity (%)		
Drainage	Number of sites	Within sites	Between sites	Between drainages	Total	Within sites	Between sites	Between drainages
Babbage River	3	0.0256	0.0020		0.0276	92.9	7.1	
Hulahula River	2	0.0463	0.0009		0.0472	98.1	1.9	
Sagavanirktok	3	0.0429	0.0013		0.0442	97.1	2.9	
Average	16	0.0383	0.0004	0.0033	0.0420	91.1	0.9	8.0

became anadromous. Life history strategy can also be related to feeding and thus partially explained by a growth-dependent maturity (Jonsson and Hindar 1982).

Management Implications

The observed pattern of differentiation among populations within and among river systems refuted the null hypothesis that Beaufort Sea Dolly Varden are a single panmictic population. Nearly all collections of Dolly Varden from different river systems were genetically distinct, and several systems supported more than one population. Human activity affecting a critical habitat in localized areas, such as an overwintering area or access to Beaufort Sea coastal feeding areas in summer, could threaten individual populations of Dolly Varden from this region. For anadromous populations, changes in their distribution and abundance will affect species abundance within a broad coastal area that serves as an important fishing area. Loss of populations would reduce the overall genetic diversity of Dolly Varden in this region.

Acknowledgments

This study was funded by the Minerals Management Service, Department of the Interior, through an interagency agreement with the National Oceanic and Atmospheric Administration (NOAA), Department of Commerce, as part of the Alaska Outer Continental Shelf Environmental Assessment Program. We especially thank Randy Bailey, formerly with the U.S. Fish and Wildlife Service, and Lyman Thorsteinson, NOAA, who supported this study. We also thank the many service personnel of the Fairbanks Fishery Assistance Office who collected samples. Patty Gallagher, Kristin Denning, and Jonathan Pyatskowit produced the figures. Noelle Hardt and Lisa Bishop–Oltz helped compile the manuscript.

References

Aebersold, P. B., G. A. Winans, D. J. Teel, G. B. Milner, and F. M. Utter. 1987. Manual for starch gel electrophoresis: a method for the detection of genetic variation. NOAA (National Oceanic and Atmospheric Administration) NMFS (National Marine Fisheries Service) Technical Report 61.

Allendorf, F. W., and G. H. Thorgaard. 1984. Tetraploidy and the evolution of salmonid fishes. Pages 1–53 *in* B. J. Turner, editor. Evolutionary genetics of fishes. Plenum, New York.

Allendorf, F. W., and F. M. Utter. 1979. Population genetics. Pages 407–454 *in* W. S. Hoar, S. S. Randall, and J. R. Brett, editors. Fish physiology, volume 8. Academic Press, New York.

Altukhov, Y. P., and E. A. Salmenkova. 1991. The genetic structure of salmon populations. Aquaculture 98:11–40.

Andersson, L., N. Ryman, and G. Ståhl. 1983. Protein loci in the Arctic charr, *Salvelinus alpinus* L.: electrophoretic expression and genetic variability patterns. Journal of Fish Biology 23:75–94.

Armstrong, R. H. 1984. Migration of anadromous Dolly Varden charr in southeastern Alaska—a manager's nightmare. Pages 559–570 *in* L. Johnson and B. L. Burns, editors. Biology of the Arctic charr. Proceedings of the international symposium on Arctic charr. University of Manitoba Press, Winnipeg.

Bain, L. H. 1974. Life histories and systematics of Arctic char (*Salvelinus alpinus., L.*) in the Babbage River system, Yukon Territory. Pages 1–156 *in* P. J. McCart, editor. Life histories of three species of freshwater fishes in Beaufort Sea Drainages, Yukon Territory. Canadian Arctic Gas Study Limited, Biological Report Series 18, Calgary, Alberta.

Boyer, S. H., D. C. Fainer, and M. A. Naughton. 1963. Myoglobin: inherited structural variation in man. Science 140:1228–1231.

Chakraborty, R. 1980. Gene diversity analysis in nested subdivided populations. Genetics 96:721–726.

Clayton, J. W., and D. N. Tretiak. 1972. Amine citrate buffers for pH control in starch gel electrophoresis. Journal of the Fisheries Research Board of Canada 29:1169–1172.

Cooper, D. W. 1968. The significance level in multiple tests made simultaneously. Heredity 23:614–617.

Craig, P. C. 1977a. Ecological studies of anadromous and resident populations of Arctic char in the Canning River drainage and adjacent coastal waters of the Beaufort Sea, Alaska. Pages 1–116 *in* P. J. McCart, editor. Fisheries investigations along the North Slope and Beaufort Sea coast in Alaska with emphasis on Arctic char. Canadian Arctic Gas Study Limited, Biological Report Series 41, Calgary, Alberta.

Craig, P. C. 1977b. Arctic char in Sadlerochit Spring, Arctic National Wildlife Refuge. Pages 1–28 *in* P. J. McCart, editor. Fisheries investigations along the North Slope and Beaufort Sea coast in Alaska with emphasis on Arctic char. Canadian Arctic Gas Study Limited, Biological Report Series 41, Calgary, Alberta.

Craig, P. C. 1989a. Subsistence fisheries at coastal villages in the Alaskan Arctic, 1970–1986. Biological Papers of the University of Alaska 24:131–152.

Craig, P. C. 1989b. An introduction to anadromous fishes in the Alaskan Arctic. Biological Papers of the University of Alaska 24:27–54.

Craig, P. C., and P. J. McCart. 1975. Fish utilization of nearshore coastal waters between the Colville and Mackenzie Rivers with an emphasis on anadromous species. Pages 172–219 *in* P. C. Craig, editor. Fisheries investigations in a coastal region of the Beaufort Sea. Canadian Arctic Gas Study Limited/Alaskan Arctic Gas Study Company, Biological Report Series 34, Calgary, Alberta.

Ferguson, A. 1981. Systematics of Irish charr as indicated by electrophoretic analysis of tissue proteins. Biochemical Systematics and Ecology 9:225–232.

Furniss, R. A. 1974. Inventory and cataloguing of Arctic area waters. Alaska Department of Fish and Game Annual Report 15:1–45, Juneau.

Furniss, R. A. 1975. Inventory and cataloguing of Arctic area waters. Annual report of progress, 1974–1975. Alaska Department of Fish and Game, Federal Aid in Fish Restoration, Project F-9-6, Study G-I, Juneau.

Gyllensten, U. 1985. The genetic structure of fish: differences in the intraspecific distribution of biochemical genetic variation between marine, anadromous, and freshwater species. Journal of Fish Biology 26:691–699.

Hindar, K., N. Ryman, and G. Ståhl. 1986. Genetic differentiation among local populations and morphotypes of Arctic charr, *Salvelinus alpinus*. Biological Journal of the Linnean Society 27:269–285.

Jonsson, B., and K. Hindar. 1982. Reproductive strategy of dwarf and normal Arctic charr (*Salvelinus alpinus*) from Vangsvatnet Lake, western Norway. Canadian Journal of Fisheries and Aquatic Sciences 39:1404–1413.

Kornfield, I., K. F. Beland, J. R. Moring, and F. Kircheis. 1981. Genetic similarity among endemic Arctic char (*Salvelinus alpinus*) and implications for their management. Canadian Journal of Fisheries and Aquatic Sciences 38:32–39.

McCart, P. 1980. A review of the systematics and ecology of Arctic char, *Salvelinus alpinus*, in the western Arctic. Canadian Technical Report of Fisheries and Aquatic Sciences 935.

McCart, P. J., and P. C. Craig. 1973. Life history of two isolated populations of Arctic charr (*Salvelinus alpinus*) in spring-fed tributaries of the Canning River, Alaska. Journal of the Fisheries Research Board of Canada 30:1215–1220.

Nei, M. 1972. Genetic distance between populations. American Naturalist 196:283–292.

Nei, M. 1973. Analysis of gene diversity in subdivided populations. Proceedings of the National Academy of Sciences of the United States of America 70:3321–3323.

Nevo, E., A. Beiles, and R. Ben-Schlomo. 1984. The evolutionary significance of genetic diversity: ecological, demographic and life history correlates. Lecture Notes in Biomathematics 53:12–213.

Nordeng, H. 1983. Solution to the "char problem" based on Arctic char (*Salvelinus alpinus*) in Norway. Canadian Journal of Fisheries and Aquatic Sciences 40: 1372–1387.

Reist, J. D. 1989. Genetic structuring of allopatric populations and sympatric life history types of charr, *Salvelinus alpinus/malma*, in the western Arctic, Canada. Physiology and Ecology Japan, Special Volume 1:405–420.

Reist, J. D., J. D. Johnson, and T. J. Carmichael. 1997. Variation and specific identity of char from northwestern arctic Canada and Alaska. Pages 250–261 *in* J. Reynolds, editor. Fish ecology in Arctic North America. American Fisheries Society Symposium 19, Bethesda, Maryland.

Ridgway, G. J., S. W. Sherburne, and R. D. Lewis. 1970. Polymorphisms in the esterases of Atlantic herring. Transactions of the American Fisheries Society 99: 147–151.

Ryman, N. 1983. Patterns of distribution of biochemical genetic variation in salmonids: differences between species. Aquaculture 33:1–21.

Ryman, N., and G. Ståhl. 1981. Genetic perspectives of the identification and conservation of Scandinavian stocks of fish. Canadian Journal of Fisheries and Aquatic Sciences 38:1562–1575.

Shaklee, J. B., F. W. Allendorf, D. C. Morizot, and G. S. Whitt. 1990. Gene nomenclature for protein-coding loci in fish. Transactions of the American Fisheries Society 119:2–15.

Shaklee, J. B., and S. R. Phelps. 1990. Operation of a large-scale, multiagency program for genetic stock identification. Pages 817–830 *in* N. C. Parker and five coeditors. Fish-marking techniques. American Fisheries Society Symposium 7, Bethesda, Maryland.

Sneath, P. H. A., and R. R. Sokal. 1973. Numerical taxonomy. Freeman, San Francisco.

Sokal, R. R., and F. J. Rohlf. 1981. Biometry, 2nd edition. Freeman, San Francisco.

Wright, S. 1978. Evolution and the genetics of populations, volume 4. Variability within and among populations. University of Chicago Press, Chicago.

Yoshihara, H. T. 1973. Monitoring and evaluation of Arctic waters with emphasis on the North Slope drainages in annual report of progress, 1972–1973. Alaska Department of Fish and Game, Federal Aid in Fish Restoration Project F-9-5, Job G-III-A, Juneau.

American Fisheries Society Symposium 19:250–261, 1997

Variation and Specific Identity of Char from Northwestern Arctic Canada and Alaska

J. D. REIST, J. D. JOHNSON, AND T. J. CARMICHAEL
Department of Fisheries and Oceans
501 University Crescent, Winnipeg, Manitoba R3T 2N6, Canada

Abstract.—Char from east of the Mackenzie River system in Canada are considered to be Arctic char *Salvelinus alpinus*. Anadromous char south of the Alaska peninsula are, with few exceptions, considered to be Dolly Varden *S. malma*. Considerable debate has occurred with respect to specific affinities of the char found between these two regions along the continental north slope from Point Barrow, Alaska, eastward to the Mackenzie River and along the Alaskan coast from Point Barrow south to the Alaska peninsula. Several authors have suggested that these are taxonomic forms of either Arctic char or Dolly Varden, and this issue has not been resolved in the literature or in common usage. Morphological and genetic variation were examined for char from locations in the western Arctic in order to assign specific identification to several char forms found on the Yukon north slope. Anadromous and isolated riverine resident char from the Yukon north slope were similar to each other and to Dolly Varden from southeastern Alaska. Both were different from lacustrine Yukon north slope char and anadromous char from east of the Mackenzie River. Lacustrine north slope char exhibited greater similarity to Arctic char than to Dolly Varden. These results substantiate the view that riverine char (anadromous, residual, and isolated stream-resident forms) from the continental north slope west of the Mackenzie River are Dolly Varden, whereas lacustrine char from this area are relict Arctic char.

Fishes of the genus *Salvelinus* represent a confusing array of variation that has led to the description of numerous taxonomic forms throughout their Holarctic range. In northwestern arctic North America, three species have been formally identified or described: Arctic char *S. alpinus*, assumed to be synonymous with the Eurasian taxon described by Linnaeus; Dolly Varden *S. malma*, assumed to be synonymous with the Asian taxon described by Walbaum; and Angayukaksurak char *S. anaktuvukensis*, described by Morrow (1973). Additionally, numerous authors have described various geographically or ecologically defined subspecific taxa and forms and allied these with the above species. Although general agreement prevails among authors with respect to the existence of the various subspecific groups, considerable disagreement exists as to the affinities of both defined subspecific taxa and certain geographically defined groups. The taxonomic identity of several forms of chars found in the Beaufort Sea area is the subject of this article.

Salvelinus Taxonomy in Western Arctic North America

Two taxa of char co-occur in southwestern Alaska: Arctic char with high counts of pyloric caeca and gill rakers and Dolly Varden with low counts for these characters (DeLacy and Morton 1942; McPhail 1961). McPhail (1961) considered these to be good biological species because of a lack of hybridization where they co-occurred. McPhail (1961) summarized the observed morphological variation for Dolly Varden as follows: a northern form with more vertebrae and gill rakers occurred north of the Alaska peninsula and on the Seward Peninsula; a southern form occurred south of the Alaska Peninsula, throughout the Aleutians and the remaining southern portion of the North American range; and a form occurred in central Alaska with low counts of pyloric caeca. Morphological variation for Arctic char was summarized as follows (McPhail 1961): the Bering Sea–western Arctic form with fewer pyloric caeca and lower gill rakers was distributed from the lower Kuskokwim River north and east to the Mackenzie River, Canada; the eastern form with higher counts was distributed east of the Mackenzie River; and the Bristol Bay–Gulf of Alaska form occurred south of the Kuskokwim River and had counts similar to those of the eastern form. Two forms of char co-occurring in the Sagavanirktok River basin, Alaskan north slope, were equated with the Bering Sea–western Arctic and eastern forms of Arctic char by McCart and Craig (1971).

Morrow (1973) described the Angayukaksurak char, *S. anaktuvukensis*, as a distinct species. McCart (1980) reexamined the taxonomy of chars from the Prudhoe Bay area east to the Mackenzie River and considered the Angayukaksurak char to be a variant form of Dolly Varden. McCart (1980) concluded that only two forms of char were present in this area: all stream-resident and isolated stream-resident char, as well as a single lake-resident population, were equated with the Bering Sea–western Arctic form of Arctic char (McPhail 1961), and all remaining lake populations were the eastern form of Arctic char. Genetic evidence supported this, in that allozyme variation in liver esterases of the eastern form was fixed for a fast allele and 98.6% of the Bering Sea–western Arctic form exhibited the alternative slow allele.

Morrow (1980) examined variation in Alaskan chars with the express aim of establishing the specific identity of the Bering Sea–western Arctic form. Multivariate comparison using meristic data indicated that lake-resident Arctic char from southern Alaska were distinct from both southern and northern forms of Dolly Varden. The latter pair overlapped each other, but some degree of separation was evident. Almost complete overlap was observed between the Bering Sea–western Arctic form of Arctic char and the northern form of Dolly Varden, leading Morrow (1980) to conclude that these constitute the same taxon, Dolly Varden.

Thus, we have two species-level taxa with conflicting composition. Usage of char nomenclature in studies since 1980 reflects this confusion. For example, in comparing aging structures of char from the Wood River in interior Alaska, Baker and Timmons (1991) explicitly indicated that the taxon under consideration was the Bering Sea–western Arctic form of Arctic char of McPhail (1961), which Morrow (1980) would consider to be northern form Dolly Varden. Barber and McFarlane (1987) compared aging structures from two populations of eastern form Arctic char (sensu McPhail 1961) from the central Northwest Territories and two populations from the North Slope, coastal Alaska (northern Dolly Varden sensu Morrow 1980 or Bering Sea–western Arctic form of Arctic char sensu McPhail 1961) and equated all with Arctic char. DeCicco (1989) considered the char from the Kotzebue area of western Alaska to be northern form Dolly Varden (sensu Morrow 1980) but equivalent to the Bering Sea–western Arctic form of Arctic char of McPhail (1961). Other examples of such confusion of taxa exist in the literature.

Regardless of the specific designation of these taxonomic forms of char, all exhibit a wide range of alternative life history strategies. These include co-occurring anadromous and residual (nonanadromous) forms, stream-resident forms isolated above impassible falls as well as nonisolated stream-resident forms, and lacustrine life history types (Bain 1974; Armstrong and Morrow 1980; McCart 1980; Johnson 1989; Reist 1989; Reynolds and Gregory 1989).

In summary, three taxonomic forms of char are widespread throughout northwestern North America and probably represent two species. Agreement prevails in that char occurring in lacustrine habitats west of the Mackenzie River are Arctic char (with one exception). For southern Dolly Varden, agreement prevails in that this form occurs in riverine habitats south of the Alaska Peninsula. For char distributed in riverine habitats north of the Alaska Peninsula to the Mackenzie River, agreement occurs as to taxonomic distinctness, but disagreement prevails as to the specific affinity of this taxon.

Objectives of the Study

The objectives of this study are (1) to determine whether chars from northwestern arctic areas of North America represent two or more taxonomically recognizable forms; (2) to determine whether ecologically defined, life history types represent either distinct taxa or differentiation within taxa defined by tests of the first objective; (3) (assuming more than one taxon is present) to develop criteria for the recognition of the taxa and assign individuals from various locations and life history types to the correct taxon; and (4) to establish the limits of distribution in the northwestern Arctic of the taxonomically recognizable forms present.

These objectives can be reformulated into the general null hypothesis that all samples from the study area represent the same taxon. The testable prediction is that no discontinuity in variation should be present across the samples. That is, for samples exhibiting similar life histories, variation should be clinal regardless of the type of data examined. Between life history types, variation may be clinal or discontinuous, but it should be attributable to ecological rather than taxonomic sources. If the null hypothesis is rejected and thus two or more taxa exist, the next hypothesis to be tested is that all life history types of char (anadromous, nonanadromous residual, and isolated) from riverine habitats along the continental north coast west of the Mackenzie River are the northern form of Dolly Varden. The testable prediction is that these will exhibit greater similarity to Dolly Varden from elsewhere than to Arctic char. The final hypothesis to be examined is whether lacustrine char from this same area are eastern form Arctic char, with the prediction that these will be similar to Arctic char from elsewhere.

Materials and Methods

Logic of the Study

The initial step was to establish two reference sample groups: group A, Klutina River, Alaska; and group H, Cambridge Bay, Northwest Territories—to be used as indicators of the possible taxa present. Because the type specimens could not be examined, the reference samples were established on the basis of their geographical situation and comparison to literature descriptions of char from the same general area. The second step was to identify the reference samples relative to preexisting taxonomic criteria. Of necessity, this was accomplished using meristic criteria only. To simplify presentation of results, the remaining individual samples (see Appendix Table A.1) were grouped on the basis of similar geographic location, life history, and habitat to form six groups of unknowns: group B, Nome River, Alaska; groups C (anadromous), D (residual), and E (isolated), life history types of char from Canadian riverine habitats west of the Mackenzie River; group F, lacustrine char from isolated lakes west of the Mackenzie River; and group G, anadromous char from locations east of the Mackenzie River. Third, using a linear discriminant function for meristic data from the two reference samples, the six groups of unknowns were classified to specific morphological groupings. Fourth, the results of steps 1–3 were examined in reference to the null hypothesis outlined previously and discontinuities representing taxonomic boundaries sought. Because the establishment of reference samples, the classification of unknowns, and the existence of taxonomic boundaries all represent a hypothesis based upon morphological data, the final step was to retest these morphological results using different data—in this case, genetic variation at isozyme loci.

This analytical logic establishes that groups exist and to some extent establishes the membership of those groups; however, it does not specify the taxonomic level of those groups (e.g., species rather than some subspecific taxon). Existence of groups is a fact that can be established by discontinuities in the data. However, the assignment of a particular taxonomic level is a matter of philosophical interpretation. Such interpretation is open to debate, as has been the norm in the taxonomy of chars.

Data

Eight meristic variables were counted for all fish: upper (UGR) and lower (LGR) gill rakers, with the middle raker included in the latter; pyloric caeca (PYL); branchiostegal rays (BRC); dorsal (DRC), anal (ARC), pectoral (PRC), and pelvic (VRC) principal fin rays. Total gill raker and pyloric caecum counts have traditionally been used to differentiate the taxa of char in this area. As a result of the influence of environment on meristic characters, clinal variation in such characters is possible. Therefore, these meristic results were corroborated with a parallel set of allozyme genetic data. Two enzyme systems and a total of three loci exhibited variation, repeatable results, and adequate models: superoxide dismutase (SOD, Enzyme Commission number 1.15.1.1) showed one locus and phosphoglucomutase (PGM, Enzyme Commission number 2.7.5.1) was resolved for two loci (PGM-1 and PGM-2).

Analyses

Meristic variation in the two reference samples was analyzed using multivariate discriminant analysis (Norusis 1990). Such analysis is a three-step process: (1) the null hypothesis of centroid equality is tested, and if rejected; (2) a linear discriminant function is constructed between the groups; (3) a posteriori, the individuals used to construct the function can be scored and classified as to group membership by proximity to a group centroid or distribution or both, thus providing a measure of the discriminating power of the function. Alternatively, using the unstandardized coefficients for the function, individuals extrinsic to the analysis (i.e., true unknowns) can be scored and similarly classified. To ensure that the results obtained were representative of real variation in the data and not imposed by the constraints of the analysis, principal component analysis and scoring was employed that paralleled this discriminant analysis.

Several techniques were employed to summarize the variation observed and to address questions of similarity between groups. These included the unweighted pair-group method using arithmetic averages (UPGMA) for clustering, principal coordinates analysis, and multidimensional scaling of Euclidean distances for meristic data and Nei's genetic distance for genetic data (Rohlf 1990). Each of these techniques has idiosyncratic assumptions that may affect interpretation of results; thus, similarity of results between techniques indicates true relationships inherent in the data.

Results

Identification of the Reference Samples

Comparison of total gill raker and pyloric caecum counts was used to establish the identity of the reference samples (Figure 1). Char from the area near Klutina River (group A) were characterized by low counts for total gill rakers and pyloric caeca, similar to counts for nominate Dolly Varden (McPhail 1961; Morrow 1980). Char from Cambridge Bay were characterized by high counts of both variables and were similar to counts for nominate Arctic char from the central Canadian Arctic (McPhail 1961). Thus, we concluded that these reference samples respectively represent the two major char taxa present in the study area.

Meristic Discriminant Analysis of Reference Samples

The discriminant analysis of meristic variation between the two reference groups indicated a highly significant multivariate difference ($P < 0.0001$) between the representatives of Dolly Varden (Klutina River) and Arctic char (Cambridge Bay). The best discriminating variables were pyloric caeca and lower and upper gill rakers, with minimal contribution from the remaining meristic variables (Table 1). The two groups were completely separate and a posteriori classification accuracy was 100% (Figure 2A, top panel). Fish representative of Dolly Varden all exhibited negative scores, whereas those representative of Arctic char exhibited positive scores. Principal component analysis in which no a priori grouping structure was imposed gave similar results. With the exception of four individuals from Cambridge Bay that occupied the intermediate area, two groups of fish were evident and this separation was primarily attributable to gill raker and pyloric caecum variation.

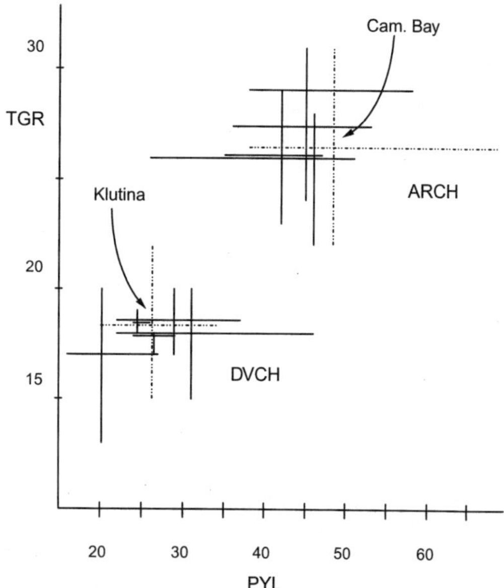

FIGURE 1.—Identification of reference samples—Klutina River Dolly Varden *Salvelinus malma* (DVCH) and Cambridge (Cam.) Bay Arctic char *S. alpinus* (ARCH)—to taxon by comparison of meristic variation with samples from the same area. Individual samples are centered on the bivariate mean values for total gill rakers (TGR) and pyloric caeca (PYL) and the range of each character is given by the respective line. Arctic char values are derived from McPhail (1961) for Sapuladjuk, Kathewachaga, Bloody Falls, and Bernard Harbor localities. Dolly Varden values are derived from McPhail (1961) for Summit Lake and Turnagain Arm and from Morrow (1980) for Tutka Bay, Bear Creek, and Ketchikan area.

Discriminant Analysis Classification of Unknowns

Using the unstandardized discriminant coefficients from the above analysis (Table 1), fish from all other samples were treated as unknowns and classified. This procedure involves computing the discriminant score as the sum of the unstandardized coefficients multiplied by the respective character values from each individual including the constant. Thus, the scores for the unknowns are expressed relative to the centroids, and the distributions of scores for the reference samples are used to construct the function.

A bimodal distribution of scores resulted when all fish extrinsic to the discriminant analysis were plotted simultaneously (Figure 2A, bottom panel). Such bimodality is primary evidence of the existence of two taxa in the samples and further indicates that the reference samples do not simply represent extremes of clinal variation. Additional examination of scoring of groups within these unknowns revealed the following pattern of differences.

Fish from the Nome River all scored negatively and overlapped the distribution of scores of the Klutina River reference sample but were displaced somewhat toward zero (Figure 2B). Anadromous char from locations west of the Mackenzie River (group C including Joe Creek, Firth River, Canoe River, Cache Creek, and Rat River as well as coastal locations near Herschel Island [Ptarmigan Bay, Thetis Bay, and Pauline Cove]) primarily exhibited negative scores between −4 and −1 (Figure 2B). Thus, like the Nome River fish, these were not exactly coincident with, but substantively overlapped, the Klutina reference sample. A few individuals from some of these samples also exhibited scores in the 0 to +1 range of the function, thus overlapping minimally the distribution of scores for

Cambridge Bay fish. However, these scores represent outliers distant from both the sample centroid and the majority of the distribution of scores.

Anadromous char collected from locations east of the Mackenzie River (group G including mainland coastal samples [Wood Bay], mainland riverine samples [Hornaday River and Horton River], and arctic island samples [unnamed lake on Banks Island, Kuujjua River, Kuuk River, Kagloryuak River, Kagluk River, and Naloagyuak River]) scored positively and exhibited overlap with the scores of the Cambridge Bay reference sample (Figure 2B). A few individuals from these samples scored in the 0 to −1 area of the discriminant function, thus overlapping minimally the distribution of scores for the Klutina River fish. These scores represent outliers distant from both the sample centroids and the primary distribution of scores.

Fish representing the residual life history type from rivers west of the Mackenzie River (group D including Joe Creek, Firth River, Babbage River, Canoe River, and Cache Creek) scored negatively and overlapped the distribution of scores of Klutina River fish (Figure 2C). Fish from riverine habitats west of the Mackenzie River that were isolated above impassable falls (group E—Babbage River and Cache Creek) exhibited similar scores (Figure 2C). However, lacustrine fish from the north slope west of the Mackenzie River (group F—Lakes 103 and 104) exhibited complete overlap of discriminant scores with those of the Cambridge Bay reference sample (Figure 2C).

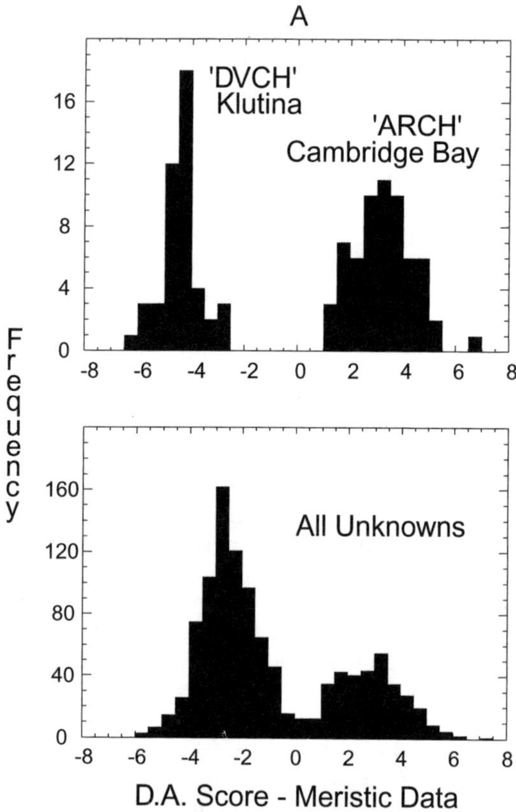

FIGURE 2.—Discriminant analysis (D.A.) scores of groups. (**A**) Reference samples—group A, Klutina River Dolly Varden *Salvelinus malma* (DVCH) and group H Cambridge Bay Arctic char *S. alpinus* (ARCH) (top panel) and all other, unknown char (bottom panel).

Numerical Taxonomy of Meristic Variation

Clustering, principal coordinates, and multidimensional scaling analyses of the Euclidean distances between the eight groups of fish for meristic mean values gave similar results (Figure 3). The most obvious feature on the phenogram (Figure 3A) was the formation of two groups of fish: (1) all riverine char from west of the Mackenzie River and (2) all char from east of the Mackenzie River plus the lacustrine fish from the North Slope west of the Mackenzie River. Within both groups two subgroups were apparent. For group 1, two subgroups were associated at a Euclidean distance of 3.2—the first consisted of the Klutina River fish and the isolated life history type from the Yukon North Slope rivers and the other consisted of the Nome River, anadromous, and residual life history types from west of the Mackenzie River. For group 2, the two subgroups were anadromous and Cambridge Bay fish associated with lacustrine fish at a Euclidean distance of 4.0.

The principal coordinates analysis resulted in a similar arrangement in which axis I (60% of total variation) separated chars from riverine habitats west of the Mackenzie River from the lacustrine fish and chars east of the Mackenzie River (Figure 3B). Axis II (30%) separated the subgroups within these two groups similarly to that seen for the clustering. Axis III (6%, not illustrated) indicated that the isolated fish were somewhat distinct from all others, thus the implied relationships of these with the Klutina River sample observed in the phenogram is probably artifactual. The multidimensional scaling analysis reinforced the results observed in the previously noted phenetic analyses and also suggested that the association between the isolated group and Klutina River fish as seen in the phenogram was artifactual (Figure 3C).

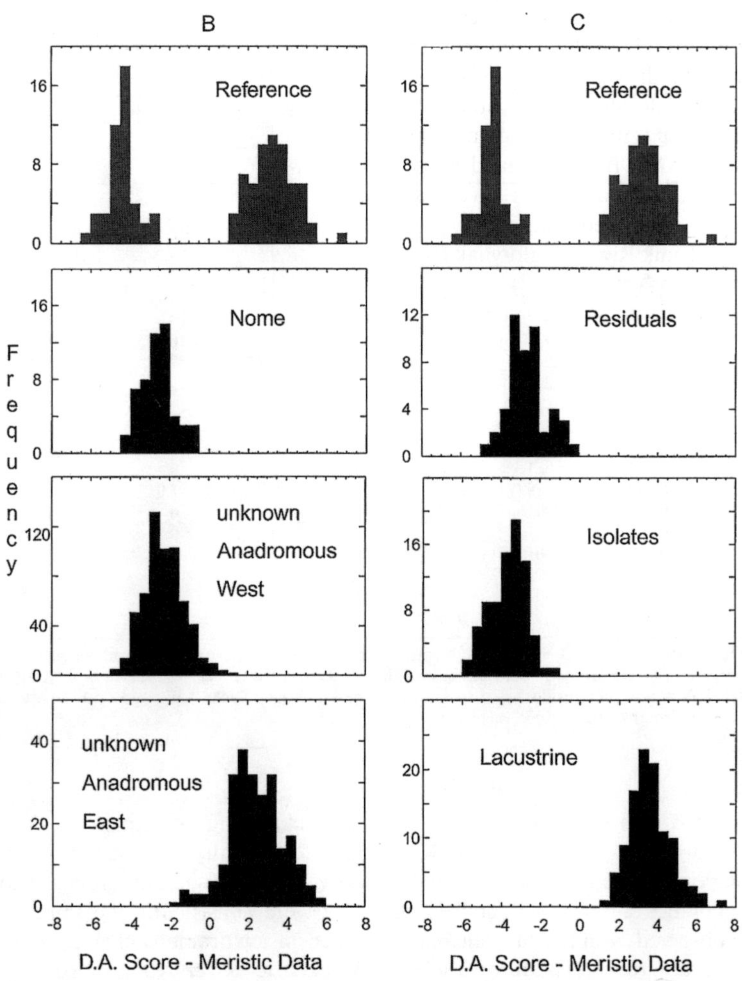

FIGURE 2.—Continued. (**B**) Groups of anadromous unknowns compared to the reference samples (top panel)—Nome River group B (second panel), anadromous fish from west of the Mackenzie River group C (third panel), and anadromous fish from east of the Mackenzie River group G (bottom panel). (**C**) Other life history types from the Yukon North Slope compared to the reference samples (top panel)—residual riverine group D (second panel), isolated riverine group E (third panel), and lacustrine group F (bottom panel) forms.

Genetic Results

For genetic variation, the lacustrine fish and both groups from east of the Mackenzie River generally exhibited low variation relative to the samples from west of the Mackenzie River except for the isolated fish (Table 1). Superoxide dismutase-1 exhibited two alleles, **a* and **b*, with mobilities of 100 and 84, respectively (allelic mobilities were established relative to the Klutina River sample). Between groups, the pattern of variation indicated that for all lacustrine char, Cambridge Bay char, and with the exception of two heterozygous fish of 342 anadromous fish, complete fixation occurred for the **b* allele, which was most prevalent overall. The isolated riverine group from west of the Mackenzie River was also fixed for **b*. The **a* allele was present at a moderate frequency (5%+) in the residual and anadromous fish from west of the Mackenzie River but was very frequent in both the Nome and Klutina samples. For PGM-1 three alleles (**a*, **b*, **c*) with relative mobilities of −100, −117, and −92, respectively, were observed. With the exception of one heterozygous fish of 360, all char east of the Mackenzie River as well as the lacustrine char from west of the Mackenzie River and Cambridge Bay char were fixed for the **c* allele.

TABLE 1.—Meristic mean values for the char groups, coefficients (Std = standardized and Unstd = unstandardized) for discriminant analysis between Klutina and Cambridge Bay reference samples, and allele frequencies for allozymes for the groups. Meristic acronyms are explained in "Materials and Methods." Enzyme acronyms are SOD = superoxide dismutase and PGM = phosphoglucomutase.

	Mean for group[a]								Coefficient	
Variable	KL	NO	AW	AE	IW	RW	LW	CB	Std	Unstd
					Meristic data					
DRC	10.7	11.1	10.9	10.9	11.1	10.9	10.4	10.8	−0.005	−0.007
ARC	9.1	9.8	9.7	9.7	9.2	9.5	9.2	10.1	0.113	0.169
PRC	13.2	13.9	13.8	14.0	13.0	13.8	13.3	14.2	−0.097	−0.150
VRC	8.9	9.1	9.1	9.5	8.8	9.0	9.7	9.7	−0.068	−0.129
BRC	10.6	11.0	11.4	11.2	11.0	11.2	10.9	11.5	0.106	0.148
UGR	7.6	9.0	8.9	11.2	8.4	8.6	10.3	10.9	0.438	0.486
LGR	10.8	12.7	12.6	15.9	12.1	12.7	17.0	15.5	0.368	0.358
PYL	26.3	28.1	30.2	42.8	24.7	29.1	49.1	48.4	0.767	0.194
Constant										−16.925
					Genetic data					
*SOD-1*a*	0.535	0.411	0.081	0.001	0	0.049	0	0		
**b*	0.465	0.589	0.919	0.999	1.0	0.951	1.0	1.0		
*PGM-1*a*	0.812	0.961	0.999	0.002	1.0	1.0	0	0		
**b*	0.188	0	0	0	0	0	0	0		
**c*	0	0.039	0.001	0.998	0	0	1.0	1.0		
*PGM-2*a*	0.925	1.0	0.999	0.998	1.0	1.0	1.0	1.0		
**b*	0.075	0	0.001	0	0	0	0	0		
**c*	0	0	0	0.002	0	0	0	0		

[a]KL = group A, Klutina River char. NO = group B, Nome River char. AW = group C, anadromous fish from rivers west of the Mackenzie River. AE = group G, anadromous fish from east of the Mackenzie River. IW = group E, isolated riverine char from west of the Mackenzie River. RW = group D, residual char from west of the Mackenzie River. LW = group F, lacustrine char from west of the Mackenzie River. CB = group H, Cambridge Bay char.

All other fish exhibited high frequencies of the **a* allele. The **b* allele was only observed in the Klutina River sample. For PGM-2 three alleles (**a*, **b*, **c*) with relative mobilities of −100, −70, and −119, respectively, were observed, with the **a* allele being the most frequent in all groups. The rare **b* allele was observed only in anadromous fish from west of the Mackenzie River and in the Klutina River fish, and the **c* allele was observed only in one heterozygous fish from east of the Mackenzie River.

The UPGMA phenogram of Nei genetic distances exhibited two major groups associated at a distance of 0.5: (1) anadromous fish east of the Mackenzie River, lacustrine fish from west of the Mackenzie River, and the Cambridge Bay sample; and (2) all remaining groups (Figure 4A). Within the latter cluster, two groups that were associated at a distance of 0.1 were apparent: (a) Klutina–Nome fish and (b) Yukon North Slope anadromous, residual, and isolated groups. Principal coordinates analysis and multidimensional scaling duplicated the results observed in the phenogram (Figure 4B, 4C).

Summary

In summary, all analyses, whether morphological or genetic, as well as all types of analysis (discriminant, clustering, principal coordinate, and multidimensional scaling) indicated the presence of two taxa in these fish. The groups belonging to these taxa were consistently the same: One taxon consisted of anadromous char from areas east of the Mackenzie River, Cambridge Bay fish, and the lacustrine samples from the Yukon North Slope west of the Mackenzie River. The other taxon included riverine fish from west of the Mackenzie River, including the two samples from Alaska.

Discussion

Presence of Two Taxa

The presence of a discontinuity in meristic variation (i.e., the bimodality of discriminant scores) indicates that two taxa of char are present in northwestern arctic North America. The association of ecologically defined life history types with their taxon of presumptive origin indicates that the variation is primarily the result of evolutionary rather than ecological causes. Thus, as originally formulated, the null hypothesis that all chars in this area represent the same taxon is rejected. All life history types of riverine char (anadromous, nonanadromous, and residual) from the continental coast west of the Mackenzie River and the Nome River anadromous char both exhibited greater similarity to the reference Dolly Varden sample than either did to the reference Arctic char sample. Thus, it can be concluded that morphologically these are all representatives of the Dolly Varden taxon. Anadromous char from east of the Mackenzie River and lacustrine char from isolated lakes west of the Mackenzie River exhibited greater morphological similarity to the reference Arctic char sample than either did to the reference Dolly Varden sample. Thus, these should all be considered as representatives of the Arctic char taxon.

The support of this morphological hypothesis by genetic results, in particular the almost complete fixation of alternative alleles for two enzyme loci, further strengthens the case for the existence of the two taxa. Analyses of these genetic data confirmed the presence of at least two taxa; however, some results suggested the presence of two lower-level taxa within the Dolly Varden group. Resolution of this issue awaits future work.

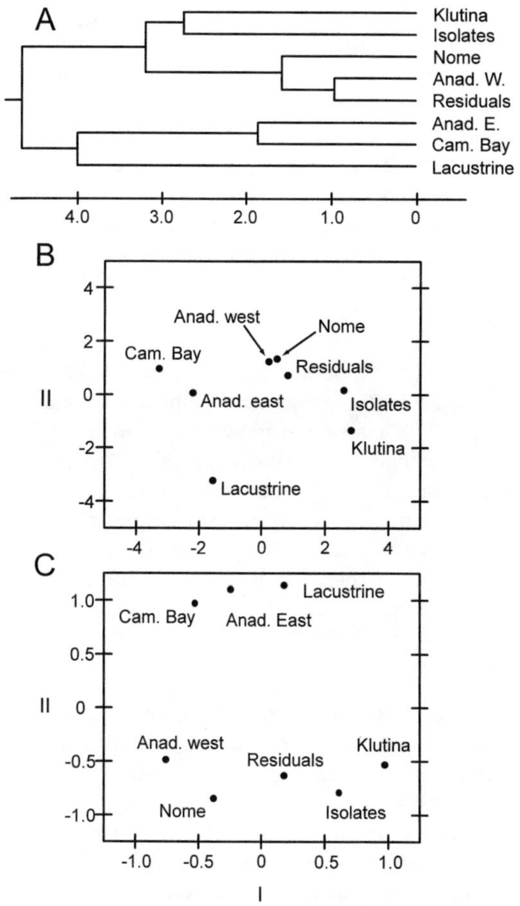

FIGURE 3.—Phenetic results obtained for meristic data. (**A**) Clustering using unweighted pair-group method with arithmetic averages. (**B**) Principal coordinates analysis. (**C**) Multidimensional scaling analysis of Euclidean distances for standardized meristic mean values. Groups are reference Dolly Varden *Salvelinus malma* (Klutina River, group A), isolated (Isolates) riverine char from west of the Mackenzie River (group E), Nome River anadromous char (group B), anadromous (Anad.) char from west (W.) of the Mackenzie River (group C), residual char from west of the Mackenzie River (group D), anadromous char from east (E.) of the Mackenzie River (group G), reference Cambridge (Cam.) Bay Arctic char *S. alpinus* (group H), and lacustrine char from west of the Mackenzie River (group F).

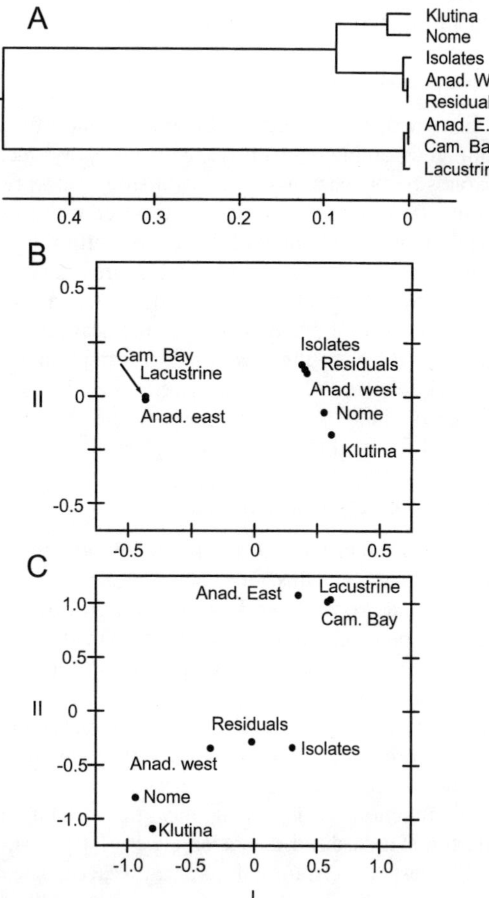

FIGURE 4.—Phenetic results obtained for genetic data. (**A**) Clustering using unweighted pair-group method with arithmetic averages. (**B**) Principal coordinates analysis. (**C**) Multidimensional scaling analysis of Nei's genetic distance. Groups and abbreviations are as in Figure 3.

This evidence supports the contention of Morrow (1980) that all riverine-dwelling char west of the Mackenzie River and throughout Alaska are forms of Dolly Varden. There is no evidence for the presence of a riverine, often anadromous, form of Arctic char (i.e., Bering Sea–western Arctic form of McPhail 1961) anywhere west of the Mackenzie River. Thus, this taxon as originally formulated by McPhail (1961), reiterated by McCart (1980), and in prominent usage since 1980 should be equated with Dolly Varden. Future usage of this taxon as a form for Arctic char is not warranted. Alternatively, there is excellent evidence for the presence of a lacustrine form of char that is very similar, morphologically and genetically, to eastern form Arctic char (McPhail 1961; Morrow 1980; this study). These are distributed in several lakes across the continental North Slope (McCart 1980) and throughout western and southern Alaska (Morrow 1980).

Identification of the Taxa

In view of this taxonomic distinction and of the continuing ambiguity with respect to exactly which taxon is being examined in a given study, interpretation of previous literature must proceed with caution. To avoid ambiguity in the future, information regarding the habitat occupied, the degree of isolation if any, and total gill raker and pyloric caecum counts should be presented in all publications regarding char from the area. Such information will provide relevant identifying criteria for the taxa being examined. To aid this process, the following protocol for field identification is given.

Field Identification of Individuals

To establish criteria for the rapid field identification of char from this area, a linear discriminant function was constructed between two groups: (1) all anadromous and residual North Slope Dolly Varden and (2) all anadromous char east of the Mackenzie River. Two meristic variables, total gill raker (TGR) count and pyloric caecum (PYL) count, were used because of their discriminatory power and for field convenience.

The function was highly significant ($P < 0.0000$), and the standardized coefficients were 0.8111 and 0.6195, respectively, for gill rakers and caeca. The a posteriori classification accuracy was very high, although not perfect; for Dolly Varden, 99.4% (687 of 691) were correctly classified, and for Arctic char, 96.8% (337 of 348) were correctly classified. Thus, in most cases, this discriminant function results in accurate classification of unknown individuals from the area of the continental north slope and coastal regions to the east of the Mackenzie River.

To classify unknowns, the unstandardized discriminant coefficients must be used to compute the discriminant score as 0.5297 multiplied by the TGR count plus 0.1167 multiplied by the PYL count minus the constant 16.3680. These scores are compared to the group centroids and ranges of scores given herein to hypothesize the identity of the individual. For correctly classified Dolly Varden and Arctic char, these values respectively were centroid, −1.534 and range, −3.97 to +0.66; and centroid, 3.045 and range, +1.13 to +6.06. The highest score for Dolly Varden and the lowest score for Arctic char represent individuals that were not misclassified to an inappropriate group. The scores of misclassified individuals ranged from 1.19 to 1.49 for Dolly Varden (4 individuals) and from 0.14 to 0.85 for 10 of 11 misclassified Arctic char. One of the misclassified Arctic char scored at −0.51, representing an extreme outlier. Thus, the scores between 0.14 and 1.49 represent an ambiguous zone for purposes of classifying individuals, and identification of unknowns in that range should be considered suspect and corroborated by other criteria. Discriminant scores less than 0.14, especially negative values, represent Dolly Varden. Discriminant scores greater than 1.49 represent Arctic char.

Level of the Taxa

The discovery of two taxa does not automatically determine the taxonomic level to which those taxa should be assigned. Rather, as pointed out previously, determination of taxonomic level is a philosophical point of view imposed on the system by human interpretation. However, in applying the biological species concept, coexistence without widespread hybridization, as seen for Dolly Varden and Arctic char in southwest Alaska (McPhail 1961), provides strong evidence for the taxa described herein to be considered distinct at the species level. This interpretation is reinforced by the apparent ecological and biological differences: riverine spawning for Dolly Varden and lacustrine spawning for Arctic char; age of first seaward migration of 3–4 years for Dolly Varden (McCart 1980) and 4–7 years for Arctic char (Johnson 1980); greatest ages attained of approximately 17 years for North Slope anadromous Dolly Varden (McCart 1980) compared to greater than 24 years for Nauyuk Lake Arctic char (Johnson 1980); and development of a strong kype in Dolly Varden but less so in Arctic char (J. D. Reist, personal observation).

Distribution and Subspecific Diversity of the Taxa

From the evidence presented herein and insofar as coastal char populations are concerned, the Mackenzie River apparently forms a demarcation in the distribution of the taxa. In particular, Dolly Varden apparently do not occur east of the Mackenzie River, and the only Arctic char populations in North America that occur west of the Mackenzie River are lacustrine.

These distributions suggest that Arctic char was once more widely distributed in western North America. This species is cold-adapted, being found in northernmost freshwaters (Johnson 1980), and seems to require environments that include lacustrine habitats. Thus, in areas that are cold, where lakes are present, and in areas of a former more-widespread distribution, Arctic char presently occur. Conversely, Dolly Varden is a species adapted to warmer environments (e.g., distribution south to Washington State, Haas and McPhail 1991), is found primarily in riverine environments, and exhibits limited use of lacustrine habitats. Range expansion of this taxon to areas east of the Mackenzie River appears to be limited either by ecological factors (e.g., unsuitable habitat, in that cool, nonturbid mountain rivers are not found or environments are colder) or by biological factors (e.g., coexistence with anadromous forms of Arctic char is unsuccessful).

Dolly Varden distribution in the study area is virtually coincident with the habitat available during the Wisconsin glaciation (Lindsey and McPhail 1986). Presumably, isolation in northern and southern Beringian refuges has contributed to subspecific diversification into the northern and southern forms. The reconstruction of Beringian physiography during the Wisconsin glaciation includes major rivers draining now-submerged lands under the Chukchi and Bering seas (Lindsey and McPhail 1986). Areas on the north side and western tip of the Seward Peninsula and those farther north were drained by the Chukchee Sea River. Areas immediately south were drained by the Gulf of Anadyr River, and those farther south by the Yukon and Kuskokwim rivers. Thus, in contrast to the findings of Morrow (1980), the Seward Peninsula appears to be the area of demarcation between these two forms. The association of Nome River fish with North Slope Dolly Varden in some analyses and with Klutina River fish in others indicates the possibility of greater diversity within this taxon. Perhaps three forms are present: (1) southern—from the Yukon River southward; (2) Anadyrian—southern portion of the Seward Peninsula and areas on the south side of Norton Sound, as well as southern drainages of the Chukotka Peninsula in Siberia; and (3) a northern form distributed to the north of the Seward Peninsula and along both the Siberian and North American continental north coasts. More examination of subspecific diversity within Dolly Varden is warranted.

The distribution of relictual populations of Arctic char within the study area is poorly known but apparently widespread. Presumably this indicates a previous, probably pre-Illinoian glaciation, more widespread distribution. The glacial extent was greater and thus a cooler climate prevailed during the Illinoian glaciation (Lindsey and McPhail 1986). Because Arctic char can survive at glacial fronts (Priede 1989), there would be ample opportunity for trans-basin movement during ice retreat either throughout interior Beringia or along coastal margins, as well as for isolation in mountainous lakes as deglaciation occurred. Detailed examination of subspecific diversity in Arctic char populations throughout the study area is needed.

Thus, like the conclusions of Morrow (1980), our results indicate that river-dwelling char from areas west of the Mackenzie River are representatives of Dolly Varden. Lacustrine fish from this area, with one exception, all represent relictual forms of Arctic char (Morrow 1980). The exception is a lacustrine population of Dolly Varden reported from the Canning River, Alaska (McCart 1980). The existence of the Bering Sea–western Arctic taxon of Arctic char (McPhail 1961; McCart 1980) present in riverine habitats in this area is not supported by the results of this study.

Acknowledgments

We acknowledge the financial support for this study from implementation funds of the Inuvialuit Settlement Agreement administered by the Department of Fisheries and Oceans (DFO)–Inuvialuit Fisheries Joint Management Committee and by regular DFO funding. Substantive logistical support was received from the Science Institute of the Northwest Territories, Inuvik Research Centre. The assistance of the following individuals is greatly appreciated: for providing samples from their own studies—R. Baker, W. Bond, A. Kristofferson, and B. Stewart; for aid in collecting Canadian samples—D. Bodaly, R. Fudge, and B. Macdonald; for aid in collecting Alaskan samples—K. Alt and F. DeCicco; and for arranging for sample collections—V. Gillman. The efforts of M. Koshinsky and her crew in processing the fish are appreciated. D. Ostlere, T. Stevens, and B. Tilwari all contributed to the electrophoretic work. Permission to collect in Northern Yukon National Park was given by Parks Canada. The editorial efforts of G. Decterow are greatly appreciated.

References

Armstrong, R. H., and J. E. Morrow. 1980. The Dolly Varden charr, *Salvelinus malma*. Pages 99–140 *in* E. K. Balon, editor. Charrs: salmonid fishes of the genus *Salvelinus*. Dr. W. Junk, the Hague, Netherlands.

Bain, L. H. 1974. Life histories and systematics of Arctic charr (*Salvelinus alpinus* L.) in the Babbage River system, Yukon Territory. Arctic Gas Biological Report Series 18(1), Calgary, Alberta.

Baker, T. T., and L. S. Timmons. 1991. Precision of ages estimated from five bony structures of Arctic char (*Salvelinus alpinus*) from the Wood River system, Alaska. Canadian Journal of Fisheries and Aquatic Sciences 48:1007–1014.

Barber, W. E., and G. A. McFarlane. 1987. Evaluation of three techniques to age Arctic char from Alaskan and Canadian waters. Transactions of the American Fisheries Society 116:874–881.

DeCicco, A. L. 1989. Movements and spawning of adult Dolly Varden charr (*S. malma*) in Chukchi Sea drainages of northwestern Alaska: evidence for summer and fall spawning populations. Pages 229–238 *in* Kawanabe et al. (1989).

DeLacy, A. C., and W. M. Morton. 1942. Taxonomy and habits of the chars, *Salvelinus alpinus* and *Salvelinus malma* of the Karluk Drainage System. Transactions of the American Fisheries Society 72:79–91.

Haas, G. R., and J. D. McPhail. 1991. Systematics and distributions of Dolly Varden (*Salvelinus malma*) and bull trout (*Salvelinus confluentus*) in North America. Canadian Journal of Fisheries and Aquatic Sciences 48:2191–2211.

Johnson, L. 1980. The arctic charr, *Salvelinus alpinus*. Pages 15–98 *in* E. K. Balon, editor. Charrs: salmonid fishes of the genus *Salvelinus*. Dr. W. Junk, the Hague, Netherlands.

Johnson, L. 1989. The anadromous Arctic charr, *Salvelinus alpinus*, of Nauyuk Lake, N.W.T., Canada. Pages 201–228 *in* Kawanabe et al. (1989).

Kawanabe, H., F. Yamazaki, and D. L. G. Noakes, editors. 1989. Biology of chars and masu salmon. Physiology and Ecology Japan, Special Volume 1.

Lindsey, C. C., and J. D. McPhail. 1986. Zoogeography of fishes of the Yukon and Mackenzie basins. Pages 639–674 *in* C. H. Hocutt and E. O. Wiley, editors. The zoogeography of North American freshwater fishes. Wiley, New York.

McCart, P. J. 1980. A review of the systematics and ecology of Arctic charr, *Salvelinus alpinus*, in the western Arctic. Canadian Technical Report of Fisheries and Aquatic Sciences 935.

McCart, P., and P. Craig. 1971. Meristic differences between anadromous and freshwater-resident Arctic char (*Salvelinus alpinus*) in the Sagavanirktok River Drainage, Alaska. Journal of the Fisheries Research Board of Canada 28:115–118.

McPhail, J. D. 1961. A systematic study of the *Salvelinus alpinus* complex in North America. Journal of the Fisheries Research Board of Canada 18:793–816.

Morrow, J. E. 1973. A new species of *Salvelinus* from the Brooks Range, northern Alaska. Biological Papers of the University of Alaska 13:1–8.

Morrow, J. E. 1980. Analysis of the Dolly Varden charr, *Salvelinus malma*, of northwestern North America and northeastern Siberia. Pages 323–338 *in* E. K. Balon, editor. Charrs: salmonid fishes of the genus *Salvelinus*. Dr. W. Junk, the Hague, Netherlands.

Norusis, M. J. 1990. SPSS/PC+ advanced statistics 4.0 for the IBM PC/XT/AT and PS/2. SPSS Inc., Chicago.

Priede, I. G. 1989. Observations on landlocked and migratory charr in Arctic north east Greenland. Pages 107–108 *in* Kawanabe et al. (1989).

Reist, J. D. 1989. Genetic structuring of allopatric populations and sympatric life history types of charr, *Salvelinus alpinus/malma*, in the western Arctic, Canada. Pages 405–420 *in* Kawanabe et al. (1989).

Reynolds, J. B., and L. S. Gregory. 1989. Dolly Varden and their habitat, Tiekel River, Copper River basin, Alaska. Page 250 *in* Kawanabe et al. (1989).

Rohlf, F. J. 1990. NTSYS-pc, numerical taxonomy and multivariate analysis system, version 1.60. Exeter Software, Setauket, New York.

Appendix: Samples Used in Study

TABLE A.1.—Locations of samples used in this study of char from northwestern arctic Canada and Alaska and the group to which they were assigned (see text). Latitude and longitude refer to collection location for the sample and number represents the number of fish used in the morphological portion of the study. Location code is the original code to be used in cross-referencing to other work. Abbreviations in location names are AK = Alaska, NWT = Northwest Territories, YT = Yukon Territory, R = River, L = Lake, and Ck = Creek.

Group	Location name	Latitude (°N)	Longitude (°W)	Date	Life history type[a]	Habitat[b]	Number	Presumptive taxon[c]	Location code
A	Klutina R, AK	61° 45′	145° 45′	Sep 1991	?	S?	49	DVCH-s	91–12
B	Nome R, AK	64° 40′	165° 20′	Sep 1991	A	S	54	DVCH-n	91–13
C	Joe Creek, YT	68° 56′	140° 58′	Sep 1986	A	S	52	DVCH-n	86–58
D	Joe Creek, YT	68° 56′	140° 58′	Sep 1986	R	S	26	DVCH-n	86–58
C	Firth R, YT	68° 40′	140° 55′	Sep 1986	A	S	54	DVCH-n	86–59
C	Firth R, YT	68° 40′	140° 55′	Sep 1988	A	S	17	DVCH-n	88–07
D	Firth R, YT	68° 40′	140° 55′	Sep 1988	R	S	9	DVCH-n	88–07
F	Lake 103, YT	69° 26′	139° 34′	Sep 1988	L	L	60	ARCH	88–06/1
F	Lake 104, YT	69° 26′	139° 36′	Sep 1988	L	L	49	ARCH	88–06/2,3,4
C	Babbage R, YT	68° 40′	139° 13′	Sep 1988	A?	S	4	DVCH-n	88–10/1
E	Babbage R, YT	68° 34′	139° 19′	Sep 1986	I	S	26	DVCH-n	86–60
E	Babbage R, YT	68° 38′	139° 22′	Sep 1988	I	S	26	DVCH-n	88–10/3
C	Canoe R, YT	68° 46′	138° 44′	Sep 1986	A	S	68	DVCH-n	86–61
C	Canoe R, YT	68° 46′	138° 45′	Sep 1988	A	S	56	DVCH-n	88–09
D	Canoe R, YT	68° 46′	138° 45′	Sep 1988	R	S	11	DVCH-n	88–09
C	Cache Creek, YT	68° 17′	136° 21′	Sep 1986	A	S	62	DVCH-n	86–62
D	Cache Creek, YT	68° 17′	136° 21′	Sep 1986	R	S	4	DVCH-n	88–62
C	Cache Creek, YT	68° 17′	136° 21′	Sep 1988	A	S	66	DVCH-n	88–05/1,2,3
D	Cache Creek, YT	68° 17′	136° 21′	Sep 1988	R	S	3	DVCH-n	88–05/1,2,3
E	Cache Creek, YT	68° 16′	136° 23′	Sep 1988	I	S	31	DVCH-n	88–05/4
C	Ptarmigan Bay, YT	69° 26′	139° 01′	Aug 1988	A	S	32	DVCH-n	88–57
C	Ptarmigan Bay, YT	69° 28′	139° 01′	Aug 1989	A	S	23	DVCH-n	88–05
C	Thetis Bay, YT	69° 33′	139° 02′	Aug 1989	A	S	13	DVCH-n	89–03
C	Pauline Cove, YT	69° 35′	138° 52′	Aug 1989	A	S	59	DVCH-n	89–04
C	Shingle Point, NWT	68° 59′	137° 31′	Aug 1989	A	S	16	DVCH-n	89–20
C	Rat River, NWT	67° 47′	136° 19′	Sep 1986	A	S	58	DVCH-n	86–63
C	Rat River, NWT	67° 47′	136° 19′	Sep 1988	A	S	51	DVCH-n	88–11
G	Wood Bay, NWT	69° 48′	129° 42′	Aug 1989	A	?	8	ARCH	89–08
G	Horton R, NWT	69° 56′	126° 48′	Sep 1988	A	?	12	ARCH	88–31
G	Hornaday R, NWT	69° 24′	123° 35′	Aug 1986	A	?	50	ARCH	86–64
G	Unnamed L, NWT	73° 05′	118° 15′	Aug 1987	A?	L?	52	ARCH	87–83
G	Kuuk R, NWT	70° 34′	112° 38′	Sep 1987	A	L?	31	ARCH	87–82
G	Kagluk R, NWT	70° 13′	112° 58′	Sep 1988	A	L?	30	ARCH	88–36
G	Kagloryuak R, NWT	70° 18′	111° 24′	Aug 1989	A	L?	28	ARCH	89–18
G	Naloagyuak R, NWT	70° 13′	112° 13′	Aug 1989	A	L?	31	ARCH	89–19
G	Kuujjua R, NWT	71° 15′	116° 30′	Oct 1991	A	L?	35	ARCH	91–25
H	Ferguson L, NWT	69° 22′	105° 03′	Dec 1987	A	L?	51	ARCH	87–85
H	30 Mile Ck, NWT	69° 16′	108° 00′	Dec 1987	A	L	13	ARCH	87–86
H	Byron Bay, NWT	69° 04′	109° 14′	Dec 1987	A	L	4	ARCH	87–87
H	Surrey, L, NWT	69° 40′	106° 40′	Dec 1987	A	L	5	ARCH	87–88

[a]Life history types are A = anadromous, R = residual member of anadromous populations, I = isolated river resident (nonanadromous), L = isolated lake resident (nonanadromous), and ? = uncertain type.

[b]Habitats where spawning or residence occurs are S = streams and rivers, L = lakes, and ? = uncertainty in habitat.

[c]Presumptive taxon as established herein: DVCH-s = southern form of Dolly Varden char, DVCH-n = northern form of Dolly Varden char, and ARCH = Arctic char.

American Fisheries Society Symposium 19:262–273, 1997

Use of a Stress Index to Estimate Temperature and Salinity Stress in Arctic Ciscoes

JAMES D. BRYAN AND ROBERT G. FECHHELM
LGL Alaska Research Associates, Inc.
4175 Tudor Centre Drive, Anchorage, Alaska 99508, USA

Abstract.—Causeways along Alaska's Beaufort Sea coast could cause nearshore fishes to be exposed to stressful combinations of temperature and salinity. We developed an index to characterize the level of stress that could be produced in juvenile Arctic ciscoes *Coregonus autumnalis* by extreme temperatures and salinities. Our index is based upon the observed tolerances and responses of Arctic ciscoes to various temperature and salinity regimes. The index was used to categorize the stress potential of conditions in the nearshore Beaufort Sea and to determine the degree of exposure of age-1 Arctic ciscoes to potentially stressful temperature and salinity combinations. Results of index evaluation suggest that the stress potential in early summer is caused by low temperature, but the potential increases because of the likelihood of a sharp increase in salinity in mid-July and a rapid cooling of coastal waters in late summer. The risk of stress in age-1 Arctic ciscoes is greatest from late July to mid-August, when nonstressful combinations of temperature and salinity become rare, substantial numbers of fish remain in the coastal zone, and fish parasite loads are high. However, Arctic ciscoes can reduce stress by entering lower-salinity overwintering sites in late summer, thus avoiding the cold, saline water along the coast and counteracting halophilic parasites.

River discharge along Alaska's Beaufort Sea coast creates a narrow band of relatively warm, low-salinity water during the summer (Niedoroda and Colonell 1990). This coastal band is an important summer feeding habitat for Arctic ciscoes *Coregonus autumnalis* and other amphidromous fish species (Craig 1984). Causeways associated with oil and gas production on Alaska's North Slope alter the coastal hydrography (Niedoroda and Colonell 1990), at times reducing the extent of warm, freshwater habitat and potentially exposing fish to cold marine water.

Biotic or abiotic challenges that require a compensatory response on the part of a biological system are said to produce stress (Pickering 1981b; Wedemeyer et al. 1990); extreme or rapidly changing temperature and salinity are among the many factors that can act as stressors (Schreck 1981; Wedemeyer et al. 1990). Causeway-induced hydrographic changes could expose fish to potentially stressful combinations of temperature and salinity and could exacerbate the effects of naturally occurring stressors.

Few studies of Prudhoe Bay fishes have attempted to detect the characteristic hematological and histological symptoms of stress. Although fish growth, condition, and abundance are affected by stress and have been monitored extensively in Prudhoe Bay, they are limited in their utility as stress indicators (Ryan and Harvey 1977; Rice 1990; Shuter 1990).

To evaluate the role that causeway-induced hydrographic changes may play in producing stress in Prudhoe Bay fishes, we developed a stress index. Our index is based on the classification of environmental variables by Fry (1947, 1971), the classification of stress by Brett (1958), and incorporates the observed effects of temperature and salinity on age-1 Arctic ciscoes. The index, then, represents a simple method of interpreting the temperature and salinity combinations recorded in the nearshore Beaufort Sea in terms of the observed responses of juvenile Arctic ciscoes to temperature and salinity.

The stress index was also used to examine relationships between the temperatures and salinities reported at fyke-net locations and the observed catch per unit effort (CPUE, fish $\cdot$ net^{-1} $\cdot$ d^{-1}) of age-1 Arctic ciscoes. We estimated the degree to which these fish have been exposed to potentially stressful conditions and examined the conditions that may affect exposure.

Methods

If fish acclimated to a known temperature and salinity are exposed to more extreme ambient conditions, the level of stress produced by those conditions might be estimated if the relationships among stress, ambient, and acclimation conditions were known. If the acclimation temperature and salinity are not known, as is generally the case in the field, the severity of the ambient conditions could be categorized by the level of stress that would be produced in fish acclimated to a reference temperature and salinity. Accordingly, our stress index categorizes the level and type of stress that would be produced by extreme ambient temperatures and salinities based upon the observed responses of juvenile Arctic ciscoes acclimated in the laboratory to warm freshwater.

Lethal Limits

Given sufficient time, fish can acclimate to a wide range of temperatures and salinities. However, fish cannot acclimate to temperatures beyond their ultimate incipient lethal temperatures (Fry 1947, 1971). At low temperatures, high salinity may often act as a "masking factor" (Fry 1947, 1971), increasing the lethality of cold water (Umminger 1971; Hazel and Prosser 1974; Eddy 1981). To separate conditions beyond the compensatory limits of Arctic ciscoes from those within, we divided temperature and salinity (T–S) combinations into those that permit relatively long-term survival and acclimation (tolerance) and those that produce lethal stress (resistance, i.e., lethal category in Figure 1). The boundary between tolerance and resistance was determined from T–S combinations to which juvenile Arctic ciscoes could and could not be acclimated (Fechhelm et al. 1983, 1991) and from measurements of the 96-h "dose" lethal to 50% of test organisms (LD_{50}) (Fechhelm et al. 1991), a more conventional estimate of the incipient lethal level (Fry 1971).

No-stress Zone

Nonlethal T–S combinations were divided into those that do not produce symptoms of sublethal stress (no-stress category in Figure 1) and those that do. Elliott (1981:212) defined the "optimum temperature range" as "... the range over which feeding occurs and there are no external signs of abnormal behavior, i.e., thermal stress is not obvious" and further stated that his optimum temperature range is similar to the preferred temperature range. Our no-stress category should then include temperatures and salinities that are within fishes' normal physiological range and would not be avoided in gradients of temperature or salinity.

The no-stress temperature range was determined from the preferred temperature range recorded by Fechhelm et al. (1983) (Figure 1). However, available data do not provide a clear upper limit for nonstressful salinities. It is unlikely that Arctic ciscoes acclimated to freshwater would be seriously stressed by moderate salinities of 10–20‰, because dilute salt water has often been found to reduce stress in freshwater fish (Schreck 1981). Given that the highest growth rates of anadromous fishes tend to occur at salinities of 10 ± 2‰, (i.e., salinities nearly isosmotic with fish blood; Brett [1979]) and that Arctic ciscoes are generally considered to be among the more euryhaline of the nearshore fishes (Craig 1984), we believe that salinities below 15‰ probably were not stressful within the preferred temperature zone (Figure 1).

Sublethal Stress Zone

The T–S combinations beyond the boundaries of the no-stress zone but within the zone of tolerance were categorized as combinations with the potential for producing sublethal stress. The transition from T–S combinations that produce lower sublethal stress with primarily bioenergetic consequences (i.e., the "limiting" or "loading" stress of Brett [1958]) to those that produce higher sublethal stress where vital processes may become disrupted (Brett's "inhibiting" stress, as well as limiting and loading stress) is undoubtedly gradual, without a distinct boundary. Yet fish acclimated to warm freshwater may survive milder sublethal T–S combinations with little or no further acclimation (low sublethal, Figure 1), whereas combinations nearer their lethal limits may require acclimation for survival (high sublethal, Figure 1). Because Arctic ciscoes prefer, and normally occur in, relatively warm water of lower salinity, an important division among sublethal T–S combinations separates the colder, more-marine combinations that require previous acclimation or gradual exposure to avoid mortality from the less-severe combinations that may produce stress but no mortality.

We defined the boundary between high- and low-sublethal conditions as the most extreme T–S combinations that produced no more than 5% mortality (after 96 h) in juvenile Arctic ciscoes when directly transferred to those combinations from 12°C and 0‰ (see Figure 1 of Fechhelm et al. 1991).

Stress Categories

Stress index categories (no stress, low sublethal, high sublethal, and lethal, as defined above) were assigned numeric values to allow arithmetic descriptions of average stress-index values (Figure 1). We also defined a temperature index and a salinity index, each with numeric values, to describe variation of the stress index along the salinity and temperature axes (see Figure 1).

Exposure to Stressful Conditions

The risk that potentially stressful conditions represent to the population of age-1 Arctic ciscoes

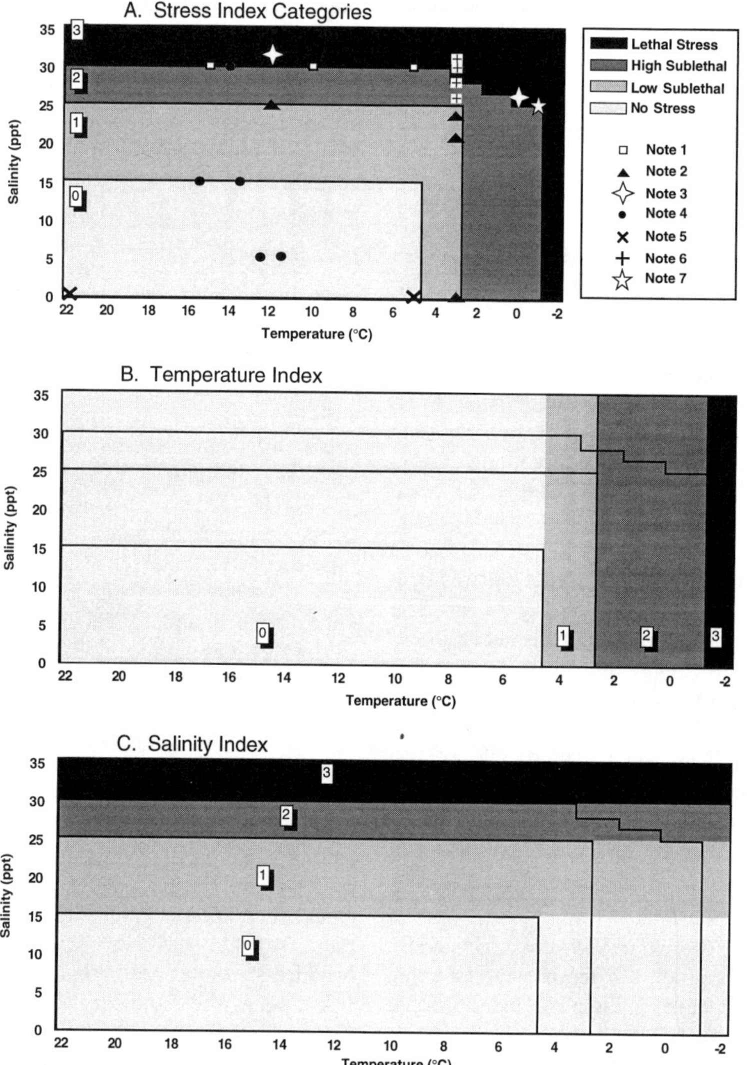

FIGURE 1.—Classification of temperature and salinity combinations by stress index (**A**), temperature index (**B**), and salinity index (**C**) categories for age-1 Arctic ciscoes *Coregonus autumnalis*. See text for explanation of categories. Numbers shown are the numeric values assigned to each category for the purpose of describing average conditions. Note 1: Salinities to which Arctic ciscoes could be acclimated at 15, 10, and 5°C (Fechhelm et al. 1983). Note 2: Temperatures and salinities that produced no more than 5% mortality after direct transfer from 12°C and 0‰ (Fechhelm et al. 1991). Note 3: Estimates of temperature and salinity combinations beyond the zone of tolerance (Fechhelm et al. 1991 and W. J. Wilson, LGL Alaska Research Associates, Inc., personal communication). Note 4: Preferred temperatures at indicated acclimation and test salinities (Fechhelm et al. 1983). Note 5: Estimated boundaries of preferred temperature range (Fechhelm et al. 1983). Note 6: Temperatures and salinities that produced more than 5% mortality (in 96 h) after direct transfer from 12°C and 0‰ (Fechhelm et al. 1991). Note 7: Overwintering temperature and salinity observed by Schmidt et al. (1989).

depends upon the degree to which the population is exposed to those conditions, that is, upon the proportion of the population that occurs within areas that may include potentially stressful conditions and upon the distribution of that proportion among ambient temperatures and salinities. We estimated the degree of exposure by comparing catch rates of age-1 Arctic ciscoes among stress index categories.

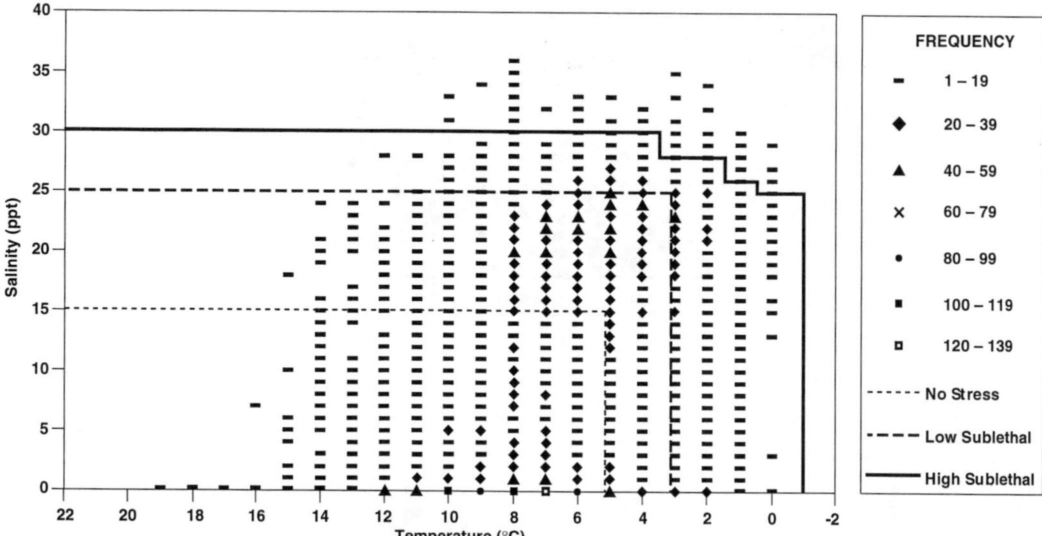

FIGURE 2.—Frequency of occurrence of bottom temperature and salinity combinations recorded in the Prudhoe Bay area nearshore zone in field studies during summers 1986–1991. Symbols in the legend indicate the number of occurrences of each combination and heavy lines indicate the boundaries of the stress index categories. The lethal zone is above the high-sublethal boundary.

Stress Indicators

There are limited data on liver glycogen levels and parasite loads, more conventional indicators of stress, that may be used to corroborate the results of the stress index analysis. These data are discussed after the stress index analysis.

Results and Discussion

Stress Potential

Figure 2 displays the frequencies of T–S combinations recorded for nearshore bottom conditions in the Prudhoe Bay area during summer studies from 1986 to 1991 (Glass et al. 1990; LGL 1990, 1991, 1992a, 1992b; Reub et al. 1991). The most frequent combinations were within the no-stress zone, near 0‰ salinity and from 5 to 12°C. A second peak occurred near the low-sublethal–high-sublethal boundary from 3 to 8°C and from 20 to 25‰, conditions that Arctic ciscoes can survive when acclimated to warm freshwater (the most commonly occurring nearshore condition). High-sublethal combinations were less frequent, and few combinations were within the lethal zone.

Figure 3 illustrates the weekly averages of the stress, temperature, and salinity indexes for bottom conditions during summers 1986–1991. Also included are average index values for surface conditions in the Prudhoe Bay area that were recorded during summers 1982–1985 (Critchlow 1983; Griffiths et al. 1983; WWC 1983; Biosonics 1984; Moulton et al. 1986; Cannon et al. 1987). Values of the stress index were relatively low and quite variable early in the season but generally increased throughout the summer. The temperature index exhibited a generally similar trend but with relatively low values (higher temperatures) during late July and early August. Early in the season the stress potential was attributable almost exclusively to low temperature because almost all salinities were within the no-stress range. Values of the salinity index tended to increase abruptly in late July, ultimately compounding the effects of low temperature later in the season.

Exposure

To determine the relative degree of exposure of age-1 Arctic ciscoes to potentially stressful conditions, we compared the daily medians of fyke-net catch rates (CPUE) with the daily averages of the stress index for those years (since 1982) reporting a strong age-1 Arctic cisco year-class (1986, 1987, 1988, and 1991; raw data reported in Glass et al. 1990; LGL 1990, 1992b; Reub et al. 1991). We chose the median catch rate of all reporting fyke nets as our estimate of overall daily abundance

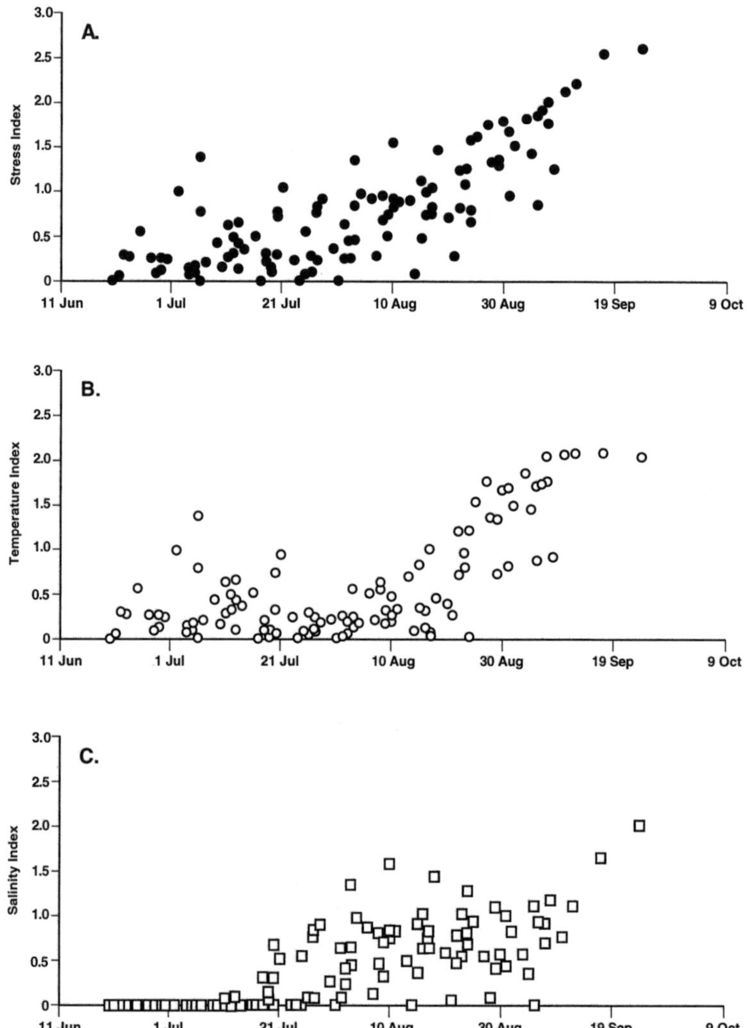

FIGURE 3.—Weekly averages of the stress index (**A**), temperature index (**B**), and salinity index (**C**) for bottom conditions (1986–1991) and surface conditions (1982–1985) recorded in the Prudhoe Bay area nearshore zone.

because of the highly skewed nature of the catch rate distributions. Locations of fyke-net stations reporting the catch rates, temperatures, and salinities used to determine CPUE and the average stress potential are shown in Figure 4.

The lower daily catch rates occurred late in the season, when average values of the stress index were at their highest (Figure 5). As a result, daily median CPUE and the daily average of the stress index were inversely correlated (Spearman's rank correlation coefficient, Mendenhall et al. [1986]) in each of the 4 years examined (1986, −0.231; 1987, −0.564; 1988, −0.591; 1991, −0.111). The degree of correlation was significant ($\alpha = 0.05$) for each year except 1991, when the sampling season ended before the more widespread development of stressful conditions in September. Although fish were caught until the end of the season each year, the period of greater abundance (to the degree that catch represents abundance) occurred before the period of higher stress potential.

Stress levels in age-1 Arctic ciscoes may have been lower than those suggested by the daily averages of the stress index if the fish distribution was biased toward lower-stress categories. We calculated the expected daily catch within each stress category by multiplying the total number of fish caught (per day) in all nets by the fraction of all nets

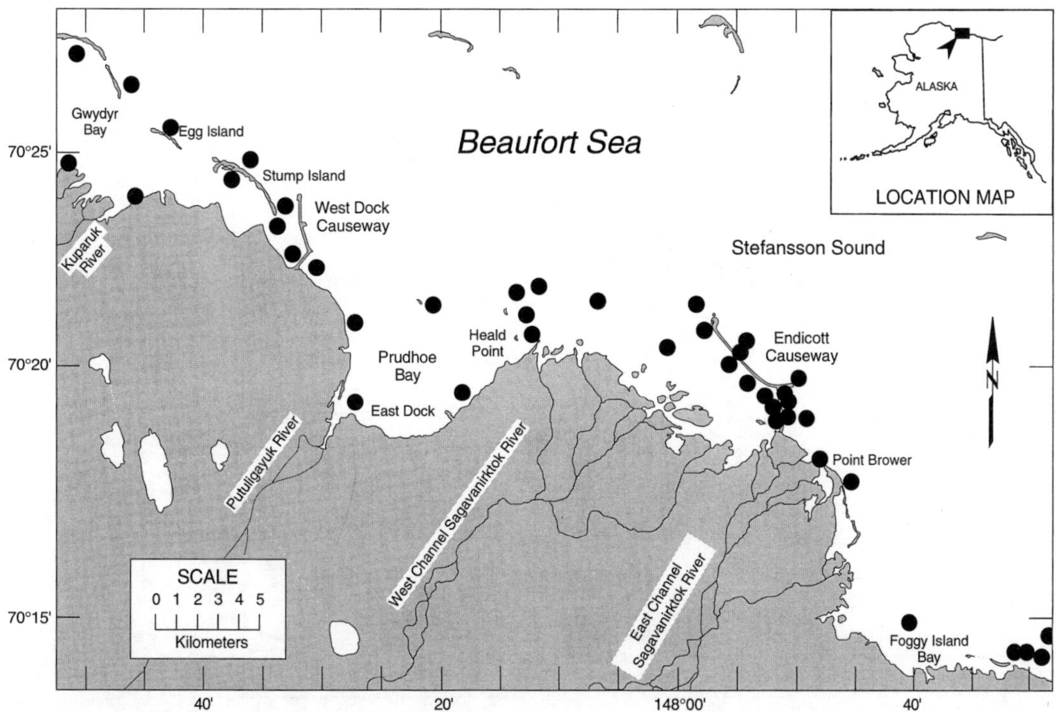

FIGURE 4.—General locations of fyke nets along the Beaufort Sea coast reporting the catch rates, temperatures, and salinities that were used in this study.

that reported conditions within the individual categories, and then compared the expected and observed catch for each day within each category tested (Wilcoxon signed rank test, Ott [1988]). In 1986, 1987, and 1988 significantly more fish were caught in the no-stress category than would be expected on the basis of environmental availability alone (Bonferroni correction: $\alpha = 0.05/4$ years; see Figure 6).

There was no statistically significant bias among stress categories in 1991 ($\alpha = 0.05/3$ categories; Figure 6), when conditions remained mild in the nearshore zone until the end of the (abbreviated) sampling season, with average salinities generally remaining within the no-stress category (Figure 7). Low salinities suggest reduced mixing of nearshore and offshore waters that could reduce food availability (LGL 1992b). If so, fish may have been forced to forage nearer the nearshore–offshore interface, offsetting the water quality bias observed in previous years.

To estimate the degree of exposure of the age-1 Arctic cisco population to extreme temperature and salinity, we estimated total fish abundance within each stress index category by taking the seasonal totals of daily catch (adjusted to 24-h effort) in each category, then dividing each category total by the total effort for the season (Figure 8). In 1986, 1987, and 1991 total catch rate in the high-sublethal and lethal categories remained low and nearly the same among years, with differences in annual abundance manifested as changes in the low-sublethal and no-stress categories. Consequently, in years with a greater overall abundance, the additional fish appear to have been "funnelled" into the lower-stress categories, reducing the fraction of the seasonal catch that occurred in the higher categories. Accordingly, a relatively greater fraction of the total catch was recorded in the higher-stress categories in years with lower abundance.

In 1988 a substantially higher overall catch occurred in the high-sublethal category than in other years as a result of relatively high daily catch rates late in the season, when there were no lower-stress T–S combinations available. Consequently, a relatively greater fraction of the overall seasonal catch occurred in higher categories in 1988 than in the other 3 years. In 1986, 1987, and to a lesser degree in 1991, salinities increased abruptly in mid-July and temperatures decreased relatively quickly near

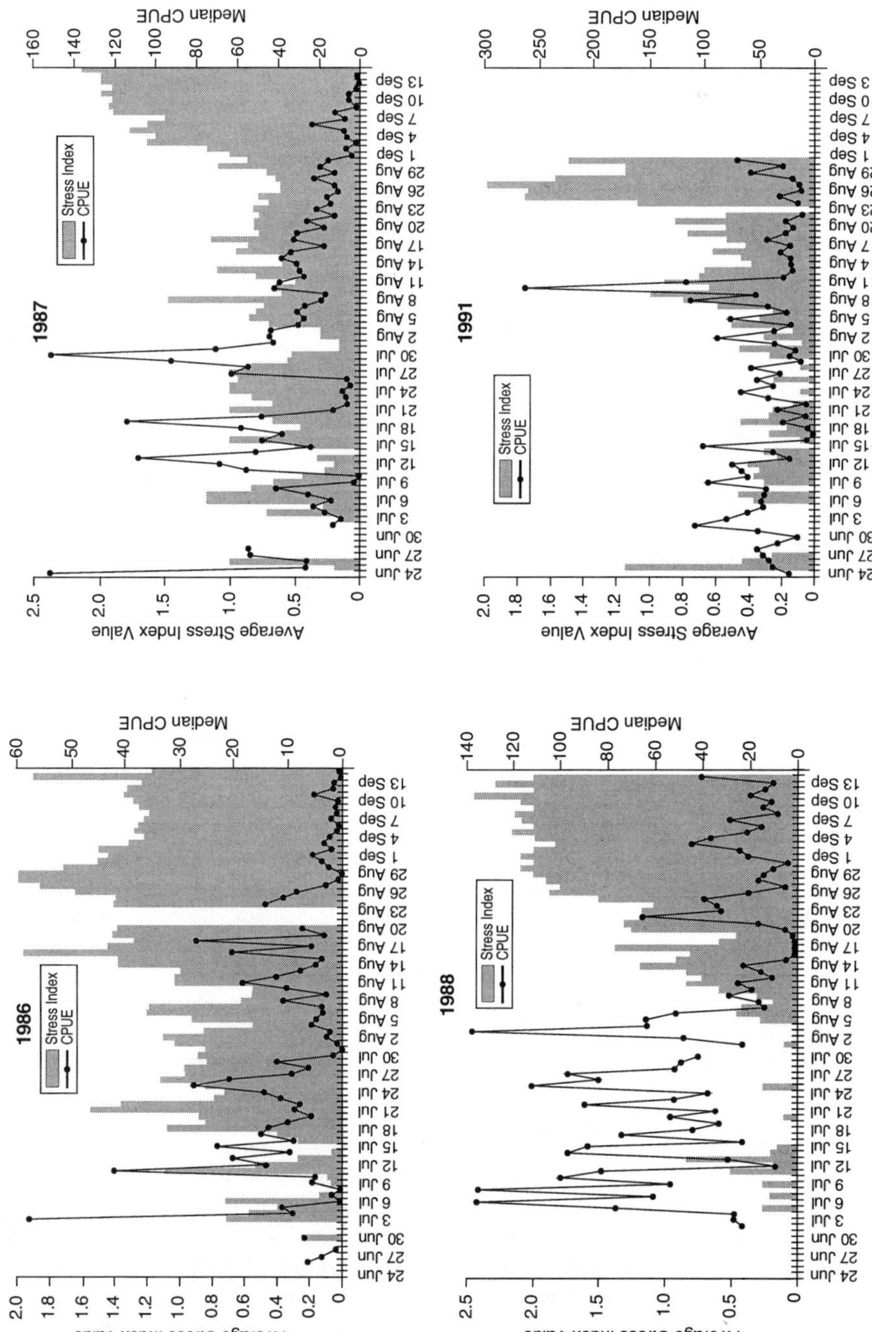

FIGURE 5.—Daily medians of catch per unit effort (CPUE, fish $\cdot$ net^{-1} $\cdot$ d^{-1}) of age-1 Arctic ciscoes *Coregonus autumnalis* and daily averages of the stress index for all reporting fyke-net stations in 1986, 1987, 1988, and 1991.

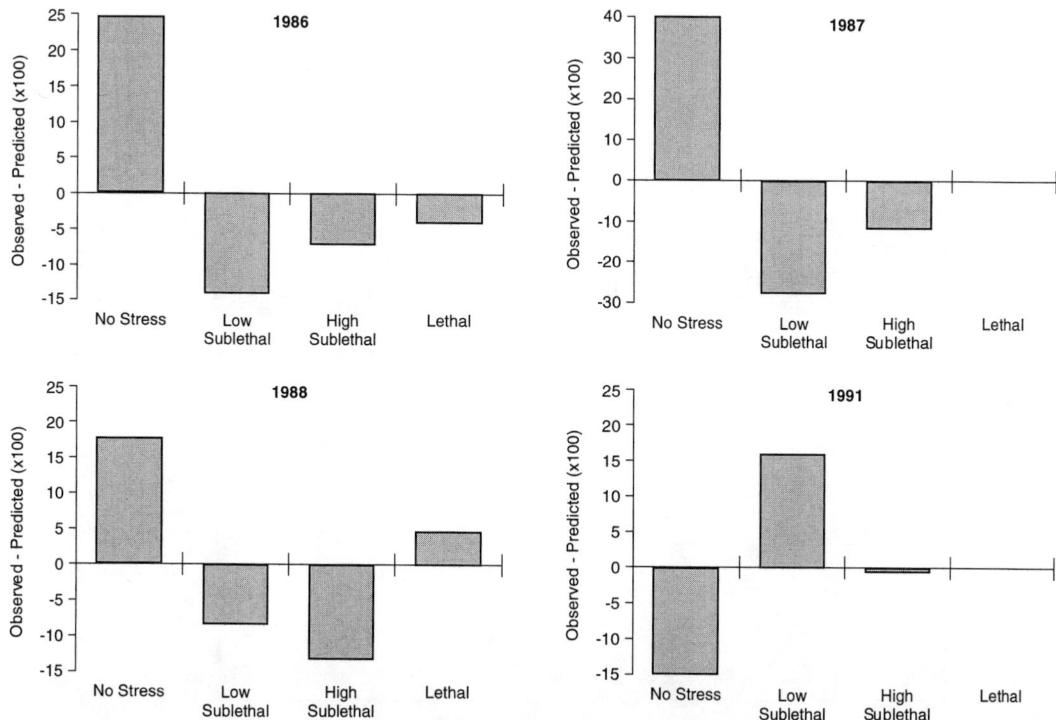

FIGURE 6.—Cumulative daily (observed − predicted) catch of age-1 Arctic ciscoes *Coregonus autumnalis* by stress index category for 1986, 1987, 1988, and 1991. See text for an explanation of predicted catch. Results of two-tailed signed rank tests on the daily differences between observed and predicted catch for no-stress, low-sublethal, high-sublethal, and lethal categories, respectively, were as follows: 1986, P = 0.0000043, 0.016, 0.0068, and 0.0084; 1987, P = 0.0010, 0.063, 0.0014, and NA; 1988, P = 0.0069, 0.025, 0.33, and 0.94; and 1991, P = 0.44, 0.34, 0.46, and NA. The acronymn NA indicates that too few daily values were available for a valid test ($N < 10$).

the end of August (Figure 7 and 3, respectively). In 1988, however, salinities increased slowly over July and August with water temperature dropping at a relatively constant rate from mid-August until the end of the summer. The gradual change in conditions may have allowed fish to more fully acclimate to the more-severe T–S combinations, making those combinations less stressful, or if changing water quality is among the cues that affect the regional movements of Arctic ciscoes, gradually worsening conditions may not have produced a clear environmental signal to return to overwintering grounds.

Liver Glycogen Levels

Reduced liver glycogen levels may be used as an indicator of stress because the characteristic stress-induced increase in blood glucose level is attributable, in part, to elevated glycogenolysis in the liver and muscles (Schreck 1981; Eckert et al. 1988; Wedemeyer et al. 1990). Glass (1991) reported average liver glycogen levels in Arctic ciscoes (<150 mm fork length) collected from 18 August to 15 September 1987 at Oliktok Point (approximately 60 km west of Prudhoe Bay) and the Sagavanirktok River delta (Figure 4) and on 25 and 26 June 1988 from the Colville (approximately 80 km west of Prudhoe Bay) and Sagavanirktok river deltas. Glass (1991) found that liver glycogen levels in Sagavanirktok samples were higher than in Colville–Oliktok samples, although the difference was not statistically significant. However, liver glycogen levels in samples from both regions were significantly lower in the fall than in the early season, suggesting that the fish of both regions were stressed after conditions deteriorated in the fall.

Parasite Infestation

Although a number of factors undoubtedly affect the level of parasitism in fish, increases in the degree of parasitism may be indicative of a stress-

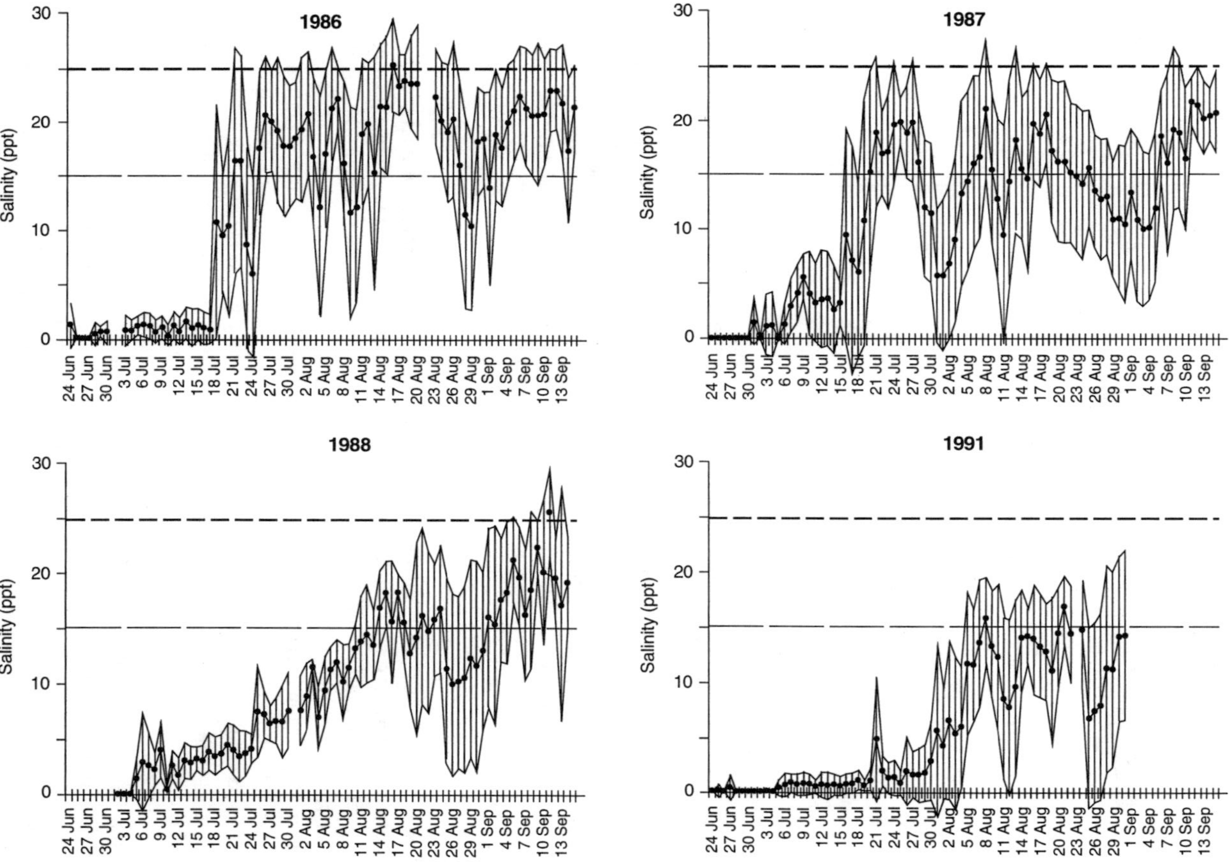

FIGURE 7.—Daily average of bottom salinities for all reporting fyke nets (solid circles) during 1986, 1987, 1988, and 1991. Vertical lines indicate 1 SD above and below the average. Heavy horizontal dashed lines indicate the highest no-stress salinity (lower line) and the highest low sublethal salinity (upper line).

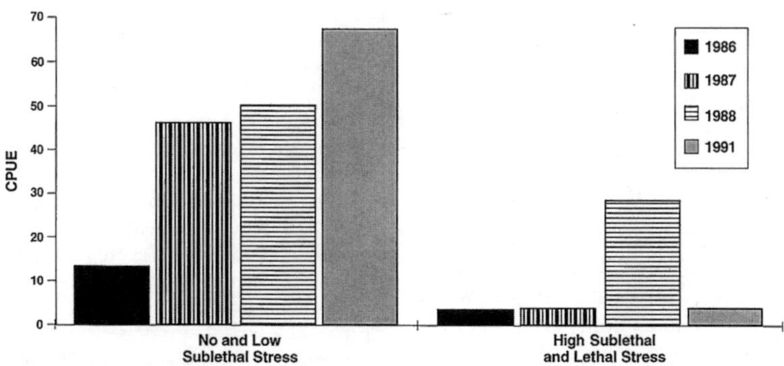

FIGURE 8.—Overall seasonal catch per unit effort (CPUE, fish · net^{-1} · d^{-1}) of age-1 Arctic ciscoes *Coregonus autumnalis* by combined stress index category.

induced reduction in the effectiveness of the immune system (Esch et al. 1975; Anderson 1990). Craig and Haldorson (1981) examined Arctic cisco parasites in samples taken west of Prudhoe Bay in Simpson Lagoon on 30 June, 27 July, and 2 September in 1977 and from under the ice on the Colville River delta in November of 1976 and 1977. The dominant parasite found was the cestode *Diplocotyle olrikii* (probably synonymous with *Bothrimonus sturionus*; Margolis and Arthur 1979). *D. olrikii* is thought to be intolerant of low salinity and has been reported to die when its host enters fresh water (Bauer 1970). Craig and Haldorson (1981) found no *D. olrikii* in the freshwater November samples and very few in the June samples, but they found a high infestation rate in July that decreased only slightly by early September. The incidence of other parasites also peaked in their July sample.

Parasite infestation rates increased as salinities in Simpson Lagoon rose to potentially stressful levels in late July (Figure 9), when salinities characteristically increase in the nearshore zone (Figure 3). An increase in parasitism after exposure to stressful salinities is consistent with a stress-induced reduction in immune function in the host by conditions that benefit the parasite. The lower infestation rates

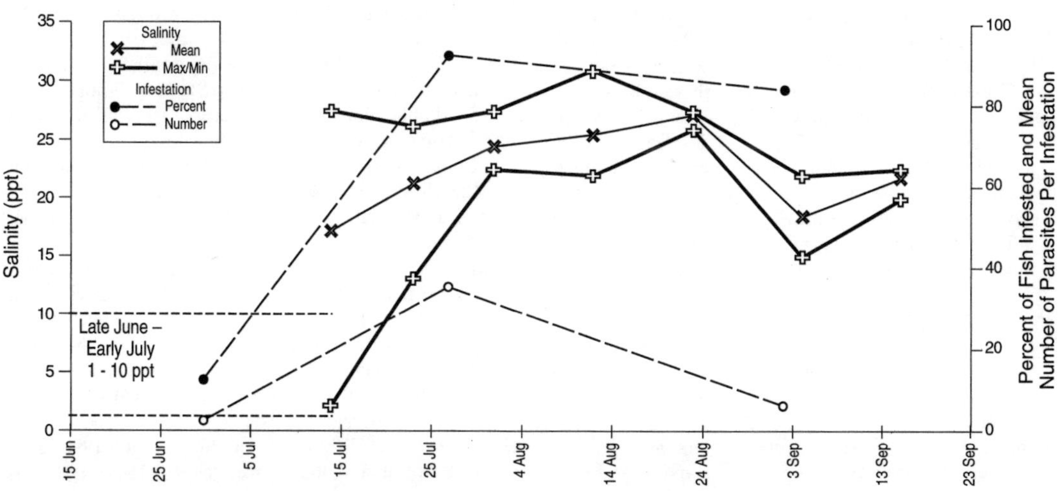

FIGURE 9.—Level of infestation of Arctic ciscoes *Coregonus autumnalis* with *Diplocotyle olrikii* and ambient salinities recorded by Craig and Haldorson (1981) for samples taken in Simpson Lagoon in 1977. Solid circles indicate percent of fish infested; open circles indicate the average number of *D. olrikii* per fish. Heavy solid lines (with open plus signs) indicate maximum and minimum recorded salinities; regular solid line (with open x's) indicates average salinities. Craig and Haldorson (1981) did not report salinities for June and early July, but they described salinities as ranging from 1 to 10‰.

recorded by Craig and Haldorson (1981) in the freshwater November samples suggest that Arctic ciscoes avoided stressful salinities, shifting the balance between parasite and host in their favor, by returning to overwintering sites of lower salinity.

Conclusions

In general, the period of greatest risk from nearshore conditions in the Prudhoe Bay area to age-1 Arctic ciscoes is from late July to mid-August, when nonstressful temperature and salinity combinations become rare and substantial numbers of fish remain in the coastal area, possibly with heavy parasite loads. However, juvenile Arctic ciscoes can reduce stress levels during late summer by entering overwintering sites of lower salinity, thereby avoiding the cold, saline water along the coast and eliminating halophilic parasites.

Acknowledgments

We thank J. Erwin, B. Gallaway, B. Griffiths, and B. Wilson for their critical review of the manuscript. Environmental-monitoring programs were funded by the Duck Island Unit owners, which is operated by BP Exploration (Alaska) Inc. The conclusions and opinions expressed in this article do not necessarily represent those of any of the above-mentioned organizations or individuals.

References

Anderson, D. P. 1990. Immunological indicators: effects of environmental stress on immune protection and disease outbreaks. Pages 38–50 *in* S. M. Adams, editor. Biological indicators of stress in fish. American Fisheries Society Symposium 8, Bethesda, Maryland.

Bauer, O. N. 1970. Parasites and diseases of U.S.S.R. coregonids. Pages 267–278 *in* C. C. Lindsey and C. S. Woods, editors. Biology of coregonid fishes. University of Manitoba Press, Winnipeg.

Biosonics, Inc. 1984. Prudhoe Bay waterflood project fish monitoring program, 1983. Report of Biosonics, Inc. to U.S. Army Corps of Engineers, Anchorage, Alaska.

Brett, J. R. 1958. Implications and assessments of environmental stress. Pages 69–83 *in* P. A. Larkin, editor. Investigations of fish–power problems. H. R. MacMillan Lectures in Fisheries. University of British Columbia, Vancouver.

Brett, J. R. 1979. Environmental factors and growth. Pages 599–675 *in* W. S. Hoar, D. J. Randall, and J. R. Brett, editors. Fish physiology, volume 8. Academic Press, New York.

Cannon, T. C., B. Adams, D. Glass, and N. Troy. 1987. Fish distribution and abundance. Pages 1–129 *in* 1985 Endicott Environmental Monitoring Program, volume 6. Report of Envirosphere Company to U.S. Army Corps of Engineers, Anchorage, Alaska.

Craig, P. C. 1984. Fish use of coastal waters of the Alaskan Beaufort Sea: a review. Transactions of the American Fisheries Society 113:265–282.

Craig, P. C., and L. Haldorson. 1981. Fish. Pages 384–678 *in* Beaufort Sea barrier island–lagoon ecological process studies: final report, Simpson Lagoon. Environmental assessment of the Alaskan continental shelf, final report of the principal investigators, volume 7. U.S. Bureau of Land Management and National Oceanic and Atmospheric Administration, Outer Continental Shelf Environmental Assessment Program, Boulder, Colorado.

Critchlow, K. 1983. Prudhoe Bay waterflood project fish monitoring program, 1982. Report of Woodward-Clyde Consultants to U.S. Army Corps of Engineers, Anchorage, Alaska.

Eckert, R., D. R. Randall, and G. Augustine. 1988. Animal physiology, third edition. Freeman, New York.

Eddy, F. B. 1981. Effects of stress on osmotic and ionic regulation in fish. Pages 77–102 *in* Pickering (1981a).

Elliott, J. M. 1981. Some aspects of thermal stress on freshwater teleosts. Pages 209–245 *in* Pickering (1981a).

Esch, G. W., J. W. Gibbons, and J. E. Bourque. 1975. An analysis of the relationship between stress and parasitism. American Midland Naturalist 93:339–353.

Fechhelm, R. G., J. D. Fechhelm, C. J. Herlugson, and D. K. Beaubien. 1991. Bioassay tests of acute temperature and salinity shock to Arctic cisco. Pages 109–118 *in* C. S. Benner and R. W. Middleton, editors. Fisheries and oil development on the continental shelf. American Fisheries Society Symposium 11, Bethesda, Maryland.

Fechhelm, R. G., W. H. Neill, and B. J. Gallaway. 1983. Temperature preference of juvenile Arctic cisco (*Coregonus autumnalis*) from the Alaskan Beaufort Sea. Biological Papers of the University of Alaska 21:24–38.

Fry, F. E. J. 1947. Effects of the environment on animal activity. University of Toronto Studies, Biology Series 55, Publications of the Ontario Fisheries Research Laboratory, Toronto.

Fry, F. E. J. 1971. The effect of environmental factors on the physiology of fish. Pages 1–98 *in* W. S. Hoar and D. J. Randall, editors. Fish physiology, volume 6. Academic Press, New York.

Glass, D. R. 1991. Changes in the condition and physiological parameters of anadromous fish species in the central Beaufort Sea during the overwintering period, part 1. Pages 1–45 *in* 1987 final report for the Endicott Environmental Monitoring Program. Report of Envirosphere Company to U.S. Army Corps of Engineers, Anchorage, Alaska.

Glass, D. R., C. Whitmus, and M. Prewitt. 1990. Fish distribution and abundance. Pages 1–154 *in* 1986 Endicott Environmental Monitoring Program, volume 5. Report of Envirosphere Company to U.S. Army Corps of Engineers, Anchorage, Alaska.

Griffiths, W. B., D. R. Schmidt, R. G. Fechhelm, and B. J. Gallaway. 1983. Fish ecology. Pages 1–323 *in* B. J. Gallaway and R. P. Britch, editors. Environmental summer studies for the Endicott Development, volume 5. Report of LGL Alaska Research Associates, Inc. and Northern Technical Services to Sohio Alaska Petroleum Company, Anchorage, Alaska.

Hazel, J. R., and C. L. Prosser. 1974. Molecular mechanisms in temperature compensation in poikilotherms. Physiological Reviews 54:620–677.

LGL (LGL Alaska Research Associates, Inc.). 1990. Recruitment and population studies, analysis of fyke net data. Pages 1–317 *in* the 1988 Endicott Development fish monitoring program, volume 2. Report of LGL Alaska Research Associates, Inc., to BP Exploration (Alaska) Inc., Anchorage, Alaska, and the North Slope Borough, Point Barrow, Alaska.

LGL (LGL Alaska Research Associates, Inc.). 1991. Analysis of fyke net data. Pages 1–143 *in* the 1989 Endicott Development fish monitoring program, volume 2. Report of LGL Alaska Research Associates, Inc., to BP Exploration (Alaska) Inc., Anchorage, Alaska, and the North Slope Borough, Point Barrow, Alaska.

LGL (LGL Alaska Research Associates, Inc.). 1992a. Analysis of fyke net data. Pages 1–160 *in* the 1990 Endicott Development fish monitoring program, volume 2. Report of LGL Alaska Research Associates, Inc., to BP Exploration (Alaska) Inc., Anchorage, Alaska, and the North Slope Borough, Point Barrow, Alaska.

LGL (LGL Alaska Research Associates, Inc.). 1992b. Analysis of fyke net data. Pages 1–188 *in* the 1991 Endicott Development fish monitoring program, volume 2. Report of LGL Alaska Research Associates, Inc., to BP Exploration (Alaska) Inc., Anchorage, Alaska, and the North Slope Borough, Point Barrow, Alaska.

Margolis, L., and J. R. Arthur. 1979. Synopsis of parasites of fishes of Canada. Canada Department of Fisheries and Oceans, Bulletin 199, Ottawa.

Mendenhall, W., R. L. Scheaffer, and D. D. Wackerly. 1986. Mathematical statistics with applications, third edition. Duxbury Press, Boston.

Moulton, L. L., and seven coauthors. 1986. 1984 Central Beaufort Sea fish study. Waterflood Monitoring Program Fish Study. Report of Entrix, Inc., LGL Ecological Research Associates, Inc., and Woodward-Clyde Consultants to Envirosphere Company, Anchorage, Alaska, and U.S. Army Corps of Engineers, Anchorage, Alaska.

Niedoroda, A. W., and J. M. Colonell. 1990. Beaufort Sea causeways and coastal ocean dynamics. Pages 509–516 *in* Proceedings of the 9th international conference of offshore mechanics and arctic engineering. American Society of Mechanical Engineers, New York.

Ott, L. 1988. An introduction to statistical methods and data analysis, third edition. PWS-Kent Publishing Company, Boston.

Pickering A. D., editor. 1981a. Stress and fish. Academic Press, New York.

Pickering, A. D. 1981b. Introduction: the concept of biological stress. Pages 1–9 *in* Pickering (1981a).

Reub, G. S., J. D. Durst, and D. R. Glass. 1991. Fish distribution and abundance. Pages 1–60 *in* 1987 Endicott Environmental Monitoring Program, volume 6. Report of Envirosphere Company to U.S. Army Corps of Engineers, Anchorage, Alaska.

Rice, J. A. 1990. Bioenergetics modeling approaches to evaluation of stress in fishes. Pages 80–92 *in* S. M. Adams, editor. Biological indicators of stress in fish. American Fisheries Society Symposium 8, Bethesda, Maryland.

Ryan, P. M., and H. H. Harvey. 1977. Growth of rock bass, *Ambloplites rupestris*, in relation to the morphoedaphic index as an indicator of an environmental stress. Journal of the Fisheries Research Board of Canada 34:2079–2088.

Schmidt, D. R., W. B. Griffiths, and L. R. Martin. 1989. Overwintering biology of anadromous fish in the Sagavanirktok River delta, Alaska. Biological Papers of the University of Alaska 24:55–74.

Schreck, C. B. 1981. Stress and compensation in teleostean fishes: response to social and physical factors. Pages 295–321 *in* Pickering (1981a).

Shuter, B. J. 1990. Population-level indicators of stress. Pages 145–166 *in* S. M. Adams, editor. Biological indicators of stress in fish. American Fisheries Society Symposium 8, Bethesda, Maryland.

Umminger, B. L. 1971. Chemical studies of cold death in the gulf killifish, *Fundulus grandis*. Comparative Biochemistry and Physiology A Comparative Physiology 39:625–632.

Wedemeyer, G. A., B. A. Barton, and D. J. McLeay. 1990. Stress and acclimation. Pages 451–489 *in* C. B. Schreck and P. B. Moyle, editors. Methods for fish biology. American Fisheries Society, Bethesda, Maryland.

WWC (Woodward-Clyde Consultants). 1983. Lisburne development area: 1983 environmental studies—final report. Report of Woodward-Clyde Consultants to ARCO-Alaska, Inc., Anchorage.

American Fisheries Society Symposium 19:274–286, 1997

Estimating Abundance of Arctic Anadromous Fish in the Coastal Beaufort Sea during Summer: Is an Open Population Model Necessary?

Benny J. Gallaway and Robert G. Fechhelm
LGL Alaska Research Associates, Inc.
4175 Tudor Centre Drive, #202, Anchorage, Alaska 99508, USA

William J. Gazey
W. J. Gazey Research
1214 Camas Court, Victoria, British Columbia V8X 4R1, Canada

Abstract.—Arctic anadromous fishes exhibit extensive movements along the Beaufort Sea coast during their summer feeding season. Long-term monitoring programs at specific locations within this distributional range usually use catch-per-unit-effort (CPUE) data to index annual abundance of these fishes, but the sampling gears that are effective in capturing the fish are passive. This can lead to a strong bias because high catches may reflect strong movement, not high abundance. Multiple mark–recapture models have been used to supplement CPUE data for better understanding of fluctuations in abundance. The models routinely used (Schumacher–Eschmeyer, Schnabel, and Gazey–Staley) depend on the unlikely assumption that the population size does not change during the experiment. We report the degree to which the estimates for one species, least cisco *Coregonus sardinella*, are compromised by this invalid assumption. Estimates obtained from an open population model (Jolly–Seber) are compared to estimates obtained from the closed model (Gazey–Staley) using the same data. In 1990, only a 3% difference during periods of abundance was observed between the mean of the daily estimates obtained using the open population model and the maximum likelihood estimate provided by the closed population model. In 1991, the difference between the two estimates was 21% but was not statistically significant. We show that the "population" is closed to immigration during the experimental period, although some emigration occurs. Under these conditions a valid estimate can be obtained from the closed population model.

The construction of causeways in the nearshore Beaufort Sea (Figure 1) and their potential effects on anadromous coregonid fish have generated large, long-term monitoring programs (Colonell and Gallaway 1990; Gallaway et al. 1991; Hachmeister et al. 1991). Fundamental questions arise regarding our capability to evaluate effects of coastal development on these fish: How many fish are present? What proportion of these fish are exposed to potential effects? To answer these questions requires an ability to accurately estimate or reliably index population size.

Catch per unit effort (CPUE) at index stations is the most common means by which this objective is achieved, but the sampling gears that are effective in capturing the fish are passive. This can lead to bias because high catches may reflect strong movement, not high abundance. In recent years, multiple mark–recapture studies have been conducted to supplement CPUE data to gain a better understanding of fluctuations in coregonid abundance.

Four basic assumptions must be met to obtain valid mark–recapture estimates of abundance. Stated formally, these are as follows (Gazey and Staley 1986):

1. The population is closed, so the population size does not change over the period of the experiment.
2. The probability of capturing a marked individual at any given time is equal to the proportion of marked members in the population at that time.
3. Animals do not lose their marks over the period of the study.
4. All marked fish are reported when recaptured.

For small coregonid fishes that have been studied in the developed region (Arctic ciscoes *Coregonus autumnalis* and broad whitefish *C. nasus* between 120 and 250 mm fork length), the mark–recapture approach appears valid; that is, there is no reason to expect any major violations of the underlying assumptions.

The small coregonid fishes in question mostly overwinter in the Sagavanirktok River and exhibit a restricted distribution around its delta (Fechhelm et al. 1992; Griffiths et al. 1992). The developed region within which these small fish are restricted is totally encompassed by the sampling effort (Figure 1). Even though the sampling is systematic rather than random, the small fish exhibit distributional and

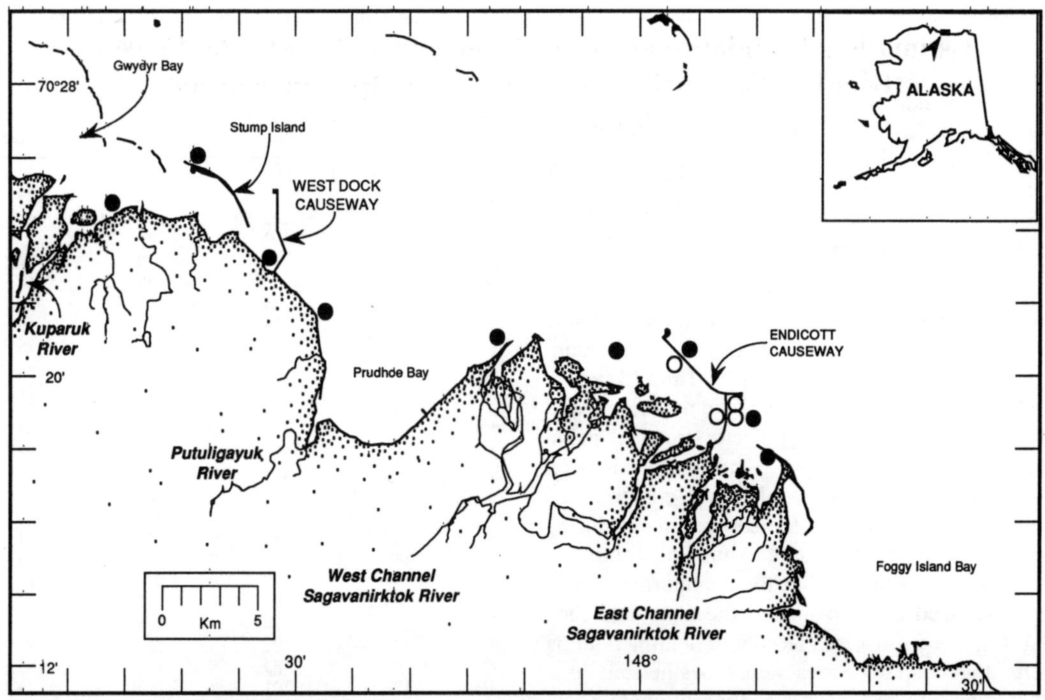

FIGURE 1.—Map of study area in the Beaufort Sea showing causeways and fyke-net station locations in 1990 (closed circles) and 1991 (closed and open circles).

movement characteristics that would suggest uniform mixing of marked and unmarked animals. The populations are believed to be reasonably constant (Griffiths et al. 1992) because (1) catch patterns show that no immigration to or emigration from other areas occurs, (2) mortality during the brief summer period is believed to be low, and (3) losses from and gains into the size interval of interest because of growth are considered negligible relative to the total numbers of fish present.

Most of the large (>250 mm fork length) coregonids that use our study area (e.g., least ciscoes *C. sardinella*) overwinter in the Colville River. Because of the distance between the Colville River and our study area near Prudhoe Bay (approximately 100 km), obtaining accurate estimates of abundance poses a formidable problem: sampling is restricted to only the eastern one-third of the overall distributional range used during the summer feeding period. Large numbers of least ciscoes may move into our area during summer. If so, they may remain in our area for some period, move farther eastward, or move back to the west, depending upon hydrographic conditions and prey abundance encountered along the way. A model that assumes an open population for our area seems more appropriate than a closed population model.

The suitability of a closed population model for estimating abundance is important. In these models, only two categories of fish need be known—those that are marked and those that are not marked. In an open population model, an additional piece of information must be known—when each marked animal was last captured. Fish must be classified into their capture and recapture period combination, thus substantially increasing the total number of animals required to obtain reliable estimates in comparison to the closed population model. A unique mark for each individual would be needed. In some instances a unique mark for each fish or the increased sampling effort is either impossible on a practical basis or cost prohibitive.

In 1990 and 1991, unusually high numbers of large least ciscoes from the Colville River were caught and tagged in our study area. Significant numbers of these fish were subsequently recaptured, and these data enabled us to estimate abundance using models developed for both open and closed populations. We tested the hypothesis that abundance estimates using a closed population model would not differ significantly from abundance estimates with an open population model. Rejection of the hypothesis would suggest that open population models are required. Failure to reject the hypothesis would imply that closed models might be acceptable, although such a finding would require explanation.

TABLE 1.—Number of least ciscoes *Coregonus sardinella* caught and marks released by sampling day (1 = 6 July) during 1990 and 1991.

	1990		1991	
Day	Caught	Released	Caught	Released
1			33	30
2	227	219	49	47
3	227	202	64	60
4	4	3	43	39
5	10	9	29	27
6	9	9	19	18
7	26	25	27	25
8	17	16	67	56
9	39	35	94	78
10	31	30	62	52
11	323	203	158	142
12	540	475	292	255
13	500	438	215	194
14	672	593	317	277
15	479	419	479	353
16	273	233	325	271
17	227	200	627	560
18	165	132	1,425	1,314
19	182	160	862	770
20	24	17	1,090	1,015
21	284	249	547	522
22	202	175	316	299
23	118	102	519	472
24	64	53	192	165
25	123	108	192	173
26	122	102	249	230
27	145	134	387	354
28	264	215	317	296
29	197	169	170	153
30	114	104	175	166
31	40	28	236	222
32	30	22	83	41
33	50	24	294	146
34	70	59	342	315
35	208	187	118	0
36	206	185	244	0
37	89	79	379	0
38	91	70	267	0
39	42	36	375	0
40	22	7	386	0
41			390	0
42	44	0	325	0
43	11	0	637	477
44	5	2	805	752
45	3	0	498	456
46	30	13	209	0
47	46	40	244	0
48	64	52	79	0
49	34	13		
50	10	8	215	0
51	16	14	161	0
52	4	4	92	0
53	15	5	142	0
54	47	24	169	0
55	37	35	263	0
56	31	27	139	1
57	23	18	197	0
58	62	0		

Study Area and Methods

Our study area encompassed the coastal zone extending from the Kuparuk River to the east side of the Sagavanirktok River delta (Figure 1). Sampling was conducted at 9 sites in 1990 and 13 sites in 1991. Fish were captured using fyke nets comprised of double cod-end traps (1.27-cm stretched nylon mesh), two 15.2-m wings (25-cm stretched nylon mesh), and a common 60-m lead (25-cm stretched nylon mesh) attached to the shore. Except during periods of inclement weather, sampling continued on a 24-h basis throughout the summer, with nets being serviced at approximately the same time each day. Captured fish were placed into floating holding pens. Large least ciscoes were anesthetized in a dilute solution of tricaine methansulfonate (MS-222), identified, measured for fork length (FL), and tagged with individually numbered Floy anchor tags. Individuals were held in the floating pens until they recovered from the effects of the anesthetic, whereupon they were released. Data (tag number, FL) for recaptured tagged fish were recorded by date and station location. Table 1 displays the number of fish caught and marks released by day during 1990–1991. Tables 2 and 3 list the subsequent recapture of marks by day of release and recapture for 1990 and 1991, respectively.

Established tests of model assumptions for a closed population (e.g., Otis et al. 1978) could not be applied because of the sparse recapture history (see Tables 2 and 3). However, the expected total number of recaptures each day could be calculated under the assumptions of a closed population and capture probabilities that are proportional to the number of marks released to the population, that is,

$$E[R_t] = \frac{M_t C_t}{\sum_t M_t C_t} \sum_t R_t, \tag{1}$$

where M_t is the number of marks released to the population before day t, and R_t is the total number of recaptures taken during day t from a sample of C_t fish. Because the expected number of recaptures was small (<2) for many days, we pooled the expected recoveries into 11 periods, each of which consisted of five sampling days, and compared them to the actual recoveries (column totals in Tables 2 and 3). Although this goodness-of-fit procedure provides a direct, albeit statistically weak, test of the

TABLE 2.—Numbers of recaptured least ciscoes *Coregonus sardinella* in the Beaufort Sea study area by release and recapture sampling day (1 = 6 July) during 1990.

Release day	Recapture day																													Total
	3	12	13	14	15	16	17	18	19	21	22	23	25	26	27	28	29	30	31	33	34	36	37	38	39	42	48	51	57	
1	0	0	1	0	0	0	0	0	0	0	0	0	0	0	0	0	0	0	0	0	0	0	0	0	0	0	0	0	0	1
2	1	0	1	0	1	0	0	0	0	0	0	0	0	1	0	0	0	0	0	0	0	0	0	1	0	0	0	0	0	5
3		1	0	0	0	0	0	0	0	1	0	0	0	0	0	0	0	0	0	0	0	0	0	0	0	0	1	0	0	3
5		0	0	0	0	0	0	0	0	0	0	0	0	0	0	0	0	0	0	0	0	0	0	0	0	0	0	0	1	1
6		0	1	0	0	0	0	0	0	0	0	0	0	0	0	0	0	0	0	0	0	0	0	0	0	0	0	0	0	1
7		0	0	0	0	0	0	0	0	1	0	0	0	0	0	0	0	0	0	0	0	0	0	0	0	0	0	0	0	1
9		0	0	0	0	0	0	0	0	0	0	0	0	0	0	0	0	0	0	0	0	0	0	0	0	0	0	0	1	1
11		1	0	0	1	0	0	0	0	0	1	0	1	0	0	0	0	0	0	0	0	0	0	1	0	0	0	0	0	5
12			1	0	0	1	1	1	0	1	0	0	0	0	0	0	1	1	0	0	0	0	1	0	0	1	0	0	0	9
13				1	0	0	0	0	0	0	0	0	0	0	0	0	1	0	0	0	0	0	0	0	0	1	0	1	0	4
14					1	0	1	1	1	0	0	1	0	0	0	0	0	0	0	0	0	1	0	0	0	0	0	0	0	6
15						0	0	1	0	2	0	0	0	0	0	1	1	0	0	0	0	0	0	0	0	0	1	0	0	6
16							0	0	0	0	1	0	0	0	0	0	0	0	0	0	0	0	0	0	0	0	1	0	0	2
18									0	0	0	0	0	1	0	0	0	0	0	0	0	0	0	0	0	0	0	0	0	1
19										1	0	0	0	0	0	0	0	0	0	0	0	0	0	0	0	1	0	0	0	2
20										0	0	0	0	0	0	0	0	0	0	0	0	0	0	0	0	0	1	0	0	1
21											0	0	0	0	1	0	0	0	0	0	0	0	0	0	0	0	0	0	0	1
22												1	0	0	0	1	0	0	0	0	0	0	0	1	0	0	0	0	0	3
26															0	0	0	0	0	0	1	0	0	0	0	0	0	0	0	1
27																3	0	0	0	0	0	0	0	0	0	0	0	0	0	3
29																		0	0	1	0	0	0	0	0	0	0	0	0	1
30																			1	0	0	0	0	0	0	0	0	0	0	1
34																						1	0	0	1	0	0	0	0	2
Total	1	2	4	1	3	1	2	3	1	6	2	2	1	2	1	5	3	1	1	1	1	2	1	3	1	3	4	1	2	61

suitability of the closed population model, we believe that a comparison of population estimates with open or closed assumptions is more germane.

The sequential Bayes algorithm described by Gazey and Staley (1986) was selected as the closed population estimator. The mode of the posterior distribution, the maximum-likelihood estimate of the sampling distribution, was used as the point estimate; it is equivalent to the iterative methods recommended by Darroch (1958) to obtain a maximum-likelihood population estimate. The 95% confidence interval quoted subsequently herein was the 95% highest probability density of the posterior distribution. The Jolly–Seber method (program JOLLY in Krebs 1989) was used to estimate daily abundance and its variance assuming an open population. The mean of the daily estimates was used as the population estimate over any given period of time. The Jolly–Seber 95% confidence interval for this mean was calculated assuming that the daily estimates are independent and the mean is normally distributed:

$$\frac{\sum N_t}{n} \pm t_{(n-1)} \frac{\sqrt{\sum \mathrm{Var}[N_t]}}{n}, \tag{2}$$

where N_t is the daily population estimate, n is the number of days in the given period, and $t_{(n-1)}$ is the cumulative value of the t-distribution with $n-1$ degrees of freedom. According to the central limit theorem, the summation of the daily variances for the calculation of the confidence interval should ensure good conformity to the normality assumption.

Daily data were obtained for the period 7 July to 31 August 1990 and for the period 6 July to 30 August 1991. The maximum-likelihood estimates obtained from the Gazey–Staley method for these periods were compared to the means of the daily Jolly–Seber abundance estimates. Mean estimates derived from the alternative models were considered significantly different in instances where the respective 95% confidence internals did not overlap.

Least ciscoes from the Colville River were not present in the study area over the entire period. A mean of the daily estimates for the entire summer, therefore, gives a biased picture of how many least ciscoes were present on average in the system because occurrence and residence times are not considered. The Jolly–Seber daily estimates were inspected to determine the periods of sustained abundance in the study area. Means of the daily estimates within these periods were considered representative of the actual abundance of large least ciscoes in the study area each year, taking occurrence and residence time into account.

TABLE 3.—Numbers of recaptured least ciscoes *Coregonus sardinella* in the Beaufort Sea study area by release and recapture sampling day (1 = 6 July) during 1991.

Release day	Recapture day																							
	3	5	8	11	12	13	14	15	16	17	18	19	20	21	22	23	24	25	26	27	28	29	30	31
1	0	1	0	0	0	0	0	0	0	0	0	0	0	0	0	0	0	0	0	0	0	0	0	0
2	1	0	0	0	0	0	0	0	0	0	0	0	0	0	0	0	0	0	0	0	0	0	0	0
3	0	0	0	0	1	0	0	0	0	0	0	0	0	0	0	0	0	0	0	0	0	0	0	0
5			0	0	0	0	0	0	0	0	1	1	0	0	0	1	0	0	0	0	0	0	0	0
6			0	0	0	0	0	0	0	0	0	0	0	0	0	0	0	0	2	0	0	0	0	0
7			1	0	0	0	1	0	0	0	0	0	0	0	0	0	0	0	0	0	0	0	0	0
8				1	0	0	0	0	0	0	0	0	0	0	0	0	0	0	0	1	0	0	0	0
9				0	0	0	0	0	0	1	0	0	0	0	0	3	0	0	0	0	0	0	0	0
10				0	0	0	0	0	0	0	0	0	0	0	0	0	0	0	0	1	0	0	0	0
11					0	0	0	0	1	1	1	1	0	0	0	0	0	0	0	0	0	0	0	0
12						1	0	0	1	2	1	0	0	0	0	1	0	0	0	0	0	0	0	0
13							0	0	1	0	1	0	1	0	0	0	0	0	0	0	0	0	0	0
14								2	0	0	4	0	1	0	0	0	0	0	0	0	0	0	0	0
15									1	0	1	0	1	0	0	0	0	0	0	0	0	1	0	0
16										0	3	1	2	0	0	0	0	0	0	0	0	0	0	0
17											2	0	0	0	1	1	0	1	2	1	0	0	0	0
18												0	1	1	0	1	0	1	0	1	2	2	1	1
19													0	0	0	1	0	0	0	1	0	0	0	0
20														0	1	2	0	1	1	1	1	1	0	0
21															4	1	0	0	0	0	0	0	0	0
22																0	1	0	0	1	0	0	0	0
23																	0	0	0	0	0	0	1	0
24																		0	0	0	2	0	0	0
25																			0	0	0	0	0	0
26																				0	0	1	0	0
27																					1	1	0	0
28																						2	0	0
29																							0	0
30																								0
31																								
32																								
33																								
34																								
43																								
44																								
45																								
Total	1	1	1	1	1	1	1	2	4	4	14	3	6	1	6	11	1	3	5	7	6	8	2	1

One reason the Gazey–Staley method was selected is that it yields sequential plots of posterior distributions (one for each sampling occasion) that can be used as a visual diagnostic feature enabling evaluation of assumption violations. Ideally, the posterior distributions should tend to stabilize or mass about a single population value. A continuous trend toward a larger or smaller population size is strong evidence that the closed population assumption has been violated. Gazey and Staley (1986) note that the sensitivity of the posterior is greater for an increasing population than for a decreasing population because only the unmarked population is affected in the former.

The posterior distributions were examined in the sequence in which the daily samples were obtained and in the reverse order (because the samples are independent, the order of calculation is immaterial to the final estimate) to determine the approximate dates on which the estimates converge around a single estimate, without suggestion of any increasing or decreasing trends. These estimates were considered directly comparable to the similar estimates derived from the Jolly–Seber data adjusted for residence period.

Results

Observed and expected recaptures (assuming a closed population and capture probabilities proportional to marks released) were pooled into 11 periods, each consisting of five sampling days, for large least ciscoes during 1990 and 1991 (Figure 2). A chi-square goodness-of-fit test between the observed and expected recoveries was not statistically significant ($P > 0.05$) for either year.

The daily patterns of least cisco abundance reflected by the Jolly–Seber estimates (Figure 3) dif-

TABLE 3.—Continued.

Recapture day																									
32	33	34	35	36	37	38	39	40	41	42	43	44	45	46	47	48	50	51	52	53	54	55	56	57	Total
0	0	1	0	0	0	0	0	0	0	0	0	0	0	0	0	0	0	0	0	0	0	0	0	0	2
0	0	0	0	0	0	0	0	0	0	0	0	0	0	0	0	0	0	0	0	0	0	0	0	0	1
0	0	0	0	0	0	0	0	0	1	0	0	0	0	0	0	0	0	0	0	0	0	0	0	0	2
0	0	0	0	0	0	0	0	0	0	0	0	0	0	0	0	0	0	0	0	0	0	0	0	0	3
1	0	0	0	0	0	0	0	0	0	0	0	0	0	0	0	0	0	0	0	0	0	0	0	0	3
0	0	0	0	0	0	0	0	0	0	0	0	0	0	0	0	0	0	0	0	0	0	0	0	0	2
0	0	0	0	0	0	0	0	0	0	0	0	0	0	0	0	0	0	0	0	0	0	0	0	0	2
0	0	0	0	0	0	0	0	0	0	0	0	1	1	0	0	0	0	1	0	0	1	0	0	0	8
0	0	0	0	0	0	0	0	0	0	0	1	0	0	0	0	0	0	0	0	0	0	0	0	0	2
0	0	0	0	1	1	1	0	0	1	0	1	0	0	1	0	0	0	0	0	0	0	1	0	0	11
0	0	0	0	0	0	0	0	0	0	1	0	0	0	0	0	0	0	0	0	0	1	0	0	1	9
0	0	0	0	0	0	0	0	0	0	0	0	0	0	0	0	0	0	0	0	0	0	0	1	0	4
0	0	1	0	1	0	0	0	0	0	0	1	0	0	0	0	0	0	1	0	0	0	0	0	0	11
0	0	0	0	0	0	0	0	0	0	0	0	0	1	0	0	0	0	1	0	0	0	0	0	0	6
0	0	0	0	1	0	0	0	0	0	0	0	0	0	0	1	0	0	0	0	2	0	0	0	0	10
0	0	0	0	0	0	1	0	0	0	1	0	1	1	0	2	0	0	0	0	0	0	0	1	0	15
0	2	0	0	0	0	1	0	1	0	0	0	2	3	0	1	0	2	0	0	0	0	1	0	2	26
1	0	0	0	0	0	1	0	0	0	1	2	0	0	1	0	1	0	0	0	0	0	0	0	0	9
0	0	1	0	0	1	0	1	0	1	1	2	1	0	0	2	0	0	0	1	1	0	1	0	0	21
0	0	0	2	0	0	1	1	0	0	2	0	1	0	0	0	0	1	0	0	0	0	0	0	0	13
0	2	1	0	0	0	0	0	0	0	0	0	0	1	1	0	0	1	0	0	0	0	0	0	0	8
0	0	1	0	0	0	0	0	0	0	0	0	0	0	1	1	0	0	0	0	0	0	0	0	0	4
0	0	0	0	0	0	0	0	1	1	0	0	0	0	0	0	0	0	0	0	0	0	0	0	0	4
0	0	0	0	0	1	1	1	0	0	0	0	0	0	0	0	0	0	0	0	0	0	0	0	0	3
0	0	0	0	0	0	0	0	1	0	0	1	0	1	0	0	0	0	1	0	0	1	0	0	0	6
0	0	0	0	0	0	0	1	0	0	0	0	0	0	0	0	0	1	0	1	0	0	0	0	0	5
0	0	0	0	1	0	0	0	0	0	0	0	0	0	0	0	0	0	0	0	0	0	0	0	0	3
0	0	0	0	0	0	1	0	0	0	0	0	0	0	0	0	0	0	0	0	0	0	0	0	0	1
0	0	0	0	1	0	0	0	0	0	0	0	0	0	0	0	0	0	0	0	0	0	0	0	0	1
2	0	0	0	0	1	0	0	0	2	0	1	0	0	0	0	0	0	0	0	0	0	0	0	0	6
	0	0	0	0	0	0	0	0	0	0	0	0	0	1	0	0	0	0	0	0	0	0	0	0	1
		0	0	1	0	0	0	1	0	1	0	0	1	0	0	0	0	0	0	0	0	0	0	0	4
			0	0	0	1	0	0	0	0	0	0	1	0	0	0	0	0	0	0	0	0	0	0	2
												0	1	1	1	0	0	0	0	0	0	0	0	0	3
													4	0	0	0	0	0	0	1	0	0	0	0	5
														6	4	1	0	0	0	0	0	0	0	0	11
4	4	5	2	6	4	8	4	4	6	7	9	6	15	12	12	2	5	4	2	4	3	3	2	3	227

fered between years, although the maximum numbers of fish present on a given day were similar each year (about 1.0 million in 1990 versus 1.3 million in 1991). Early in 1990 (9 July), a large pulse of least ciscoes moved east of West Dock, but the main group of fish did not arrive until 18 July. These fish were abundant in the area until 29 July. After this date, emigration appears evident, based on greatly reduced catches. A late spike in abundance, representing fish entering the area from the east and moving westward through the study area, occurred on 10–11 August; this may represent the return migration of the early vanguard of least ciscoes that had moved rapidly eastward through our study area in early July. For the entire season in 1990, the mean of the daily Jolly–Seber estimates was 105,000; the 95% confidence interval was 53,000–157,000 (hereafter reported in parentheses after mean abundance estimates). Mean abundance during the 12-d "residence period" was about three times as large, 349,000 (183,000–515,000).

In 1991, least ciscoes arrived in our study area on about 16 July, only 2 d earlier than in 1990 (Figure 3). However, they remained abundant through 7 August, about 1 week longer than they remained in 1990. After this date, they were virtually absent until 16–17 August, when a substantial group of least ciscoes moved east to west through the study area. Again, these probably represent the return migration of fish that had previously moved farther eastward than our study area. The mean daily Jolly–Seber abundance estimate of least ciscoes during 1991 was 177,000 (102,000–252,000). The mean estimate for the 23-d period of residence observed in 1991 was 361,348 (297,000–425,000). These data show that a similar number of large least ciscoes were in the system each year, but the fish remained longer in 1991 than in 1990.

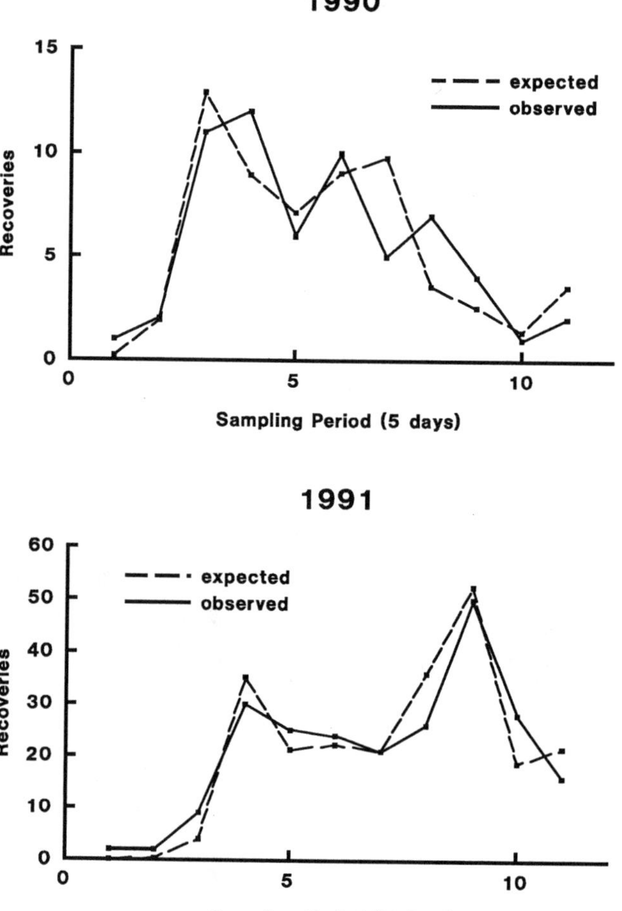

FIGURE 2.—Observed and expected recaptures (assuming a closed population and capture probabilities proportional to marks released) pooled into 11 periods, each consisting of five sampling days, for large least ciscoes *Coregonus sardinella* during 1990 and 1991.

The Jolly–Seber model also produced estimates of recruitment and survival that are not reported here. In all cases, recruitment was not significantly different from 0 and survival was not significantly different from 1.

Figure 4 shows the sequence (in order of sampling) of posterior distributions derived using the Gazey–Staley method. The posteriors with modes smaller than 250,000 in 1990 and 350,000 in 1991 were all from the early season. In 1990, most of the posteriors mass or stabilize about a single population value. On the other hand, in 1991 there was substantial change in the population size, particularly in the early season where there was continuous movement in the posteriors, with modes of 100,000–350,000. The maximum-likelihood estimates of least cisco abundance based on the full 1990 and 1991 seasons were 337,000 (256,000–428,000) and 457,000 (402,000–521,000), respectively (statistics obtained from the final posterior sequence, identified as the thick line in Figure 4). An examination of the posterior distributions sequenced forward and backward in time (not shown) to determine a stable period yielded the dates 20 July to 4 August in 1990 and 23 July to 7 August in 1991. The resulting estimates based on data from these periods in 1990 and 1991 were 331,000 (237,000–481,000) and 458,000 (371,000–574,000), respectively.

Gazey–Staley estimates were also generated for the same residence periods delineated from the Jolly–Seber estimates (18–29 July 1990 and 16 July to 7 August 1991). The resulting estimates were not

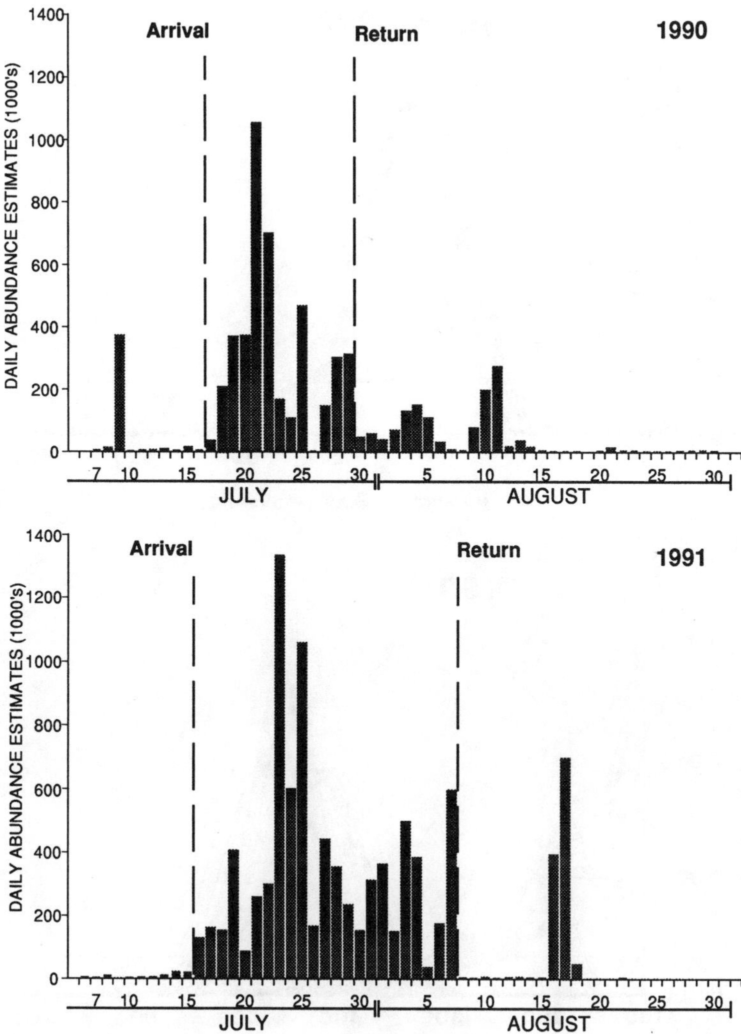

FIGURE 3.—Daily abundance of large least ciscoes *Coregonus sardinella* in the Prudhoe Bay region in 1990 and 1991, estimated using the Jolly–Seber model.

significantly different from any of the previous ones generated by this approach. In 1990, an estimated 297,000 (205,000–451,000) large least ciscoes were present, and in 1991, the estimate was 424,000 (349,000–522,000).

Discussion

Open versus Closed Models

A closed population model cannot be rejected for the 1990 and 1991 least cisco mark–recapture data. Recoveries expected under the closed model were not significantly different from the observed recoveries. Furthermore, population estimates using either an open or closed model during the resident periods were not significantly different. Of the five estimates calculated in this study, the only one (in each year) significantly different from any of the others was the daily mean of the Jolly–Seber estimates based on the full season's data (Figure 5). It is lower than the other estimates only because there were many days when least ciscoes were not present in the study area. However, the 1991 Gazey–Staley posteriors indicated quite clearly that immigration to the study area occurred before 16 July. Moreover, mean daily abundance of least ciscoes derived from the Jolly–Seber estimates for the period of residence differed from the full-season Gazey–

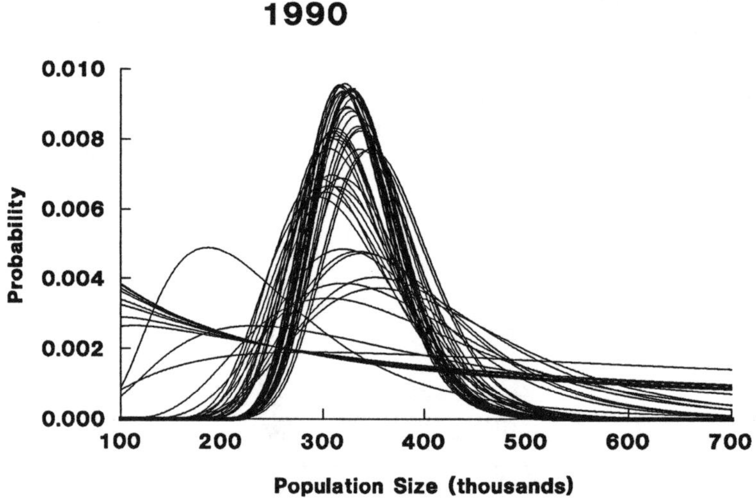

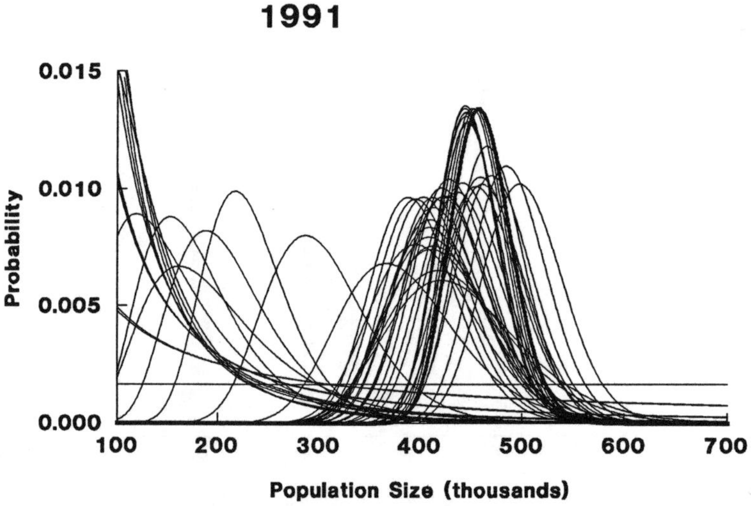

FIGURE 4.—Estimated probability of a given population size for large least ciscoes *Coregonus sardinella* in the study area in the Beaufort Sea, 1990 and 1991. The maximum-likelihood estimate of the final sequence was 337,000 in 1990 and 457,000 in 1991.

Staley estimates by only 3% in 1990 but by 21% in 1991. The greater disparity between the estimates in 1991 than in 1990 may have resulted from oceanographic differences between the 2 years.

Oceanographic Effects

Oceanographic conditions in our study area during 1991 were markedly different from any conditions we observed in more than a decade of study. Air and water temperatures were relatively low throughout the season, and ice remained near shore throughout the season. Salinity levels during the year were markedly lower than usual, and regional upwellings of cold, marine bottom water into the nearshore zone did not occur. This probably greatly reduced prey levels in the system. The condition of some species (e.g., Arctic cisco) deteriorated over the summer (Fechhelm et al. 1996), which did not bode well for their survival during winter 1991–1992.

Although we do not have condition data for least ciscoes, their response to these conditions may have included a longer residence time in the coastal zone

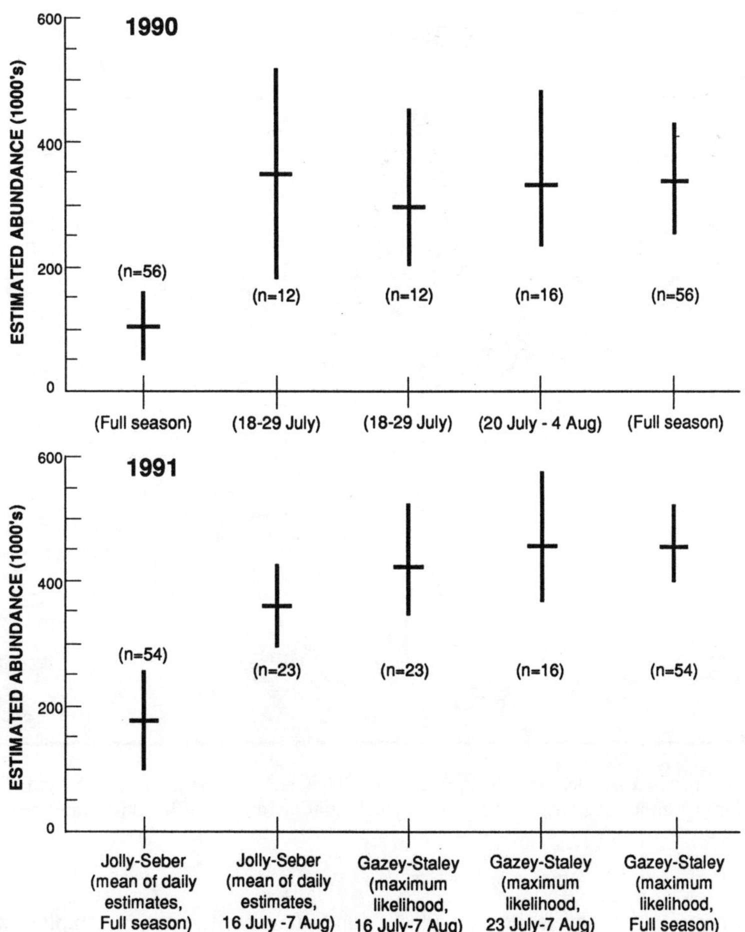

FIGURE 5.—Comparisons of abundance estimates for least ciscoes *Coregonus sardinella* for the study area in the Beaufort Sea in 1990 and 1991 based on the Jolly–Seber model (which assumes an open population) and the Gazey–Staley model (which assumes a closed population). Horizontal lines show the mean (Jolly–Seber) or maximum-likelihood (Gazey–Staley) estimates; vertical lines show the 95% confidence interval.

to try to acquire a greater ration. Their movement behavior may have resulted in a more open population in our area in 1991 as compared with 1990, which would account for the greater disparity between the abundance estimates for 1991 as compared with 1990.

We believe the optimum use of open or closed estimation models will be determined by understanding the oceanographic and biological dynamics for the year in consideration.

Large least ciscoes that use our study area overwinter in the Colville River delta (Figure 6). At breakup (late June to early July) they enter the coastal zone, moving eastward into Simpson Lagoon for feeding. Simpson Lagoon is a narrow, long and shallow barrier island–lagoon system located between Prudhoe Bay and the Colville River (Figure 6). The major channels are the broad opening at the western end and the deep channels on either side of Egg Island at the eastern end. Freshwater discharges from the Colville and Kuparuk rivers directly affect the lagoon, with the Colville River (about an order of magnitude larger than the Kuparuk) exhibiting the major influence (Colonell and Niedoroda 1990). Warm, brackish water from the more distant Sagavanirktok River also reaches the eastern portion of the lagoon on east winds after passing outside the shoals across the mouth of Prudhoe Bay or, to a lesser degree, passing through Prudhoe Bay (Colonell and Niedoroda 1990). The

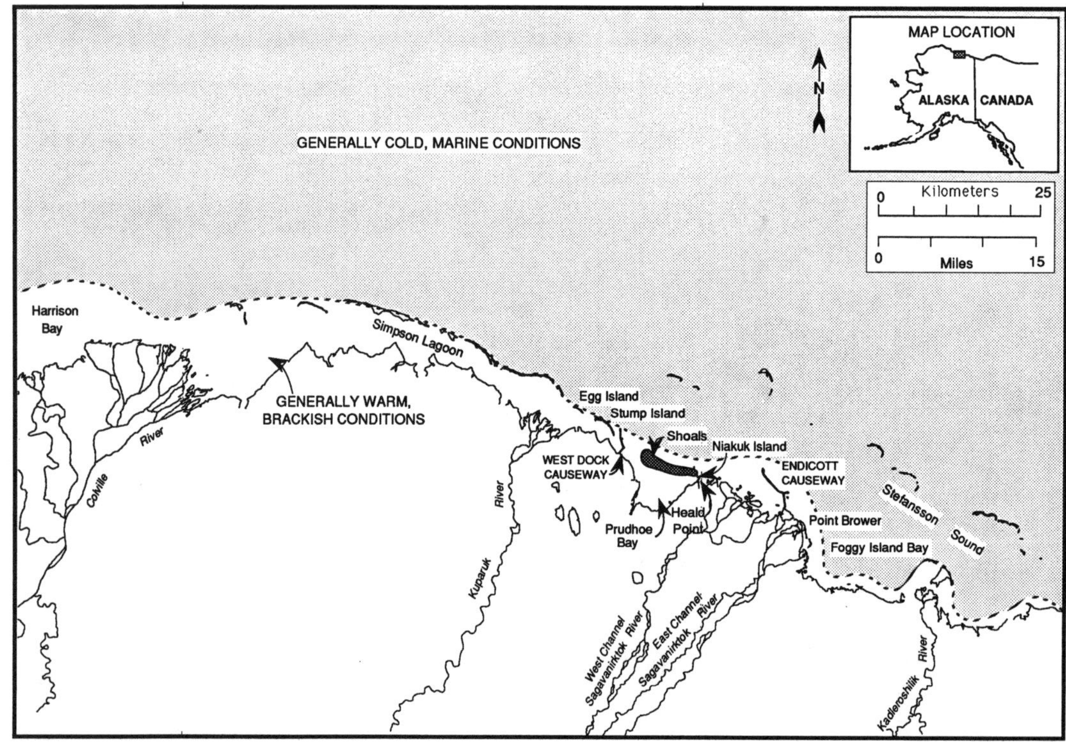

FIGURE 6.—The central-Alaska Beaufort Sea, from the Colville River to Stefansson Sound. The dashed line (approximate 2-m depth contour) delineates the generally cold marine and warm brackish water conditions.

shallow, enclosed character of the lagoon enables insolation to be effective in warming the water, thus providing a favorable feeding habitat for anadromous fish for most of the summer.

Prudhoe Bay, the Sagavanirktok River delta, and Foggy Island Bay east of the river are all east of Simpson Lagoon and part of Stefansson Sound (Figure 6). The sound extends eastward from the Sagavanirktok River delta for about 75 km to the mouth of the Staines River. Except at its eastern end, the sound is much broader than Simpson Lagoon, and the barrier-island chain protecting it is much less complete than the barrier-island chain associated with Simpson Lagoon. The sound is also deeper than the lagoon, reaching maximum depths of about 8–9 m. Consequently, most of the sound is essentially a marine environment for most of the summer, except for Prudhoe Bay, the Sagavanirktok River delta, and a narrow, brackish margin along the mainland shore east of the delta (Figure 6).

At least some large ciscoes moved through Simpson Lagoon and into the brackish areas of Stefansson Sound each year that sampling was conducted. However, large numbers of least ciscoes moved from Simpson Lagoon into the sound less than one year in two. For example, in the first 7 years (1985–1991) of the Endicott fish-monitoring program, large numbers of least ciscoes were encountered in the Prudhoe Bay–Sagavanirktok River region in 1988, 1990, and 1991. How far eastward from the Sagavanirktok River these fish travel along the brackish water shoreline is not well documented, but the eastern end of Stefansson Sound is generally considered to be the end of the summer range of Colville River least ciscoes (Craig 1989; Gallaway et al. 1991).

Fechhelm et al. (1989) observed that there is a short time window available for large least ciscoes to move eastward beyond West Dock. Large least ciscoes from the Colville River usually do not arrive in this area until about mid-July, and regional upwellings of cold, marine bottom waters typically begin under east-wind conditions about 1 week later (Colonell and Niedoroda 1990). In particular, the eastern end of Simpson Lagoon becomes colder and more saline. These circumstances typically signal the end of eastward movement out of Simpson Lagoon by large least ciscoes. Thus, the period during which fish immigrate into areas east of West Dock is short.

Fechhelm et al. (1989) also show that the actual immigration periods are typically even shorter than the brief periods during which the hydrography around West Dock would allow movement further eastward if the fish were present. The bulk of the fish usually enter the Prudhoe Bay region in one to several large waves of short (1–2 d) duration. When the hydrography shifts, immigration ceases. At this point, the population in the study area is closed to further immigration (Fechhelm et al. 1989).

Fish that moved into habitats east of West Dock may begin to return to the Colville River as early as the end of July, but often the return migration is not initiated until after the first week in August. The return migrations appear to be keyed to the severity and number of marine intrusions that occur, beginning in late July. These events are governed by annual wind patterns. Despite annual variation, after 15 August few fish are usually present east of West Dock, and the area is completely vacated by the end of August (Fechhelm et al. 1989).

Summary

When large least ciscoes enter the study area, they do so quickly, are resident for some period of weeks, and then leave under the influence of oceanographic processes. Only the late, seasonal emigration violates the closed population assumption. The lack of a profound effect on the population estimates can be explained by the fact that marked and unmarked fish have the same likelihood of emigrating. Thus, marks applied before the onset of emigration have the same capture probability even after the onset of emigration. Only marks applied after the onset of emigration have the potential to bias the population estimates. Moreover, fewer marked fish are released or recaptured because fewer fish are available after the onset of emigration. Even then, as Gazey and Staley (1986) note, the sensitivity of the posterior is greater for an increasing population than for a decreasing population, because only the unmarked population is affected in the former.

Acknowledgments

We thank C. W. Wilkinson and V. J. Chhipa for programming and data analysis assistance and J. Erwin and G. F. Hubbard for manuscript production. Inputs and critical review were received from our colleagues at LGL, namely W. J. Wilson and W. B. Griffiths. J. M. Colonell (an oceanographer with latent biological tendencies) has contributed for many years to our understanding of physical oceanography. The overall study from which this work resulted was funded by BP Exploration (Alaska) Inc., conducted in part to fulfill Endicott Development Monitoring requirements pursuant to Title 19 of the North Slope Borough (NSB) Municipal Code. We particularly thank members of the NSB's Scientific Advisory Committee who provide oversight reviews of our studies. The conclusions and opinions expressed in this paper represent those of the authors and do not necessarily represent those of the above-named individuals or organizations.

References

Colonell, J. M., and B. J. Gallaway, editors. 1990. An assessment of marine environmental impacts of West Dock causeway. Report for Prudhoe Bay unit owners represented by ARCO-Alaska, Inc. Prepared by LGL Alaska Research Associates, Inc. and Environmental Science and Engineering, Inc., Anchorage, Alaska.

Colonell, J. M., and A. W. Niedoroda. 1990. Appendix B. Coastal oceanography of the Alaskan Beaufort Sea. Pages B1–B74 *in* J. M. Colonell and B. J. Gallaway, editors. An assessment of marine environmental impacts of West Dock causeway. Report for Prudhoe Bay unit owners represented by ARCO-Alaska, Inc. Prepared by LGL Alaska Research Associates, Inc. and Environmental Science and Engineering, Inc., Anchorage, Alaska.

Craig, P. C. 1989. An introduction to anadromous fishes in the Alaska Arctic. Biological Papers of the University of Alaska 24:27–54.

Darroch, J. N. 1958. The multiple-recapture census: I. Estimation of a closed population. Biometrika 46: 343–359.

Fechhelm, R. G., J. S. Baker, W. B. Griffiths, and D. R. Schmidt. 1989. Localized movement patterns of least cisco (*Coregonus sardinella*) and Arctic cisco (*C. autumnalis*) in the vicinity of a solid-fill causeway. Biological Papers of the University of Alaska 24:75–106.

Fechhelm, R. G., R. E. Dillinger, Jr., B. J. Gallaway, and W. B. Griffiths. 1992. Modeling of in situ growth and temperature relationships of yearling broad whitefish in Prudhoe Bay, Alaska. Transactions of the American Fisheries Society 121:1–12.

Fechhelm, R. G., W. B. Griffiths, L. R. Martin, and B. G. Gallaway. 1996. Intra- and interannual variation in the relative condition and proximate body composition of Arctic ciscoes from the Prudhoe Bay region of Alaska. Transactions of the American Fisheries Society 125:600–612.

Gallaway, B. J., W. J. Gazey, J. M. Colonell, A. W.

Niedoroda, and C. J. Herlugson. 1991. The Endicott Development Project—preliminary assessment of impacts from the first major offshore oil development in the Alaskan Arctic. Pages 42–80 *in* C. S. Benner and R. W. Middleton, editors. Fisheries and oil development on the continental shelf. American Fisheries Society Symposium 11, Bethesda, Maryland.

Gazey, W. J., and M. J. Staley. 1986. Population estimation from mark–recapture experiments using a sequential Bayes algorithm. Ecology 67:941–951.

Griffiths, W. B., B. J. Gallaway, W. J. Gazey, and R. E. Dillinger. 1992. Growth and condition of Arctic cisco and broad whitefish as indicators of causeway-induced effects in the Prudhoe Bay region, Alaska. Transactions of the American Fisheries Society 121: 557–577.

Hachmeister, L. E., D. R. Glass, and T. C. Cannon. 1991. Effects of solid-fill gravel causeways on the coastal central Beaufort Sea environment. Pages 81–96 *in* C. S. Benner and R. W. Middleton, editors. Fisheries and oil development on the continental shelf. American Fisheries Society Symposium 11, Bethesda, Maryland.

Krebs, C. J. 1989. Ecological methodology. Harper and Rowe, New York.

Otis, D. L., K. P. Burnham, G. C. White, and D. R. Anderson. 1978. Statistical inference from capture data on closed animal populations. Wildlife Monographs 62.

American Fisheries Society Symposium 19:287–294, 1997

Reproductive Biology and Distribution of the Snow Crab from the Northeastern Chukchi Sea

J. M. PAUL AND A. J. PAUL

University of Alaska, Institute of Marine Science, Seward Marine Center Laboratory
Post Office Box 730, Seward, Alaska 99664, USA

WILLARD E. BARBER

University of Alaska, School of Fisheries and Ocean Sciences,
Fairbanks, Alaska 99775, USA

Abstract.—We determined size at maturity, fecundity, distribution, and abundance of the snow crab *Chionoecetes opilio* during 1990 and 1991 in the northeastern Chukchi Sea, Alaska. Snow crab abundance and biomass varied extensively between stations but tended to be greatest in the southern part of the study area and offshore rather than inshore. Biomass estimates varied extensively, ranging from 1.2 to 4,000 kg/km^2. At all latitudes most females with mature-shaped abdominal flaps were gravid. The smallest ovigerous snow crab was 34-mm carapace width (CW). Average CW of ovigerous females was 46 mm (SD = 4) and carried 19,900 (SD = 6,500) eggs. The equation describing the relationship between CW and the number of eggs (*Y*) per clutch was $Y = 0.672\ (\mathrm{CW})^{2.668}$; $r^2 = 0.54$. The vasa deferentia of males were examined for the presence of spermatophores. All males 35-mm CW and greater had spermatophores; only 19% with CW of 25- to 29-mm were so classified. Spermatophore diameters increased from about 44 μm in 25- to 29-mm CW males to about 64 μm in males 35- to 44-mm CW. Spermatophore diameters for crabs between 45- and 70-mm CW were similar, generally between 75 and 80 μm.

The snow crab *Chionoecetes opilio* is a circumpolar species for which there are substantial fisheries in the Atlantic and the Pacific oceans. In the northwest Pacific Ocean, snow crabs occur in the northern Sea of Japan, the Bering and Chukchi seas from Wrangel Island to Point Barrow, and the Beaufort Sea at the mouth of the Mackenzie River (Slizkin 1989). Most of the existing information on the biology of this species has been developed for the northwestern Atlantic Ocean or the Sea of Japan stocks (Bowerman and Melteff 1984).

In the northeastern Chukchi Sea, snow crabs are a dominant benthic species, but because they are not harvested their basic biology is poorly described. This preliminary survey provides new information on distribution trends and reproduction for a seldom-sampled population.

Methods

The sampling sites were dictated by oil lease sales in the area north and east of Point Hope (north of approximately 68°20′), east of the international boundary (168°58′), and limited in the north by sea ice. Before sampling, 11 transects and 56 stations were located perpendicular to the coast on a nautical chart. Stations were positioned approximately 55–110 km apart depending on transect length. Nearshore stations were established closer to one another so that within a transect there were at least two stations bounded by the coastal current. In 1990 there were 48 stations sampled and in 1991 there were 8 additional stations sampled to examine snow crab abundance and biomass (Figure 1). Sea ice or bad weather precluded actual collection at some of the predetermined station positions, so samples were taken as close as possible to the desired site.

During 1990 the bottom temperature averaged 3.8°C (SD = 3.3°C) and ranged from −1.0 to 12°C; salinity was 29–33 ppt. Contour maps of bottom temperature and salinity for 1990 occur in Weingartner (1997, this volume). During 1991 the bottom temperature averaged 0.1°C (SD = 2.0°C), and ranged from −1.0 to 7°C; salinity was 29–33 ppt.

A chartered 34-m commercial trawler, *Ocean Hope III*, was used during 16 August to 17 September 1990 and 10–23 September 1991 to collect snow crab abundance and distribution data. Otter trawling from this vessel was conducted with the National Marine Fisheries Service's standard 83-112 survey trawl, which was fished directly on the bottom. The trawl had a 34.1-m footrope set back 7 cm from a tickler chain and a 25.2-m headrope. The mesh of the cod end was 90 mm and contained a 32-mm stretched mesh liner. The effective opening of the net was monitored by a Scanmar electronic mensuration unit hung from one wing; the net fished as expected. At most stations there were two 30-min tows in sequence along the same tract.

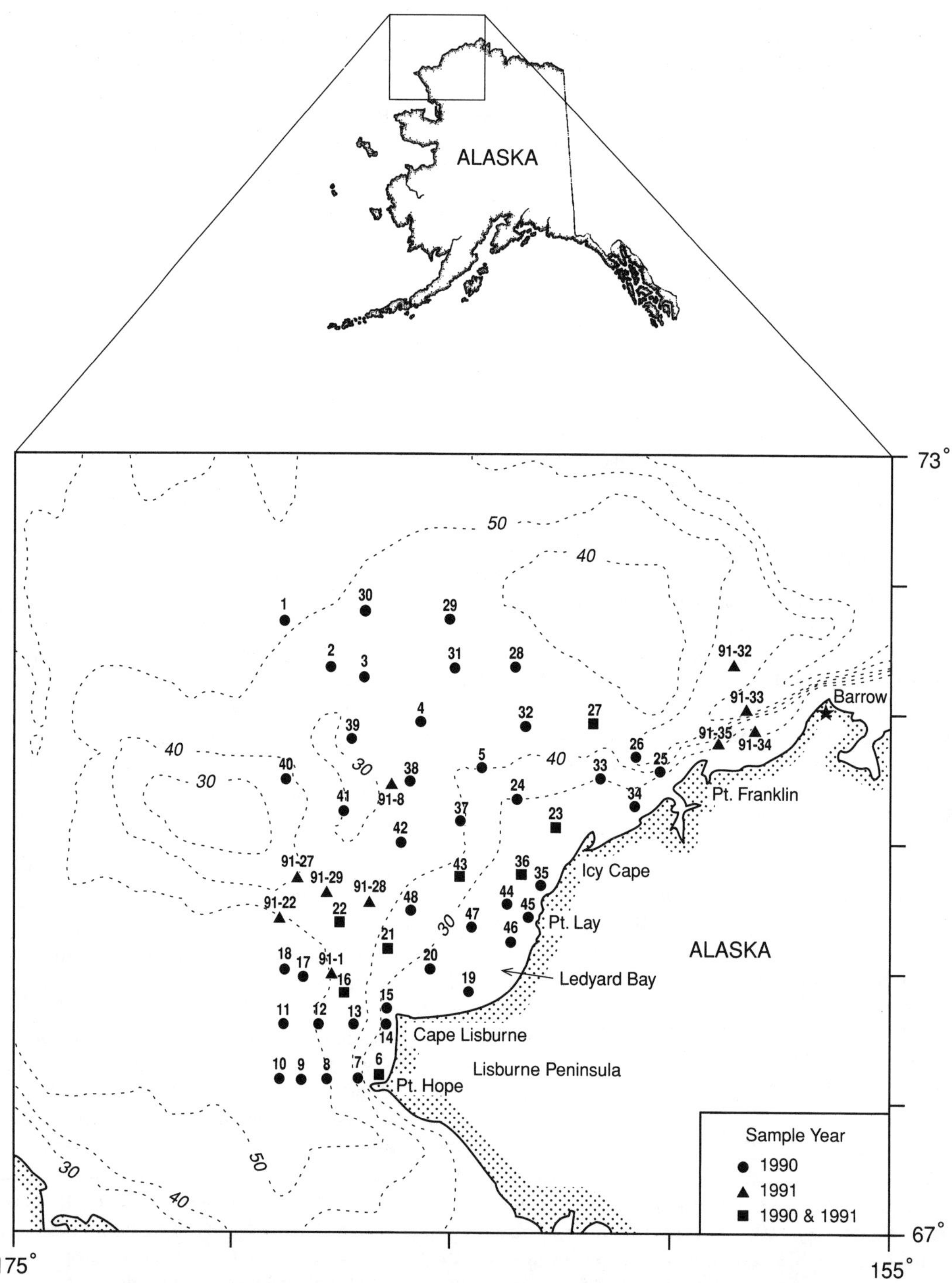

FIGURE 1.—Stations sampled for distribution and abundance of the snow crab *Chionoecetes opilio* in the northeastern Chukchi Sea during 1990 and 1991. Dotted lines indicate depth contours.

At most stations a gross weight of all snow crabs combined was measured; then all snow crabs were counted. Periodically, catches had more crabs or other species than could be processed before the next trawl sample. In those cases, all the snow crabs in the catch were put into baskets. Baskets were filled one at a time as the crew cleared the sorting area and no bias such as size selection was apparent. All baskets were then weighed and some were randomly selected to obtain the number and weight of snow crabs. The average weight of snow crabs at a station was estimated by dividing the total weight of all individuals in the sample by the number of crabs. Weighing individuals separately was not feasible aboard ship.

Abundance (number/km^2) and biomass (kg/km^2) estimates were based on trawl-opening width, the distance trawled, and the numbers and biomass of the catch. The distance trawled was determined from the ship's location at the beginning and end of each set by a Global Positioning System. The data from both tows at a station were averaged to estimate the abundance and biomass at each station.

The 73-m *Oshoro Maru* was used in July 1991 to collect snow crabs for additional reproductive observations; its otter trawl net had a 43.3-m headrope, a 48.6-m footrope with roller gear, and 90-mm stretched mesh with a 32-mm stretched mesh liner. These samples were not used to estimate abundance or biomass.

To examine the relationship between crab size and latitude, the carapace width (CW) of all the females collected at station 21 (N = 25) in July 1991 and at stations 23 (N = 51) and 91–32 (N = 88) in September 1991 (Figure 1) were measured to the nearest 1 mm. The CW of all males from stations 91-1 (N = 125) and 91–8 (N = 131) collected in July 1991 were also measured to the nearest 1 mm. Station locations are shown in Figure 1.

To determine if mature female snow crabs occurred throughout the sampling area, 503 females from 20- to 64-mm CW (Table 1) were collected at 14 sites (Table 2) in July and September 1991. They were examined for the presence of a mature-shaped abdominal flap and eggs.

In the August and September collections, eggs were too developed for fecundity estimates, but 93 females collected at station 1 in July had clutches of bright orange eggs that appeared to be recently extruded. Two females with new eggs had damaged abdominal flaps, so the eggs from only 91 clutches were counted. The dry weight of 100 eggs subsampled from each of the 91 clutches was measured. Eggs were dried at 60°C in a convection oven until a constant weight was reached. The remaining eggs on the pleopods of each female were removed and dried as previously stated. Fecundity estimates were determined by dividing the dry weight of the total egg mass by the average dry weight of eggs in the subsample. The number of eggs (Y) was related to carapace width with a power curve: $Y = a(CW)^b$, where $a > 0$).

Males (N = 318) used for maturity studies were all captured in September 1991 at stations 16, 21, 22, 27, 91-29, and 91-32 (Figure 1). Males over 20-mm CW were divided into groups based on 5-mm increments and had to be taken from several stations to get a minimum of 25 males of the desired

TABLE 1.—Number of snow crabs *Chionoecetes opilio* from the northeastern Chukchi Sea examined for the presence of eggs (in July and September) and spermatophores (in September), 1991.

Carapace width (mm)	Females		Males	
	Total number	Percent with eggs	Total number	Percent with spermatophores
15–19	0	0	1	0
20–24	17	0	36	0
25–29	27	0	37	19
30–34	54	2	49	49
35–39	29	52	25	100
40–44	156	73	25	100
45–49	169	86	26	100
50–54	41	93	29	100
55–59	9	75	25	100
60–64	1	100	25	100
65–69	0		26	100
70–74	0		14	100
Total	503		318	

TABLE 2.—Number of northeastern Chukchi Sea female snow crabs *Chionoecetes opilio* with mature-shaped abdominal flaps carrying eggs, July and September 1991.

Station location		Females		
Latitude N	Longitude W	Total number	Number with mature-shaped abdominal flaps	Percent with eggs
69°02′	167°38′	100	97	98
69°03′	166°43′	12	2	100
69°23′	166°28′	22	16	100
69°32′	165°59′	8	7	100
69°38′	167°41′	50	48	98
69°40′	168°31′	50	44	100
69°54′	168°42′	35	34	100
70°00′	165°03′	4	2	100
70°13′	167°05′	27	24	96
70°21′	162°53′	51	0	0
70°31′	166°08′	50	42	98
70°33′	162°20′	4	0	0
70°58′	163°39′	2	1	100
71°37′	159°02′	88	7	100

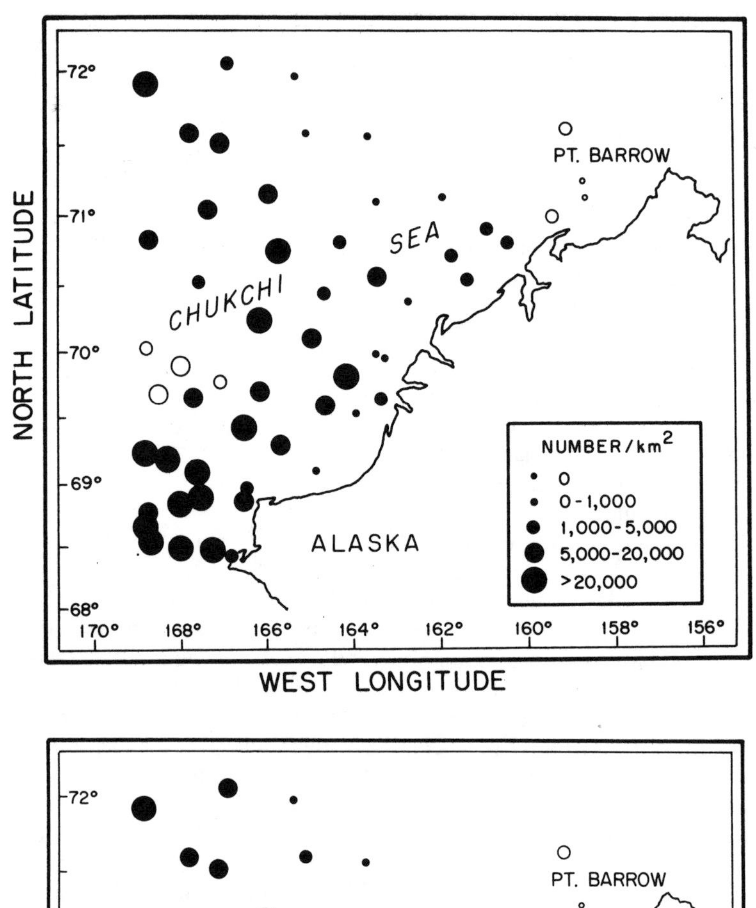

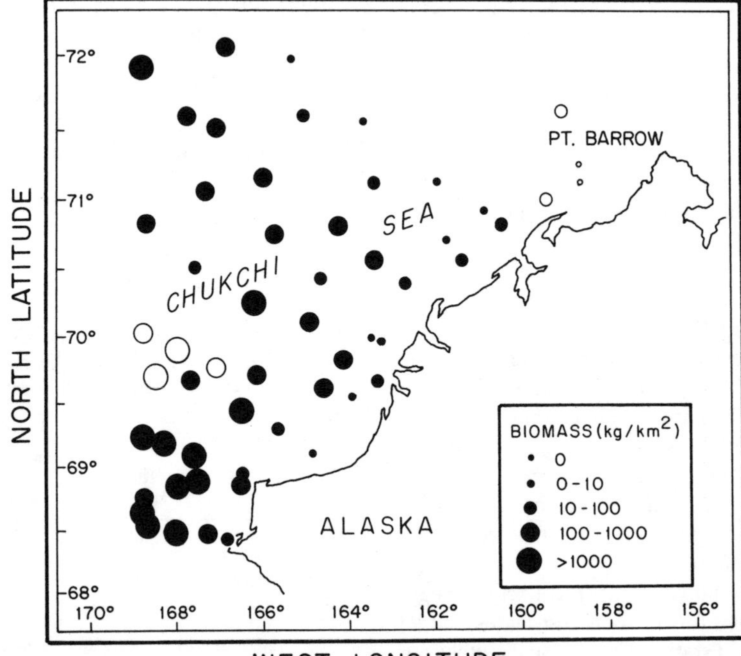

FIGURE 2.—Abundance (upper panel) and biomass (lower panel) of the snow crab *Chionoecetes opilio* captured by otter trawl at stations occupied in the northeastern Chukchi Sea during 1990 (closed circles) and 1991 (open circles).

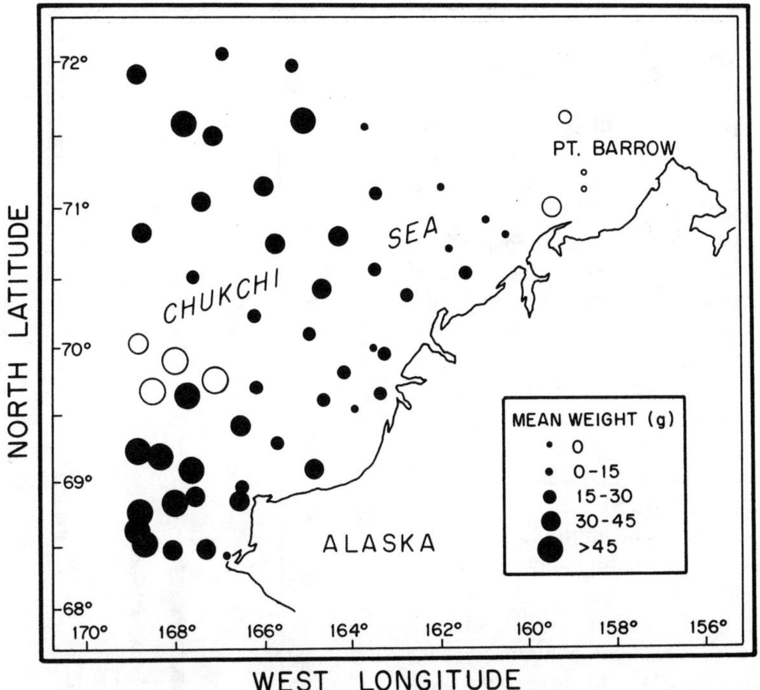

FIGURE 3.—Mean weight of the snow crab *Chionoecetes opilio* captured by otter trawl at stations occupied in the northeastern Chukchi Sea during 1990 (closed circles) and 1991 (open circles).

sizes (Table 1). In the group with the largest males (70- to 74-mm CW) only 14 individuals were captured. Size at the onset of physiological maturity was determined by histological examination of the vas deferens for spermatophores. The vasa deferentia were fixed in Bouin's solution, embedded, sectioned to 10–12 μm thick, and mounted following the techniques of Paul and Paul (1989). Standard Ehrlich's hematoxylin and eosin-y staining sequence (Clarke 1973) was used to enhance morphological identification of spermatophores.

Results

Snow crabs were present at all stations, with the largest abundance and biomass tending to be in the southern part of the study area, but varying extensively between stations (Figure 2). Abundance and biomass estimates also varied considerably between trawls at most stations. The highest estimated mean abundance was at station 1 (100,000/km^2). The largest mean biomass (4,000 kg/km^2) was at station 8, although station 1 biomass was nearly equivalent (3,100 kg/km^2). Lowest mean abundance was at station 28 (190/km^2), whereas the lowest biomass estimate occurred at station 35 (1.2 kg/km^2). Average crab weight (total weight of catch/number of crabs) was generally greater in the southern area than in the northern area and lower inshore than offshore (Figure 3). Individuals from stations south of about 70°N generally had greater average weight, but there was extensive variation. This trend was also reflected in carapace width. The smallest female crabs occurred at the northern sample sites (Figure 4). At the most northern station at which they were captured (91-32), the average size of all females (N = 88) was 33-mm CW (SD = 9 mm, range = 20–54 mm). At the more southern stations 23 (N = 51) and 21 (N = 22), female CW averaged 35 mm (SD = 5 mm, range = 28–44 mm) and 45 mm (SD = 4 mm, range = 36–50 mm), respectively. Male size at the north (91-8) and south (91-1) stations follow this same pattern (Figure 4). Gravid females were found at all latitudes sampled and 99% of those with mature-shaped abdominal flaps carried eggs (Table 2). Egg-bearing females ranged from 34- to 60-mm CW and averaged 46 mm (SD = 4 mm). Most females greater than 35-mm CW were carrying eggs (Table 1). The average size of the gravid females from southern station 21 was 45 mm (N = 16, SD = 4 mm, range = 36–50 mm). At the more northern station (91-32) gravid females had a mean CW of 50 mm (N = 7, SD = 4 mm, range = 44–54 mm). No females less than 30-mm CW had mature-shaped abdominal flaps or eggs. Of those in the 30- to 34-mm CW size group, only 2% had mature-shaped abdominal flaps and eggs. In the size groups of females that were larger than 35-mm CW, no less than 52% were gravid (Table 1). Of the gravid females examined for fecundity during July, 93 carried recently extruded bright orange eggs, while 4 females had eyed eggs that appeared ready to hatch. The average number of eggs carried by the 91 females was 19,900 (SD = 6,500). The equation describing the relationship between CW and the number of eggs (Y) per clutch was $Y = 0.672 (CW)^{2.668}$; $r^2 = 0.54$ (Figure 5).

Presence of spermatophores in the vasa deferentia suggests males begin to mature between 25- and 29-mm CW (Figure 6). No males smaller than 25-mm CW contained spermatophores, whereas only 19% of males in the 25- to 29-mm CW range did. All males 35-mm CW and larger contained spermatophores (Table 1). Spermatophore diameters increased from about 44 μm in the 25- to 29-mm CW size group to 64 μm in males greater than 35-mm CW. In males between 45- and 70-mm CW, spermatophore diameters were approximately 75–80 μm (Figure 6).

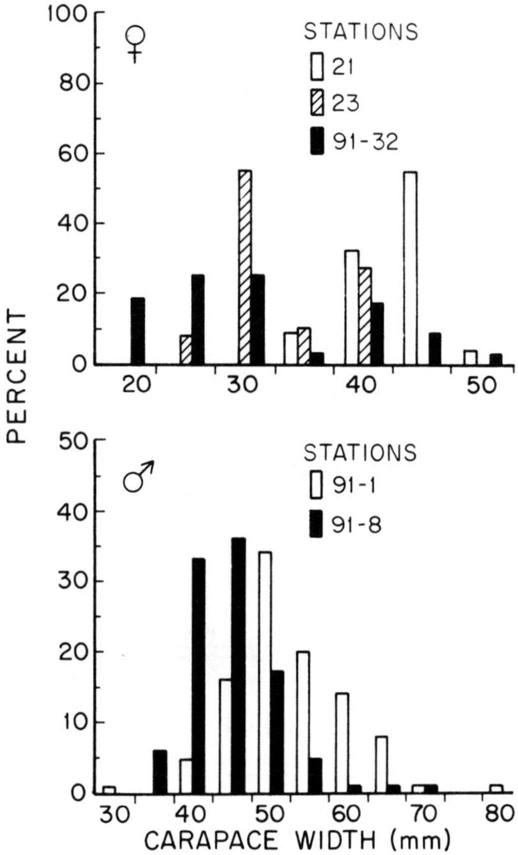

FIGURE 4.—Carapace widths of northeastern Chukchi Sea female snow crabs *Chionoecetes opilio* (upper panel) from stations 21 (69°26′N, 166°31′W), 23 (70°22′N, 162°43′W), and 91-32 (71°37′N, 159°02′W), and from male snow crabs (lower panel) from stations 91-1 (69°02′N, 167°38′W) and 91-8 (70°31′N, 166°08′W).

Discussion

Although snow crabs are widely distributed throughout the northwestern Atlantic Ocean (Elner 1982), Bering Sea (Otto 1982; Slizkin 1989), and northeastern Chukchi Sea (this study) little is known about the factors influencing their distribution and abundance. These factors must include larval recruitment dynamics, benthic habitat requirements, thermal tolerance, water depth preferences, predation, competition, and cannibalism, but the relative importance of each of these factors is unknown.

Prey preference for specific sediment types or depths may explain some of the distribution pattern observed for the snow crab. In the northwestern Gulf of St. Lawrence, where depths exceeded 135 m, Derosiers et al. (1982) found increases in snow crab size with increasing depth. The northeastern Chukchi Sea stations sampled (Figure 1), with one exception (91-33), had depths with a range of only 14–52 m, and snow crabs were found at every station. In the northeastern Chukchi Sea, the benthic sediments (Sharma 1979) form a general pattern of long, narrow gravel belts along the shore and in a few isolated patches in offshore regions. Sand predominates in nearshore areas, and silts and clays predominate offshore. Within this broad pattern, however, there is a mosaic of sediment types (Naidu 1988). Thus, during sampling we probably trawled across several bottom types, each with its own prey community. To better identify sediment habitats preferred by snow crabs, smaller areas must be sampled.

Chukchi Sea snow crabs tend to be smaller than those from the Bering Sea or the northern Atlantic Ocean. Chukchi Sea females averaged 46-mm (this study) to 50-mm CW (Jewett 1981) versus 63- to 72-mm CW in the Bering Sea (Somerton 1981). In

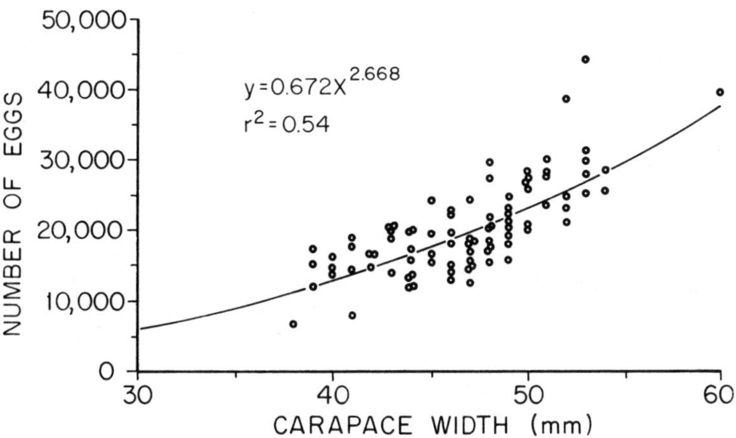

FIGURE 5.—Number of eggs in clutches of various-sized snow crabs *Chionoecetes opilio* captured by otter trawl in the northeastern Chukchi Sea during July 1991. The equation describing the relationship between carapace width (CW) and the number of eggs (Y) per clutch was $Y = 0.672\ (X)^{2.668}$ where X = CW.

commercial fisheries the legal size limit for males ranges from 78-mm CW in Japan (Sinoda 1982) to greater than 94-mm CW in Atlantic Canada (Elner 1982). Chukchi Sea males seldom reach 78-mm CW.

Fecundity estimates for snow crabs from this study are similar to other estimates (Jewett 1981). Fecundity of snow crabs is positively correlated with body size (Haynes et al. 1976; Paul and Fuji 1989). In the southeastern Bering Sea, a 55-mm CW female would have about 24,500 eggs (Haynes et al. 1976) versus 29,000 (Figure 6) for Chukchi Sea specimens 55-mm CW. In Atlantic Canada multiparous females typically carry 52,000–80,000 eggs, depending on geographic region (Davidson et al. 1985). There they reach maturity around 50-mm CW (Watson 1970) and a 55-mm female would carry about 26,600 eggs (Haynes et al. 1976).

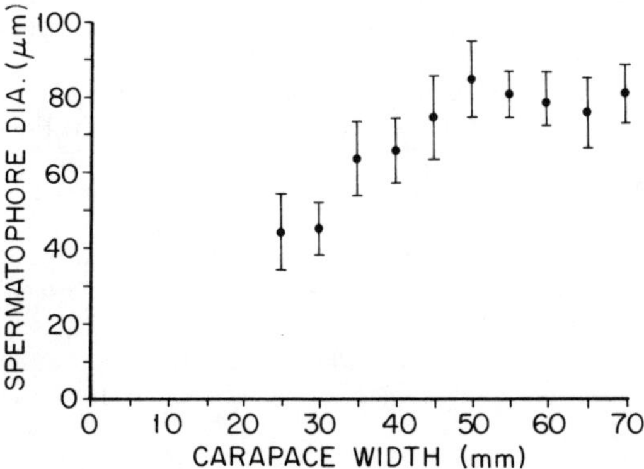

FIGURE 6.—Diameter of spermatophores in relation to size of male snow crabs *Chionoecetes opilio* captured by otter trawl in the northeastern Chukchi Sea during September 1991. Vertical bar represents ± 1 SD.

In 1976 Jewett (1981) found that only 3.3% of mature Chukchi Sea females were ovigerous versus 99% in this study. He collected in September and October with a benthic trawl, so the sampling period and collection gear were similar. He reported that 93% (N = 63) of mature females without external eggs had developing orange ova, so it is possible that they had not yet ovulated. In 1991 snow crabs had extruded eggs before 25 July. These contrasting observations suggest that the dates of snow crab ovulation vary interannually.

Based on claw morphometrics, Somerton (1981) estimated that 50% of male snow crabs in the Bering Sea that reached 65-mm CW were mature. In the northeastern Chukchi Sea, males greater than 65 mm are rare and all those greater than 35 mm have sperm. However, Hartnoll (1965) emphasized that spermatophore presence in small males is only circumstantial evidence of maturity. Laboratory studies and in situ observations are needed to identify the size at which male Chukchi Sea snow crabs mate with primiparous and multiparous mates.

Recently there has been considerable speculation on the reproductive habits of the snow crab, especially the importance of carapace and claw sizes of males (Conan and Comeau 1986). Comparison of size at maturity of female snow crabs from the

Chukchi Sea, Bering Sea, and the north Atlantic suggests that this species has a considerable capacity to modify size at maturity in response to environmental factors. A comprehensive comparison of snow crab mating behavior from several geographical areas, including the Chukchi Sea where there is no fishery, could improve our understanding of the reproductive biology of this valuable circumpolar species.

Acknowledgments

We thank the crews of the *Oshoro Maru* and *Ocean Hope III* and T. Sample, C. Armistead, T. Dark, and R. Bakkala for equipment and assistance. The Alaska Outer Continental Shelf Region of the Minerals Management Service, U.S. Department of the Interior, Anchorage, Alaska, provided funding under contract 14-35-0001-30559.

References

Bowerman, J. H., and B. R. Melteff. 1984. A bibliography of references on the genus *Chionoecetes*. University of Alaska, Alaska Sea Grant Report 84-7, Fairbanks.

Clarke, G., editor. 1973. Staining procedures used by the biological stain commission, 3rd edition. Williams and Wilkins, Baltimore, Maryland.

Conan, G. Y., and M. Comeau. 1986. Functional maturity and terminal molt of male snow crab, *Chionoecetes opilio*. Canadian Journal of Fisheries and Aquatic Sciences 43:1710–1719.

Davidson, K., J. C. Roff, and R. W. Elner. 1985. Morphological, electrophoretic, and fecundity characteristics of Atlantic snow crab, *Chionoecetes opilio*, and implications for fisheries management. Canadian Journal of Fisheries and Aquatic Sciences 42:474–482.

Derosiers, G., J.-C. F. Brethes, and F. Coulombe. 1982. Edaphic segregation within a population of *C. opilio* in the southwestern part of the Gulf of St. Lawrence (Chaleur Bayarea). Pages 353–379 *in* Melteff (1982).

Elner, R. W. 1982. Overview of the snow crab *Chionoecetes opilio* fishery in Atlantic Canada. Pages 3–19 *in* Melteff (1982).

Hartnoll, R. G. 1965. The biology of spider crabs: a comparison of British and Jamaican species. Crustaceana 9:1–16.

Haynes, E., J. F. Karinen, J. Watson, and D. J. Hopson. 1976. Relation of number of eggs and egg length to carapace width in the brachyuran crabs *Chionoecetes bairdi* and *C. opilio* from the southeastern Bering Sea and the Gulf of St. Lawrence. Journal of the Fisheries Research Board of Canada 33:2592–2959.

Jewett, S. C. 1981. Variations in some reproductive aspects of female snow crabs *Chionoecetes opilio*. Journal of Shellfish Research 1:95–99.

Melteff, B., editor. 1982. Proceedings of the international symposium on the genus *Chionoecetes*. University of Alaska, Alaska Sea Grant College Report 82-10, Fairbanks.

Naidu, A. S. 1988. Marine surficial sediments. Section 1.4 *in* Bering, Chukchi, and Beaufort seas: coastal and ocean zones strategic assessment data atlas. NOAA (National Oceanic and Atmospheric Administration) Strategic Assessment Branch, Ocean Assessment Division, Washington, DC.

Otto, R. S. 1982. An overview of the eastern Bering Sea Tanner crab fisheries. Pages 83–118 *in* Melteff (1982).

Paul, A. J., and A. Fuji. 1989. Bioenergetics of the Alaskan crab *Chionoecetes bairdi* (Decapoda: Majidae). Journal of Crustacean Biology 9:25–36.

Paul, A. J., and J. M. Paul. 1989. The size at the onset of maturity in male *Chionoecetes bairdi* (Decapoda, Majidae). Pages 95–103 *in* B. Melteff, editor. Proceedings of the international symposium on king and Tanner crabs. University of Alaska, Alaska Sea Grant College Program Report 90-04, Fairbanks.

Sharma, G. D. 1979. The Alaskan shelf: hydrographic, sedimentary and geochemical environment. Springer-Verlag, New York.

Sinoda, M. 1982. Fisheries for the genus *Chionoecetes* in southwest Japan Sea. Pages 21–29 *in* Melteff (1982).

Slizkin, A. G. 1989. Tanner crabs (*Chionoecetes opilio*, *C. bairdi*) of the northwest Pacific: distribution, biological peculiarities, and population structure. Pages 27–33 *in* B. Melteff, editor. Proceedings of the international symposium on king and Tanner Crabs. University of Alaska, Alaska Sea Grant College Program Report 90-04, Fairbanks.

Somerton, D. A. 1981. Regional variation in the size and maturity of two species of Tanner crab (*Chionoecetes bairdi* and *C. opilio*) in the eastern Bering Sea, and its use in defining management subareas. Canadian Journal of Fisheries and Aquatic Sciences 38:163–174.

Watson, J. 1970. Maturity, mating, and egg laying in the spider crab, *Chionoecetes opilio*. Journal of the Fisheries Research Board of Canada 27:1607–1616.

Weingartner, T. J. 1997. A review of the physical oceanography of the northeastern Chukchi Sea. Pages 40–59 *in* J. Reynolds, editor. Fish ecology in Arctic North America. American Fisheries Society Symposium 19, Bethesda, Maryland.

American Fisheries Society Symposium 19:295–309, 1997

Weight–Length Relationships and Condition of Dolly Varden in Coastal Waters of the Arctic National Wildlife Refuge, Alaska

TEVIS J. UNDERWOOD, DOUGLAS E. PALMER,[1] LAURA A. THORPE, AND BRUCE M. OSBORNE

U.S. Fish and Wildlife Service, Fairbanks Fishery Resource Office
101 12th Avenue, Box 17, Fairbanks, Alaska 99701, USA

Abstract.—Few baseline parameters have been established for arctic fish populations, despite the need for comparison of baseline data to collections from developed areas. Using least-squares regression, we established body condition baselines for Dolly Varden *Salvelinus malma* inhabiting the nearshore coastal waters of the Arctic National Wildlife Refuge, Alaska. Biologists collected length and weight data during two time periods in three areas (Simpson Cove, Barter Island, and Beaufort Lagoon) over 4 years (1988–1991). Using covariance analyses, we detected differences in body condition within season, among areas, and among years, but not between sexes. Low body condition corresponded to years when the arctic ice pack did not recede appreciably from the mainland. Heavy ice pack probably reduced coastal upwelling and hence primary production. In this study, all fish samples were collected within 18-d time periods, and significant differences in condition were found that previous investigations were unable to detect. Seasonal differences in body condition indicated a need to design experiments with short collection periods during the same time periods each year.

Fisheries resources of the Arctic National Wildlife Refuge are an important part of Alaska's arctic ecosystem. Numerous fish species depend on nearshore waters of the Beaufort Sea. The nearshore aquatic habitat is affected by a complex process, incorporating factors from the air, sea, and land to form a unique habitat with many estuarine characteristics (Craig 1984). This unique habitat funnels energy into a food web used by fish, migratory birds, marine mammals, and subsistence fishers. The refuge was set aside to conserve these resources and interactions (USFWS 1988).

The 1980 Alaska National Interest Lands Conservation Act temporarily prohibited oil development on the refuge and required a comprehensive and continuing assessment of the fish and wildlife resources. Several biological studies were subsequently initiated, including small-fishery studies in some coastal lagoons. In 1987, a review of fisheries data indicated the need for further study to provide an adequate baseline to monitor, evaluate, and mitigate impacts in case potential congressional action opened the area to oil development. Additional measures of environmental change were needed.

Several parameters can be used to describe and monitor fish populations. This article focuses on body condition as an index of environmental change using weight–length data for Dolly Varden *Salvelinus malma* that inhabit the refuge's coastal waters. Le Cren (1951) described weight–length relationships as a measure of condition using regression analysis. Recent articles and comments by Cone (1989) and Cone et al. (1990) have clarified the use of this method and the distinction between fish condition and fish form. Colonell and Gallaway (1990) provided an arctic Alaska overview of the use of weight–length data to measure condition. In arctic Alaska the primary species of interest with regard to condition have been Arctic cisco *Coregonus autumnalis* and broad whitefish *C. nasus* (Griffiths et al. 1992), but least cisco *C. sardinella* and Dolly Varden have also been investigated (Gallaway et al. 1991). Few differences in fish condition have been reported among regions and years for arctic species (Whitmus et al. 1987; LGL et al. 1990; Gallaway et al. 1991).

Study Area

Three areas were selected for sampling within the study area (Figure 1). The first area, Simpson Cove, is a sheltered location near the western edge of the refuge. The second area, Barter Island, includes Kaktovik and Jago lagoons. Kaktovik Lagoon is a pulsing lagoon that exchanges water through changes in water surface elevation (i.e., tidal or storm events). Jago Lagoon is a limited-exchange lagoon connecting to the Beaufort Sea and Kaktovik Lagoon. Limited-exchange lagoons exchange water through ocean currents as well as tidal and storm events. Beaufort Lagoon, the third sampling

[1]Present address: U.S. Fish and Wildlife Service, Kenai Fisheries Assistance Office, Post Office Box 1670, Kenai, Alaska 99611, USA.

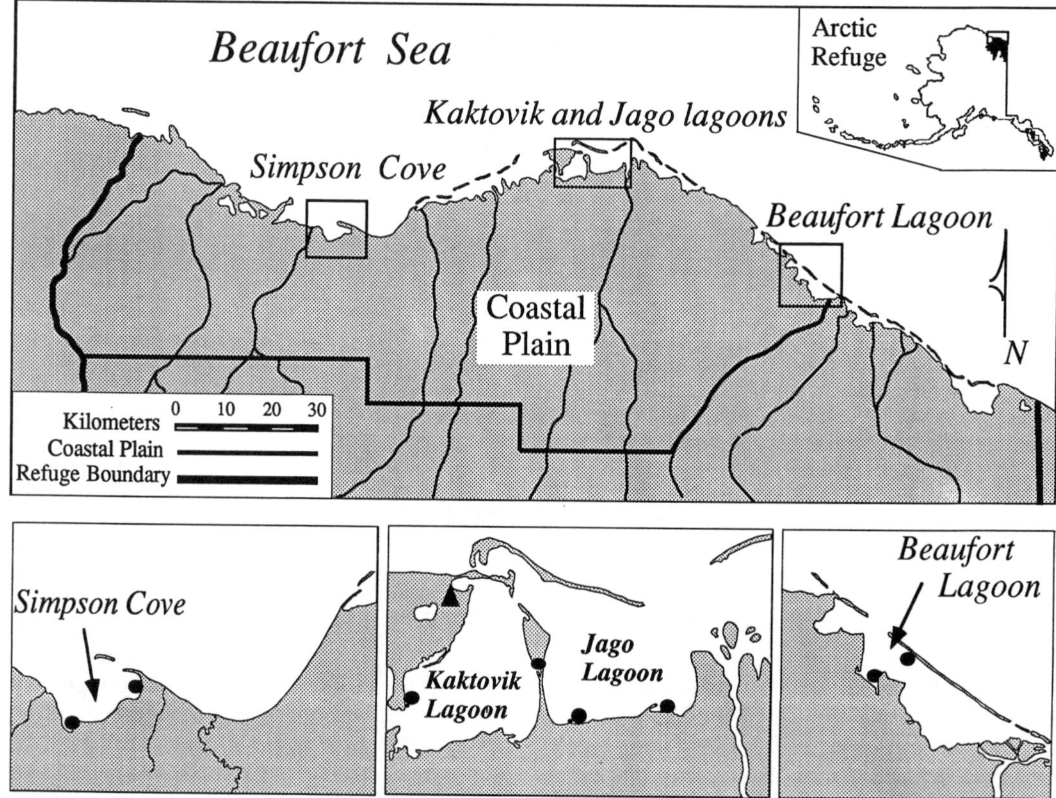

FIGURE 1.—The coastal plain and nearshore area of the Arctic National Wildlife Refuge. The refuge is in the northeastern corner of Alaska adjacent to the Alaska–Canada border. The three sampling areas were Camden Bay (Simpson Cove) to the west, Barter Island, which included the Kaktovik and Jago lagoon system, and Beaufort Lagoon to the east. Solid circles indicate net sites and the solid triangle indicates the village of Kaktovik on Barter Island.

area, is also a limited-exchange lagoon (Hachmeister and Vinelli 1984). Barrier islands, gravel spits 10–100 m or more in width, protect the lagoons and Simpson Cove from the Beaufort Sea. Sheltered waters within these coves and lagoons have a maximum water depth of 3–4 m. In contrast, the Beaufort Sea can exceed 6 m within 0.5 km of shore. Inland, narrow sand and gravel beaches rim the lagoons and are associated with tundra bluffs 1–3 m high.

Coastal waters of the refuge are frozen from October to June or July. Melting snow from coastal plain rivers accelerates the melting of coastal pack ice in late spring. The melting ice and river runoff form a low salinity (10–25‰), relatively warm (5 to 11°C) band of brackish water along the coast. This band of brackish water resists mixing with high salinity (27–32‰), cold (−1 to 3°C) marine waters of the Beaufort Sea (Craig 1984). The band persists throughout much of the open-water season. Formation of the brackish water band was described by Truett (1981); Craig (1984) described its importance to Beaufort Sea fishes.

Methods

Fish were captured in dual-trap fyke nets during two time periods over 4 years in three areas described. The dual-trap fyke nets consisted of two adjacent traps each constructed with 1.5-m wide and 1.2-m high frames at the mouth (Figure 2). The mesh sizes were 12.5-mm stretch for the traps and 25-mm stretch for the wings and leads. Two fyke nets were placed in each water body fished (Table 1). The fyke nets were fished in water depths of 1.3 m or less (Figure 1). Fyke nets were fished with leads fully extended except at one Kaktovik Lagoon station (KL10), where only 30 m of lead was used because of the steep bottom gradient. Traps, lead, and wings were anchored in place using solid steel

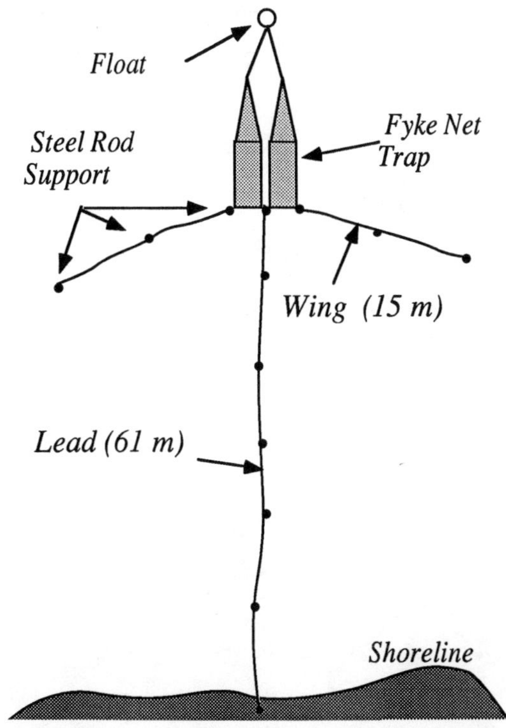

FIGURE 2.—A dual-trap fyke net used for sampling nearshore coastal waters of the Arctic National Wildlife Refuge.

rods 3 m in length and 1.5 cm in diameter. Fyke nets were checked once daily except when severe weather precluded boat travel.

The sampling period was 12–28 July or 27 August to 14 September, or both, depending on the analysis. Sampling regimes changed slightly over the 4 years of data collection, which mainly affected sample sizes. In 1988, Beaufort Lagoon was not sampled; in the other areas, stratified sampling by length groups was not used, although an effort was made to select fish from the entire length range. After 1988, five to eight fish were collected from each 25-mm length interval to avoid bias in estimates. In 1988 and 1989, one target sample size was set and collection was coordinated between sampling areas. In 1990 and 1991, target sample sizes were set for each sampling area, making each area independent (three separate samples) and increasing the total number of samples collected.

Dolly Varden were measured to the nearest 1 mm (fork length). Fish weighing less than 500 g (<1 kg in 1991) were weighed to the nearest 1 g with an electronic balance. Fish weights greater than or equal to 500 g were determined using Pesola spring scales. Weights between 500 g and 1 kg were measured to the nearest 10 g. Weights between 1 and 2 kg were measured to the nearest 50 g. Weights greater than 2 kg were measured to the nearest 100 g.

Fish condition (weight of a fish at a given length) and form (rate of increase in weight as a function of length [Cone 1989]) were compared following Cone's (1989) suggested method using least-squares regression. We regressed $\log_e$ weight on $\log_e$ length. The least-squares regression was calculated using the SAS General Linear Model analysis of covariance procedure (SAS 1988; Ramon et al. 1991). Natural logarithm transformations were used to linearize the data and stabilize variances. This procedure allows separate evaluations of slopes and intercepts. Outliers were defined as observations with absolute studentized values greater than 4 (Neter et al. 1990). Residuals were tested by the Shapiro–Wilk test to examine data normality, and the data were plotted for variance evaluation (Shapiro and Wilk 1965; SAS 1988).

Differences in condition were recognized only when body form was constant (i.e., slopes were not significantly different) and differences in intercept were significant (Cone et al. 1990). When slope differences were evident, the data were reanalyzed and length ranges were truncated so that groups of fish with similar length ranges were tested. When differences in condition were found among treatment groups, pairwise tests (α = 0.05) of least-squares means (SAS 1988) were conducted to determine which lines differed in elevation. Test classes with less then 32 observations were omitted from our analyses.

TABLE 1.—Fyke-net station locations in the nearshore coastal waters of the Arctic National Wildlife Refuge, Alaska. The first two letters of the station designation indicate the water body where the net was located: Simpson Cove (SC), Kaktovik Lagoon (KL), Jago Lagoon (JL), and Beaufort Lagoon (BL).

Station	Latitude	Longitude
	Simpson Cove	
SC01	69°58.98′N	144°50.20′W
SC04	69°57.66′N	144°57.00′W
	Barter Island	
KL05	70°05.44′N	143°39.56′W
KL10	70°06.59′N	143°31.00′W
JL12	70°05.22′N	143°28.50′W
JL14	70°05.51′N	143°22.23′W
	Beaufort Lagoon	
BL02	69°53.28′N	142°18.59′W
BL04	69°54.35′N	142°17.23′W

TABLE 2.—Comparisons of condition between female and male Dolly Varden *Salvelinus malma* collected in July from coastal waters of the Arctic National Wildlife Refuge. Analyses were for combined years (1988–1991) and within an individual year (1988 and 1991).

		Slopes		Intercepts		
Group	*N*	*b*(SE)	*P*-values	*a*(SE)	*P*-values	r^2
			All years			
Females	281	3.07 (0.02)		−12.08 (0.09)		0.99
Males	146	3.06 (0.02)		−12.03 (0.13)		0.99
			$P = 0.72$		$P = 0.87$	
			1988			
Females	121	3.11 (0.03)		−12.33 (0.17)		0.99
Males	86	3.12 (0.04)		−12.37 (0.22)		0.99
			$P = 0.85$		$P = 0.051$	
			1991			
Females	129	3.05 (0.02)		−11.95 (0.10)		0.99
Males	51	3.01 (0.02)		−11.80 (0.14)		0.99
			$P = 0.36$		$P = 0.11$	

We tested weight–length data on Dolly Varden for (1) sex-based differences in condition by comparing female and male fish collected in July using data from all years pooled; (2) seasonal differences in condition by comparing fish collected early in the open-water season (10–24 July) with those collected late in the season (27 August to 12 September) for all years pooled and within individual years; (3) overwintering differences in condition by comparing fish collected after 27 August of one year with fish collected in July of the following year (we analyzed data from the winters of 1988–1989, 1989–1990, and 1990–1991 separately); (4) among-area differences in condition by comparing fish collected within each sampling area for all years pooled and within each individual year (we conducted tests separately on fish collected in July and those collected after 27 August); and (5) among-year differences in condition by comparing fish collected in each year for all areas combined and within each individual sampling area (we conducted separate tests on fish collected in July and those collected after 27 August).

The pooled-data analyses require acknowledgement that spurious differences could result if unequal sample sizes make incorrect the assumption that the data can be pooled reliably. These assumptions were made based on life history information indicating wide dispersal and mixing of char from coastal rivers among areas and years (Underwood et al. 1995). Analysis of nonpooled data provided insight as to the reliability of pooling. In addition, a two-way analysis of covariance was done to simultaneously test area, year, and interaction effects. This procedure was conducted to examine possible spurious results as discussed above and to better partition sources of variation. Our linear model was

$$\log W_{ijk} = \mu + \alpha_i + \beta_j + (\alpha\beta)_{ij} + \gamma(\log L_{ijk}) + \varepsilon_{ijk}, \quad (1)$$

where W is weight, L is length, k is the observation number, μ is the overall mean, α_i is the main effect due to the ith area, β_j is the main effect due to the jth year, $(\alpha\beta)_{ij}$ is the interaction effect at the ith area and the jth year, and γ is the regression coefficient for the relationship between log L and log W. Higher-order interaction terms were sequentially removed from the model when found to be insignificant. In this manner all variation as measured by the sum of squares was partitioned to the significant effects.

Results

Condition Comparisons

Sex differences.—No significant differences in condition were detected between female and male Dolly Varden (Table 2). Results of the pooled-year and individual-year analyses were similar; however, for the 1988 analysis, the difference between the *P*-value of 0.051 and the a priori α of 0.05 was biologically insignificant. Plots of the log-transformed data indicate that females were more common than males in the smaller-length intervals (Figure 3). Analysis of 1989 and 1990 data was inappropriate because of low sample sizes.

Seasonal differences.—Fish condition increased during the field season based on the pooled analysis and analyses within 1990 and 1991 (Table 3). Within

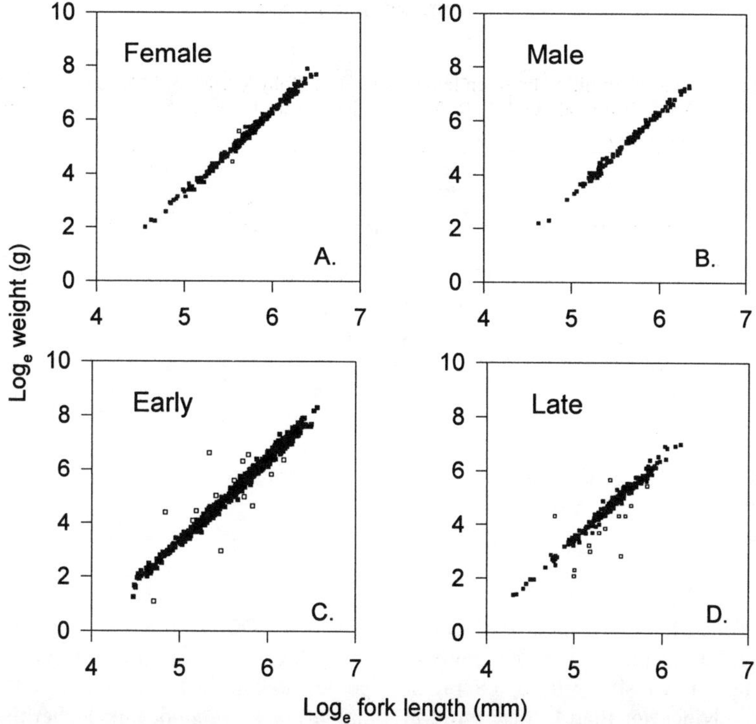

FIGURE 3.—Log-transformed weight–length data, pooled analysis (open squares = outliers), used for comparisons of condition of Dolly Varden *Salvelinus malma* between sexes in July (**A** and **B**: 1988, 1989, and 1991 data) and between seasons (**C** and **D**: 1989–1991 data). Seasonal data correspond to early (9–31 July) and late (1–14 September) sampling periods. Dolly Varden were collected from coastal waters of the Arctic National Wildlife Refuge.

TABLE 3.—Seasonal fish condition comparisons for Dolly Varden *Salvelinus malma* (sexes combined) from coastal waters of the Arctic National Wildlife Refuge for all years combined and for individually sampled years, 1988–1991. Asterisks indicate significant differences in condition.

		Slopes		Intercepts		
Group	*N*	*b*(SE)	*P*-values	*a*(SE)	*P*-values	r^2
			All years			
Early	401	3.12 (0.03)		−12.37 (0.24)		0.97
Late	208	3.02 (0.04)		−11.72 (0.23)		0.96
			P = 0.064		P = 0.0001*	
			1988: 110–360 mm			
Early	193	3.13 (0.03)		−12.45 (0.16)		0.98
Late	36	3.28 (0.06)		−13.13 (0.34)		0.99
			P = 0.04		P = 0.0001	
			1989: 100–360 mm			
Early	51	3.06 (0.03)		−11.94 (0.17)		0.99
Late	46	3.17 (0.05)		−12.40 (0.28)		0.99
			P = 0.06		P = 0.0001*	
			1990			
Early	246	3.18 (0.02)		−12.61 (0.10)		0.99
Late	57	3.08 (0.08)		−11.92 (0.42)		0.96
			P = 0.087		P = 0.001*	
			1991: 160–310 mm			
Early	103	3.10 (0.04)		−12.31 (0.23)		0.98
Late	83	2.97 (0.06)		−11.44 (0.32)		0.97
			P = 0.090		P = 0.0001*	

TABLE 4.—Condition comparisons in overwintering Dolly Varden *Salvelinus malma* from coastal waters of the Arctic National Wildlife Refuge for the winters of 1988–1989, 1989–1990, and 1990–1991. Asterisks indicate significant differences in condition.

Group	N	Slopes		Intercepts		r^2
		b(SE)	P-values	a(SE)	P-values	
			1988–1989: 145–360 mm			
Fall	34	3.32 (0.09)		−13.37 (0.48)		0.97
Spring	40	3.12 (0.05)		−12.24 (0.30)		0.99
			P = 0.063		P = 0.340	
			1989–1990			
Fall	55	3.17 (0.03)		−12.35 (0.18)		0.99
Spring	246	3.18 (0.02)		−12.61 (0.21)		0.99
			P = 0.68		P = 0.0001*	
			1990–1991: 160–427 mm			
Fall	56	3.17 (0.16)		−12.49 (0.88)		0.88
Spring	205	3.14 (0.02)		−12.52 (0.12)		0.99
			P = 0.76		P = 0.0001*	

1988, the slope coefficients remained significantly different and therefore a statement about fish condition could not be made. Fish captured in September were consistently heavier than fish captured in July. Small and large fish were not equally represented in the early and late samples (Figure 3).

Overwintering.—Fish condition decreased during the winter in two of the three winters (Table 4). Differences in condition between the fall of one year and the following spring were significant in 1989–1990 and 1990–1991 but not during 1988–1989 (Table 4). Plots of transformed data show different size group availability (Figure 4).

Spatial differences.—Fish condition differed among areas during July (Table 5). Fish from Barter Island had higher condition than those from Beaufort Lagoon and Simpson Cove in the pooled analysis; however, the differences were small because their significance varies from one analysis to the next. No fish condition differences were found at Barter Island and Beaufort Lagoon in the individual-year analyses. Plots of transformed data indicate some differences in the availability of size groups (Figure 5). The visible differences are small in plots of the three modeled linear relationships (Figure 6A). Sample sizes in 1988 and 1989 were too low to be analyzed separately.

After 27 August, statements about condition could only be made about those fish collected in 1991 because of slope heterogeneity (Table 6). In late 1991 no differences in condition were significant among areas. Plots of transformed data indicate differences among the areas (Figure 5). Sample sizes in the falls of 1988 and 1990 were too small to analyze separately.

Annual differences.—Fish condition was significantly higher in some years (Table 7). Based on the pooled analysis, fish condition in 1989 and 1990 was similar but was significantly higher than that in 1988 and 1991. Fish condition in 1988 was significantly higher than that in 1991 in the pooled analysis, but did not differ in comparisons containing only fish from Barter Island. Plots of transformed data indicate that few large fish and few small fish (<145 mm) were captured in 1988 (Figure 7). Plots of the calculated weight–length relationships indicate noticeable differences in line elevations (i.e., condition; Figure 6B). Data from 1988 and 1989 in Beaufort Lagoon and Simpson Cove were not included in those analyses because of small sample sizes.

Body condition of fish collected after 27 August differed by year based on the pooled analysis (Table 8). Fish condition was higher in 1989 than in 1990 and 1991. The latter 2 years were similar to each other with respect to condition. Statements about condition within Barter Island were invalid because of heterogeneous slopes. Plots of transformed data indicate differences in the distribution or availability of certain sizes of char (Figure 8). Data from 1988 were excluded from the analysis because of low sample sizes. Also, sample sizes in Beaufort Lagoon and Simpson Cove were too small to be analyzed separately.

Partitioning Variation

Two-way analysis of covariance results indicate that variation of each term is significant for the three effects: area, year, and their interaction (Table 9). Little can be said about the main effects

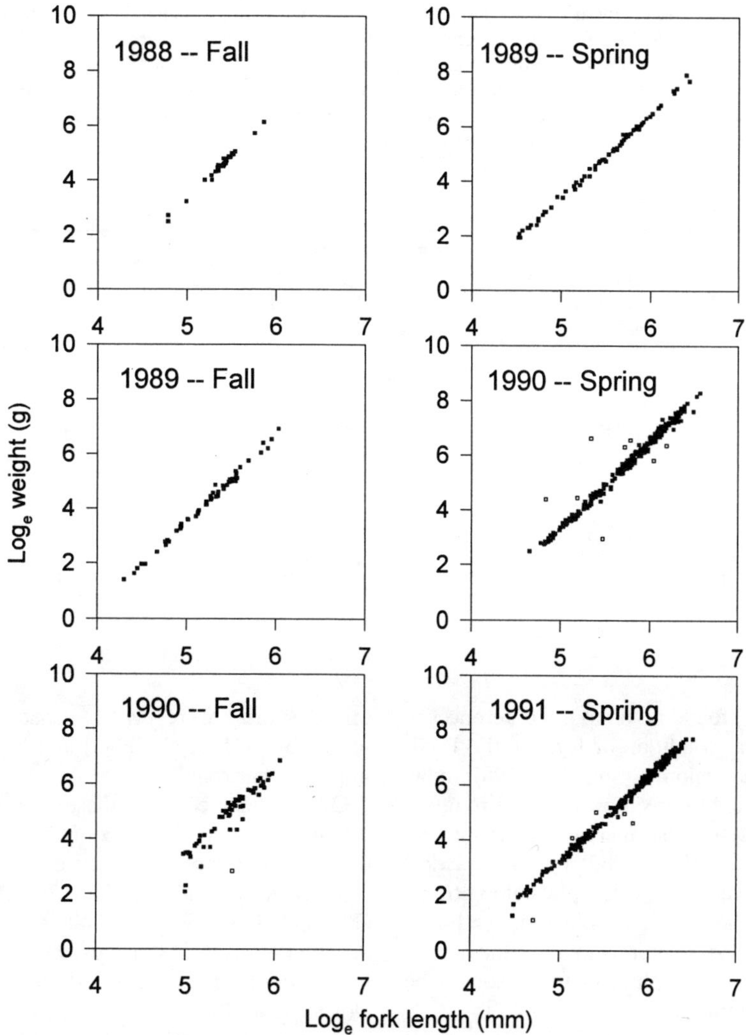

FIGURE 4.—Log-transformed weight–length data (open squares = outliers) used for comparisons of condition of Dolly Varden *Salvelinus malma* from coastal waters of the Arctic National Wildlife Refuge during three winters, 1988–1989, 1989–1990, and 1990–1991.

because of the significant interaction term. However, it is clear that the year effect adds the majority of variation. The area effect is significant and not attributable to the unequal sample sizes. Additional evidence of the greater variation among years, relative to that among areas, can be seen by comparing the distance between the lines for the modeled relationships (Figure 6).

Discussion

Results presented here represent a departure from previously published reports regarding Dolly Varden body condition. Seasonal data provide new insight into the characteristics of body condition as a measurable population variable. Our findings contrast with those of Whitmus et al. (1987), who reported no temporal or spatial differences in Dolly Varden condition; they indicated that because of sampling regimes, condition analyses in their study (and in other studies) may have been "seriously biased." Colonell and Gallaway (1990) concluded that few statistical differences were observed in anadromous fish in arctic Alaska and that even fewer differences were biologically significant. Gal-

TABLE 5.—Spatial condition comparisons for Dolly Varden *Salvelinus malma* collected in July from coastal waters of the Arctic National Wildlife Refuge for all years combined (1988–1991) and sampled years individually (sample sizes in 1988 and 1989 were too low to be analyzed separately but were included in the overall analyses). Asterisks indicate significant differences in condition. Capital letters in the "Pairwise results" column are the results of the pairwise test of least-squares mean values. Rows with similar letters within a group are not statistically different ($P > 0.05$). Letters represent the highest value to the lowest value alphabetically.

Group	N	Slopes		Intercepts		r^2	Pairwise results
		b(SE)	P-values	a(SE)	P-values		
All years: 125–615 mm							
Beaufort Lagoon	191	3.16 (0.02)		−12.61 (0.12)		0.99	B
Barter Island	398	3.12 (0.01)		−12.34 (0.10)		0.99	A
Simpson Cove	221	3.18 (0.02)		−12.71 (0.14)		0.99	B
			$P = 0.125$		$P = 0.01$*		
1990: 140–550 mm							
Beaufort Lagoon	46	3.28 (0.03)		−13.15 (0.19)		0.99	A
Barter Island	85	3.17 (0.03)		−12.52 (0.06)		0.99	A
Simpson Cove	83	3.26 (0.04)		−13.10 (0.22)		0.99	B
			$P = 0.063$		$P = 0.0001$*		
1991							
Beaufort Lagoon	145	3.11 (0.02)		−12.34 (0.10)		0.99	A, B
Barter Island	71	3.12 (0.02)		−12.38 (0.14)		0.99	A
Simpson Cove	103	3.09 (0.03)		−12.24 (0.19)		0.99	B
			$P = 0.72$		$P = 0.006$*		

laway et al. (1991) found no temporal differences in Dolly Varden char condition or form in the Prudhoe Bay, Alaska, region during 1985–1987; however, these fish were sampled over the entire open-water season, a design that increases variation and reduces precision. Cone's (1989) list of "practical considerations" (i.e., sample size, plots of log-transformed data, coefficient of determination, and standard errors of the parameter estimates) and a complete description of sampling methods should be reported in the future.

Body form was not expected to differ among areas and years. Gallaway et al. (1991) reported no change of form for Dolly Varden char between 1985 and 1987 in the nearshore coastal waters around Prudhoe Bay, Alaska. Dolly Varden generally exhibit wide dispersal of overwintering populations from rivers to nearshore coastal waters in the summer. This dispersal effectively mixes various anadromous stocks, and similar mixed stocks were expected among areas and years. The detected changes in the slope coefficient of Dolly Varden may not be the result of different stocks; more likely it occurred because equivalent stanzas—groups of fish at particular life stages with common body form (Bagenal and Tesch 1978)—were not available for each sample. Plots of linearized data indicate slight dissimilarity (Figures 3–5, 7, and 8); size groups available to be sampled differed by area and year. The effects of varying fish availability were not clear. Mixed stanzas were evident in scatter plots presented by LGL et al. (1990) and had no effect on slope comparisons.

Our findings contrast those of previous studies that indicate real differences in weight–length relationships and condition between female and male fish (Bagenal and Tesch 1978; Anderson and Gutreuter 1983). Sex-based differences in condition may be the result of dimorphism (Eschmeyer 1954; Rosen et al. 1982), gonadal development (Van Oosten and Hile 1947; Ruelle 1977; Olmsted and Cloutman 1979; Ehrhardt and Die 1988), or stage of maturity (Shireman 1974; Gabelhouse 1991). These possible differences were not statistically significant in our study.

Seasonal changes in condition highlight the need to limit the duration of data collection because of the added variation (an increase in condition) as summer progresses. Intuitively, we reason that fish condition should increase during summer, when food is abundant in the Arctic and temperatures are favorable for growth. Gonad production and the storage of energy reserves presumably occur at this time. Results similar to ours have been demonstrated in other species (Benson 1953; Shireman 1974; Gabelhouse 1991). In future comparative studies, samples should be taken in short time periods within and between years to increase test precision.

In contrast to increases in condition during sum-

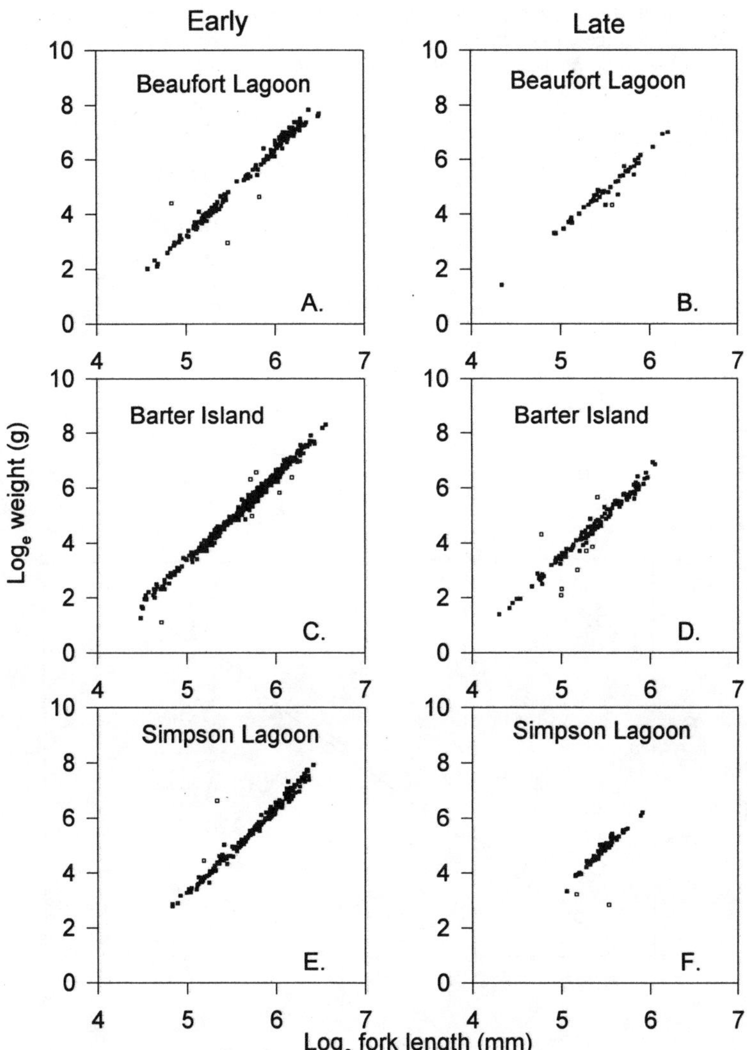

FIGURE 5.—Log-transformed weight–length data, pooled analysis (open squares = outliers), used for comparisons of condition of Dolly Varden *Salvelinus malma* among three areas in coastal waters of the Arctic National Wildlife Refuge in early (July, first column) and late (after 27 August, second column) sampling periods. The data in **A** and **B** represent 1990 and 1991; in **C–F** the data represent 1988–1991.

mer, decreases in energy reserves during winter would be expected. Such reductions were detected in the winters of 1989–1990 and 1990–1991 but not in 1988–1989. Reduced fish condition during winter has been demonstrated in brown trout *Salmo trutta* (Beyerle and Cooper 1960) and largemouth bass *Micropterus salmoides* (Adams et al. 1982). The failure to detect a decrease in condition for 1988–1989 data may indicate that spring sampling did not begin early enough to measure the period of lowest condition. By July, feeding has already begun and fish condition may have rebounded from winter minima. Alternatively, condition in fall 1988 may have been low, making further reductions during winter minimal.

High annual variation of environmental conditions in arctic regions is well documented, and variation in condition among years might be expected. Weight–length relationships in 1988 and 1991 had low line elevations that equate to low fish condition relative to 1989 and 1990. In 1988 and 1991, the ice pack did not recede appreciably from the mainland. Fish condition could be related to pack ice through direct or indirect factors. First, lower water temper-

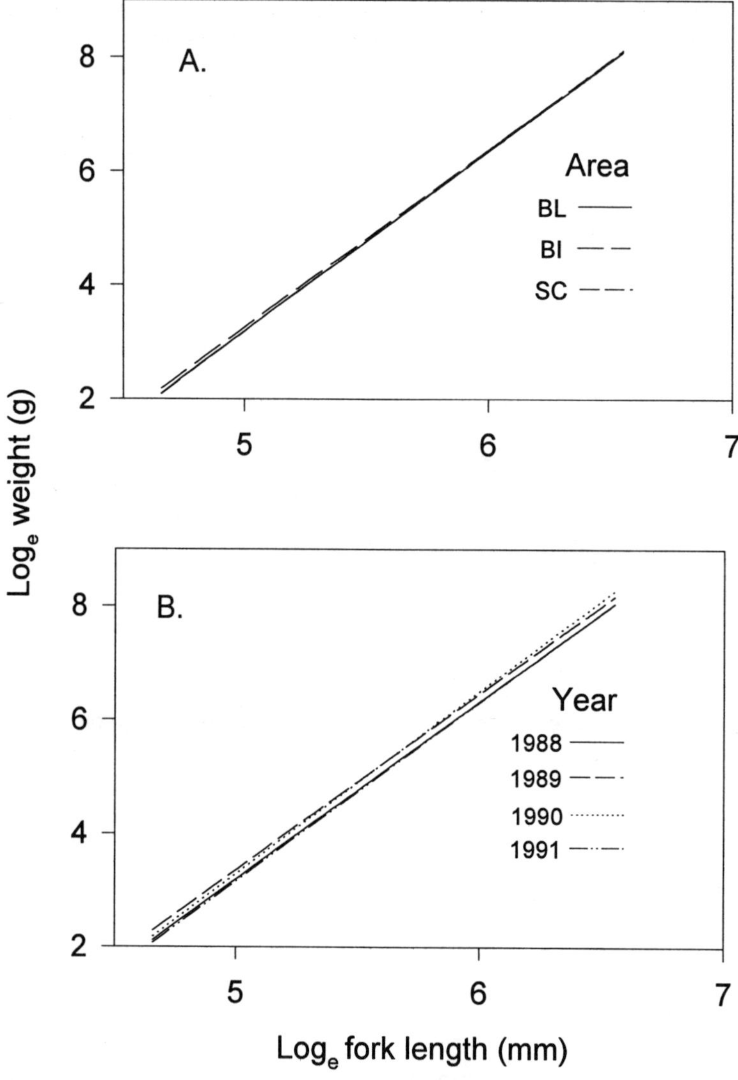

FIGURE 6.—Plots of modeled weight–length relationships for Dolly Varden *Salvelinus malma* for the three areas (**A**) and 4 years (**B**) during July. Abbreviations in (**A**) are BL = Beaufort Lagoon, BI = Barter Island, and SC = Simpson Cove.

ature might affect metabolism and activity level. Water temperature data in Jago Lagoon on 13 August in 1988 ranged from 2 to 6°C (Hale 1990) compared with 6–8°C on 16 August in 1989 (Hale 1991). In Kaktovik Lagoon, differences were slightly larger for those dates, 6–7°C compared with 9 to greater than 11°C, respectively (Hale 1990, 1991). Second, the ice pack inhibits wind-driven upwelling, which lowers food transport from marine to nearshore brackish water areas and lowers food abundance (Craig 1984; Gallaway et al. 1991). Third, shading from the pack ice would reduce primary production in offshore areas. Finally, cold weather associated with a lasting pack ice could affect the timing of river breakup. Delayed river breakup in years of severe ice could reduce the time available to feed in productive nearshore waters.

Plots of the modeled weight–length relationship indicate that less line separation was found among areas than was found among years: variation among years was greater than among areas (Figures 6A and 6B). The two-way analysis of covariance also

TABLE 6.—Spatial condition comparisons for Dolly Varden *Salvelinus malma* collected after 27 August from coastal waters of the Arctic National Wildlife Refuge for all years combined (1988–1991) and for 1991 separately (sample sizes in the falls of 1988–1990 were too small to analyze separately).

Group	N	Slopes b(SE)	Slopes P-values	Intercepts a(SE)	Intercepts P-values	r^2
			All years			
Beaufort Lagoon	50	2.90 (0.06)		−11.11 (0.35)		0.97
Barter Island	141	3.07 (0.03)		−11.89 (0.17)		0.98
Simpson Cove	77	3.11 (0.07)		−12.15 (0.41)		0.95
			P = 0.021		P = 0.0001	
			1991			
Beaufort Lagoon	43	2.92 (0.05)		−11.19 (0.26)		0.98
Barter Island	43	2.87 (0.04)		−10.84 (0.25)		0.90
Simpson Cove	32	2.96 (0.07)		−11.32 (0.41)		0.94
Without outliers			P = 0.57		P = 0.08	

indicates that annual variation was much greater than variation among areas, and the significant result of the area component helps alleviate fears of spurious results, given the unbalanced design.

Reduced variation of modeled relationships among areas relative to years might be expected. Dolly Varden disperse widely from overwintering areas along nearshore coastal waters during the Beaufort Sea open-water season (Craig and McCart 1976; McCart 1980). In addition, tagged Dolly Varden char moved between our sampling areas within several days of tagging (Underwood et al. 1995). With such mixing, little if any variation among areas was expected. The variation seen may have been caused by environmental conditions within the areas, but other sources may have contributed.

TABLE 7.—Annual condition comparisons for Dolly Varden *Salvelinus malma* collected in July from coastal waters of the Arctic National Wildlife Refuge for all sampling areas combined (1988–1991) and for sampling areas individually; within individual sampling areas, 1988 and 1989 sample numbers were too low to be analyzed. Asterisks indicate significant differences in condition. Capital letters in the "Pairwise results" column are the results of the pairwise test of least-squares mean values. Rows with similar letters within a group are not statistically different ($P > 0.05$). Letters represent the highest value to the lowest value alphabetically.

Group	N	Slopes b(SE)	Slopes P-values	Intercepts a(SE)	Intercepts P-values	r^2	Pairwise results
			All areas: 145–450 mm				
1988	204	3.12 (0.03)		−12.40 (0.15)		0.99	B
1989	47	3.10 (0.04)		−12.14 (0.22)		0.99	A
1990	172	3.21 (0.03)		−12.76 (0.16)		0.99	A
1991	228	3.15 (0.02)		−12.59 (0.11)		0.99	C
			P = 0.056		P = 0.0001*		
			Beaufort Lagoon				
1990	54	3.21 (0.03)		−12.76 (0.20)		0.99	
1991	144	3.12 (0.02)		−12.38 (0.10)		0.99	
			P = 0.006		P = 0.0001		
			Barter Island				
1988	196	3.10 (0.02)		−12.25 (0.13)		0.99	B
1989	64	3.06 (0.02)		−11.91 (0.12)		0.99	A
1990	98	3.13 (0.02)		−12.25 (0.12)		0.99	A
1991	71	3.12 (0.02)		−12.38 (0.14)		0.99	B
			P = 0.17		P = 0.001*		
			Simpson Cove: 160–450 mm				
1990	64	3.16 (0.05)		−12.52 (0.31)		0.98	
1991	86	3.11 (0.03)		−12.41 (0.19)		0.99	
			P = 0.52		P = 0.0001*		

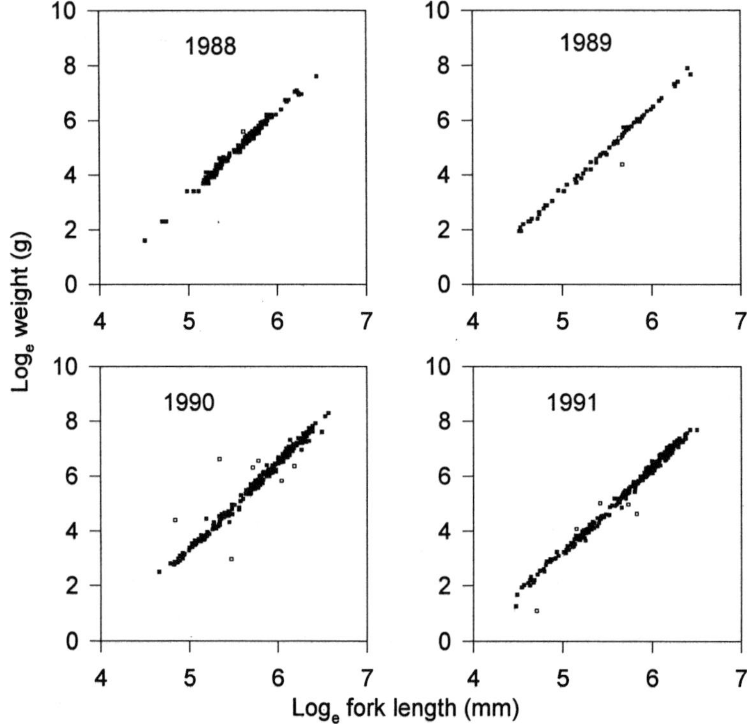

FIGURE 7.—Log-transformed weight–length data, pooled analysis (open squares = outliers), used for comparisons of condition of Dolly Varden *Salvelinus malma* from coastal waters of the Arctic National Wildlife Refuge among years during July, 1988–1991.

Future research in the Arctic should focus on clarifying and minimizing sources of variation and discovering appropriate stanzas to sample. Sampling for a limited number of days during the same time each season would minimize variation (changes in condition) demonstrated in this study and in studies of other arctic fish species during summer (LGL et al. 1990). Sampling similar locations from year to year appears to be less important, but may reduce variation slightly. Minimizing known sources of variation within a stanza will provide sensitive tests of changes in fish condition without changes in

TABLE 8.—Annual condition comparisons for Dolly Varden *Salvelinus malma* collected after 27 August from coastal waters of the Arctic National Wildlife Refuge for all sampling areas combined and for Barter Island separately, 1989–1991. Asterisks indicate significant differences in condition. Capital letters in the "Pairwise results" column are the results of the pairwise test of least-squares mean values. Rows with similar letters within a group are not statistically different ($P > 0.05$). Letters represent the highest value to the lowest value alphabetically.

Group	N	Slopes		Intercepts		r^2	Pairwise results
		b(SE)	P-values	a(SE)	P-values		
All areas: 160–341 mm							
1989	33	3.14 (0.11)		−12.23 (0.61)		0.96	A
1990	46	3.28 (0.23)		−13.12 (1.27)		0.82	B
1991	95	2.90 (0.05)		−11.08 (0.26)		0.97	B
			$P = 0.06$		$P = 0.002$*		
Barter Island							
1989	40	3.22 (0.04)		−12.60 (0.19)		0.99	
1990	46	3.47 (0.15)		−14.17 (0.85)		0.92	
1991	43	2.87 (0.04)		−10.84 (0.25)		0.98	
			$P = 0.0003$		$P = 0.0005$		

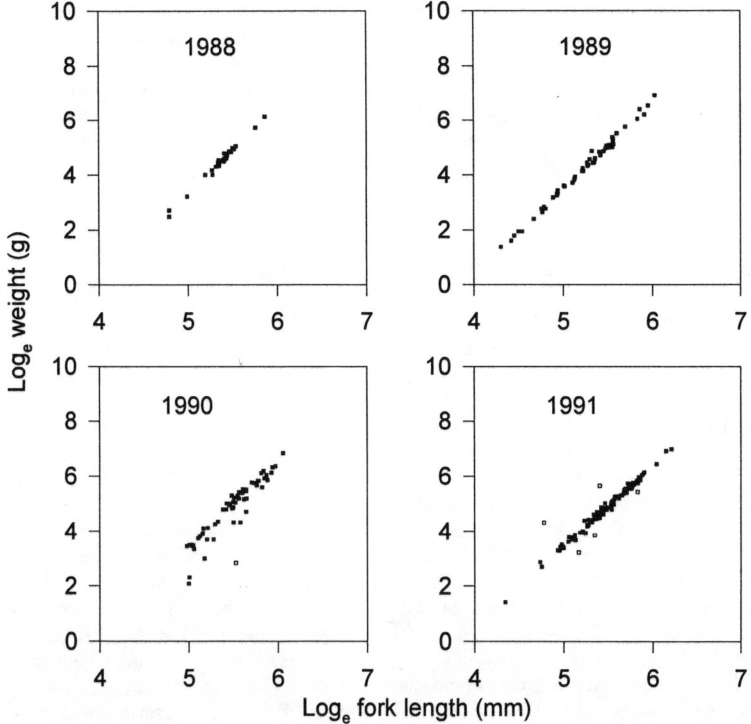

FIGURE 8.—Log-transformed weight–length data, pooled analysis (open squares = outliers), used for comparisons of condition of Dolly Varden *Salvelinus malma* from coastal waters of the Arctic National Wildlife Refuge among years after 27 August, 1988–1991.

slope that add ambiguity. If condition is to serve as a long-term monitoring tool to evaluate environmental impacts, study design must account for and minimize variation within the desired comparison.

TABLE 9.—Results of the two-way analysis of covariance model. Higher-order nonsignificant interaction terms were removed from the model sequentially. Model term "FL" represents the log$_e$ of the fork length.

Source	df	Sum of squares	Mean square	F-value	$P > F$
Model					
Year	3	72.77	24.26	796.62	0.0001
Area	2	5.57	2.78	91.43	0.0001
Year × area	4	9.19	2.30	75.47	0.0001
FL	1	1,448.96	1,448.96	47,585.08	0.0
Error	850	25.88	0.03		
Corrected total	860	1,562.37			

Acknowledgments

We thank Monty Millard, Steve Klein, Judy Gordon, and Randy Bailey of the U.S. Fish and Wildlife Service (USFWS) for reviewing the manuscript. Mike Millard of USFWS and Pham Quang of the University of Alaska Fairbanks rendered valuable assistance regarding statistical analyses. Douglas J. Fruge and David Wiswar provided leadership during the project's first year.

References

Adams, S. M., R. B. McLean, and J. A. Parrotta. 1982. Energy partitioning in largemouth bass under conditions of seasonally fluctuating prey availability. Transactions of the American Fisheries Society 111:549–558.

Anderson, R. O., and S. J. Gutreuter. 1983. Length, weight, and associated structural indices. Pages 283–300 *in* L. A. Nielsen and D. L. Johnson, editors. Fisheries techniques. American Fisheries Society, Bethesda, Maryland.

Bagenal, T. B., and F. W. Tesch. 1978. Age and growth. Pages 101–136 *in* T. Bagenal, editor. Methods for assessment of fish production in fresh waters, 3rd edition. Blackwell Scientific Publications, London.

Benson, N. G. 1953. Seasonal fluctuations in the feeding of brook trout in the Pigeon River, Michigan. Transactions of the American Fisheries Society 83:76–83.

Beyerle, G. B., and E. L. Cooper. 1960. Growth of brown trout in selected Pennsylvania streams. Transactions of the American Fisheries Society 89:255–262.

Colonell, J. M., and B. J. Gallaway, editors. 1990. An assessment of marine environmental impacts of West Dock causeway. Report for Prudhoe Bay unit owners represented by ARCO-Alaska, Inc. Prepared by Woodward-Clyde Consultants and LGL Alaska Research Associates, Inc., Anchorage, Alaska.

Cone, R. S. 1989. The need to reconsider the use of condition indices in fishery science. Transactions of the American Fisheries Society 118:510–514.

Cone, R. S., and five coauthors. 1990. Properties of relative weight and other condition indices. Transactions of the American Fisheries Society 119:1048–1058.

Craig, P. C. 1984. Fish use of coastal waters of the Alaskan Beaufort Sea: a review. Transactions of the American Fisheries Society 113:265–282.

Craig, P. C., and F. J. McCart. 1976. Fish use of nearshore coastal waters in the western Arctic: emphasis on anadromous species. Pages 361–388 *in* D. W. Hood and D. C. Burrell, editors. Assessment of the Arctic marine environment. University of Alaska, Institute of Marine Science, Fairbanks.

Ehrhardt, N. M., and D. J. Die. 1988. Selectivity of gill nets used in the commercial Spanish mackerel fishery of Florida. Transactions of the American Fisheries Society 117:574–580.

Eschmeyer, P. H. 1954. The reproduction of lake trout in southern Lake Superior. Transactions of the American Fisheries Society 84:47–74.

Gabelhouse, D. W., Jr. 1991. Seasonal changes in body condition of white crappies and relations to length and growth in Melvern Reservoir, Kansas. North American Journal of Fisheries Management 11:50–56.

Gallaway, B. J., W. J. Gazey, J. M. Colonell, A. W. Niedoroda, and C. J. Herlugson. 1991. The Endicott development project—preliminary assessment of impacts from the first major offshore oil development in the Alaskan Arctic. Pages 42–80 *in* C. S. Benner and R. W. Middleton, editors. Fisheries and oil development on the continental shelf. American Fisheries Society Symposium 11, Bethesda, Maryland.

Griffiths, W. B., B. J. Gallaway, W. J. Gazey, and R. E. Dillinger. 1992. Growth and condition of Arctic cisco and broad whitefish as indicators of causeway-induced effects in the Prudhoe Bay region, Alaska. Transactions of the American Fisheries Society 121:557–577.

Hachmeister, L. E., and J. B. Vinelli. 1984. Physical oceanography. Pages 501–579 *in* J. C. Truett, editor. Environmental characterization and biological use of lagoons in the eastern Beaufort Sea. NOAA (National Oceanic and Atmospheric Administration), OCSEAP (Outer Continental Shelf Environmental Assessment Program), final report, volume 24, Anchorage, Alaska.

Hale, D. A. 1990. A description of the physical characteristics of nearshore and lagoonal waters in the eastern Beaufort Sea. NOAA (National Oceanic and Atmospheric Administration), Office of Oceanography and Marine Assessment. Report to U.S. Fish and Wildlife Service (Contract 14-16-0007-89-7718-1), Anchorage, Alaska.

Hale, D. A. 1991. A description of the physical characteristics of nearshore and lagoonal waters in the eastern Beaufort Sea, 1989. NOAA (National Oceanic and Atmospheric Administration), Office of Oceanography and Marine Assessment. Report to U.S. Fish and Wildlife Service (Contract 14-16-0007-89-7718-2), Anchorage, Alaska.

Le Cren, E. D. 1951. The length–weight relationship and seasonal cycle in gonad weight and condition in the perch *Perca fluviatilis*. Journal of Animal Ecology 20:201–219.

LGL (LGL Alaska Research Associates, Inc.), Hunter/ESE Environmental Services, Inc., and W. J. Gazey Research. 1990. Analysis of 1985–1987 data collected by the Endicott development monitoring program, volumes 1–3. Final Report to BP Exploration (Alaska) Inc., Anchorage, and the North Slope Borough, Barrow, Alaska.

McCart, F. J. 1980. A review of the systematics and ecology of Arctic char (*Salvelinus alpinus*) in the western Arctic. Canadian Technical Report of Fisheries and Aquatic Sciences 935.

Neter, J., W. Wasserman, and M. H. Kutner. 1990. Applied linear statistical models: regression, analysis of variance, and experimental designs, 3rd edition. Irwin, Homewood, Illinois.

Olmsted, L. L., and D. G. Cloutman. 1979. Life history of the flat bullhead, *Ictalurus platycephalus*, in Lake Norman, North Carolina. Transactions of the American Fisheries Society 108:38–42.

Ramon, C. L., R. J. Freund, and P. C. Spector. 1991. SAS system for linear models, 3rd edition. SAS Institute Inc., Cary, North Carolina.

Rosen, R. A., D. C. Hales, and D. G. Unkenholz. 1982. Biology and exploitation of paddlefish in the Missouri River below Gavins Point Dam. Transactions of the American Fisheries Society 111:216–222.

Ruelle, R. 1977. Reproductive cycle and fecundity of white bass in Lewis and Clark Lake. Transactions of the American Fisheries Society 106:67–76.

SAS (SAS Institute Inc.). 1988. SAS/STAT™ user's guide, release 6.03 edition. SAS Institute Inc., Cary, North Carolina.

Shapiro, S. S., and M. B. Wilk. 1965. An analysis of variance test for normality (complete samples). Biometrika 52:591–611.

Shireman, J. V. 1974. Serum protein variability in bluegills, *Lepomis macrochirus* Rafinesque, from Iowa farm ponds. Transactions of the American Fisheries Society 103:630–632.

Truett, J. 1981. Physical processes. Pages 52–108 *in* LGL Ecological Research Associates, Inc. and LGL Ltd., editors. Beaufort Sea barrier island–lagoon ecological process studies: final report, Simpson Lagoon. Environmental assessment of the Alaskan continental shelf, volume 7. Bureau of Land Management and National Oceanic and Atmospheric Administration, Outer Continental Shelf Environmental Assessment Program, Anchorage, Alaska.

Underwood, T. J., J. A. Gordon, L. A. Thorpe, M. J. Millard, and B. M. Osborne. 1995. Characteristics of selected fish populations of Arctic National Wildlife Refuge coastal waters, final report, 1988–1991. U.S. Fish and Wildlife Service, Fairbanks Fisheries Resource Office, Alaska Fisheries Technical Report 28, Fairbanks.

USFWS (U.S. Fish and Wildlife Service). 1988. Arctic National Wildlife Refuge final comprehensive conservation plan, environmental impact statement, wilderness review, and wild river plan. Final report of the U.S. Fish and Wildlife Service, Anchorage, Alaska.

Van Oosten, J. V., and R. Hile. 1947. Age and growth of the Lake Whitefish, *Coregonus clupeaformis* (Mitchill), in Lake Erie. Transactions of the American Fisheries Society 77:178–249.

Whitmus, C. J., T. C. Cannon, and S. S. Parker. 1987. Age, growth, and condition of anadromous fish. Pages 1–34 *in* 1985 Final report for the Endicott Environmental Monitoring Program, volume 7. Report of Envirosphere Company to U.S. Army Corps of Engineers, Anchorage, Alaska.

American Fisheries Society Symposium 19:310–318, 1997

Food Habits of Four Demersal Chukchi Sea Fishes

K. O. Coyle, J. A. Gillispie, R. L. Smith, and W. E. Barber
University of Alaska Fairbanks, School of Fisheries and Ocean Sciences
Fairbanks, Alaska 99775, USA

Abstract.—Four common Chukchi Sea fishes taken during August–September 1990–1991 were examined for food habits. The species studied were Arctic cod *Boreogadus saida*, Arctic staghorn sculpin *Gymnocanthus tricuspis*, Bering flounder *Hippoglossoides robustus*, and saffron cod *Eleginus gracilis*. All four species occurred in abundance in the northeastern Chukchi Sea. Schoener dietary overlap indexes on the same species from different stations indicated considerable differences in the diets between stations. Therefore, examination of interspecific dietary overlap was limited to those stations where two or more species were captured in the same tows. Highest dietary overlap occurred between Arctic and saffron cod, and negligible overlap occurred between the above two species and the Arctic staghorn sculpin near Point Hope. Arctic cod preyed primarily on planktonic and epibenthic organisms; saffron cod consumed epibenthic and benthic fauna; Arctic staghorn sculpin consumed polychaetes and mollusks; and Bering flounder took fish and Crustacea. The diets of these fishes were not substantially different from those reported for closely related species studied elsewhere. Differences in the diets between stations are probably at least partially related to the distributions of water masses originating elsewhere—in the northern Bering Sea and Arctic Ocean.

Fish are a major link in the arctic food chain between herbivores and apex consumers. Arctic cod *Boreogadus saida* is one of the most abundant fish in the Arctic (Borkin et al. 1987; Lonne and Gulliksen 1989) and is an important food for a variety of marine mammals and birds (Frost and Lowry 1984; Lonne and Gulliksen 1989). Saffron cod *Eleginus gracilis*, Bering flounder *Hippoglossoides robustus*, and Arctic staghorn sculpin *Gymnocanthus tricuspis* are also consumed by marine mammals in the Chukchi Sea (Lowry et al. 1981). Thus, information on the diets of these fishes in the Chukchi Sea is central to understanding trophic relationships between primary and apex consumers in the arctic food web. This paper presents information on the diets of the above four fishes in the eastern Chukchi Sea north of Point Hope.

Methods

The fish samples were collected in August–September 1990 and 1991, as outlined in Smith et al. (1997, this volume). Station locations where fish for stomach analysis were collected are shown in Figure 1. The fish were identified, weighed to the nearest 1 g, measured (total length) to the nearest 1 mm, and dissected for stomach removal. Material in the stomachs was identified to the lowest taxonomic category possible, and wet weights of each taxa were obtained to the nearest 1 mg on a Cahn electrobalance.

The index of relative importance (IRI), as defined by Brodeur and Pearcy (1990), was calculated for each prey taxon in all stomachs of each fish species at each station:

$$\mathrm{IRI} = F_i(N_i + W_i), \qquad (1)$$

where F_i is the percent frequency of occurrence of prey taxon i in nonempty stomachs; N_i is the number of individuals of prey taxon i of the total number of prey organisms, expressed as a percent; and W_i is the weight of individuals of prey taxon i of the total weight of prey organisms, expressed as a percent. Values of IRI were converted to percent of total IRI for each fish species.

The Shannon–Weaver measure of niche breadth (dietary evenness) was computed for each taxon (Brodeur and Pearcy 1990); it is essentially a measure of diversity expressed by the formula:

$$H' = -\Sigma p_i \log_2(p_i), \qquad (2)$$

where H' is the diversity index and p_i is the proportion by weight or number of prey taxon i in the stomachs. Dietary evenness is the ratio of the diversity in the stomachs to the maximum possible diversity:

$$H'\mathrm{max} = \log_2(r), \qquad (3)$$

$$J = H'/H'\mathrm{max} \qquad (4)$$

where J is the evenness, H'max is the maximum possible diversity, and r is the total number of taxa.

Niche overlap was computed using the Schoener's (1968) percent similarity index (PSI) as discussed in Linton et al. (1981):

$$\mathrm{PSI} = [1 - 0.5\,(\Sigma|p_{ij} - p_{ik}|)] \times 100, \qquad (5)$$

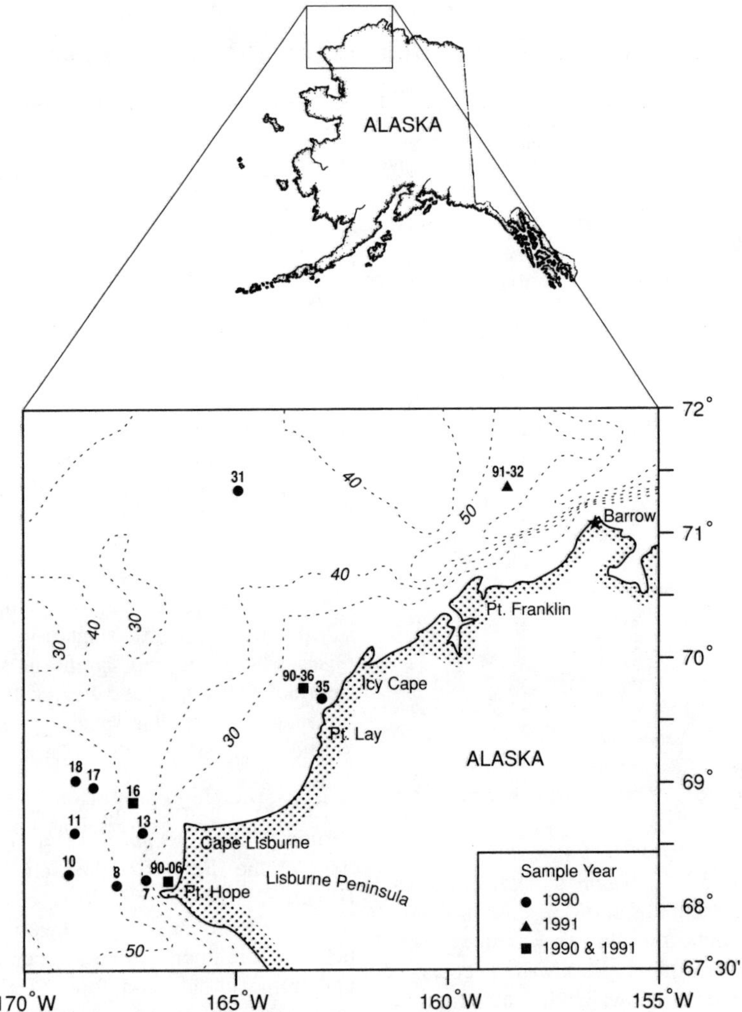

FIGURE 1.—Station locations in the northeastern Chukchi Sea where fishes were collected for stomach analysis. Dashed lines are depth contours in meters.

where p_{ij} is the proportion by weight of prey taxon i in predator j and p_{ik} is the proportion by weight of prey taxon i in predator k.

Results

The average length of fish at each station ranged from 76 to 160 mm and average weight from 9 to 53 g (Table 1). Fish for stomach analysis were selected to include the entire size range of specimens collected by the trawls. If the trawls contained high numbers of small specimens, the average size of individuals in the subsample for stomach analysis was significantly greater than that of the whole sample because of selective subsampling. The average length of Arctic cod in the whole sample at stations 7 and 91-32 was 110 mm and 106 mm, respectively; the average length of Arctic staghorn sculpin at stations 10, 35, 90-06, and 90-36 was 84 mm, 76 mm, 66 mm, and 79 mm, respectively; and the average length of saffron cod at station 90-06 was 131 mm. There were no significant differences between the average lengths of Bering flounder taken in the tows and those of flounder selected for stomach analysis. The majority of fish had been feeding before capture.

Schoener's (1968) test for species overlap was applied to single species to assess the effect of different locations on the diets. The overlap values

TABLE 1.—The average length and weight and the number of fish with full and empty stomachs in the subsamples of four demersal fish taken in August–September 1990 and 1991 at various stations in the northeastern Chukchi Sea.

Station number	Average length (mm)	Average weight (g)	Number with food	Number empty
		Arctic cod		
13	125	20	20	0
18	160	40	19	4
31	117	17	21	1
7	124	21	24	2
90-06	144	29	30	2
91-32	142	26	9	0
		Arctic staghorn sculpin		
8	99	17	18	0
10	103	26	25	1
11	87	17	20	0
17	85	15	20	0
18	102	22	9	0
35	107	28	25	7
90-06	76	9	25	2
90-36	94	24	24	0
		Bering flounder		
8	119	35	49	23
11	140	53	22	2
16	129	42	38	14
17	126	36	29	0
18	128	39	19	1
		Saffron cod		
90-06	153	47	24	0

TABLE 3.—Matrix of percent similarity index for Arctic staghorn sculpin *Gymnocanthus tricuspis* from various stations in the northeastern Chukchi Sea during 1990 and 1991.

	Station						
Station	8	10	11	17	35	90-06	90-36
8	100.0	33.3	24.2	29.3	2.8	11.7	0.7
10		100.0	21.7	22.0	5.2	8.5	2.1
11			100.0	29.7	2.6	19.4	3.3
17				100.0	7.0	30.5	4.1
35					100.0	30.2	18.6
90-06						100.0	1.1
90-36							100.0

varied from 0.3 to 58.9% (Tables 2–4), with highest values tending to occur at stations in close proximity to one another, indicating that these fish are generalists, capable of exploiting a broad spectrum of prey, depending on prey availability at particular locations. Because the overlap indexes are so different and vary markedly from 100%, overlap measures to compare different species can be applied only in those cases where the species were taken in sufficient numbers at the same stations. Thus, interspecific overlap was examined only between Arctic cod, Arctic staghorn sculpin, and saffron cod at Station 90-06 and between Bering flounder and Arctic staghorn sculpin at Stations 8, 11, and 17. Overlap between Bering flounder and Arctic staghorn sculpin at Stations 8, 11, and 17 was 5.0, 2.0, and 13.0%, respectively. Overlap between Arctic cod and saffron cod was 42.5%; overlap between Arctic cod and Arctic staghorn sculpin and between saffron cod and Arctic staghorn sculpin was 2.4%. Overlap is considered significant at values greater than 60% (Brodeur and Pearcy 1990). Thus, even between such similar species as saffron cod and Arctic cod, overlap at the same station was insignificant.

Relative importance indexes indicate that humpy shrimp *Pandalus goniurus* was a substantial component of the diets of both Arctic and saffron cod (Figure 2). However, saffron cod were taking a substantial number of crangonid shrimps and benthic amphipods while Arctic cod were taking epibenthic mysids and fishes. *Calanus* sp. was a dominant component of Arctic cod diet at stations 7, 13, and 18; the pelagic amphipod *Parathemisto libellula* was dominant at station 91-32 (Figure 3). The infaunal cumacean genera *Eudorella* and *Leucon* were important in the diet at stations 7 and 13; almost all of the cumaceans consisted of pelagic-stage, adult males. The unknown food category (sta-

TABLE 2.—Matrix of percent similarity index for Bering flounder *Hippoglossoides robustus* from various stations in the northeastern Chukchi Sea during 1990.

	Station				
Station	8	11	16	17	18
8	100.0	28.0	58.9	54.7	45.8
11		100.0	15.7	16.9	15.8
16			100.0	48.3	48.0
17				100.0	36.6
18					100.0

TABLE 4.—Matrix of percent similarity index for Arctic cod *Boreogadus saida* from various stations in the northeastern Chukchi Sea during 1990 and 1991.

	Station					
Station	91-32	90-06	13	18	31	7
91-32	100.0	3.5	0.3	13.8	6.5	9.6
90-06		100.0	1.0	3.4	2.7	3.4
13			100.0	26.4	0.5	49.6
18				100.0	25.3	30.5
31					100.0	18.1
7						100.0

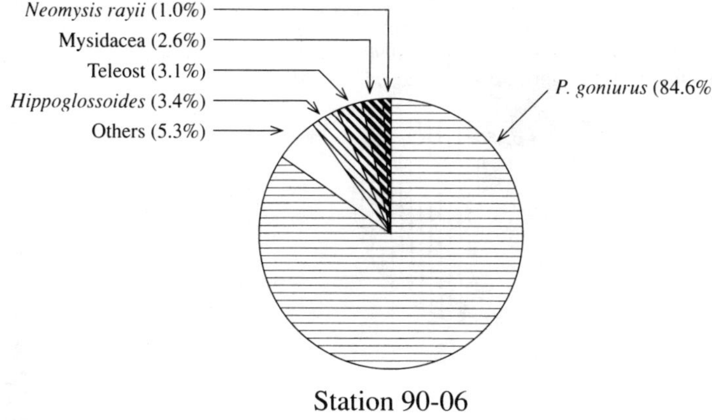

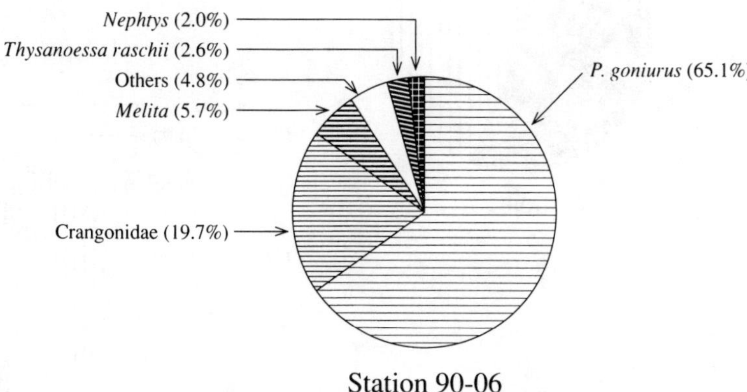

FIGURE 2.—Index of relative importance (percent of total IRI) of major taxa in the diet of Arctic cod *Boreogadus saida* and saffron cod *Eleginus gracilis* collected from station 90-06 in the northeastern Chukchi Sea during 1991.

tions 13 and 18) consisted of tissue, possibly belonging to larvaceans or pteropods. Decapod zoeae were important to the diets of Arctic cod at stations 7 and 31. The "Others" category consisted of many taxa that either were taken infrequently or were too small to contribute substantially to the diet. In Arctic cod they included benthic amphipods (*Protomedeia* spp., *Rhachotropis oculata*, *Anonyx* spp., *Boeckosimus* sp., and *Maera* sp.), pelagic amphipods (*Hyperia* sp. and *Hyperoche medusarum*), mysids (*Mysis oculata* and *Pseudomma truncatum*), and fishes (*Pungitius*, Pleuronectidae). The "Others" category in the saffron cod diet included polychaetes (Polynoidae, *Nephtys* sp.) and infaunal stages of cumaceans (*Diastylis* spp.). Thus, while both cod species consumed epibenthic taxa, the saffron cod tended to take more benthic prey and the Arctic cod more pelagic prey.

Arctic staghorn sculpin at station 90-06 were primarily consuming polychaetes and gastropods (Figure 4). The gastropods were not identified; identifiable polychaetes consisted of Flabelligeridae, Ampharetidae, and *Pectinaria* sp. The "Others" category consisted of benthic amphipods (*Ericthonius* sp. and *Ischyrocerus* sp.), a cumacean (*Diastylis bidentata*), an isopod, and some bivalves. Arctic staghorn sculpin at station 90-36 were primarily consuming euphausiids (*Thysanoessa raschii*), a shrimp

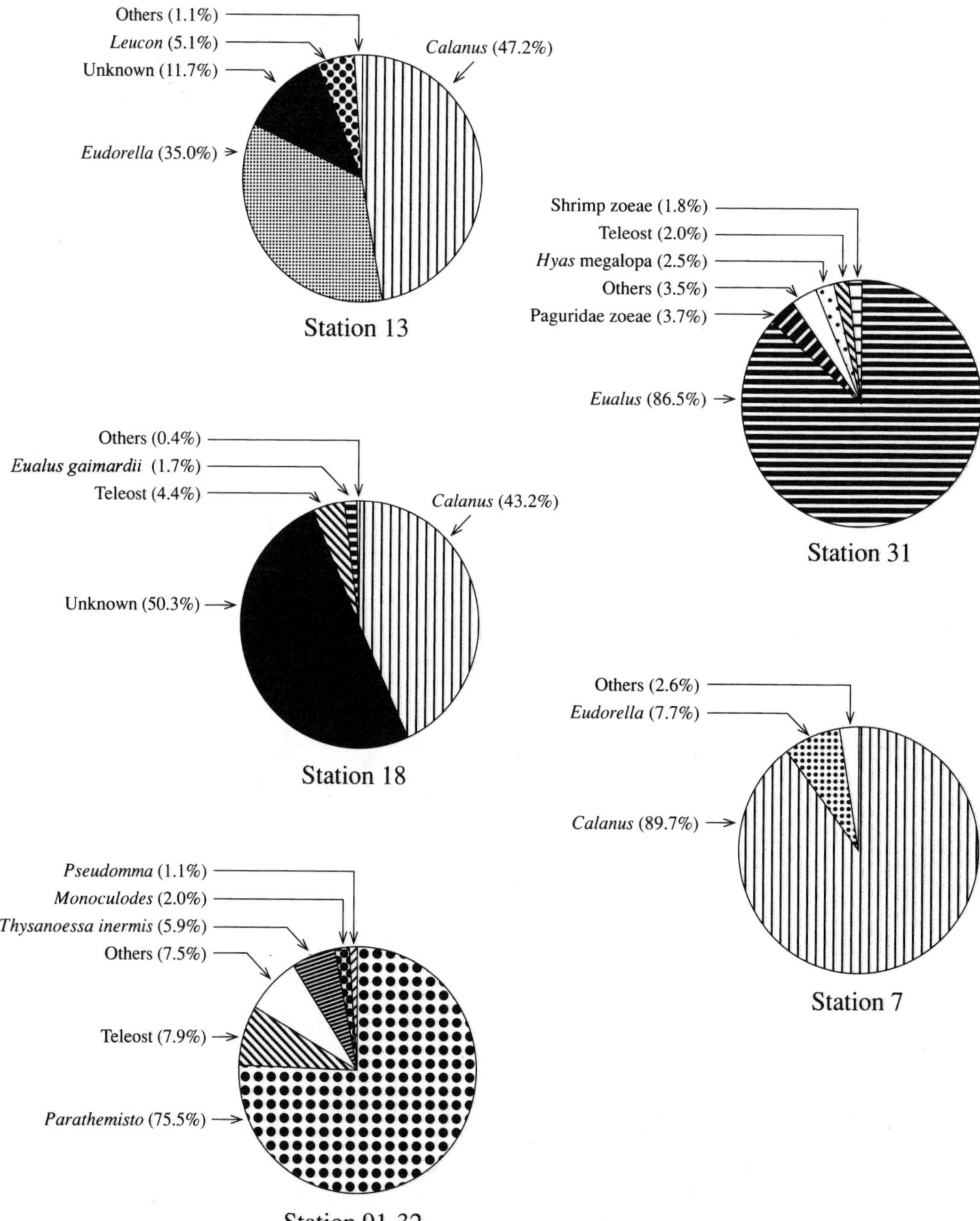

FIGURE 3.—Index of relative importance (percent of total IRI) of major taxa in the diet of Arctic cod *Boreogadus saida* collected from five stations in the northeastern Chukchi Sea during 1990 and 1991.

(circumpolar eualid *Eualus gaimardii*), and a benthic amphipod. This is the only station where they consumed primarily Crustacea. At the remaining two stations, stations 10 and 35, the primary diet included polychaetes (*Nephtys* sp., Opheliidae and Flabelligeridae) and *Echiurus echiurus*.

The Bering flounder selected for stomach analysis were all collected at stations near Point Hope.

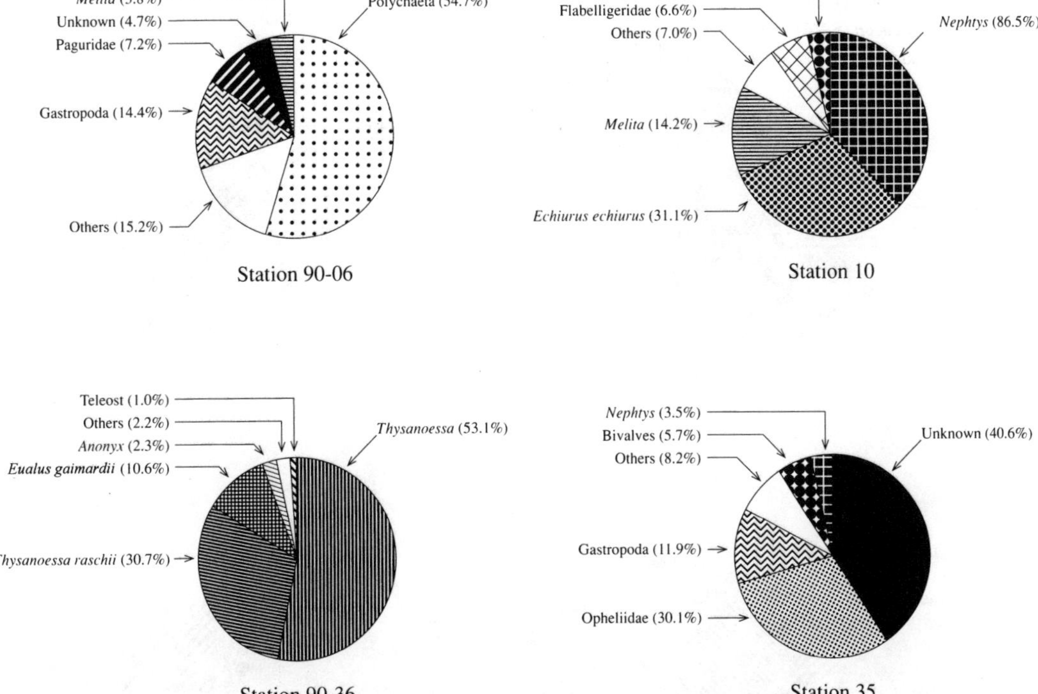

FIGURE 4.—Index of relative importance (percent of total IRI) of major taxa in the diet of Arctic staghorn sculpin *Gymnocanthus tricuspis* collected from four stations in the northeastern Chukchi Sea during 1990.

Therefore, our data on the relative importance of prey taxa reflect prey availability over a small portion of the total study area. Bering flounder mainly were consuming fish; the most important identifiable fish was *Lumpenus* sp. (Figure 5). Other teleost families included zoarcids, agonids, cottids and gadids. Benthic and epibenthic crustaceans constituted most of the rest of the diet. The infaunal amphipod *Byblis* was important at station 8. The infaunal crangonid kuro shrimp *Argis lar* and epibenthic hippolytids (*Spirontocaris* sp. and circumpolar eualid) were the dominant shrimp in the diet. Pagurid crabs were important at station 16.

The Shannon–Weaver statistic is a measure of dietary evenness. In the absence of data on prey populations, the Shannon–Weaver statistic is sometimes considered a measure of niche breadth (Brodeur and Pearcy 1990). The values for Bering flounder and Arctic staghorn sculpin are higher than for Arctic and saffron cod because the former were consuming fewer species in more similar amounts by number and weight than the latter (Table 5).

Discussion

In general, the diets of the fishes studied here do not differ substantially from those reported elsewhere. Arctic cod consume primarily epontic amphipods and plankton during the ice-covered period, thus obtaining much of their energy through primary consumers feeding on ice algae blooms (Lonne and Gulliksen 1989). During summer they feed primarily on plankton in the Beaufort and northern Chukchi seas and on epibenthic shrimp and gammaridean amphipods in the northern Bering Sea (Lowry and Frost 1981). The diet of Arctic staghorn sculpin in our study area is similar to that of threaded sculpin *Gymnocanthus pistilliger*, near the Kamchatka Peninsula, which feeds primarily on polychaetes and *Echiurus* (Tokranov 1985). American plaice *Hippoglossoides platessoides* feed primarily on mollusks, fish, crustaceans, and polychaetes (MacDonald and Green 1986); however, ophiuroids can be a dominant prey item in arctic regions where alternative prey is in low abundance or absent (Berestovskiy 1989). Flathead soles *H. elassodon*

Hippoglossoides robustus

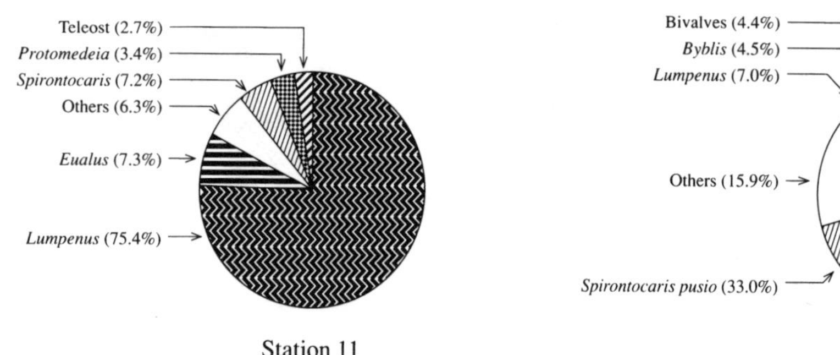

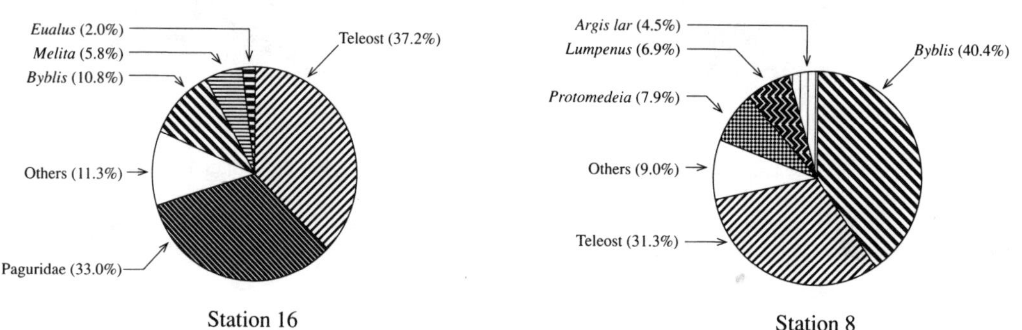

FIGURE 5.—Index of relative importance (percent of total IRI) of major taxa in the diet of Bering flounder *Hippoglossoides robustus* collected from five stations in the northeastern Chukchi Sea during 1990 and 1991.

less than 25 cm long from the eastern Bering Sea feed primarily on small crustaceans such as mysids, gammaridean amphipods, and crangonid shrimp (Livingston et al. 1986).

Previous research in the northern Bering Sea and southern Chukchi Sea has demonstrated the profound influence of different water masses and their associated properties on carbon production rates and the composition of both benthic and planktonic communities (Naumenko and Dzhangil'din 1987; Grebmeier et al. 1988; Springer et al. 1989; Walsh et al. 1989). The organic carbon and plankton advected northward with the major water masses from the northern Bering Sea (Coachman et al. 1975) influence species distribution, abundance, and biomass, thus affecting the diets of fishes, perhaps as far north as Point Franklin. Thus, an understanding of the factors influencing the diets of Arctic cod, Bering flounder, and Arctic staghorn sculpin in the Chukchi Sea requires an understanding of processes occurring in the Chirikov Basin, a region of the northern Bering Sea bounded by Siberia and Alaska, Saint Lawrence Island, and the Bering Strait.

Water in the Chirikov basin is divided into three

TABLE 5.—Shannon-Weaver evenness statistic for diets of four predator species collected from the northeastern Chukchi Sea in August–September 1990 and 1991.

Predator taxon	*N*	Total prey taxa in diet	Evenness by weight	Evenness by number
Bering flounder	157	18	0.61	0.76
Arctic staghorn sculpin	166	21	0.71	0.29
Arctic cod	123	32	0.02	0.16
Saffron cod	24	15	0.02	0.02

water masses: the Anadyr, Bering Shelf, and Alaska coastal water (Coachman et al. 1975). The characteristics and distribution of these water masses are described by Weingartner (1997, this volume). Intense summer phytoplankton blooms are often observed in the western and central Chirikov basin, where upwelling and mixing of Anadyr and Bering Shelf water result in elevated nutrient levels in the photic zone (Hansell et al. 1989). The advection of carbon originating from blooms in the Chirikov basin through the Bering Strait appears to result in elevated chlorophyll, zooplankton, and benthic stocks in the southern Chukchi Sea (Naumenko and Dzhangil'din 1987; Grebmeier et al. 1988; Springer et al. 1989). The frontal system between Bering Shelf and Alaska coastal water can be followed northward toward Cape Lisburne (Weingartner 1997) and is particularly apparent when examining the horizontal distribution of isohaline and isothermal contours in the bottom water off Point Hope and Cape Lisburne. Presumably, zooplankton, particulate organic carbon, and the remaining nutrients are advected northward along with coastal and Bering Shelf water, thus affecting feeding conditions for fish north of Point Hope.

The impact of northward transport of food is suggested by the diets of Arctic cod near Cape Lisburne and Point Hope. *Calanus* was a dominant taxon in Arctic cod stomachs at stations 7, 13, and 18. The highest concentrations of *Calanus marshallae* in the southern Chukchi Sea have been associated with the frontal region near the 32.4‰ isohaline line (Springer et al. 1989). These copepods were probably being concentrated in the frontal region and transported northward past Point Hope and Cape Lisburne, where they were consumed by cod. *Calanus* were absent from Arctic cod stomachs at station 90-06, which is east of the frontal zone in Alaska coastal water, where *Calanus* are in low concentrations or absent. The presence of saffron cod, a neritic species, at station 90-06 is another indication that Alaska coastal water is influencing community composition there.

The frontal region near Point Hope and Cape Lisburne (Weingartner 1997) apparently also impacts the benthic community. Cluster and principal coordinate analysis on abundance data reveal two distinct benthic communities, a coastal and an offshore group, in the Point Hope–Cape Lisburne region (Feder et al. 1989). The offshore group occurred in the frontal region near Point Hope and Cape Lisburne and includes stations 7, 8, 10, 11, 13, 16, and 18 in our study (Figure 1). The offshore community was dominated by benthic suspension feeders, in particular the amphipod *Byblis gaimardi* and juvenile specimens of the barnacle *Balanus crenulatus* (Feder et al. 1989). Such organisms tend to be most abundant in places such as frontal regions, where elevated primary production results in a high flux of organic carbon to benthic consumers (Grebmeier et al. 1988). *Byblis* was an important prey organism in Bering flounder stomachs at stations 16 and 18, and was dominant at station 8. Some of the fish consumed by Bering flounder may also have been feeding on *Byblis* or *Balanus*. Because we have no information on the distribution and abundance of shrimp in the study area, little can be said concerning water column processes that may be affecting their importance in diets of Bering flounder. Benthic biomass in the offshore group was dominated by polychaetes and *Echiurus echiurus* (Feder et al. 1989), the dominant prey items in Arctic staghorn sculpin at station 10.

The potential effects of water masses on the source of carbon in the diets of fish north of Cape Lisburne is problematic because stomach analysis was done on fishes from only four stations and almost nothing is known of the abundance and distribution of epifaunal invertebrates. Arctic staghorn sculpin at stations 90-36 and 35, near Icy Cape, were feeding primarily on *T. raschii*, Opheliidae, gastropods, and bivalves. *Parathemisto libellula*, dominating Arctic cod diets at station 91-35, may be a resident population in Chukchi Sea water or may have been advected south with arctic water. The frontal system in the Point Franklin area (Weingartner 1997) may be a fairly persistent feature separating water masses originating in the northern Bering and southern Chukchi seas from resident Chukchi or arctic water masses; if so, substantially different epibenthic and planktonic communities would be expected on either side of the frontal zone and such differences would undoubtedly be reflected in the diets of fishes inhabiting the different water masses.

Acknowledgments

We thank the crew of the *Ocean Hope III* and T. Sample, R. G. Bakkala, C. Armistead, and T. Dark for their assistance. We particularly thank R. M. Meyer for his enthusiastic support and encouragement throughout the study. This study was funded by the Alaska Outer Continental Shelf Region of the Minerals Management Service, U.S. Department of the Interior, Anchorage, Alaska, under contract 14-35-0001-3-559.

References

Berestovskiy, E. G. 1989. The diet of *Hippoglossoides platessoides limandoides* (Bloch) in some regions of the Barents and Norwegian Seas. Pages 109–123 *in* S. G. Podrazhanskaya, editor. Daily rhythms in feeding and ration of commercial fishes of the world's oceans. (In Russian.) Ministerstvo Rybnogo Khozyaystva Soyuz Sovietskikh Sotsialisticheskikh Respublik, Vsesoyuznyi Nauchno-Issledovatelskii Institut Morskogo Rybnogo Khozyaistva i Okeanograffi, Moscow.

Borkin, I. V., V. P. Ponomarenko, V. L. Tret'yak, and V. N. Shleynik. 1987. Arctic cod *Boreogadus saida* (Lepechin)—fish of the polar seas; stocks and utilization. Pages 183–207 *in* O. S. Skarlato, editor. Biological resources of the Arctic and Antarctic. (In Russian.) Nauka Press, Moscow.

Brodeur, R. D., and W. G. Pearcy. 1990. Trophic relationships of juvenile Pacific salmon off the Oregon and Washington coast. U.S. National Marine Fisheries Service Fishery Bulletin 88:617–636.

Coachman, L. K., K. Aagard, and R. B. Tripp. 1975. Bering Strait: the regional physical oceanography. University of Washington Press, Seattle.

Feder, H. M., A. S. Naidu, M. J. Hameedi, S. C. Jewett, and W. R. Johnson. 1989. The Chukchi Sea continental shelf: benthos and environmental interactions. NOAA (National Oceanic and Atmospheric Administration) OCSEAP (Outer Continental Shelf Environmental Assessment Program) Final Report 68:25–311, Anchorage, Alaska.

Frost, K. J., and L. F. Lowry. 1984. Trophic relationships of vertebrate consumers in the Alaskan Beaufort Sea. Pages 381–401 *in* P. W. Barnes, D. M. Schell, and E. Reimnifz, editors. The Alaskan Beaufort Sea: ecosystems and environments. Academic Press, New York.

Grebmeier, J. M., C. P. McRoy, and H. M. Feder. 1988. Pelagic–benthic coupling on the shelf of the northern Bering and Chukchi seas. I. Food supply source and benthic biomass. Marine Ecology Progress Series 48: 57–67.

Hansell, D. A., and five coauthors. 1989. Summer phytoplankton production and transport along the shelf break in the Bering Sea. Continental Shelf Research 9:1085–1104.

Linton, L. R., R. W. Davies, and F. J. Wrona. 1981. Resource utilization indices: an assessment. Journal of Animal Ecology 50:283–292.

Livingston, P. A., D. A. Dwyer, D. L. Wenker, M. S. Yang, and G. M. Lang. 1986. Trophic interactions of key species in the eastern Bering Sea. International North Pacific Fisheries Commission Bulletin 47:49–65.

Lonne, O. J., and B. Gulliksen. 1989. Size, age and diet of polar cod, *Boreogadus saida* (Lepechin 1773), in ice covered waters. Polar Biology 9:187–191.

Lowry, L. F., and K. J. Frost. 1981. Distribution, growth and foods of Arctic cod (*Boreogadus saida*) in the Bering, Chukchi and Beaufort seas. Canadian Field-Naturalist 95:186–191.

Lowry, L. F., K. J. Frost, and J. J. Burns. 1981. Trophic relationships among ice-inhabiting phocid seals and functionally related marine mammals in the Chukchi Sea. Environmental Assessment of the Alaskan Continental Shelf, Final Reports of Principal Investigators. National Oceanic and Atmospheric Administration Biological Studies 11:37–95.

MacDonald, J. S., and R. H. Green. 1986. Food utilization by five species of benthic feeding fish in Passamaquoddy Bay, New Brunswick. Canadian Journal of Fisheries and Aquatic Sciences 43:1534–1546.

Naumenko, N. I., and C. A. Dzhangil'din. 1987. Distribution of plankton and some fish species in the southern Chukchi Sea. Pages 224–238 *in* O. S. Skarlato, editor. Biological resources of the Arctic and Antarctic. (In Russian.) Nauka Press, Moscow.

Schoener, T. W. 1968. The *Anolis* lizards of Bimini: resource partitioning in a complex fauna. Ecology 49: 704–726.

Smith, R. L., W. E. Barber, M. Vallarino, J. Gillispie, and A. Ritchie. 1997. Population biology of the Arctic staghorn sculpin in the northeastern Chukchi Sea. Pages 133–139 *in* J. Reynolds, editor. Fish ecology in Arctic North America. American Fisheries Society Symposium 19, Bethesda, Maryland.

Springer, A. M., C. P. McRoy, and K. R. Turco. 1989. The paradox of pelagic food webs in the northern Bering Sea. II. Zooplankton communities. Continental Shelf Research 9:359–386.

Tokranov, A. M. 1985. Feeding by species of sculpins of the genus *Gymnocanthus* (Cottidae) from Kamchatka waters. Journal of Ichthyology 25:46–51.

Walsh, J. J., and 20 coauthors. 1989. Carbon and nitrogen cycling within the Bering/Chukchi seas: sources of organic matter affecting AOU demand of the Arctic Ocean. Progress in Oceanography 22:277–359.

Weingartner, T. J. 1997. A review of the physical oceanography of the northeastern Chukchi Sea. Pages 40–59 *in* J. Reynolds, editor. Fish ecology in Arctic North America. American Fisheries Society Symposium 19, Bethesda, Maryland.

American Fisheries Society Symposium 19:319–325, 1997

Energy Content of Arctic Cod and Saffron Cod in the Northeastern Chukchi Sea

R. L. Smith, J. M. Paul, and J. Gillispie
University of Alaska Fairbanks, Institute of Marine Science
Fairbanks, Alaska 99775, USA

Abstract.—The energy content of Arctic cod *Boreogadus saida* and saffron cod *Eleginus gracilis* captured during September 1991 in the northeastern Chukchi Sea was examined using bomb calorimetry. Mean specific energy contents of 4,201 and 4,159 J/g (wet weight of fish) were obtained for Arctic and saffron cod, respectively. These values are within 8 and 40% of two other gadid energy values previously used to model marine mammal bioenergetics. No relationship existed between specific energy content and body weight. Thus, for both species, energy density remained constant over the size ranges examined. Total energy content per fish (E) was regressed against wet weight (W) for both species. The relationship for Arctic cod was $E = 4.47W - 5.11$ ($r^2 = 0.976$); for saffron cod, $E = 4.58W - 4.98$ ($r^2 = 0.997$). There was no difference in energy content of males and females in either Arctic or saffron cod. Regressions of total body energy on otolith length (ϕ) for Arctic and saffron cod were $E = 0.582\phi^{2.71}$ ($r^2 = 0.934$) and $E = 0.022\phi^{3.92}$ ($r^2 = 0.907$), respectively. These equations allow estimation of energy intake of marine mammals and seabirds by measuring the gadid otoliths recovered from stomach contents.

Trophic studies of marine mammals and seabirds in Alaskan waters have documented the importance of gadid fishes as food for a variety of species. In the Bering, Chukchi and Beaufort seas, Arctic cod *Boreogadus saida* and saffron cod *Eleginus gracilis* occur in the diets of 13 marine mammal and 20 seabird species. Walleye pollock *Theraga chalcogramma* contributes to the diets of 11 marine mammals and 13 seabird species (Frost and Lowry 1981a, 1981b). In the Canadian high Arctic, Arctic cod is the dominant food of Glaucous gulls *Larus hyperboreus*, Arctic terns *Sterna paradisaea*, black guillemots *Cepphus grylle* (Divoky 1984), ringed seals *Phoca hispida*, narwhals *Monodon monoceros*, white whales *Delphinapterus leucas*, harp seals *P. groenlandica*, fulmars *Fulmarus*, and murres *Uria* (Bradstreet et al. 1986). Areal or interannual fluctuations in the quality and quantity of prey may have significant effects on the condition of predators, their ability to successfully reproduce and, ultimately, on population structure and size.

Recent efforts to model the energetic requirements of predators on fish have allowed estimates of those predators' impacts on prey stocks. For example, Ashwell-Erickson and Elsner (1981) investigated the dietary conversion of fish prey and the metabolic requirements of Bering Sea harbor *P. vitulina* and spotted *P. largha* seals and predicted an annual consumption of about 82,000 metric tons of walleye pollock, 52,000 metric tons of capelin *Mallotus villosus*, 37,000 metric tons of herring *Clupea pallasi*, and 46,000 metric tons of invertebrates by the populations of these two species. Similar projections might be possible for the Chukchi Sea if the metabolic requirements of the predators and the energy content of the prey were known. The purpose of this study was to provide information on energy content of two of the principal prey fishes in the Chukchi and northern Bering seas, Arctic cod and saffron cod.

Methods

Saffron cod were collected at station 90-06 (68°26′N, 166°38′W), near Point Hope, Alaska, on 23 September 1991. Arctic cod were collected at stations 90-06, 91-32 (20 September 1991; 71°37′N, 159°02′W; northwest of Barrow) and 91-22OH (17 September 1991; 69°40′N, 168°31′W; northwest of Cape Lisburne). Fish were removed from the trawl net, placed in whirl-pak bags, and immediately frozen. Fish remained frozen until dissection and analysis.

Each fish was weighed (nearest 1 g), measured (nearest 1 mm, fork length), and dissected to obtain total weight (corrected for the weight of stomach contents) and gonad weight. Otoliths were removed and stored in 50% glycerol. After dissection, each fish was freeze-dried to constant weight (dry weight, g) and homogenized in a tissue grinder. A subsample was analyzed on a Parr 1241 adiabatic calorimeter for energy content (not adjusted for nitrogen formation during bomb calorimetry). The above weights and energy measures allowed calculation of condition, specific energy content (J/g wet

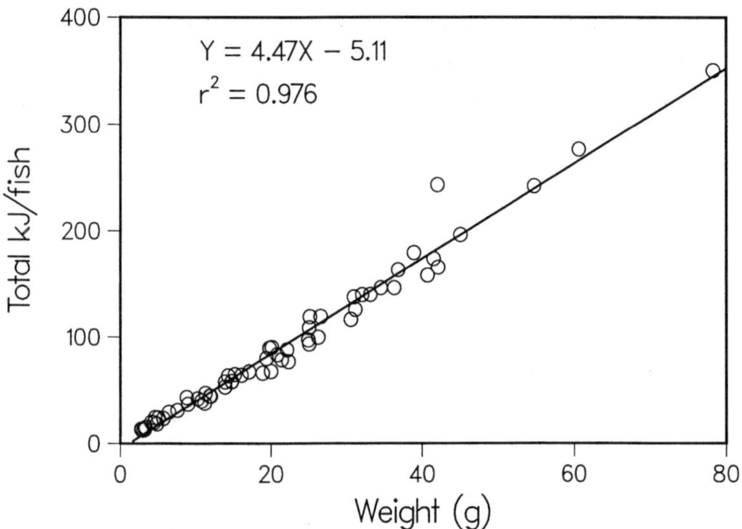

FIGURE 1.—Total body energy as a function of fish wet weight for Arctic cod *Boreogadus saida* from the northeastern Chukchi Sea, September 1991.

weight) and total energy content (kJ/fish) for each specimen.

Otoliths were measured (long axis, nearest 0.1 mm) on a dissecting microscope fitted with an ocular micrometer calibrated with a stage micrometer. Length–weight regressions for both species were calculated for comparison with length–weight regressions from other studies. Energy–weight regressions were generated for males and females of each species. Slopes and elevations of regressions of the two sexes were compared using *t*-statistics outlined in Zar (1984).

Results

Sixty Arctic cod and 51 saffron cod contributed to this study. Arctic cod, excluding stomach contents, weighed 3–78 g (77–225 mm fork length); saffron cod weighed 3–214 g (77–298 mm). Weight (W) as a function of fork length (L) in Arctic cod was $W = 4.78 \cdot 10^{-6}L^{3.08}$ ($r^2 = 0.993$). The corresponding equation for saffron cod was $W = 2.01 \cdot 10^{-6}L^{3.26}$ ($r^2 = 0.983$). Fork length as a function of otolith length (ϕ) was $L = 20.4\phi + 13.6$ for Arctic cod ($r^2 = 0.954$) and $L = 20.0\phi - 23.4$ for saffron cod ($r^2 = 0.922$).

Mean specific energy content for Arctic cod was 4,201 J/g wet weight (SD = 447); the value for saffron cod was 4,159 J/g wet weight (SD = 326). Regression of specific energy content against fish weight produced a scatter of points for both species: $r^2 = 0.015$ and 0.061 for Arctic and saffron cod, respectively. These values suggest that, within the size range tested, there is no relationship between these two variables.

Total energy content (E) per fish was expressed as a function of fish weight (kJ/g · W). For Arctic cod, females were represented by the equation $E = 4.45W - 8.29$ ($r^2 = 0.99$); the equation for males was $E = 4.65W - 3.95$ ($r^2 = 0.96$). Neither slopes nor elevations of these regressions differed significantly, with *t*-values of 1.25 ($P > 0.20$) and 0.71 ($P > 0.20$), respectively. The combined regression (both sexes) was $E = 4.47W - 5.11$ ($r^2 = 0.976$; Figure 1). Energy–weight regressions for female and male saffron cod were $E = 4.59W - 5.74$ ($r^2 = 0.99$) and $E = 4.54W - 5.33$ ($r^2 = 0.99$), respectively. Neither slopes ($t = 0.90$, $P > 0.20$) nor elevations ($t = 0.39$, $P > 0.50$) differed in these regressions. Combining sexes yielded the regression $E = 4.58W - 4.98$ ($r^2 = 0.997$; Figure 2).

Total energy per fish was regressed against otolith length (ϕ). The relationship derived for Arctic cod (sexes combined) conformed to the equation $E = 0.582\phi^{2.71}$ ($r^2 = 0.934$; Figure 3). The equivalent relationship for saffron cod was $E = 0.022\phi^{3.92}$ ($r^2 = 0.907$; Figure 4). In Arctic cod, total energy content as a function of fork length was $E = 2.42 \cdot 10^{-5}L^{3.04}$ ($r^2 = 0.975$; Figure 5). The relationship for saffron cod was $E = 3.18 \cdot 10^{-6}L^{3.46}$ ($r^2 = 0.978$; Figure 6).

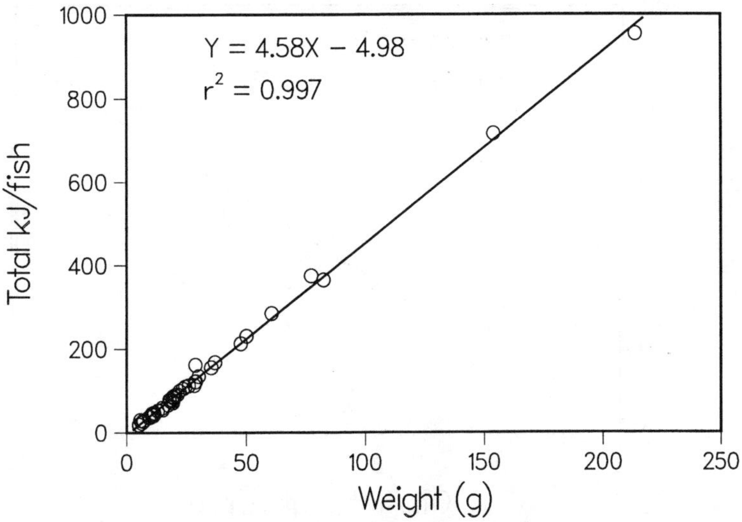

FIGURE 2.—Total body energy as a function of fish wet weight for saffron cod *Eleginus gracilis* from the northeastern Chukchi Sea, September 1991.

Discussion

The Arctic cod sample used in this study, although small ($N = 60$), appears to be representative of Chukchi Sea Arctic cod. Our length–weight equation (sexes combined) is almost identical to that of Gillispie et al. (1997, this volume). The two equations predict weights of 6.95 g and 6.85 g, respectively, for a 100-mm fork length fish. The equation in Frost and Lowry (1981a) predicts a 5.69-g fish at 100 mm ($N = 118$); Wolotira et al.'s (1977) equation predicts a 6.62-g fish. Craig et al. (1984) gave a length–weight equation for summer-caught Arctic cod from Simpson Lagoon on the Beaufort Sea coast that predicts a weight of 7.35 g for a 100-mm fish. The extreme values differ by about 30%. Variation in length–weight relation-

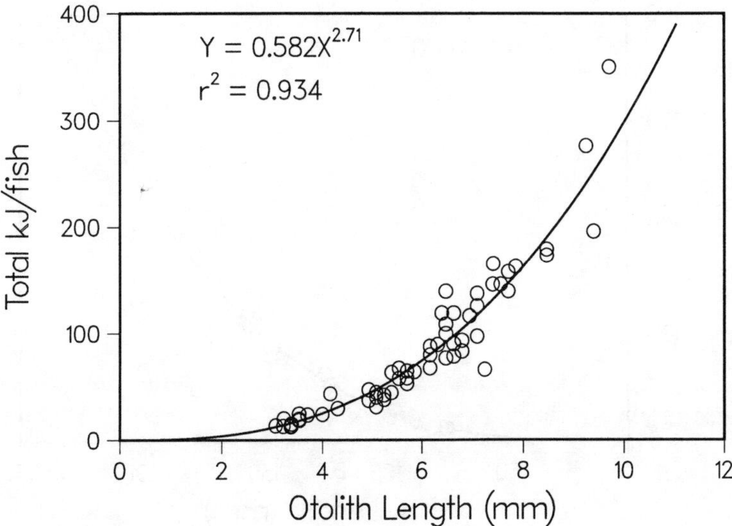

FIGURE 3.—Total body energy as a function of otolith length for Arctic cod *Boreogadus saida* from the northeastern Chukchi Sea, September 1991.

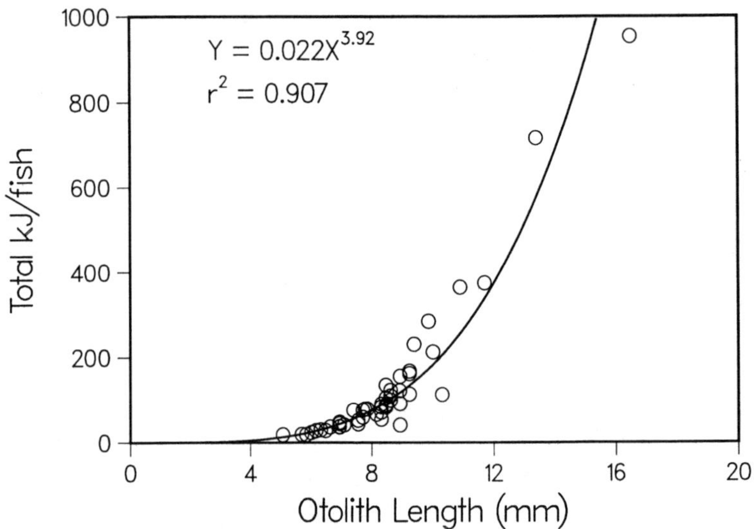

FIGURE 4.—Total body energy as a function of otolith length for saffron cod *Eleginus gracilis* from the northeastern Chukchi Sea, September 1991.

ships is undoubtedly attributable to collection season, maturation of gonads, presence of food in stomach, and other variables. Our combined length–weight regression for saffron cod agrees well with those of Wolotira et al. (1977) and Frost and Lowry (1981a). The equations predict a 100-mm fish weighing 6.66, 6.70, and 6.22 g, respectively.

Our otolith length–fork length regression for Arctic cod is in good agreement with those of Frost and Lowry (1981a) and Gillispie et al. (1997). Given a 6.0-mm otolith, our equation predicts a 136-mm Arctic cod. The same otolith length would predict a 142-mm fish according to Gillispie et al. (1997), a 142-mm fish using Bradstreet et al.'s (1986) data from Point Lay, and a 148-mm Arctic cod according to Frost and Lowry (1981a). Equations derived

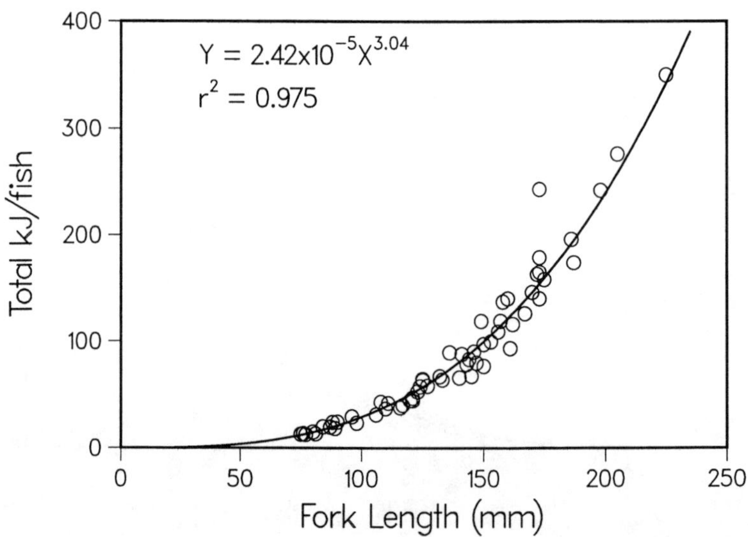

FIGURE 5.—Total body energy as a function of fork length for Arctic cod *Boreogadus saida* from the northeastern Chukchi Sea, September 1991.

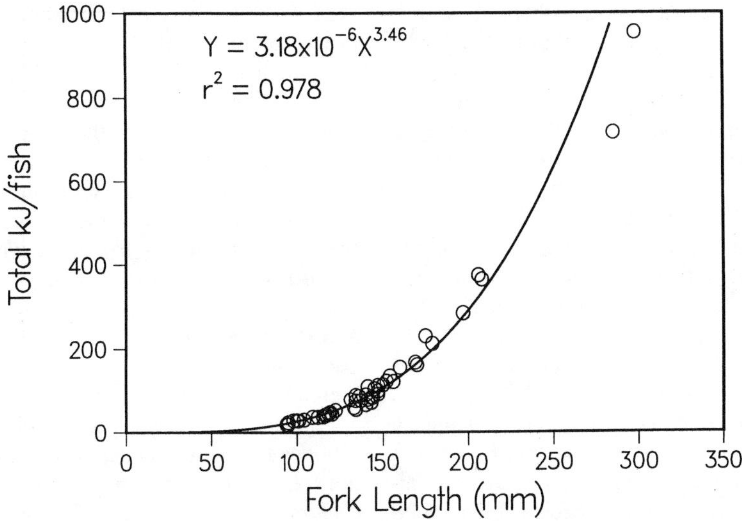

FIGURE 6.—Total body energy as a function of fork length for saffron cod *Eleginus gracilis* from the northeastern Chukchi Sea, September 1991.

from three different sampling years in Simpson Lagoon, Beaufort Sea, yield lengths of 142, 143, and 148 mm, respectively, for a fish with otolith of 6.0-mm length (Bradstreet et al. 1986). For a saffron cod with otolith length of 6.0 mm, our equation predicts a 97-mm fish; the equation from Frost and Lowry (1981a) predicts a 91-mm fish. Thus, collectively, these regressions differ in their predictions by a maximum of 8% in Arctic cod and by 6% in saffron cod.

Bradstreet et al. (1986) reported significant differences in the otolith length–body length regressions for Arctic cod populations from across its North American distribution. It may be necessary to construct energy–otolith length regressions from several populations of Arctic cod to accurately assess the energy intake of cod predators.

The specific energy values reported here are within 10% of the values reported elsewhere (Table 1): adult Pacific cod *Gadus macrocephalus* (Smith et al. 1990), adult walleye pollock (Ashwell-Erickson and Elsner 1981; Smith et al. 1988), and low-condition juvenile walleye pollock (Harris et al. 1986). However, the energy content for walleye pollock reported by Miller (1978) is 40–50% higher, and the energy value for high-condition juvenile walleye pollock reported by Harris et al. (1986) is even larger.

The power function relationships of total fish energy to otolith length and fork length allow esti-

TABLE 1.—Mean energy content of Alaskan gadid fishes. Standard deviation is in parentheses.

Species and stage	Energy/g wet weight: Calories	Energy/g wet weight: Joules	Source
Arctic cod	1,005 (107)	4,201 (447)	This study
Saffron cod	995 (78)	4,159 (326)	This study
Pacific cod			Smith et al. (1990)
Ripe	994	4,155	
Spent	926	3,871	
Walleye pollock (juvenile)			Harris et al. (1986)
Low condition	993 (55)	4,151 (230)	
High condition	1,813 (439)	7,578 (1,835)	
Walleye pollock			Smith et al. (1988)
Ripe	890 (80)	3,720 (334)	
Spent	790 (70)	3,302 (293)	
Walleye pollock	1,088 (59)	4,548 (247)	Ashwell-Erickson and Elsner (1981)
Walleye pollock	1,408 (80)	5,885 (334)	Miller (1978)

mation of the total energy per fish of these two gadids by measuring length, either of the otolith or of the entire fish. Otoliths of these two species may be distinguished by reference to the keys in Frost (1981). A 100-mm Arctic cod has 29 kJ of energy; a saffron cod of that length contains 26.5 kJ. Thus, an error in identification of a 100-mm fish would produce an error in total energy content of about 9%. Similarly, a misidentification of an 8-mm otolith would result in an error of about 50% in total energy. Considerable care should be exercised in assumptions based on otoliths alone. Furthermore, Welch et al. (1992) report that otoliths from the stomachs of predators are often abraded or broken. Dissolution of part of the otolith in the stomach would reduce the length of the otolith, causing an underestimate of energy intake by predators. Difficulty relating energy intake to broken or abraded otoliths could be, in part, resolved by developing regressions of energy to otolith width.

The Arctic cod in this study were beginning to develop gonads for spawning in winter, with gonad indexes of 3% for females and 8.5% for males. We do not have data on ripe versus spent fish and thus cannot depict the entire range of energy contents these fish might exhibit. However, two other Alaskan gadids have been evaluated for energy content relative to spawning. Adult Pacific cod exhibit a 7% decrease in specific energy content because of spawning (Smith et al. 1990); walleye pollock exhibit an 11% decrease (Smith et al. 1988). Thus, one might expect a similar fluctuation in specific energy (J/g) around the values we report for Arctic and saffron cod. Of greater significance is the influence of spawning on total energy (kJ/fish) in the gadids previously studied. The 7% decrease in specific energy is accompanied by a weight decrease, so spent Pacific cod have 30% less total energy than ripe fish (Smith et al. 1990). Walleye pollock exhibit a similar energy depletion during spawning, amounting to about 13% less than the prespawning value (Smith et al. 1988).

Many predators are seasonal in their reliance on Arctic and saffron cod as prey. For example, the seabird colonies at Cape Thompson in the southeastern Chukchi Sea (Swartz 1966) are occupied during the late spring (early May for common murres *Uria aalge*, mid-June for black-legged kittiwakes *Rissa tridactyla*), summer, and early fall (mid-October in black-legged kittiwakes). In September and October the vast majority of these birds depart for lower latitudes. One clear example of a predator relying on Arctic cod for the entire year is the ringed seal in the high Canadian Arctic (Bradstreet et al. 1986). This seal can successfully maintain breathing holes in the pack ice. Use of our relationships depicting the energy content of late-summer fish could be inappropriate for modeling the entire annual energy budget of the ringed seal because of the potential for error in the winter months associated with cod spawning. Additional determinations of body energy for these two gadids are needed for the months centering around spawning.

Springer (1992:92) reviewed the significance of walleye pollock in the ecology of the Gulf of Alaska and Bering Sea, noting the "unsettling coincidence between changing pollock biomass, changing commercial harvests and changing numbers of pollock consumers." Declining pollock biomass correlates significantly with declines in the biomasses of sea lions *Eumetopias* sp. and Pribilof Islands fur seal *Callorhinus ursinus* pups. Conversely, an inverse correlation exists between declining pollock biomass and Pribilof Islands kittiwake productivity. In the Lancaster Sound region of the Canadian Arctic, Welch et al. (1992) documented the critical role played by Arctic cod in terms of energy flow in the ecosystem. Most of the seabirds and marine mammals rely directly on Arctic cod for their subsistence. White whales rely 100% on Arctic cod; it is estimated that each beluga eats 2.5% of its body weight per day (22 kg), equivalent to about 101,000 kJ of energy (Figure 1). Similarly, Welch et al. (1992) estimated harp seals eat 4.2% of their body weight per day, all Arctic cod. This amounts to 4.2 kg/d for the average harp seal, equivalent to about 21,000 kJ/d. Clearly, northern gadids play extremely important roles in the ecosystems in which they are found and in the bioenergetics of their predator species (see also Bradstreet 1982; Craig et al. 1984; Divoky 1984; Bradstreet et al. 1986).

Springer et al. (1984) noted that seabird reproductive success in the Chukchi Sea seemed to depend on the presence of suitable forage fishes such as capelin and Pacific sand lance *Ammodytes hexapterus*. When these species are not present in sufficient quantity, the birds rely on gadids instead and typically, are unable to fledge their chicks. There is a notion that, at least for seabirds, gadids may not be as nutritious as other forage fishes (Springer 1992). Montevecchi and Piatt (1984) examined the energy content, amino acid composition, and other aspects of nutritional quality of capelin as prey for seabirds. Capelins are captured during their spawning season when energy density is low, about 3.8 kJ/g for males and 3.9 to 4.6 kJ/g for females. These values are less than or equivalent to the energy contents we report for both Arctic and saffron cod. It is possible that these gadids have limiting concentrations of some specific nutrients required for seabird survival and growth.

Acknowledgments

Contribution 1628 of Institute of Marine Science, University of Alaska Fairbanks. This study was supported by the Outer Continental Shelf Region of the Minerals Management Service, U.S. Department of the Interior under contract 14-35-0001-30559 and by the Alaska Sea Grant College Program through project RR/92-01.

References

Ashwell-Erickson, S., and R. Elsner. 1981. The energy cost of free existence for Bering Sea harbor and spotted seals. Pages 869–899 *in* D. W. Hood and J. A. Calder, editors. The eastern Bering Sea shelf: oceanography and resources. University of Washington Press, Seattle.

Bradstreet, M. S. W. 1982. Occurrence, habitat use, and behavior of seabirds, marine mammals and Arctic cod at the Pond Inlet ice edge. Arctic 35:28–40.

Bradstreet, M. S. W., and six coauthors. 1986. Aspects of the feeding ecology of Arctic cod (*Boreogadus saida*) and its importance in Arctic marine food chains. Canadian Technical Report of Fisheries and Aquatic Sciences 1491.

Craig, P. C., W. B. Griffiths, S. R. Johnson, and D. M. Schell. 1984. Trophic dynamics in an Arctic lagoon. Pages 347–380 *in* P. W. Barnes, D. M. Schell, and E. Reimnitz, editors. The Alaskan Beaufort Sea: ecosystems and environments. Academic Press, New York.

Divoky, G. J. 1984. The pelagic and nearshore birds of the Alaskan Beaufort Sea: biomass and trophics. Pages 417–437 *in* P. W. Barnes, D. M. Schell, and E. Reimnitz, editors. The Alaskan Beaufort Sea: ecosystems and environments. Academic Press, New York.

Frost, K. J. 1981. Descriptive key to the otoliths of gadid fishes of the Bering, Chukchi and Beaufort seas. Arctic 34:55–59.

Frost, K. J., and L. F. Lowry. 1981a. Trophic importance of some marine gadids in northern Alaska and their body–otolith size relationships. U.S. National Marine Fisheries Service Fishery Bulletin 79:187–192.

Frost, K. J., and L. F. Lowry. 1981b. Foods and trophic relationships of cetaceans in the Bering Sea. Pages 825–836 *in* D. W. Hood and J. A. Calder, editors. The eastern Bering Sea shelf: oceanography and resources. University of Washington Press, Seattle.

Gillispie, J. G., R. L. Smith, E. Barbour, and W. E. Barber. 1997. Distribution, abundance, and growth of Arctic cod in the northeastern Chukchi Sea. Pages 81–89 *in* J. Reynolds, editor. Fish ecology in Arctic North America. American Fisheries Society Symposium 19, Bethesda, Maryland.

Harris, R. K., T. Nishiyama, and A. J. Paul. 1986. Carbon, nitrogen and caloric content of eggs, larvae, and juveniles of the walleye pollock, *Theragra chalcogramma*. Journal of Fish Biology 29:87–98.

Miller, L. K. 1978. Energetics of the northern fur seal in relation to climate and food resources in the Bering Sea. U.S. Marine Mammal Commission, Report MMC-75/08, Washington, DC.

Montevecchi, W. A., and J. Piatt. 1984. Composition and energy contents of mature inshore spawning capelin (*Mallotus villosus*): implications for seabird predators. Comparative Biochemistry and Physiology 78A:15–20.

Smith, R. L., A. J. Paul, and J. M. Paul. 1988. Aspects of energetics of adult walleye pollock, *Theragra chalcogramma* (Pallas), from Alaska. Journal of Fish Biology 33:445–454.

Smith, R. L., A. J. Paul, and J. M. Paul. 1990. Seasonal changes in energy and the energy cost of spawning in Gulf of Alaska Pacific cod. Journal of Fish Biology 36:307–316.

Springer, A. M. 1992. A review: walleye pollock in the north Pacific—how much difference do they really make? Fisheries Oceanography 1:80–96.

Springer, A. M., D. G. Roseneau, E. C. Murphy, and M. I. Springer. 1984. Environmental controls of marine food webs: food habits of seabirds in the eastern Chukchi Sea. Canadian Journal of Fisheries and Aquatic Sciences 41:1202–1215.

Swartz, L. G. 1966. Sea-cliff birds. Pages 611–678 *in* N. J. Wilimovsky and J. N. Wolfe, editors. Environment of the Cape Thompson region, Alaska. U.S. Atomic Energy Commission, Oak Ridge, Tennessee.

Welch, H. E., and seven coauthors. 1992. Energy flow through the marine ecosystem of the Lancaster Sound region, Arctic Canada. Arctic 45:343–357.

Wolotira, R. J., Jr., T. M. Sample, and M. Morin, Jr. 1977. Demersal fish and shellfish resources of Norton Sound, the southeastern Chukchi Sea, and adjacent waters in the baseline year 1976. NOAA (National Oceanic and Atmospheric Administration) Processed Report NMFS (National Marine Fisheries Service), Northwest and Alaska Fisheries Center, Seattle.

Zar, J. H. 1984. Biostatistical analysis, 2nd edition. Prentice-Hall, Englewood Cliffs, New Jersey.

American Fisheries Society Symposium 19:326–339, 1997

Synthesis in Applied Fish Ecology: Twenty Years of Studies on Effects of Causeway Development on Fish Populations in the Prudhoe Bay Region, Alaska

WILLIAM J. WILSON AND BENNY J. GALLAWAY

LGL Alaska Research Associates, Inc.
4175 Tudor Centre Drive, Suite 202, Anchorage, Alaska 99508, USA

Abstract.—Causeways in the form of gravel piers have been used in the Alaskan Arctic to access offshore oil and gas fields and to facilitate movement of freight to the Prudhoe Bay development complex. Since the mid-1970s, environmental monitoring has been required for these developments, and as a consequence a large database was developed on the marine habitats and fish populations of the nearshore Beaufort Sea. Impact assessment studies were conducted for two specific causeways in the Alaskan Beaufort Sea, focusing on several fish species that are utilized in local commercial and subsistence fisheries. Fish-monitoring programs associated with the West Dock and Endicott causeways have been required by the North Slope Borough and have been sponsored by industry. These studies have been reviewed by an independent and multidisciplinary Science Advisory Committee (SAC) impaneled by the borough. In 1991 the SAC and the borough requested a comprehensive synthesis of available information to examine the nearly two decades of accumulated environmental data on the impacts of causeway development on arctic fish populations. This process also required integration of past monitoring reports and published articles into a definitive impact assessment. The synthesis effort was organized around four principal issues: (1) causeway effects on migration of young-of-the-year Arctic ciscoes *Coregonus autumnalis*, (2) causeway effects on movements of other fish species, (3) causeway-induced changes in nearshore hydrography and effects of those changes on fish populations, and (4) causeway effects on fishery catches in the Colville River delta. Resolution of each issue required answering multiple subsidiary questions, each requiring specific data analyses. The process of synthesis, integration, and assessment is described in this article.

Oil and gas exploration and development activities have been continuous in northern Alaska since the 1960s. Environmental impacts of these development activities have been monitored since the discovery of the Prudhoe Bay oil field in 1968. Accompanying the development of hydrocarbon resources has been the construction of solid fill causeways in the central Alaskan Beaufort Sea, and environmental concerns have been expressed regarding their effects on nearshore hydrography and fish populations.

Two causeways (gravel-filled piers) have been developed in the region: West Dock and the Endicott Causeway (Figure 1). West Dock was constructed in 1974–1975 to provide deepwater access to barges delivering supplies and equipment to develop the Prudhoe Bay oil field. The structure was lengthened during early 1976 to provide emergency access to barges trapped in nearshore ice, and in 1981 it was lengthened again for the installation of a water intake facility. The second and third segments are separated by a 15-m breach, located 2,800 m offshore. The total length of the West Dock pier is approximately 4.3 km.

The Endicott oil field is located about 16 km northeast of Prudhoe Bay, in the delta of the Sagavanirktok River (Figure 1). The field contains oil reserves of approximately 1 billion barrels in place, about 350 million of which are recoverable. Development of the field required a 16-km-long gravel access road and an 8-km causeway connecting two artificial islands that support the oil production complex. The causeway was constructed in 1985 and included two breaches totaling 210 m.

Impact assessment research and long-term monitoring of effects of these two causeways on the nearshore environment were sporadic in the 1970s but have been more or less continuous since 1981. Under permit from the U.S. Army Corps of Engineers (USACE) and the North Slope Borough (NSB), environmental monitoring has been specifically mandated to include habitat and fish population studies. Various studies have been completed, but continued long-term monitoring of the Endicott Causeway has been required by the borough because of concerns about impacts of the structure on fish populations important to residents of northern Alaska.

This article focuses on the process of evaluating causeway impacts on fish populations of the Alaskan Arctic and describes the historic monitoring efforts, database, and analytical framework used to

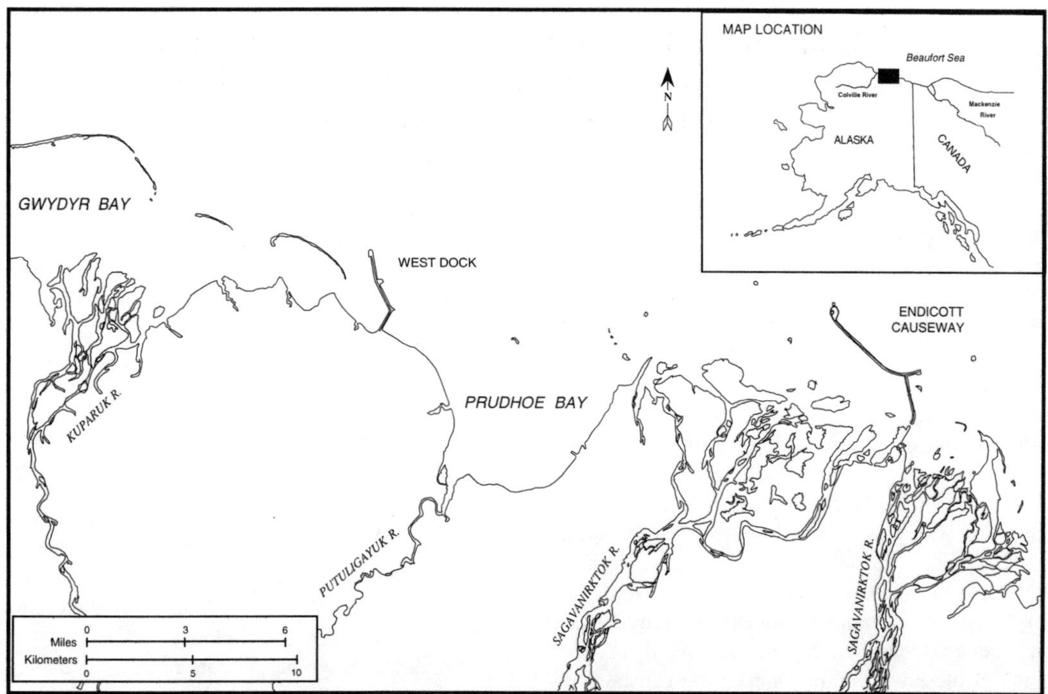

FIGURE 1.—Causeways in the Prudhoe Bay region, Alaska.

prepare a comprehensive synthesis of information for assessing impacts.

Scope and History of the Issues

During the initial planning process for the Endicott Causeway, many environmental issues were raised, including concerns about effects on fish populations:

1. What would be the effects of the causeway and causeway-induced changes in circulation and hydrography on the migration of young-of-the-year Arctic ciscoes *Coregonus autumnalis* from Canada to the Colville River of Alaska?
2. What would be the effects of the causeway and causeway-induced changes in circulation and hydrography upon the nearshore migration corridor (from the shore to the 2-m isobath) used by most species and size groups of amphidromous fish?
3. How would the temperature–salinity characteristics of the nearshore habitat be altered by the circulation and hydrographic effects resulting from the causeway, and what ramifications would these changes have on the population level of broad whitefish *C. nasus* in the Sagavanirktok River?
4. How would the Colville River fisheries of Arctic ciscoes and least ciscoes *C. sardinella* be affected?

These major concerns are restricted to the question of effects of coastal modification on physical and biological processes. The major issues are well documented, having been articulated in the early 1980s based on monitoring the West Dock structure and comments presented in a series of public forums and resource agency meetings while the Endicott Development project was being planned. These issues provided the focus for the project environmental impact statement (EIS) and the monitoring program.

Need for an Information Synthesis

Construction and operation permit stipulations imposed by the USACE and NSB for the Endicott Causeway mandated a marine ecological monitoring program, the purpose being to evaluate the predictions that had been made in the project EIS relative to the four major issues.

During 1985–1987 a joint oceanographic–fish-monitoring program was conducted under the auspices of the USACE. During 1988–1990 separate oceanographic- and fish-monitoring programs were conducted for the USACE and NSB, respectively. The USACE oceanography program continued along the lines established by the 1985–1987 program and was concluded in 1990. BP Exploration (Alaska) Inc. (BPXA), the operator of the Endicott Development, continued the oceanography program in 1991. The NSB fish-monitoring program of 1988–1991 established a fish-sampling and experimental-studies program designed to address specific questions needing resolution. The NSB impaneled an independent Science Advisory Committee (SAC) to review and provide guidance to the program.

During the first years of SAC oversight, major emphasis was placed on developing publications addressing key questions so that a synthesis of the understanding of the various elements of the research efforts eventually could be prepared based on peer-reviewed literature. Before this effort, much of the scientific information on causeway effects was unpublished. Because many of the questions were of an oceanographic nature, the fish-monitoring program also included oceanographic analyses and assessments (separate from the USACE program). Because the respective fish- and oceanographic-monitoring programs of 1988–1990 were separate, true integration and a consensus assessment were not achieved. Nevertheless, progress toward addressing the major issues was made by each of the review programs (SAC 1990a, 1990b).

During 1991 the SAC stressed the overriding importance of a comprehensive synthesis of fish and oceanographic data and a definitive assessment of impacts from the Endicott Causeway on fish and marine systems (SAC 1991). In the SAC's view, the most pressing need was to determine which impact issues (after 6 years of one of the most intensive monitoring studies ever conducted in North America) could be considered resolved and what additional information was needed to complete a reasoned assessment. This view—that there was immediate need for a comprehensive synthesis and assessment—was shared by the NSB and by the operator of the development, BPXA. Thus, major emphasis was placed on synthesis and integration of all existing data and information necessary to produce the most comprehensive assessment possible.

It is important to define the terms used in this process. The following paraphrases the SAC's concurrence on the definitions of synthesis, integration, and assessment: (1) synthesis is the process of building understanding from the separate elements and results of research and investigation; (2) integration is the act of combining elements of investigation into a whole, bringing results of unidisciplinary and multidisciplinary studies into a holistic explanation of phenomena, and seeing how conclusions of the various task elements work together and what they mean in unison; and, (3) assessment is appraising, evaluating, and coming to a conclusion.

The objectives of the synthesis, integration, and assessment program were based on the key issues:

1. To review and synthesize data relevant to assessing how the Endicott Causeway affects movement of young-of-the-year Arctic ciscoes from Canada to the Colville River of Alaska.
2. To review and synthesize data relevant to assessing how the causeway affects movement of other amphidromous fish (broad whitefish, least ciscoes, Dolly Varden *Salvelinus malma*, and Arctic ciscoes age 1 and older).
3. To review and synthesize data relevant to assessing how temperature–salinity changes attributable to the causeway affect populations of amphidromous fish.
4. To review and synthesize data relevant to assessing whether observed variations in the Colville River Arctic cisco and least cisco fisheries are attributable to the causeways.
5. To identify areas in which additional data (if any) are needed to finalize the assessments noted above in issues 1–4.

All historic and recent data from the Prudhoe Bay region were included in the synthesis, integration, and assessment program. Many investigations have been conducted in anticipation of or to directly determine the effects of industrial development on the marine environment of northern Alaska, particularly the Prudhoe Bay region. These studies generated a large database that, when used in conjunction with the more recent Endicott Causeway-monitoring database, constitutes a record of oceanographic and fish observations in the Alaskan Beaufort Sea since 1954.

The database includes data files on measurements of fish distribution and abundance, movement patterns, age and growth, food availability and feeding activities, and hydrographic conditions in the coastal zone. In addition to the archived data, gray literature has been generated as well as dozens of peer-reviewed scientific articles and publications, including a major collection of ecological papers

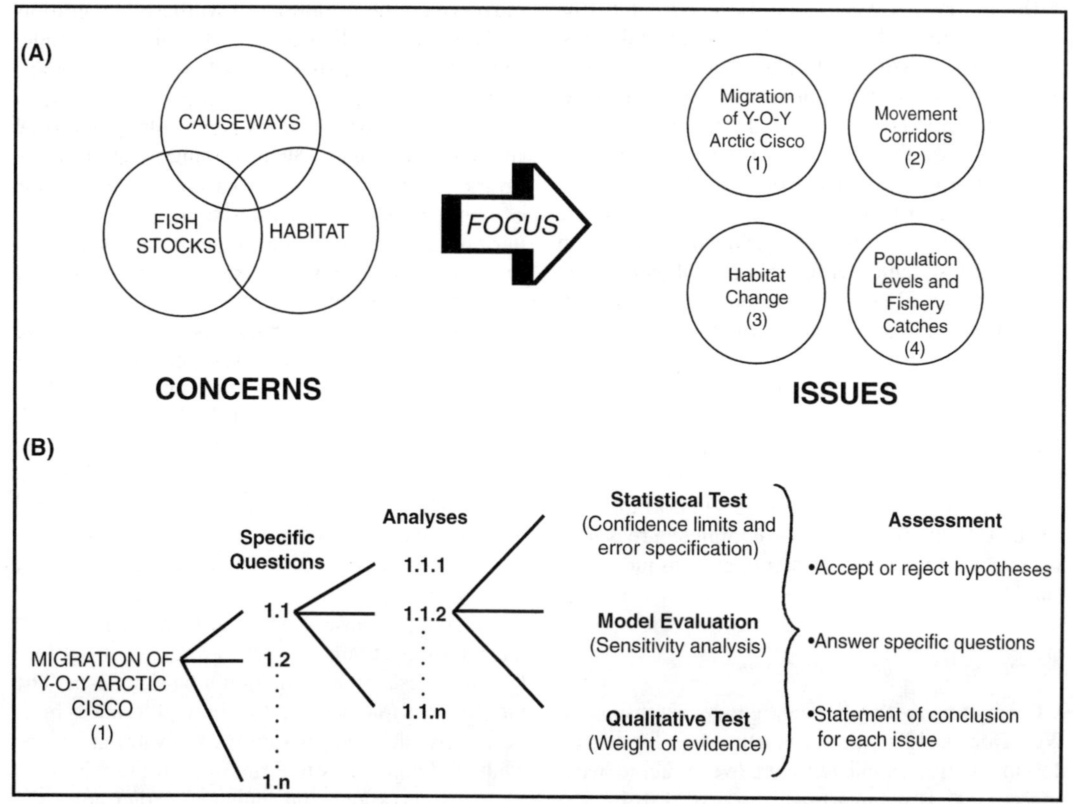

FIGURE 2.—Conceptual model of the 1991 synthesis, integration, and assessment process.

published by the University of Alaska Fairbanks (Norton 1989).

The NSB also requested that reports from all studies previously conducted—genetic-stock structure studies, basic annual environmental-monitoring studies, laboratory fish-feeding studies, and basic distribution, abundance, and movement studies—be integrated into the comprehensive synthesis report. The report was to evaluate these various documents and data sets and provide an assessment of how each of the original EIS issues has been resolved from all available data. That process of synthesis, integration, and assessment is described below.

Methods

Conceptual Approach

The conceptual design of the synthesis, integration, and assessment effort is shown in Figure 2. The concerns about causeways relate to their effects on fish and fish habitat. These general concerns were divided into four separate issues (Panel A of Figure 2), which conform to the issues originally identified in the Endicott Project EIS. To resolve each issue required answering a series of key questions (Appendix 1) related to each issue based on results of specific analyses (Panel B of Figure 1). Some of the questions are common to all the issues (effects of causeway-induced circulation changes), whereas others are specific to a single issue (has condition of broad whitefish exhibited a decline since construction of the Endicott Causeway in 1985?). Some of the questions simply required a descriptive response.

There were some points of contention regarding the interpretation of the descriptive data. In these instances, the relative merits of opposing points of view were evaluated by statistically testing well-structured hypotheses. Significant differences were stated with specified levels of confidence. If significant differences were not found, the results were evaluated in terms of the power of the test (what levels of difference could we have detected) and the possibility that a type II error was being made. Although there were relatively few general questions in this category, there were many subquestions that had to be addressed by formal hypothesis testing.

After all of the questions were addressed, the information was integrated to make a statement of conclusion regarding each issue. Statistical and simulation models were tools for achieving integration and providing the basis for the overall assessment concerning each issue. How models were used in concert with the synthesis information and results of analyses and hypothesis testing is perhaps best shown by an example, presented in the next section.

The objectives of this approach were to definitively provide an up-to-date statement of conclusion regarding each of the four issues. The fifth and secondary objective was to define important data sets that should be obtained to complete a reasoned assessment (the reasoned assessment was judged by the SAC).

Addressing the Concerns and Issues

Originating with the emergency extension of West Dock in 1976, the use of causeways for coastal developments has elicited controversy about what effects these structures might have on fish habitat and, in turn, on fish stock health and size. These three areas (causeways, fish populations, and habitat) are the major concerns that were dealt with in the synthesis, integration, and assessment effort. These three overlap considerably (see Panel A, Figure 2), although the areas of overlap are not necessarily proportional. This overlapping of issues and the complexity thus generated have led to a wide variety of opinions in the scientific community regarding the definition of impacts and the resolution of issues.

For example, the presence of causeways per se has been documented to directly block fish migrations under some circumstances (e.g., Gallaway et al. 1990). These circumstances involve the natural encroachment of marine water against the tip of a causeway and the inability of migrating fish to move around the end of the structure because they are unwilling to swim into sublethally high-salinity–low-temperature water. Blockage of fish is not attributable simply to a physical barrier; other phenomena are involved. Blockage of fish migrations at these causeways results from the hydrographic changes induced by the effect of the structures on circulation. Often referred to as the "wake eddy" effect, high-salinity–low-temperature water occasionally is advected inshore downstream of the causeway (Colonell and Gallaway 1990) and fish are temporarily unable or unwilling to swim through the water mass.

To use the example given above, the symptom of causeway effects on fish is blockage of fish movement. Therefore, the approach needed to achieve effective mitigation must be based on the mechanism producing the symptom, which depends on the location of the causeway and the nature of the surrounding environment.

Although it is clear that all causeways will have some effect on habitat, it does not necessarily follow that such changes will be detrimental to the fish. Fish species differ in terms of their habitat requirements and use patterns. And in the Arctic, the complexity and dynamics of coastal fish-habitat conditions (Neill and Gallaway 1989) render assessment of effects of habitat change a difficult challenge. What may be detrimental to one species or life history stage may be inconsequential to another. Conversely, a causeway at one location may have adverse impacts on a species, whereas at another location these impacts might not be significant. The mechanisms that account for these differences must be clearly identified if future causeways are to be evaluated objectively in terms of their likely impacts on fish populations and habitat. Furthermore, the issues associated with a particular causeway must be focused on the fish species that are considered important and potentially affected under the site-specific circumstances. Environmental-monitoring programs and directed scientific studies of the causeways have gradually unraveled natural ecological processes and how causeways interact with these local and regional processes. Defining the issues of concern about causeways has been the key element of the environmental-impact assessment process. We have attempted to guide our approach to synthesis and assessment by reliance on the original statement of issues found in the Endicott Development EIS. The next step in the program was to resolve each issue through a process of hypothesis testing.

Questions and Hypotheses

We devised a multistage approach to address each issue (Panel B, Figure 2). We began by identifying the pertinent questions for each issue. As suggested by the SAC, the questions were stated in a manner appropriate to a resource manager's needs, and the questions were restricted to those considered necessary to resolve the issues. The

TABLE 1.—An example of a group of analyses required to answer one question (see Appendix 1 for a list of all questions).

Question 1.1	What mechanisms govern the movement of young-of-the-year Arctic ciscoes from the Mackenzie River of Canada to habitats in Alaska?
Hypothesis 1.1.1	There is no significant correlation between wind speed and direction and abundance of young-of-the-year Arctic ciscoes in Prudhoe Bay.
Analysis 1.1.1	Correlation of winds to young-of-the-year abundance in Prudhoe Bay and coastal study sites between the Mackenzie River and Prudhoe Bay.
Hypothesis 1.1.2	There is no significant association between time and distance traveled by young-of-the-year Arctic ciscoes and wind speed and direction.
Analysis 1.1.2	Graphical analysis showing level of association between elapsed time and distance traveled by young-of-the-year to wind speed and direction.
Hypothesis 1.1.3	There is no significant association between historical year-class contributions to the Colville fishery and the winds that occurred during the first year of each year-class.
Analysis 1.1.3	Graphical analysis showing level of association between year-class contribution to fishery catch per unit effort and historical wind patterns.
Hypothesis 1.1.4	Abundance of young-of-the-year Arctic ciscoes in Prudhoe Bay cannot be described by simple regional-scale diffusion-transport models.
Analysis 1.1.4	Develop regional-scale diffusion-transport model based on physical processes alone and use to predict arrival times and strengths.

questions, associated with the four specific issues surrounding the two Beaufort Sea causeways, are provided in Appendix 1. This list provided the measures for determining when the program objectives had been met.

The next step was to formulate sets of testable hypotheses pertinent for each question and to document that there is an observational basis for each hypothesis; that is, we demonstrated that specific data could be collected and analyzed to test each hypothesis (an example is provided in Table 1). This documentation served to focus effort on impacts that are truly possible and within the realm of probability rather than those that "could" occur given imaginary situations. Each hypothesis was coded to a specific question, which was coded to one of the four issues.

Each hypothesis (see example in Table 1) was stated in a testable fashion. For some, an explicitly formulated statistical test was appropriate and confidence limits were specified. If the results of the test suggested that the observed differences were not significant at the specified probability level, then appropriate power analysis was conducted to provide an explicit estimate of the maximum possible difference that could have escaped detection.

It was appropriate that some hypotheses would best be tested using model evaluations, especially some of the hypotheses concerning physical processes and physical–biological relationships. In these instances, an evaluation of the sensitivity of the model result to variation in any of the input parameters (assumed or measured) was provided.

Where necessary, the suitability of the model for testing a given hypothesis was evaluated. For example, various statistical models (Gazey–Staley or "Bayes," Schnabel, and Schumacher–Eschmeyer) have been used to estimate population levels of some anadromous fish in the study area since 1988 for comparison to precauseway levels. All of these models, at least to some degree, include the assumption that the population is closed, although this is usually not the case. How seriously are the results compromised by invalidating this assumption? (See Gallaway et al. 1997, this volume.) Comparison of the results from the models based on the closed-population assumption to results obtained using an open-population model (Jolly–Seber) was suggested by the SAC, and an expanded fish tagging program was instituted in 1990 to strengthen the tests.

The results obtained for the least cisco population have enabled us to conduct the suggested comparison. The mean Bayes estimate (all three closed-population models gave similar results) suggested that 332,000 large least ciscoes were present in the Prudhoe Bay region during summer 1990. The mean daily Jolly–Seber estimate of the large least cisco population in Prudhoe Bay during the period 17 July (arrival date) to 29 July (initiation of the main return migration) was 301,000. Such results are encouraging, given that historical data are available for use in closed-population models but are not of the right type for use in open-population models.

Some hypotheses could be tested only qualitatively and relied on weight-of-evidence evaluations for their acceptance or rejection. For example, the abundance of Arctic ciscoes in the Prudhoe Bay

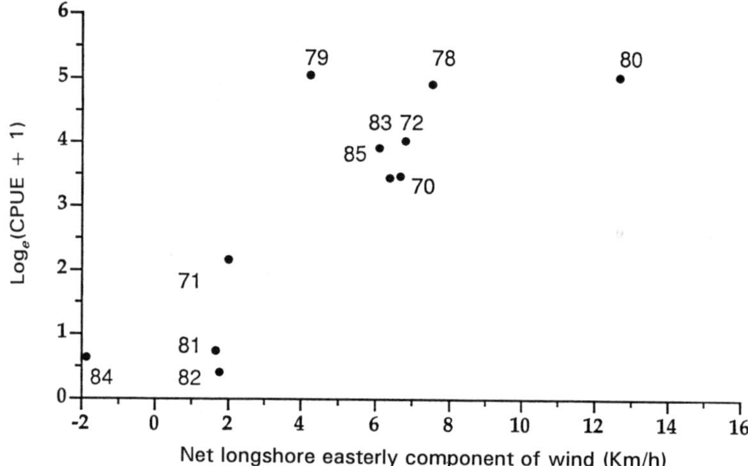

FIGURE 3.—Log$_e$ of catch per unit effort (CPUE, number of fish per net per 24 h) of year-classes (numbered) of Arctic ciscoes *Coregonus autumnalis* in the Colville River fisheries as a function of the net longshore easterly component of the wind during the year of their recruitment into the Alaskan Beaufort Sea as yearlings.

region was hypothesized to be related to wind patterns. Strong versus weak year-classes could be defined qualitatively from the historical fyke-net and fishery catch per unit effort (CPUE) and age data, but the variance associated with some of these determinations could not be estimated. However, we found that classification of CPUE by age as either strong (represented by daily fyke-net catches of 10s to 100s or 1,000s of young-of-the-year Arctic ciscoes) or weak (daily catches <10) was convincing, without knowing the variance; the differences could be attributable to chance. However, when the pattern of strong versus weak year-class determinations was compared to a quantitative determination of wind patterns, marked correspondence was observed (Figure 3). Our judgment was that the variation in CPUE was more likely attributable to wind patterns than to random variation resulting from an imprecise measure of abundance. Such an approach was used when necessary, so that managers could determine the level to which the judgments of the scientists were based on pure gestalt versus documented evidence.

After evaluating each hypothesis, a basis for answering each question was established. The combination of hypothesis testing and responding to questions led to an assessment of each issue, that is, a judgment of the degree to which each issue had been resolved (Panel B, Figure 2). This final step was to form a conclusion regarding impacts of the region's causeways on fish populations and habitats.

Results and Discussion

The draft synthesis report was completed in 1992 and reviewed by the NSB and the SAC (1993). Appendix 2 provides a brief synopsis of the degree to which each issue was resolved. The synthesis, integration, and assessment process articulated specifically how causeways interact with natural ecological processes of the Alaskan Beaufort Sea, and quantitated their effects on local fish populations. The four key issues are not fully resolved,[1] but a basis for determining how each issue will be resolved has been established (SAC 1993).

The synthesis effort also has shown that the controversy regarding causeway effects on coastal habitat and fish populations in northern Alaska has largely resulted from a poor understanding of both the biology of these fish and of the coastal oceanography. The initial approach followed by regulatory groups who had oversight responsibilities for evaluating the causeway developments had been to protect habitat, based on the premise that if the habitat is reduced, declines in the fish populations will ultimately follow (Hachmeister et al. 1991). However, more issue-based, focused investigations

[1]A supplemental synthesis, integration, and assessment was completed in early 1997 and presented to the Science Advisory Committee. In this report the remaining questions regarding causeway effects on nearshore Beaufort Sea fish populations were answered.

have shown that the summer habitat is normally not at capacity because of the limiting features of winter habitat relative to summer habitat. Furthermore, the net effect of causeways on summer habitat has been to increase locally the variability in temperature and salinity. However, the extremes are not affected and are within the limits naturally experienced by the fish. The fish are well adapted to contend with this type of environmental uncertainty (Johnson 1976, 1981, 1983; Craig 1989).

The synthesis process has also led us to believe that population parameters (e.g., population level and structure, and growth) are sensitive to habitat change and are well suited to address effects of habitat reduction. We concur with Robertson (1991) that impact predictions may be appropriately based on expected changes to habitat, but impact assessments are best made by study of the populations involved. However, long-term studies are necessary, particularly in the Arctic, to observe how the populations vary over time and to determine the key factors that govern the dynamics.

In essence, we believe that available information is adequate to determine that the effects of the Endicott Causeway on anadromous fish habitat and populations probably are not substantial.

Conclusion

We have described an approach to synthesis, integration, and assessment of a large volume of scientific data and information available for evaluating the impacts of causeways on coastal marine habitat and fish populations in the central Alaskan Beaufort Sea. We recognize other methods to accomplish impact assessment. For example, we could have defined each key question and prepared a stand-alone report on it. Alternatively, a synthesis report could have been prepared that was organized by discipline, such as by trophic level (e.g., effects on coastal hydrography, primary productivity, epibenthic invertebrates, fish, and fisheries) or by fish species (e.g., effects on Arctic ciscoes, broad whitefish, etc.).

We chose our preferred approach based largely on the history of the causeway-monitoring programs in the Alaskan Beaufort Sea. Since preparation of the Endicott Development EIS, federal and state agency concerns, and subsequently the field studies, have been issue-oriented. This mind-set continues today, so it was logical to proceed with an issue-based synthesis, which also was preferred by the SAC.

The Endicott Development is the first continuous offshore producing field in the Arctic. As the sixth largest oil field in the United States, the impacts of this development on fish and wildlife resources has been carefully monitored by regulatory agencies, conservation organizations, and the general public. The synthesis, integration, and assessment program serves as a key milestone in evaluating the impacts of the Endicott Development on coastal fish populations. The program provides a framework for decision making and a direction for future environmental studies in the North American Arctic and elsewhere where comprehensive, long-term environmental impact assessments are required.

References

Bryan, J. D., and R. G. Fechhelm. 1997. Use of a stress index to estimate temperature and salinity stress in Arctic ciscoes. Pages 262–273 *in* J. Reynolds, editor. Fish ecology in Arctic North America. American Fisheries Society Symposium 19, Bethesda, Maryland.

Colonell, J. M., and B. J. Gallaway, editors. 1990. An assessment of marine environmental impacts of West Dock Causeway. Report for Prudhoe Bay unit owners represented by ARCO-Alaska, Inc. Prepared by LGL Alaska Research Associates, Inc. and Environmental Science and Engineering, Inc., Anchorage, Alaska.

Colonell, J. M., B. J. Gallaway, and A. W. Niedoroda. 1992. Environmental effects of Beaufort Sea causeways. Pages 958–974 *in* S. A. Hughes, editor. Coastal engineering practice '92. American Society of Civil Engineers, New York.

Craig, P. C. 1989. An introduction to anadromous fishes in the Alaskan arctic. Biological Papers of the University of Alaska 24:27–54.

Fechhelm, R. G., J. S. Baker, W. B. Griffiths, and D. R. Schmidt. 1989. Localized movement patterns of least cisco (*Coregonus sardinella*) and Arctic cisco (*C. autumnalis*) in the vicinity of a solid-fill causeway. Biological Papers of the University of Alaska 24:75–106.

Fechhelm, R. G., P. S. Fitzgerald, and J. D. Bryan. 1992. Laboratory studies of the effect of salinity on the growth of yearling Arctic cisco (*Coregonus autumnalis*) in Prudhoe Bay, Alaska. The 1991 Endicott development fish monitoring program, volume 5. Report of LGL Alaska Research Associates, Inc. to BP Exploration (Alaska) Inc., Anchorage.

Gallaway, B. J., R. G. Fechhelm, W. B. Griffiths, and J. G. Cole. 1997. Population dynamics of broad whitefish in the Prudhoe Bay region, Alaska. Pages 194–207 *in* J. Reynolds, editor. Fish ecology in Arctic North America. American Fisheries Society Symposium 19, Bethesda, Maryland.

Gallaway, B. J., W. J. Gazey, J. M. Colonell, A. W. Niedoroda, and C. J. Herlugson. 1991. The Endicott development project—preliminary assessment of impacts from the first major offshore oil development in the Alaskan Arctic. Pages 42–80 *in* C. S. Benner and

R. W. Middleton, editors. Fisheries and oil development on the continental shelf. American Fisheries Society Symposium 11, Bethesda, Maryland.

Gallaway, B. J., W. J. Gazey, and L. L. Moulton. 1989. Population trends for the Arctic cisco (*Coregonus autumnalis*) in the Colville River of Alaska as reflected by the commercial fishery. Biological Papers of the University of Alaska 24:153–165.

Gallaway, B. J., W. B. Griffiths, J. M. Colonell, and A. W. Niedoroda. 1990. Anadromous fish and habitat background. Pages 14–41 *in* J. M. Colonell and B. J. Gallaway, editors. An assessment of marine environmental impacts of West Dock Causeway. Report for Prudhoe Bay unit owners represented by ARCO-Alaska, Inc. Prepared by LGL Alaska Research Associates, Inc. and Environmental Science and Engineering, Inc., Anchorage, Alaska.

Gallaway, B. J., W. B. Griffiths, P. C. Craig, W. J. Gazey, and J. W. Helmericks. 1983. An assessment of the Colville River delta stock of Arctic cisco (*Coregonus autumnalis*)—migrants from Canada? Biological Papers of the University of Alaska 21:4–23.

Griffiths, W. B., B. J. Gallaway, W. J. Gazey, and R. E. Dillinger, Jr. 1992. Growth and condition of Arctic cisco and broad whitefish as indicators of causeway-induced effects in the Prudhoe Bay region, Alaska. Transactions of the American Fisheries Society 121: 557–577.

Hachmeister, L. E., D. R. Glass, and T. C. Cannon. 1991. Effects of solid-fill gravel causeways on the coastal Beaufort Sea. Pages 81–96 *in* C. S. Benner and R. W. Middleton, editors. Fisheries and oil development on the continental shelf. American Fisheries Society Symposium 11, Bethesda, Maryland.

Johnson, L. 1976. Ecology of Arctic populations of lake trout, *Salvelinus namaycush*, lake whitefish, *Coregonus clupeaformis*, Arctic char, *S. alpinus*, and associated species in unexploited lakes of the Canadian Northwest Territories. Journal of the Fisheries Research Board of Canada 33:459–488.

Johnson, L. 1981. The thermodynamic origin of ecosystems. Canadian Journal of Fisheries and Aquatic Sciences 38:571–590.

Johnson, L. 1983. Homeostatic characteristics of single species of fish stocks in arctic lakes. Canadian Journal of Fisheries and Aquatic Sciences 40:987–1024.

Neill, W. H., and B. J. Gallaway. 1989. "Noise" in the distributional responses of fish to environment: an exercise in deterministic modeling motivated by the Beaufort Sea experience. Biological Papers of the University of Alaska 24:123–130.

Niedoroda, A. W., and J. M. Colonell. 1989. Coastal boundary layer processes in the central Beaufort Sea. Pages 41–52 *in* C. T. Mitchell, editor. A synthesis of environmental information on causeways in the nearshore Beaufort Sea, Alaska. U.S. Minerals Management Service, OCS (Outer Continental Shelf) Study, MMS 89-0038, Anchorage, Alaska.

Niedoroda, A. W., and J. M. Colonell. 1990. Beaufort Sea causeways and coastal ocean dynamics. Pages 509–516 *in* S. K. Chakrabarti, H. Maeda, C. Aage, and F. G. Nielsen, editors. Proceedings of the 9th international conference of offshore mechanics and arctic engineering. American Society of Mechanical Engineers, New York.

Norton, D. W., editor. 1989. Research advances on anadromous fish in Arctic Alaska and Canada. Nine papers contributing to an ecological synthesis. Biological Papers of the University of Alaska 24.

Robertson, S. B. 1991. Habitats versus populations: approaches for assessing impacts on fish. Pages 97–108 *in* C. S. Benner and R. W. Middleton, editors. Fisheries and oil development on the continental shelf. American Fisheries Society Symposium 11, Bethesda, Maryland.

SAC (Science Advisory Committee). 1990a. A review of the technical plan for the 1989 Endicott development fish monitoring program. Conducted by the North Slope Borough Science Advisory Committee, NSB-SAC-OR-111, Barrow, Alaska. (Available from North Slope Borough Department of Wildlife Management, Post Office Box 69, Barrow, Alaska 99723, USA.)

SAC (Science Advisory Committee). 1990b. A review of the 1990 (draft) Endicott development fish monitoring program technical plan. Conducted by the North Slope Borough Science Advisory Committee, NSB-SAC-OR-115, Barrow, Alaska. (Available from North Slope Borough Department of Wildlife Management, Post Office Box 69, Barrow, Alaska 99723, USA.)

SAC (Science Advisory Committee). 1991. A review of the 1991 (draft) Endicott development fish monitoring program technical plan. Conducted by the North Slope Borough Science Advisory Committee, NSB-SAC-OR-118, Barrow, Alaska. (Available from North Slope Borough Department of Wildlife Management, Post Office Box 69, Barrow, Alaska 99723, USA.)

SAC (Science Advisory Committee). 1993. A review of the 1991 draft Endicott development fish monitoring program report and synthesis, integration and assessment. Conducted by the North Slope Borough Science Advisory Committee, NSB-SAC-OR-124, Barrow, Alaska. (Available from North Slope Borough Department of Wildlife Management, Post Office Box 69, Barrow, Alaska 99723, USA.)

Appendix 1: Questions Relevant to Causeway Impacts on Fish Populations

Issue 1.—What would be the effects of the causeway and causeway-induced changes in circulation and hydrography on the migration of young-of-the-year Arctic ciscoes from Canada to the Colville River of Alaska?

Question 1.1 What mechanisms govern the movement of young-of-the-year Arctic ciscoes from the Mackenzie River of Canada to habitats in Alaska?

Question 1.2 Is the movement restricted to the nearshore brackish water zone?

Question 1.3 What proportion of the Canadian source population(s) of Arctic ciscoes use the Colville River and adjacent coastal zone as overwintering and rearing habitat, respectively?

Question 1.4 Does the Endicott Causeway cause the movements to be blocked in the Prudhoe Bay region, preventing the fish from reaching the Colville River?

Question 1.5 What is the fate of young-of-the-year Arctic ciscoes that attempt to overwinter in the Sagavanirktok River and to what extent will survivors ultimately contribute to the fishery?

Question 1.6 Do the area causeways or their associated hydrographic conditions increase susceptibility to disease or other environmental stresses (e.g., storms, temperature change, etc.) in young-of-the-year Arctic ciscoes?

Question 1.7 If sufficient data are not available to answer the basic questions associated with issue 1, what are the major areas in which additional data are needed?

Issue 2.—What would be the effects of the causeway and causeway-induced changes in circulation and hydrography on the nearshore migration corridor (from the shore to the 2-m isobath) used by most species and size groups of amphidromous fish?

Question 2.1 What are the mechanisms governing the movement of least ciscoes, Arctic ciscoes, and broad whitefish from the Colville River eastward into the Prudhoe Bay region?

Question 2.2 When do these species arrive at the eastern end of Simpson Lagoon and do the conditions around West Dock result in blockage?

Question 2.3 What are the movement patterns of these species into and across Prudhoe Bay and beyond, and do the conditions around the Endicott Causeway lead to blockage?

Question 2.4 Do the species using the Prudhoe Bay region during summer show distributional responses that can be related to changes in temperature and salinity?

Question 2.5 Does the West Dock Causeway contribute to blockage or delay of fish moving westward?

Question 2.6 Do the catch patterns suggest that the Endicott Causeway or hydrographic conditions around the structure contribute to blockage or delay of the resident fish moving into the Sagavanirktok River?

Question 2.7 Do the area causeways or their associated hydrographic conditions increase susceptibility to disease or other environmental stresses (e.g., storms, temperature changes, etc.) in fish moving in and through the area?

Question 2.8 When and to what extent do fish of each species use breaches in the area causeways?

Question 2.9 If sufficient data are not available to answer the basic questions associated with issue 2, what are the major areas in which additional data are needed?

Issue 3.—How would the temperature–salinity characteristics of the nearshore habitat be altered by the circulation and hydrographic effects resulting from the causeway, and what ramifications would these changes have on the population level of broad whitefish in the Sagavanirktok River?

Question 3.1 Have hydrographic and circulation effects from the Endicott Causeway exceeded the predictions of the environmental impact statement? In particular, does the causeway affect upwelling or otherwise cause shoreward incursion of a marine saltwater wedge into the delta?

Question 3.2 What temperature–salinity conditions can Arctic ciscoes and broad whitefish tolerate, and how do these tolerances compare to levels of causeway-induced changes in hydrography?

Question 3.3 Have there been any changes in abundance and age–size structure of broad whitefish or Arctic ciscoes that are attributable to the causeways?

Question 3.4 Are the fish exposed to the hydrographic effects from the causeway or do they avoid them?

Question 3.5 Is there a relationship between temperature–salinity and growth and condition of Arctic ciscoes and broad whitefish; and, if so, what are the consequences of the causeway-induced hydrographic changes on these population characteristics?

Question 3.6 What are the population-level ramifications of adverse, sublethal effects on growth and condition of Arctic ciscoes and broad whitefish, and what is the timescale required before the effects would be manifested as a decrease in the populations?

Question 3.7 Is there a relationship between nearshore hydrography and the species composition of the fish community inhabiting nearshore habitat, and to what extent do the causeway-induced changes in hydrography alter composition of this community?

Question 3.8 If sufficient data are not available to answer the basic questions associated with Issue 3, what are the major areas in which additional data are needed?

Issue 4.—What would be the impact on the Colville River Arctic and least cisco fisheries?

Question 4.1 What stocks of Arctic ciscoes are important to the Colville River fisheries and what is their source?

Question 4.2 Has the size of the catchable stocks of Arctic and least ciscoes declined since construction of the causeways?

Question 4.3 Has the size–age structure, age at sexual maturity, or growth of the catchable stocks of Arctic and least ciscoes changed since construction of the causeways?

Question 4.4 Have there been changes in natural or fishing mortality rates of the catchable cisco stocks since construction of the area causeways?

Question 4.5 How sensitive are the source population(s) of Arctic ciscoes to impacts resulting from Alaskan causeways and harvest levels?

Question 4.6 If sufficient data are not available to answer the basic questions associated with issue 4, what are the major areas in which additional data are needed?

Appendix 2: Synopsis of Causeway Impacts on Fish Populations

Issue 1.—Issue 1 concerned the potential effects of causeways on the dispersal of young-of-the-year Arctic ciscoes from the Mackenzie River, Canada, to the Colville River, Alaska.

Arctic ciscoes, predominantly from the Arctic Red and Peel River tributaries to the Mackenzie River of Canada, are transported to Alaska during their first year of life if winds are favorable. Winds are favorable on the order of 50% of the years. A substantial proportion of the population is probably affected when winds are favorable—on average about 30% or more of the population considering all years. The dispersal is mostly a passive drift process (Gallaway et al. 1983, 1989), governed by the wind effect on currents. The effects of the regional causeways on circulation do not greatly affect the dispersals, and thermal trapping as described in the project environmental impact statement has not occurred.

The dependency of the fish during the dispersal on nearshore, brackish coastal waters for survival is not well documented. There is some indication that the fish do indeed occur in offshore waters, and data for older fish show that they can acclimate to marine salinities, at least under conditions of relatively warm temperature (>3°C). Most of the fish that do find refuge in the Sagavanirktok River typically survive and overwinter there for up to 2–3 years after their initial recruitment. The number that overwinter there is determined by the number of fish that are transported this far west and whether they are transported beyond the delta before the end of summer. Maximum numbers occur when the bulk of the fish are initially carried westward beyond the area, then subjected to a wind reversal carrying them eastward into the Sagavanirktok River delta.

The area causeways do not appear to affect the numbers of fish that reach the Colville River. It should be stressed that large numbers can reach the Sagavanirktok River but not the Colville River. Large numbers can be carried beyond the Sagavanirktok River and then beyond the Colville River as well. Thus, abundance of young of the year in the Prudhoe Bay region does not necessarily index abundance that will occur in the Colville River fisheries 5–6 years later when they reach catchable size.

The available evidence is adequate to conclude that the area causeways do not affect the dispersal of young-of-the-year Arctic ciscoes. Thus, although there are remaining uncertainties concerning the year-to-year composition of the Alaskan stock, the proportion of the source stocks involved, and the factors governing their abundance in Alaska, these do not bear directly on the issue of causeway effects. The issue of causeway effects on dispersal of young-of-the-year Arctic ciscoes is judged to be resolved.

Major modifications of the two area causeways are planned for 1993, when the breaches in the Endicott Causeway will be expanded by 200 m.[1] A similar breaching expansion will occur in West Dock in 1994.[2] These alterations will, to some extent, modify existing local

[1]A 200-m breach was constructed in the Endicott Causeway during winter 1993–1994. The three Endicott breaches now total 410 m.

[2]A 200-m breach was constructed in West Dock in late 1995. The old 50-m breach has filled with sediment and no longer conveys water.

water circulation patterns. Although the effects of these modifications on circulation are expected to be minor in scale, continued monitoring will be conducted to document their effects on young-of-the-year Arctic cisco dispersal through the region.

Issue 2.—Issue 2 concerned the extent to which the causeways and their hydrographic effects influence the movements of amphidromous fish other than young-of-the-year Arctic ciscoes.

Coregonid fish that use the Prudhoe Bay area during summer overwinter in either the Colville River (Arctic cisco, least cisco, broad whitefish) or the Sagavanirktok River (Arctic cisco and broad whitefish, occasionally least cisco). Small fish of each species do not disperse as far as large fish during the summer feeding period. This is particularly true for small broad whitefish, which do not venture away from the immediate vicinity of the mouths of their overwintering areas. Some small least ciscoes from the Colville River disperse as far east as West Dock, but in less than 40% of the years. Small Arctic ciscoes from the Colville River are difficult to distinguish from small Arctic ciscoes that overwinter in the Sagavanirktok River, but the dispersal range of small Arctic ciscoes from the Colville River is expected to be similar to that observed for least ciscoes. When small Colville River fish reach as far east as West Dock, they are periodically prevented from moving farther eastward if regional upwelling has advected shelf bottom water to the vicinity of the causeway (Fechhelm et al. 1989). This phenomenon occurs infrequently at West Dock and does not appear to influence movements eastward around the Endicott structure. Also, West Dock is near the end of their eastward dispersal range. Given these and other observations, we conclude that there are no appreciable effects of the Prudhoe Bay causeways on the summer feeding dispersal of small coregonid fish from the Colville River.

All of the evidence agrees with the conclusion that migrations of large coregonid fish are not affected by either of the area causeways. There also is no evidence to support the contention that the fish moving through the Prudhoe Bay region are subjected to a higher exposure to stressful conditions than in other regions (see Bryan and Fechhelm 1997, this volume).

Enlargement of the breaches in the two area causeways does not appear to be a required mitigation measure from a fish movement standpoint. The proposed enlargement of the existing breach at West Dock may not be effective in facilitating passage of anadromous fish because of its offshore location—to be more effective, it should be located along the mainland shore. Future monitoring will probably focus on what effects the new breaching may have on localized hydrographic conditions around both causeways and on fish movement patterns in the vicinity of the causeways.

Issue 3.—The results of analyses under issue 3 quantified the effects of the area causeways on habitat and the resulting ramifications to fish populations. Whether habitat or population parameters should be used as the basis for impact assessment has generated the most controversy concerning effects from the area causeways. We have concluded that both are important, and use of one without considering the other can lead to erroneous conclusions and unjustified mitigation attempts.

The interaction between regional-scale oceanographic processes and causeways has been defined (Niedoroda and Colonell 1989, 1990), enabling quantification of the causeway effects on habitat variables that govern the well-being of resident coregonid fish populations. The key habitat factor appears to be water temperature. Water temperature explains 77–92% of the observed variance in growth of juvenile Arctic ciscoes and broad whitefish (Fechhelm et al. 1992; Griffiths et al. 1992), the size-classes of fish least able to move away from suboptimal habitat conditions and therefore most affected by causeway-induced hydrographic perturbations.

As little as 1°C reduction in temperature has been estimated to result in 20% reduction in growth of juvenile broad whitefish (Fechhelm et al. 1992). In the nearshore area west of the Endicott Causeway, the causeway has reduced the water temperature, on average, by about 0.5–0.7°C. Although the fish still use this habitat, their movements reflect an avoidance response when adverse conditions occur. Because favorable habitat remains available, there have been no significant reductions in growth (Griffiths et al. 1992). Nevertheless, the results do show some potential habitat loss.

However, it also has been hypothesized for broad whitefish that the constraint of limited winter habitat may maintain the population below carrying capacity of the summer habitat (see Gallaway et al. 1997). If this were true, habitat reduction would not necessarily lead to a reduction in the population. Population size is probably governed by a density-dependent strategy used by the fish to cope with the constraint of limited winter habitat.

Pronounced causeway effects on fish habitat are restricted to the areas in the immediate vicinity of the causeways, and these areas are avoided by the fish (Gallaway et al. 1991). Such events occur about one to five times during summer, with most occurring near the end of summer, after most of the fish have already left the area. Although significant reductions in temperature have occurred in more-nearshore areas west of the Endicott Causeway, they are small and have not had significant effects on the populations.

The decision to require additional breaching in the area causeways was based on the conclusion that the Endicott Causeway had resulted in adverse impacts on habitat and fish populations that had exceeded the predictions of the environmental impact statement (Hachmeister et al. 1991). This conclusion is erroneous based on the available evidence (Gallaway et al. 1991; Colonell et al. 1992). The requirement for additional breaching on this basis is therefore questionable. Continued monitoring will be required to determine if the additional breaching will mitigate the observed effects that are already generally less than those predicted in the project environmental impact statement.

Issue 4.—Issue 4's questions were aimed at gaining a better understanding of the dynamics of the fishery and the role of the area causeways on these dynamics.

The Colville River cisco fisheries predominantly harvest Arctic and least ciscoes along with smaller numbers of Bering ciscoes *Coregonus laurettae* and broad whitefish. The harvested Arctic ciscoes are mainly representatives of the Peel River (Mackenzie River) population, with some contribution from the Arctic Red River (Mackenzie River). Variation in stock size results from variation in meteorological patterns. High stock levels occur in Alaska when westward transport is high and young of the year are abundant. Under these conditions, a substantial portion of the source population appears to be transported to Alaska. We have not established any firm linkage between the area causeways and the success of the Alaskan fisheries; nevertheless, a key risk analysis should be conducted before a final conclusion is reached.

American Fisheries Society Symposium 19:340–345, 1997

Symposium Summary

Living With Uncertainty

Lionel Johnson[1]
10201 Wildflower Place, Sidney, British Columbia, V8L 3R3, Canada

I beseech ye in the bowels of Christ, think it possible, that ye may be mistaken.

Oliver Cromwell (1650)

It was with some trepidation that I accepted Jim Reynolds's invitation to review the proceedings of this symposium on arctic fisheries ecology, but that is nothing compared with the trepidation that I now feel at the moment of execution. It is indeed a formidable task to review the many wide-ranging and thought-provoking presentations and discussions of the last 3 days. It is not possible in the time available to review in detail all the excellent papers, but I assure you that all contributed to forming some general thoughts on arctic fishes and fisheries that I would like to discuss. Among the fisheries topics covered are the national perspective, the native view, sportfishing, pollution, the influence of oceanography and ocean variability, and general management problems. There are so many groups and factors involved that I think it helps to view the problem from the fish's perspective. To the perspicacious fish, life is sufficiently difficult coping with the natural environment without the extra hazards of dealing with an erratic predator. To provide focus to this review I have tried to develop a number of themes that seem important to me. If we are ever to manage our natural systems with any degree of success, we must attempt to understand the extremely complex processes that together form the Earth's biosystem. Sometimes it appears that there are few generalities and that each system is composed of individual complexities and idiosyncrasies, but I believe that there are general principles that we must discover. Ecosystem structure is like a cryptogram that needs to be carefully explored and examined from all aspects, with the details being gradually teased apart and then rebuilt on a theoretical basis. We may be wrong 9 times out of 10 in our attempts to find a solution to the problem, but we must keep trying. There are three additional points of general importance: first, it is unlikely that there will ever be a perfect pattern into which everything fits neatly without overlap or contradiction; second, in this or any other biological research program, the mere accumulation of data will not of itself provide a solution to the cryptogram; and third, it is most probable that when we do get some glimmer of an answer, we may not like it because it may make management more difficult than we had imagined.

How do arctic systems fit into a world pattern? Are they different qualitatively from the rest of the world, or are they merely one end of a bell-shaped continuum stretching from pole to pole? Dr. Proenza[2], very properly, encouraged us to "Think bipolar," and this approach led to one of the first major generalizations: that arctic and antarctic marine ecosystems are not merely agglomerations of species but express a degree of similarity. This conclusion was first pointed out by Captain James Clark Ross of the British Navy in the 1840s after experience in both regions. Ross (1847) noted that the structure of the ecosystems (not his exact words) was similar, although the component species in each polar region were quite different. In both regions marine mammals occupy the position of top predator, being largely dependent on small fish living a precarious existence under the ice. Evidently similar conditions result in similar structure, with the implication that the forces determining ecosystem structure do not operate in a random manner. A second major generalization was made later in the century by Alfred Russell Wallace (Wallace 1878), who noted that the diversity of animals and plants decreased from the tropics to the polar regions. The major physical trends are a decline in the sun's energy input as latitude increases and an increase in the variability of input, both within and between seasons. Another general trend is the transfer of heat from equatorial regions toward the poles; it is the vagaries of this heat transfer by oceanic and atmospheric currents that account for much of the arctic climatic variability. Evidently in response to climatic conditions, the following general rules exist: Bergmann (1962) states that warm-blooded animals are larger in cold regions than in

[1]Retired from Canada Department of Fisheries and Oceans.

[2]See Appendix.

temperate zones; Allen (1962) finds that exposed parts—such as limbs, tail, and ears—of warm-blooded animals decline in size as mean annual temperature declines; and Jordan (1892) states that polar fish have more parts than those of tropical and temperate regions. These concepts are discussed by Power (1997, this volume) in his admirable survey of arctic fishes. There is also the little noticed but important conclusion of Mann and Brylinsky (1975) that the ratio of primary production to biomass (*P*/*B*) in lakes declines with increasing latitude. This means that, energy wise, we get more bang for our buck in arctic regions than we do in temperate or tropical regions, because of straighter and more-efficient food chains in the Arctic. That is, a greater proportion of the total energy assimilated is delivered to the terminal predator, accounting for the great abundance of certain species in arctic regions even though the annual energy input is extremely low.

The articles on oceanography (Weingartner 1997, this volume; Segar and Pollard[2]) are exciting because they seem to hold out the prospect that if we can understand the oceans (and great progress is obviously being made) then we can understand a great deal about the natural history of marine and anadromous species. However, this hope is somewhat diminished by the revelation of the variability of the oceans. Nevertheless, I do not think we will make much progress in fisheries unless we assimilate every scrap of oceanographic information available and harmonize it with the results of biological investigations; although in truth, the direct correlation between the environment and the fish populations tends to be somewhat tenuous at best (Wyllie-Echeverria et al. 1997, this volume). The Arctic Ocean presents an interesting anomaly for the highly variable environment between the shore and the summer ice edge (owing to atmospheric variability) and stands in contrast to the highly stable, deep central core under the permanent ice. As a "Northern Mediterranean," water interchange between the Arctic and other oceans is reduced, thus preventing more-uniform distribution of temperature worldwide, resulting in a permanent cover of ice. The permanent ice cover absorbs virtually all the local heat input from the sun during the melt season so that the water column remains cold and stable. The ice cover thus functions as a buffer against fluctuations, thereby reducing overturn and upwelling. This low level of interchange, overturn, and upwelling is significant in arctic fisheries for it is the regions experiencing such phenomena that are the world's most important areas of fish production. Given these conditions of stability, the bottom waters become a sink for nutrients and organic matter because the detritus, falling to the depths, is carried to more southerly regions by deep currents. Diversity in a wide range of organisms has been closely correlated with total energy input (Currie and Paquin 1987; K. Patalas, paper presented at Annual Meeting of the Canadian Federation of Biological Societies, 1992). The low diversity of the arctic biota is certainly commensurate with low energy input. However, there are three additional factors reducing energy accumulation, hence contributing to low diversity: (1) the drain of nutrients and organic matter mentioned above, (2) the export south of much of the relatively abundant production of the short arctic summer through the migration of birds and mammals, and (3) the high degree of climatic variability. The continual southward drain of nutrients and energy—which, interestingly, is in the opposite direction to the flow of heat in the physical part of the system—conspires against the buildup of stable populations on land or in the oceans. Only in lakes is this drain reduced, although even here piscivorous and insectivorous birds take their toll of the total production south.

One of the continuing problems in arctic fisheries is that of species identification (Bickham et al. 1997; Reist 1997; both this volume; Dillinger et al.[2]). Why do so many similar species occupy arctic waters and frequently the same waters? Among the coregonids and the chars, what exactly constitutes a species? It is hoped that these problems will be resolved eventually because correct identification of stocks is of great importance to management and to ecosystem research. Perhaps the most pervasive theme of the articles presented at this conference has been environmental variability and the equal variability of fish stocks. In the nearshore regions, several species, such as Arctic char *Salvelinus alpinus* (Jarvela and Thorsteinson 1997; DeCicco 1997; both this volume), Arctic cisco *Coregonus autumnalis* (Moulton et al. 1997; Bond and Erickson 1997; Green and Mitchell 1997, all this volume; Dillinger et al.[2]; Thorpe and Osborne[2]), Arctic cod *Boreogadus saida* (Gillispie et al. 1997; Hop et al. 1997; Underwood 1997; all this volume), Arctic staghorn sculpin *Gymnocanthus tricuspis* (Smith et al. 1997a, this volume) and broad whitefish *C. nasus* (Gallaway et al. 1997; Tallman 1997; both this volume; Chang-Kue[2]), show much variability in population structure as measured by their abundance and modal size. Other fish species, such as fish doctor *Gymnelus viridis*, evidently exhibit considerable stability (Green and Mitchell 1997). The greatest variability in time and locality of appearance and size of individual in the school is exhibited by Arctic cod. It is crucial to understand the life cycle of Arctic cod because this fish is generally considered to be the most important forage species for marine mammals and larger fishes. The question arises as to the nature of this variability: is it true fluctuation in population structure and abundance or is it more a case of variation in the pattern of aggregation and migration masking the overall state of the population? Has this variability in the life cycle of cod evolved, not only as a response to the environment, but also as a defense against the excessive buildup of predator stocks? In effect, this would be a contest between stability (of the predator) and instability (of the prey), where variability in prey density induces instability in the predator.

The mere survival of fluctuating stocks indicates that populations in the Arctic must be robust or resilient. In other words, they must have a considerable capacity to recover from conditions of low abundance; without possessing damping or stabilizing factors, fluctuating species would be rapidly eliminated. In fact, Julian Huxley (Huxley 1942) made the percipient suggestion that, in general, evolution can be characterized by the increasing capacity of evolutionary lineages to dampen the effects of environmental fluctuation. This seems to be true at the level of the ecosystem and of the dominant or terminal species in the food chain. Arctic cod evidently does not follow this pattern closely, but on the other hand, the capacity to respond rapidly to changing food abundance is important in an efficient and effective energy collector in a variable system. This places the responsibility on the higher trophic levels to impose stability on the system as a whole.

We can gain some indication as to the nature of these stabilizing factors by examining freshwater fish populations. Parker and Johnson (1991) described several char populations from some of the most northerly freshwater ecosystems known, and DeCicco et al. (1997, this volume) described a population of Arctic grayling *Thymallus arcticus* in a riverine habitat. By their large size and considerable mean age, both populations evince a marked degree of stability over long periods of time. In this context, stability is understood to be the continuity of structure and abundance in the face of natural environmental variability.

I believe the prime factor in stability is system autonomy. This is epitomized by Parker and Johnson's (1991) small arctic lake. System autonomy in such a lake ensures the damping of external influences: organisms, confined to one locality, must exist on the physical energy entering that system at that locality, delaying for the longest possible time the dissipation of that energy. In addition, water's physical characteristics buffer atmospheric fluctuations, resulting in a lean but relatively constant environment. Within such an autonomous ecosystem, the stabilizing factors can work themselves out to their full extent. The result is evident in the large size, uniformity, considerable biomass, and great mean age of the dominant species. It is evident also in the bimodal size structure of the population, each mode being composed of fish of similar size but variable age. A bimodal structure of this type ensures the elimination of year-to-year effects and appears to be the most stable structure theoretically. The modal structure is established by symmetrical interaction between individuals within a mode but asymmetrical interaction between individuals of different modes. This means that fish in the smaller mode are relegated to the periphery of the habitat of the larger fish.

Thus, stability is associated with a mode of large, old fish, relatively uniform in size but of variable age. In the ocean such a mode is evident in the dominant populations of large marine mammals. It is my belief that the mode really exists and that fish (and, of course, mammals) of each species travel in modes recognized as populations of relatively uniformly sized individuals. This ensures uniform distribution (equipartitioning) of the available energy, which again contributes to overall stability. From the work of Hop et al. (1997) we see that individual Arctic cod schools, although variable in abundance, tend to be primarily of one size.

Such concepts are seldom included in the popular individualistic approach to population dynamics, but they need to be investigated further. Thus, one reason for large size in arctic animals seems to be the need to adapt to year-to-year fluctuations in the environment. For example, many marine mammals and top predators in lakes are large and long-lived (*K*-selection). At the other extreme, the ability to track fluctuations closely and maximize energy input has been attained by evolving a large reproductive potential and a rapid turnover time (*r*-selection). A series of primary producers each focusing on differences in the total energy input spectrum ensures the most efficient uptake of energy; at the same time, the upper trophic levels ensure the effective dampening of the system as a whole. Together, these two factors ensure a most effective system in an adverse environment.

Field studies on the natural populations that have

been the subject of this conference are most satisfactorily complemented by laboratory work on the energetics of individual species (Smith et al. 1997a; Smith et al. 1997b, this volume; Reynolds[2]; Stuby and Schell[2]). Through energetics we may eventually gain a better understanding of ecosystem structure and the constraints on energy flow through the system. From this we may gain a much better appreciation of how exploitation affects a system. Similarly, experimental work on the rehabilitation (Hemming 1997, this volume) of artificial lakes and ponds will provide guidance for effective management of such waters and show how desirable ecosystems can be created with our assistance. We who work in the Arctic are fortunate to have this enormous laboratory at our disposal, which contains many locations that are relatively undisturbed by humans. We have destroyed the structure of so many ecosystems that it behooves us to examine closely what remains of intact systems before it is too late. On the other hand, it is evident that even vast amounts of money spent on research will not necessarily provide us with the concrete answers we so ardently desire. Despite this, research provides a great deal of background information and frequently exposes completely unexpected facets of the life history of organisms, both at the center of the investigation and at the periphery.

These concepts may help us to manage our renewable resources, because they put into perspective some of the general processes on which the resources depend, even if they indicate that management is going to be more difficult than we thought. They are part of the much-discussed "ecosystem approach." Frequently this is mere lip-service, but in its better connotation, it means thinking of ourselves as being within the system rather than above it. The ecosystem approach requires integration of all information supplied by oceanography and limnology as well as the investigation of species at all levels. For these reasons it is extremely important to support and maintain the ongoing investigation of species such as mollusks (Foster et al.[2]), Bering flounder *Hippoglossoides robustus* (Smith et al. 1997b), snow crab *Chionoecetes opilio* (Paul et al. 1997, this volume), fourhorn sculpin *Myoxocephalus quadricornis* (Underwood 1997; Gordon[2]) and Arctic staghorn sculpin (Smith et al. 1997a), along with many others.

In summary, ecosystems are energetic systems which must obey rules of energy transfer and storage. We must determine what the rules are and how they apply if we are to save ourselves from ecological disaster. We must get rid of the Newtonian way of thinking that has so bedeviled biological thought. Organisms are not billiard balls interacting in specific ways but thermodynamic engines, buzzing balls of energy, with lives, goals, and constraints. One of the most significant problems with which we are faced, and one of our own making, is that we have destroyed the autonomy of many systems by removing preexisting boundaries and constraints to energy flow. We have eliminated or reshaped system boundaries, in effect creating a single unified system that is still in a state of great internal flux; only within the boundaries of the Earth's biosphere can we now view the system as being autonomous. Only when we stop inflicting new and greater perturbations will the earth system settle down in a new configuration, but eventually it will settle down. What the new configuration will be, we have little hope of envisaging.

Thus we come to practical management procedures. Management implies prediction, that is, that we are fairly certain of the response to our actions. Scientific management thus implies that we know what we are doing. Unfortunately, it is frequently little more than another "buzz-concept" dependent on a Newtonian approach. Crudely this means find out the size of the stock, what the annual production is, estimate the harvest potential, and impose regulations accordingly. Unfortunately this does not seem to work. Ask the Newfoundlanders about cod stocks. In actuality there can never be enough information for scientific management. We must always incorporate a large uncertainty factor in our estimates of potential harvest and in our assessments of the potential effect of structures.

It was suggested in the early sessions that there should be some mechanism to support biologists when they speak out against the received wisdom of their peers and superiors, thereby endangering their chances of promotion or advancement. As a possible antidote, Power (1997) suggested the equivalent of the Hippocratic oath for fisheries scientists and others involved in the management of renewable resources. Perhaps we also need a "Freedom of Assessment Act" to protect unwelcome assessments, which must be genuine assessments rather than simple opinions. However, before taking any oath or proceeding with legislation, we should develop a set of principles to which we might swear to be true. I suggest that there are certain practical principles that might be applied to management (I am indebted to Don Ludwig of the University of British Columbia for codifying many of these constraints on management and presenting them at the

Pacific Northwest Workshop on Mathematical Biology, Seattle, Washington, April 1992):

1. Scientific management is an impossibility, for there can never be enough information available. Therefore, scientists alone should not manage the resource. Nevertheless, any decision must be made after due consideration of the most extensive information available.
2. Any discussion of resource allocation should involve an attorney who will speak exclusively on behalf of the resource itself.
3. There must be input from the harvesters, for they (or their heirs, assigns, and successors) are the first to benefit or lose by the decisions made and they are the ones in most immediate contact with the resource.
4. Given these constraints, the following strategy should be employed: (a) adopt policies that are reversible; (b) favor actions that are informative; (c) place the onus on the instigator to show that no damage will accrue if irreversible structures are to be imposed on a system; (d) consider a variety of hypotheses and do not make the data fit just one hypothesis (always consider that one may be wrong); and (e) try to relate changes and proposed changes to observed or established ecological patterns.

I recognize that this brief summary and interpretation reflects what psychologists term "cognitive dissonance"—the capacity to accept evidence and arguments that reinforce one's own established ideas but reject or ignore those that do not fit. That is, we develop what is called "common sense" (the sum of our previous experience and our parental teachings), and I believe that it is dangerous. For this reason, the definition of science as organized common sense is inadequate and becomes less and less adequate the deeper we plumb the nature of the living world. In biology, as in the depths of physics, we occasionally stumble on phenomena that appear to defy common sense—phenomena that require a reorganization of our thinking processes if they are to be harmoniously accommodated. Progress in science depends on recognizing these phenomena and making the necessary adjustments to our central paradigm to bring them into accord with observation. We must accept the fact that we live on the verge of error. Biology has suffered greatly from cognitive dissonance, or faith and conviction, and an underlying belief that Newtonian principles apply. Like the physicists, we need an "Uncertainty Principle" to reinforce the possibility that we may be wrong. Biological thought, like organisms themselves, will probably always be in a state of flux, hence we must learn to live forever with uncertainty and try to plan accordingly.

And finally, on behalf of the intellectual ecosystem that occupies the habitat of arctic fish ecology, it is my great pleasure to congratulate Jim Reynolds and his hardworking committee for organizing an excellent conference. I also thank the University of Alaska Fairbanks and the Alaska Chapter of the American Fisheries Society for their invaluable support in ensuring that the conference produced a smooth and exciting flow of cerebral energy.

References

Allen, J. A. 1962. Evolution. Page 918 *in* Encyclopedia Britannica, volume 8. William Benton, Chicago.

Bergmann, C. 1962. Evolution. Page 918 *in* Encyclopedia Britannica, volume 8. William Benton, Chicago.

Bickham, J. W., J. C. Patton, S. Minzenmayer, L. L. Moulton, and B. J. Gallaway. 1997. Identification of Arctic and Bering ciscoes in the Colville River delta, Beaufort Sea Coast, Alaska. Pages 224–228 *in* Reynolds (1997).

Bond, W. A., and R. N. Erickson. 1997. Coastal migrations of Arctic ciscoes in the eastern Beaufort Sea. Pages 155–164 *in* Reynolds (1997).

Cromwell, O. 1650. From a letter to the general assembly of the Church of Scotland dated August 3, 1650. Page 310 *in* The Columbia Dictionary of Quotations, Columbia University Press, New York.

Currie, D. J., and V. Paquin. 1987. Large-scale biogeographical patterns of species richness of trees. Nature 329:326–327.

DeCicco, A. L. 1997. Movements of postsmolt anadromous Dolly Varden in northwestern Alaska. Pages 175–183 *in* Reynolds (1997).

DeCicco, A. L., M. F. Merritt, and A. E. Bingham. 1997. Characteristics of a lightly exploited population of Arctic grayling in the Sinuk River, Seward Peninsula, Alaska. Pages 229–239 *in* Reynolds (1997).

Gallaway, B. J., R. G. Fechhelm, W. B. Griffiths, and J. G. Cole. 1997. Population dynamics of broad whitefish in the Prudhoe Bay region, Alaska. Pages 194–207 *in* Reynolds (1997).

Gillispie, J. G., R. L. Smith, E. Barbour, and W. E. Barber. 1997. Distribution, abundance, and growth of Arctic cod in the northeastern Chukchi Sea. Pages 81–89 *in* Reynolds (1997).

Green, J. M., and L. R. Mitchell. 1997. Biology of the fish doctor, an eelpout, from Cornwallis Island, Northwest Territories, Canada. Pages 140–147 *in* Reynolds (1997).

Hemming, C. R. 1997. Experimental introduction of Arctic grayling to a rehabilitated gravel extraction site, North Slope, Alaska. Pages 208–213 *in* Reynolds (1997).

Hop, H., H. E. Welch, and R. E. Crawford. 1997. Population structure and feeding ecology of Arctic cod schools in the Canadian high arctic. Pages 68–80 *in* Reynolds (1997).

Huxley, J. 1942. Evolution: the modern synthesis. George Allen and Unwin, London.

Jarvela, L. E., and L. K. Thorsteinson. 1997. Movements and temperature occupancy of sonically tracked Dolly Varden and Arctic ciscoes in Camden Bay, Alaska. Pages 165–174 *in* Reynolds (1997).

Jordan, D. S. 1892. Relations of temperature to vertebrae number in fishes. Proceedings of the United States National Museum 14:107–120.

Mann, K. R., and M. Brylinsky. 1975. Estimating the productivity of lakes and reservoirs. Pages 221–226 *in* T. W. M. Cameron and L. W. Billingsley, editors. Energy flow, its biological dimension. Royal Society of Canada, Ottawa.

Moulton, L. L., L. M. Philo, and J. C. George. 1997. Some reproductive characteristics of least ciscoes and humpback whitefish in Dease Inlet, Alaska. Pages 119–126 *in* Reynolds (1997).

Parker, H. H., and L. Johnson. 1991. Population structure, ecological segregation and reproduction in nonanadromous Arctic charr (*Salvelinus alpinus*) in a series of unexploited lakes in the Canadian Arctic. Journal of Fish Biology 38:123–147.

Paul, J. M., A. J. Paul, and W. E. Barber. 1997. Reproductive biology and distribution of the snow crab from the northeastern Chukchi Sea. Pages 287–294 *in* Reynolds (1997).

Power, G. 1997. A review of fish ecology in Arctic North America. Pages 13–39 *in* Reynolds (1997).

Reist, J. D. 1997. The Canadian perspective on issues in Arctic fisheries management and research. Pages 4–12 *in* Reynolds (1997).

Reynolds, J., editor. 1997. Fish ecology in Arctic North America. American Fisheries Society Symposium 19, Bethesda, Maryland.

Ross, J. C. 1847. A voyage of discovery to the southern and antarctic regions. J. Murray, London.

Smith, R. L., W. E. Barber, M. Vallarino, J. Gillispie, and A. Ritchie. 1997a. Population biology of the Arctic staghorn sculpin in the northeastern Chukchi Sea. Pages 133–139 *in* Reynolds (1997).

Smith, R. L., M. Vallarino, E. Barbour, E. Fitzpatrick, and W. E. Barber. 1997b. Population biology of the Bering flounder in the northeastern Chukchi Sea. Pages 127–132 *in* Reynolds (1997).

Tallman, R. F. 1997. Interpopulation variation in growth rates of broad whitefish. Pages 184–193 *in* Reynolds (1997).

Underwood, T. J., D. E. Palmer, L. A. Thorpe, and B. M. Osborne. 1997. Weight–length relationships and condition of Dolly Varden in coastal waters of the Arctic National Wildlife Refuge, Alaska. Pages 295–309 *in* Reynolds (1997).

Wallace, A. R. 1878. Tropical nature and other essays. McMillan, London.

Weingartner, T. J. 1997. A review of the physical oceanography of the northeastern Chukchi Sea. Pages 40–59 *in* Reynolds (1997).

Wyllie-Echeverria, T., W. E. Barber, and S. Wyllie-Echeverria. 1997. Water masses and transport of age-0 Arctic cod and age-0 Bering flounder into the northeastern Chukchi Sea. Pages 60–67 *in* Reynolds (1997).

Appendix: Symposium Presentations Not in this Proceedings

The following presentations were made at the symposium Fish Ecology in Arctic North America, however, do not appear in this volume.

Chang-Kue, K. T. J., and E. F. Jessop. Spawning and overwintering destinations of broad whitefish in the Mackenzie Delta and southeast Beaufort Sea coast, Northwest Territories. Canada Department of Fisheries and Oceans, 501 University Crescent, Winnipeg, Manitoba R3T 2N6, Canada.

Dillinger, R. E., Jr., J. M. Green, and T. P. Birt. The *Coregonus autumnalis* species complex (Pisces: Salmonidae) in northwestern Canada and Alaska. Dillinger: University of Dubuque, Environmental Science Department, 2000 University Avenue, Dubuque, IA 52001; Green: Memorial University, Department of Biology, St. John's, Newfoundland A1B 3X9, Canada; Birt: Metro Toronto Zoo, 361A Old Finch Avenue, Scarborough, Ontario M1B 5K7, Canada.

Foster, N. R., H. M. Feder, and R. Baxter. Distribution of mollusks in the northeastern Chukchi Sea. Foster: University of Alaska Museum, 907 Yukon Drive, Fairbanks, AK 99775-6960; Feder: University of Alaska Fairbanks, Institute of Marine Science, Fairbanks, AK 99775-7220; Baxter (deceased).

Gordon, J. A. Catch per unit effort analysis for the open water season in the Arctic National Wildlife Refuge, Alaska, 1988–1991. U.S. Fish and Wildlife Service, Fishery Resource Office, 101 12th Avenue, Fairbanks, AK 99701.

Proenza, L. M. United States perspective: issues that will affect fisheries research and management by the year 2000 and beyond. Purdue University, 107 Young Graduate House, West Lafayette, IN 47906-6208.

Reynolds, J. B. Food, energetics, and growth of lake trout in Walker Lake, Alaska, and other lakes of Arctic North America. University of Alaska Fairbanks, Alaska Cooperative Fish and Wildlife Research Unit, Fairbanks, AK 99775-7020.

Segar, D. A., and D. D. Pollard. Oceanography of arctic seas. Segar: 7609 Potrero Avenue, El Cerrito, CA 94530; Pollard: Environmental and Natural Resources Institute, University of Alaska Anchorage, 707 A Street, Anchorage, AK 99501.

Stuby, L. A., and D. M. Schell. Seasonal changes in lipid and protein contents of Arctic cisco and broad whitefish. Stuby: Alaska Department of Fish and Game, 1300 College Road, Fairbanks, AK 99701; Schell: University of Alaska, Water Research Center, Fairbanks, AK 99775-5860.

Thorpe, L., and B. Osborne. Distribution and biological characteristics of Arctic char *Salvelinus alpinus* in inland waters of the Arctic National Wildlife Refuge, Alaska. U.S. Fish and Wildlife Service, Fishery Resource Office, 101 12th Avenue, Fairbanks, AK 99701.